GENETICS

Readings from
**SCIENTIFIC
AMERICAN**

GENETICS

With Introductions by
Cedric I. Davern
The University of Utah

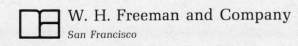 W. H. Freeman and Company
San Francisco

Some of the SCIENTIFIC AMERICAN articles in *Genetics*
are available as separate Offprints. For a complete list
of articles now available as Offprints, write to W. H.
Freeman and Company, 660 Market Street, San Francisco,
California 94104

Library of Congress Cataloging in Publication Data
Main entry under title:

Genetics: readings from Scientific American.

Includes bibliographies and index.
1. Genetics—Addresses, essays, lectures.
I. Davern, Cedric I.   II. Scientific American
[DNLM: 1. Genetics—Collected works.   QH 430 R287]
QH438.G455   575.1   80-25208
ISBN 0-7167-1200-8
ISBN 0-7167-1201-6 (pbk.)

CONTENTS

General Introduction   2

Note on cross-references. References to articles included in this book are noted by the title of the article and the page on which it begins; references to articles that are available as Offprints, but are not included here, are noted by the article's title and Off-print number; references to articles published by SCIENTIFIC AMERICAN but which are not available as Offprints are noted by the title of the article and the month and year of its publication.

GENETICS

GENERAL INTRODUCTION

GENERAL INTRODUCTION

If one were to chronicle the great achievements of the science of genetics, which seeks to describe and explain the phenomenon of heredity, such a history would show a science that has transformed the taxonomical and descriptive pursuits, with its vitalist underpinnings, of eighteenth- and nineteenth-century natural historians into the experimental and thoroughly materialist inquiries of twentieth-century research biologists.

This anthology—a selection of *Scientific American* articles together with Gregor Mendel's classic paper—documents and celebrates two seminal insights from which the field of genetic inquiry has grown: one is the synthesist tradition, and the other is the reductionist tradition.

The synthesist tradition of evolutionary studies has its roots in Darwin's theory that natural selection is the major force in the origin and evolution of species.

The reductionist tradition, which has led to identification of DNA as the substance of the gene, and has culminated in the discoveries of molecular biology concerning the structure, replication, recombination, mutation, and expression of DNA, is rooted in Mendel's positing of the gene (or "factor," as he called it) as the determinant specifying the characteristics of an organism.

The readings are meant to provide the students with a resource to complement formal genetic texts. The readings comprise a series of vignettes that capture the essence of significant problems that have characterized the field of genetics; they have been written by researchers who have directly contributed to clarification and/or solution of these problems.

The study of heredity reveals two distinct phenomena demanding explanation: constancy and change. Those familiar with the preoccupations of the pre-Socratic scientist-philosophers will recognize this problem as constituting the central paradox of their age.

Our ancient forebears were more impressed with the order presented by constancy than they were with the chaos suggested by change, and the adage "like begets like" came to have considerable currency. Yet, ancient agricultural and pastoral man consciously exploited hereditary change in plant and animal breeding while depending on hereditary constancy (or near constancy within lineages) for the success of such breeding.

Investigators of the phenomena of heredity prior to Gregor Mendel were fascinated by this problem, and they worked without success to elucidate the rules of heredity that could explain the diverse patterns of constancy and change they observed in their hybridization experiments. Mendel's insights, methods, and classifications of the data gathered from his pea hybridization experiments allowed him to perceive the order that had eluded his predecessors, and thus to elucidate the basic principles of heredity, namely, those of allele segregation and the independent assortment of allelic differences at segregating loci.

Mendel's empirical laws were published in 1865, but it was not until forty years later, when the process of meiosis was thoroughly understood, that his laws came to have a material basis in the chromosome theory of heredity as proposed independently by Walter Stanborough Sutton and Theodor Boveri in 1903.

Thus the process of meiosis provided the explanation of how the parental alleles segregated from one another during germ cell formation, and how segregating loci, when they were located on separate chromosomes (or at least far enough apart on the same chromosome), assorted independently of each other in accordance with Mendel's second law.

Exceptions to Mendel's second law of independent assortment were readily accommodated within the chromosome theory of inheritance by postulating that segregating loci showing an excess transmission of grandparental allele combinations must be located on the same chromosome.

Furthermore, as was so ably demonstrated by Alfred Sturtevant in 1913, the degrees of the linkage between different pairs of segregating loci could be used to generate an unambiguous, internally consistent linear map of a set of loci. Such maps were interpreted as reflections of the physical organization of the chromosomes as linear assemblages of genes.

These insights and discoveries ushered in the classical period of formal genetics, during which geneticists demonstrated the universality of Mendel's laws as modified by the chromosome theory of inheritance. Geneticists collected mutants, constructed genetic maps, and studied chromosome rearrangements. During this period the gene was considered to be no more than a formal element of genetic determination that occupied a specific location on the chromosome. But with the discovery that DNA is the agent of genetic transformation by Oswald Avery, C. M. MacLeod, and M. J. McCarty (1944), and with its identification as the genetic material of a bacterial virus by Alfred D. Hershey and Martha Chase (1952), the concept of the gene was transformed from an element of formal genetic analysis to a palpable entity: now it was amenable to (and deserving of) physical and chemical characterization.

Biochemists had long been interested in the phosphorus-rich acidic compound that had first been isolated from pus and salmon sperm by Johann Miescher in 1868. With the passage of time and expenditure of effort, nucleic acid was further characterized and found to contain four different nitrogenous bases, which Erwin Chargaff observed varied in relative proportions among DNAs derived from different organisms. However, as was to prove of critical importance in the solution by James D. Watson and Francis H. C. Crick of the structure of DNA, Chargaff also discovered in 1950 the now famous "complementarity" rule: despite the variation in base composition, the ratio of adenine to thymine (A:T) and the ratio of guanine to cytosine (G:C) were each always equal to unity.

This work of identifying the chemical nature of the gene culminated in Watson and Crick's proposal (1953) for the structure of DNA. The structure immediately revealed how DNA could replicate with fidelity and encode information in the linear sequence of its nucleotides.

In the meantime, a separate line of endeavor had already established how genes function. In 1940, George W. Beadle and Edward L. Tatum proposed, as a result of their studies of mutants affecting the functioning of various metabolic pathways in the fungus *Neurospora*, that genes must specify the structure of enzymes.

By 1953 it was recognized that the great diversity among proteins could be attributed to the particular sequence of the twenty different amino acids which composed their constituent polypeptide chains. Thus the gene was visualized as an extensive linear sequence of the four different nitrogenous bases (adenine, thymine, guanine, and cytosine) that in some way determined the amino acid sequence of the polypeptide specified by it.

With the advent of techniques for labeling with radioisotopes and for the separation and isolation of macromolecules and macromolecular complexes from disrupted cells, the question of what entities and processes are involved in the specification of polypeptides by genes could be pursued. Research based on these techniques led to the identification and description, in molecular terms, of the processes involved in replication, transcription, and translation, as well as the "cracking" of the genetic code.

Such a strong analytic approach could not long go without a banner and a following, and so "molecular biologists" came into being with the credo: "It is not sufficient to identify molecular entities involved in a biological process, but the mechanics of that process must be sought in the structures of those participating entities." So powerful was the seductive attraction of their reductionist platform that many scientists broke with their current disciplines to join the molecularists. Those who remained faithful to their original calling, and especially those engaged in trying to understand the complexity of nature as it is manifest in species, populations, and communities, looked upon this spectacle of reductionist enthusiasm with some distaste.

The understanding of the molecular biology of DNA—its structure, replication, mutation, recombination, and expression—provided at one and the same time a material basis for what had once seemed irreconcilable—constancy and change: constancy deriving from the high fidelity of its replication and change arising from mutation and its recombination.

Both Mendelian and molecular genetics have been assimilated into the explication of the Darwinian theory of evolution to give what Ernst Mayr describes as the modern synthesis. Like the problem of constancy and change, theories of evolution and its mechanism through natural selection also date back to the philosophical speculations of the materialist pre-Socratic scientists Anaximander and Empedocles.

One is thus left with the impression that the great problems of nature have enjoyed a remarkable Parmenidean constancy over the millenia, despite the Heraclitian flux of change and progress that appears to characterize the advance of man's intellectual heritage.

I

PRINCIPLES OF HEREDITY

I PRINCIPLES OF HEREDITY

INTRODUCTION

Inasmuch as like begets *almost*-like, the obvious first task that people interested in the phenomena of heredity set for themselves was to describe and understand the patterns of inheritance of parental traits as manifested in their hybrid progeny.

The basic technique of these scientists consisted of cross-breeding two organisms, each of which differed in one or more characteristics, and then following the manifestation of these differentiating characteristics in one or more successive generations. For the investigators prior to Mendel, the outcome of their efforts and observation did no more than to confirm the phenomenon of heredity, but its rules remained obscure, and its underlying mechanism inscrutable.

What distinguished Mendel's approach to the analysis of hereditary phenomena from those of his predecessors? The remarkable thing about Mendel's work is that many aspects distinguished it from that of his predecessors—his choice of material, his mode of observation, his method of analysis, and finally, the inductive leap he made in proposing a theory of heredity to account for his observations. So elegant was his theory that it is reasonable to conjecture that his theory may have informed his approach to the problem—in short, he may have leapt before he looked.

What did he in fact do that was so unique?

He deliberately chose a naturally inbreeding and therefore genetically homozygous plant species.

He focused on one differentiating character at a time and was careful to ensure that hybrids between different strains were not defective in their fertility.

He scored his observations by generation and thus did not confuse the generations as his predecessors had done.

He scored the frequency of each phenotypic class.

Not only was he quantitative in his approach, he also recognized the statistical nature of the manifestations of heredity that he was seeking to characterize. Thus he made provision to score adequate numbers of individuals in segregating generations to establish the validity of his inferences.

Yet, as Robert C. Olby has recently pointed out, from the data so amassed Mendel did not infer the existence of determinants of heredity that conserved their integrity in hybrids and successive generations. Instead, Mendel argued that the parental characters *per se* were conserved in the hybrid generations. While he apparently failed to distinguish between the potential and actual when it came to accounting for the manifestation of parental characteristics, his discoverers in 1900 certainly made this distinction. They explained the phenomenon of dominance and the patterns of segregation for a contrasting character pair in the hybrid generations in terms of genetic determinants or genes, thus clearly distinguishing between genotype and phenotype.

Finally, Mendel was able to show that different segregating character pairs behaved as if they were assorting independently. It was not surprising that Mendel's paper, which made such a break with the past, lay largely ignored until it was discovered almost simultaneously by three independent researchers forty-five years after it had been published. This discovery came about as a consequence of efforts at the turn of the century to find some way to test a chromosome theory of heredity put forward by August Weismann. Weismann had the controversial notion that each chromosome was the repository of the total genetic specificity of an organism. This theory arose out of Wilhelm Roux's conjecture that the processes of mitosis and meiosis—which were described some years after Mendel had reached his conclusions—must be significant cellular processes that provide the mechanical basis for the phenomenon of heredity.

Mendel's methodological heritage—namely, his rigorously quantitative and analytical approach—introduced a novel and seminal current into the largely qualitative and descriptive mainstream of biological inquiry of his time. This current was destined to become the mainstream itself, transforming biology from its genteel tradition of natural history into a rigorous science ranging from the unabashed reductionism of molecular biology to the development of sophisticated quantitative models that seek to describe and relate population dynamics, evolutionary processes, and ecological phenomena.

Alfred North Whitehead once said that modern philosophy is a footnote to Plato. A similar thing might be said about the relation of Mendel to modern biology, and it is for this reason that his 1865 paper, "Experiments in Plant Hybridization," has been chosen to open this collection.

Soon after the discovery of Mendel's work, it quickly became apparent that segregation of alleles at some pairs of genetic loci do not assort independently. These exceptions to Mendel's second law of independent assortment constituted the phenomenon of linkage. Then it was discovered that genetic loci could be assigned to linkage groups, where the number of linkage groups correspond to the haploid number of chromosomes comprising the genome. Furthermore, as Alfred Sturtevant discovered in 1913, within a linkage group, the loci could be unambiguously disposed on a linear map to reflect their physical location on the chromosome.

Now it appears that Mendel's first law does not always hold either; this is described in James F. Crow's article, "Genes That Violate Mendel's Rules." Crow describes chromosomes that act somewhat akin to cuckoo fledglings by ensuring that their less fortunate homologs end up in germ cells that degenerate.

This section closes with George W. Beadle's article, "The Genes of Men and Molds," because it documents the origin of the classic dictum of functional genetic analysis—namely, that the function of a gene is to specify an enzyme. Like Mendel's inferences about the rules of heredity, this dictum was a seminal insight whose material basis was subsequently demonstrated, as is recounted in the articles of Section III.

1 Experiments in Plant Hybridization

by Gregor Mendel

INTRODUCTORY REMARKS[1]

Experience of artificial fertilisation, such as is effected with ornamental plants in order to obtain new variations in colour, has led to the experiments which will here be discussed. The striking regularity with which the same hybrid forms always reappeared whenever fertilisation took place between the same species induced further experiments to be undertaken, the object of which was to follow up the developments of the hybrids in their progeny.

To this object numerous careful observers, such as Kölreuter, Gärtner, Herbert, Lecoq, Wichura and others, have devoted a part of their lives with inexhaustible perseverance. Gärtner especially, in his work "Die Bastarderzeugung im Pflanzenreiche" (The Production of Hybrids in the Vegetable Kingdom), has recorded very valuable observations; and quite recently Wichura published the results of some profound investigations into the hybrids of the Willow. That, so far, no generally applicable law governing the formation and development of hybrids has been successfully formulated can hardly be wondered at by anyone who is acquainted with the extent of the task, and can appreciate the difficulties with which experiments of this class have to contend. A final decision can only be arrived at when we shall have before us the results of detailed experiments made on plants belonging to the most diverse orders.

Those who survey the work in this department will arrive at the conviction that among all the numerous experiments made, not one has been carried out to such an extent and in such a way as to make it possible to determine the number of different forms under which the offspring of hybrids appear, or to arrange these forms with certainty according to their separate generations, or definitely to ascertain their statistical relations.[2]

It requires indeed some courage to undertake a labour of such far-reaching extent; this appears, however, to be the only right way by which we can finally reach the solution of a question the importance of which cannot be overestimated in connection with the history of the evolution of organic forms.

The paper now presented records the results of such a detailed experiment. This experiment was practically confined to a small plant group, and is now, after eight years' pursuit, concluded in all essentials. Whether the plan upon which the separate experiments were conducted and carried out was the best suited to attain the desired end is left to the friendly decision of the reader.

SELECTION OF THE EXPERIMENTAL PLANTS

The value and utility of any experiment are determined by the fitness of the material to the purpose for which it is used, and thus in the case before us it cannot be immaterial what plants are subjected to experiment and in what manner such experiments are conducted.

The selection of the plant group which shall serve for experiments of this kind must be made with all possible care if it be desired to avoid from the outset every risk of the questionable results.

The experimental plants must necessarily—

1. Possess constant differentiating characters.

2. The hybrids of such plants must, during the flowering period, be protected from the influence of all foreign pollen, or be easily capable of such protection.

The hybrids and their offspring should suffer no marked disturbance in their fertility in the successive generations.

Accidental impregnation by foreign pollen, if it occurred during the experiments and were not recognized, would lead to entirely erroneous conclusions. Reduced fertility or entire sterility of certain forms, such as occurs in the offspring of many hybrids, would render the experiments very difficult or entirely frustrate them. In order to discover the relations in which the hybrid forms stand towards each other and also towards their progenitors it appears to be necessary that all members of the series developed in each successive generation should be, *without exception*, subjected to observation.

At the very outset special attention was devoted to the *Leguminosae* on account of their peculiar floral structure. Experiments which were made with several members of this family led to the result that the genus *Pisum* was found to possess the necessary qualifications.

Some thoroughly distinct forms of this genus possess characters which are constant, and easily and certainly recognizable, and when their hybrids are mutually crossed they yield perfectly fertile progeny. Furthermore, a disturbance through foreign pollen cannot easily occur, since the fertilising organs are closely packed inside the keel and the anther bursts within the bud, so that the stigma becomes covered with pollen even before the flower opens. This circumstance is of special importance. As additional advantages worth mentioning, there may be cited the easy culture of these plants in the open ground and in pots, and also their relatively short period of growth.

Artificial fertilisation is certainly a somewhat elaborate process, but nearly always succeeds. For this purpose the bud is opened before it is perfectly developed, the keel is removed, and each stamen carefully extracted by means of forceps, after which the stigma can at once be dusted over with the foreign pollen.

In all, thirty-four more or less distinct varieties of Peas were obtained from several seedsmen and subjected to a two years' trial. In the case of one variety there were noticed, among a larger number of plants all alike, a few forms which were markedly different. These, however, did not vary in the following year, and agreed entirely with another variety obtained from the same seedsman; the seeds were therefore doubtless merely accidentally mixed. All the other varieties yielded perfectly constant and similar offspring; at any rate, no essential difference was observed during two trial years. For fertilisation twenty-two of these were selected and cultivated during the whole period of the experiments. They remained constant without any exception.

Their systematic classification is difficult and uncertain. If we adopt the strictest definition of a species, according to which only those individuals belong to a species which under precisely the same circumstances display precisely similar characters, no two of these varieties could be referred to one species. According to the opinion of experts, however, the majority belong to the species *Pisum sativum;* while the rest are regarded and classed, some as sub-species of *P. sativum,* and some as independent species, such as *P. quadratum, P. saccharatum,* and *P. umbellatum.* The positions, however, which may be assigned to them in a classificatory system are quite immaterial for the purposes of the experiments in question. It has so far been found to be just as impossible to draw a sharp line between the hybrids of species and varieties as between species and varieties themselves.

DIVISION AND ARRANGEMENT
OF THE EXPERIMENTS

If two plants which differ constantly in one or several characters be crossed, numerous experiments have demonstrated that the common characters are transmitted unchanged to the hybrids and their progeny; but each pair of differentiating characters, on the other hand, unite in the hybrid to form a new character, which in the progeny of the hybrid is usually variable. The object of the experiment was to observe these variations in the case of each pair of differentiating characters, and to deduce the law according to which they appear in the successive generations. The experiment resolves itself therefore into just as many separate experiments as there are constantly differentiating characters presented in the experimental plants.

The various forms of Peas selected for crossing showed differences in the length and colour of the stem; in the size and form of the leaves; in the position, colour, and size of the flowers; in the length of the flower stalk; in the colour, form, and size of the pods; in the form and size of the seeds; and in the colour of the seed-coats and of the albumen [cotyledons]. Some of the characters noted do not permit of a sharp and certain separation, since the difference is of a "more or less" nature, which is often difficult to define. Such characters could not be utilised for the separate experiments; these could only be applied to characters which stand out clearly and definitely in the plants. Lastly, the result must show whether they, in their entirety, observe a regular behaviour in their hybrid unions, and whether from these facts any conclusion can be come to regarding those characters which possess a subordinate significance in the type.

The characters which were selected for experiment relate:

1. To the *difference in the form of the ripe seeds.* These are either round or roundish, the depressions, if any, occur on the surface, being always only shallow; or they are irregularly angular and deeply wrinkled (*P. quadratum*).

2. To the *difference in the colour of the seed albumen* (endosperm).[3] The albumen of the ripe seeds is either pale yellow, bright yellow and orange coloured, or it possesses a more or less intense green tint. This difference of colour is easily seen in the seeds as [= if] their coats are transparent.

3. To the *difference in the colour of the seed-coat.* This is either white, with which character white flowers are constantly correlated; or it is grey, grey-brown, leather-brown, with or without violet spotting, in which case the colour of the standards is violet, that of the wings purple, and the stem in the axils of the leaves is of a reddish tint. The grey seed-coats become dark brown in boiling water.

4. To the *difference in the form of the ripe pods.* These are either simply inflated, not contracted in places; or they are deeply constricted between

the seeds and more or less wrinkled (*P. saccharatum*).

5. To the *difference in the colour of the unripe pods.* They are either light to dark green, or vividly yellow, in which colouring the stalks, leaf-veins, and calyx participate.[4]

6. To the *difference in the position of the flowers.* They are either axial, that is, distributed along the main stem; or they are terminal, that is, bunched at the top of the stem and arranged almost in a false umbel; in this case the upper part of the stem is more or less widened in section (*P. umbellatum*).[5]

7. To the *difference in the length of the stem.* The length of the stem[6] is very various in some forms; it is, however, a constant character for each, in so far that healthy plants, grown in the same soil, are only subject to unimportant variations in this character.

In experiments with this character, in order to be able to discriminate with certainty, the long axis of 6 to 7 ft. was always crossed with the short one of $3/4$ ft. to $1\frac{1}{2}$ ft.

Each two of the differentiating characters enumerated above were united by cross-fertilisation. There were made for the

1st trial	60	fertilisations on	15	plants.
2nd "	58	"	"	10	"
3rd "	35	"	"	10	"
4th "	40	"	"	10	"
5th "	23	"	"	5	"
6th "	34	"	"	10	"
7th "	37	"	"	10	"

From a larger number of plants of the same variety only the most vigorous were chosen for fertilisation. Weakly plants always afford uncertain results, because even in the first generation of hybrids, and still more so in the subsequent ones, many of the offspring either entirely fail to flower or only form a few and inferior seeds.

Furthermore, in all the experiments reciprocal crossings were effected in such a way that each of the two varieties which in one set of fertilisation served as seed-bearer in the other set was used as the pollen plant.

The plants were grown in garden beds, a few also in pots, and were maintained in their naturally upright position by means of sticks, branches of trees, and strings stretched between. For each experiment a number of pot plants were placed during the blooming period in a greenhouse, to serve as control plants for the main experiment in the open as regards possible disturbance by insects. Among the insects[7] which visit Peas the beetle *Bruchus pisi* might be detrimental to the experiments should it appear in numbers. The female of this species is known to lay the eggs in the flower, and in so doing opens the keel; upon the tarsi of one specimen, which was caught in a flower, some pollen grains could clearly be seen under a lens. Mention must also be made of a circumstance which possibly might lead to the introduction of foreign pollen. It occurs, for instance, in some rare cases that certain parts of an otherwise quite normally developed flower wither, resulting in a partial exposure of the fertilising organs. A defective development of the keel has also been observed, owing to which the stigma and anthers remained partially uncovered.[8] It also sometimes happens that the pollen does not reach full perfection. In this event there occurs a gradual lengthening of the pistil during the blooming period, until the stigmatic tip protrudes at the point of the keel. This remarkable appearance has been observed in hybrids of *Phaseolus* and *Lathyrus*.

The risk of false impregnation by foreign pollen is, however, a very slight one with *Pisum*, and is quite incapable of disturbing the general result. Among more than 10,000 plants which were carefully examined there were only a very few cases where an indubitable false impregnation had occurred. Since in the greenhouse such a case was never remarked, it may well be supposed that *Bruchus pisi*, and possibly also the described abnormalities in the floral structure, were to blame.

[F_1] THE FORMS OF THE HYBRIDS[9]

Experiments which in previous years were made with ornamental plants have already afforded evidence that the hybrids, as a rule, are not exactly intermediate between the parental species. With some of the more striking characters, those, for instance, which relate to the form and size of the leaves, the pubescence of the several parts, &c., the intermediate, indeed, is nearly always to be seen; in other cases, however, one of the two parental characters is so preponderant that it is difficult, or quite impossible, to detect the other in the hybrid.

This is precisely the case with the Pea hybrids. In the case of each of the seven crosses the hybrid-character resembles[10] that of one of the parental forms so closely that the other either escapes observation completely or cannot be de-

tected with certainty. This circumstance is of great importance in the determination and classification of the forms under which the offspring of the hybrids appear. Henceforth in this paper those characters which are transmitted entire, or almost unchanged in the hybridisation, and therefore in themselves constitute the characters of the hybrid, are termed the *dominant*, and those which become latent in the process *recessive*. The expression "recessive" has been chosen because the characters thereby designated withdraw or entirely disappear unchanged in their progeny, as will be demonstrated later on.

It was furthermore shown by the whole of the experiments that it is perfectly immaterial whether the dominant character belongs to the seedbearer or to the pollen-parent; the form of the hybrid remains identical in both cases. This interesting fact was also emphasised by Gärtner, with the remark that even the most practised expert is not in a position to determine in a hybrid which of the two parental species was the seed or the pollen plant.[11]

Of the differentiating characters which were used in the experiments the following are dominant:

1. The round or roundish form of the seed with or without shallow depressions.
2. The yellow colouring of the seed albumen [cotyledons].
3. The grey, grey-brown, or leather-brown colour of the seed-coat, in association with violet-red blossoms and reddish spots in the leaf axils.
4. The simply inflated form of the pod.
5. The green colouring of the unripe pod in association with the same colour in the stems, the leaf-veins and the calyx.
6. The distribution of the flowers along the stem.
7. The greater length of stem.

With regard to this last character it must be stated that the longer of the two parental stems is usually exceeded by the hybrid, a fact which is possibly only attributable to the greater luxuriance which appears in all parts of plants when stems of very different length are crossed. Thus, for instance, in repeated experiments, stems of 1 ft. and 6 ft. in length yielded without exception hybrids which varied in length between 6 ft. and 7½ ft.

The hybrid seeds in the experiments with seed-coat are often more spotted, and the spots sometimes coalesce into small bluish-violet patches. The spotting also frequently appears even when it is absent as a parental character.[12]

The hybrid forms of the seed-shape and of the albumen [colour] are developed immediately after the artificial fertilisation by the mere influence of the foreign pollen. They can, therefore, be observed even in the first year of experiment, whilst all the other characters naturally only appear in the following year in such plants as have been raised from the crossed seed.

[F_2] THE GENERATION [BRED] FROM THE HYBRIDS

In this generation there reappear, together with the dominant characters, also the recessive ones with their peculiarities fully developed, and this occurs in the definitely expressed average proportion of three to one, so that among each four plants of this generation three display the dominant character and one the recessive. This relates without exception to all the characters which were investigated in the experiments. The angular wrinkled form of the seed, the green colour of the albumen, the white colour of the seedcoats and the flowers, the constrictions of the pods, the yellow colour of the unripe pod, of the stalk, of the calyx, and of the leaf venation, the umbel-like form of the inflorescence, and the dwarfed stem, all reappear in the numerical proportion given, without any essential alteration. *Transitional forms were not observed in any experiment.*

Since the hybrids resulting from reciprocal crosses are formed alike and present no appreciable difference in their subsequent development, consequently the results [of the reciprocal crosses] can be reckoned together in each experiment. The relative numbers which were obtained for each pair of differentiating characters are as follows:

Expt. 1. Form of seed.—From 253 hybrids 7,324 seeds were obtained in the second trial year. Among them were 5,474 round or roundish ones and 1,850 angular wrinkled ones. Therefrom the ratio 2.96 to 1 is deduced.

Expt. 2. Colour of albumen.—258 plants yielded 8,023 seeds, 6,022 yellow, and 2,001 green; their ratio, therefore, is as 3.01 to 1.

In these two experiments each pod yielded usually both kinds of seeds. In well-developed pods which contained on the average six to nine seeds, it often happened that all the seeds were round (Expt. 1) or all yellow (Expt. 2); on the other hand there were never observed more than five wrinkled or five green ones in one pod. It appears to make no difference whether the pods are developed early or later in the hybrid or whether they spring from the main axis or from a lateral one. In some few plants only a few seeds developed in the first formed pods, and these possessed exclusively one of the two characters, but in the subsequently developed pods the normal proportions were maintained nevertheless.

As in separate pods, so did the distribution of the characters vary in separate plants. By way of illustration the first ten individuals from both series of experiments may serve.

	EXPERIMENT 1. Form of Seed.		EXPERIMENT 2. Color of Albumen.	
Plants	Round	Angular	Yellow	Green
1	45	12	25	11
2	27	8	32	7
3	24	7	14	5
4	19	16	70	27
5	32	11	24	13
6	26	6	20	6
7	88	24	32	13
8	22	10	44	9
9	28	6	50	14
10	25	7	44	18

As extremes in the distribution of the two seed characters in one plant, there were observed in Expt. 1 an instance of 43 round and only 2 angular, and another of 14 round and 15 angular seeds. In Expt. 2 there was a case of 32 yellow and only 1 green seed, but also one of 20 yellow and 19 green.

These two experiments are important for the determination of the average ratios, because with a smaller number of experimental plants they show that very considerable fluctuations may occur. In counting the seeds, also, especially in Expt. 2, some care is requisite, since in some of the seeds of many plants the green colour of the albumen is less developed, and at first may be easily overlooked. The cause of this partial disappearance of the green colouring has no connection with the hybrid-character of the plants, as it likewise occurs in the parental variety. This peculiarity [bleaching] is also confined to the individual and is not inherited by the offspring. In luxuriant plants this appearance was frequently noted. Seeds which are damaged by insects during their development often vary in colour and form, but, with a little practice in sorting, errors are easily avoided. It is almost superfluous to mention that the pods must remain on the plants until they are thoroughly ripened and have become dried, since it is only then that the shape and colour of the seed are fully developed.

Expt. 3. Colour of the seed-coats.—Among 929 plants 705 bore violet-red flowers and grey-brown seed-coats; 224 had white flowers and white seed-coats, giving the proportion 3.15 to 1.

Expt. 4. Form of pods.—Of 1,181 plants 882 had them simply inflated, and in 299 they were constricted. Resulting ratio, 2.95 to 1.

Expt. 5. Colour of the unripe pods,—The number of trial plants was 580, of which 428 had green pods and 152 yellow ones. Consequently these stand in the ratio 2.82 to 1.

Expt. 6. Position of flowers.—Among 858 cases 651 had inflorescences axial and 207 terminal. Ratio, 3.14 to 1.

Expt. 7. Length of stem.—Out of 1,064 plants, in 787 cases the stem was long, and in 277 short. Hence a mutual ratio of 2.84 to 1. In this experiment the dwarfed plants were carefully lifted and transferred to a special bed. This precaution was necessary, as otherwise they would have perished through being overgrown by their tall relatives. Even in their quite young state they can be easily picked out by their compact growth and thick dark-green foliage.[13]

If now the results of the whole of the experiments be brought together, there is found, as between the number of forms with the dominant and recessive characters, an average ratio of 2.98 to 1, or 3 to 1.

The dominant character can have here a *double signification*—viz. that of a parental character, or a hybrid-character.[14] In which of the two significations it appears in each separate case can only be determined by the following generation. As a parental character it must pass over unchanged to the whole of the offspring; as a hybrid-character, on the other hand, it must maintain the same behaviour as in the first generation [F_2].

[F_3] THE SECOND GENERATION [BRED] FROM THE HYBRIDS

Those forms which in the first generation [F_2] exhibit the recessive character do not further vary in the second generation [F_3] as regards this character; they remain constant in their offspring.

It is otherwise with those which possess the dominant character in the first generation [bred from the hybrids]. Of these *two*-thirds yield offspring which display the dominant and recessive characters in the proportion of 3 to 1, and thereby show exactly the same ratio as the hybrid forms, while only *one*-third remains with the dominant character constant.

The separate experiments yielded the following results:

Expt. 1. Among 565 plants which were raised from round seeds of the first generation, 193 yielded round seeds only, and remained therefore constant in this character; 372, however, gave both round and wrinkled seeds, in the proportion of 3 to 1. The number of the hybrids, therefore, as compared with the constants is 1.93 to 1.

Expt. 2. Of 519 plants which were raised from seeds whose albumen was of yellow colour in the first generation, 166 yielded exclusively yellow, while 353 yielded yellow and green seeds in the proportion of 3 to 1. There resulted, therefore, a division into hybrid and constant forms in the proportion of 2.13 to 1.

For each separate trial in the following experiments 100 plants were selected which displayed the dominant character in the first generation, and in order to ascertain the significance of this, ten seeds of each were cultivated.

Expt. 3. The offspring of 36 plants yielded exclusively grey-brown seedcoats, while of the offspring of 64 plants some had grey-brown and some had white.

Expt. 4. The offspring of 29 plants had only simply inflated pods; of the offspring of 71, on the other hand, some had inflated and some constricted.

Expt. 5. The offspring of 40 plants had only green pods; of the offspring of 60 plants some had green, some yellow ones.

Expt. 6. The offspring of 33 plants had only axial flowers; of the offspring of 67, on the other hand, some had axial and some terminal flowers.

Expt. 7. The offspring of 28 plants inherited the long axis, and those of 72 plants some the long and some the short axis.

In each of these experiments a certain number of the plants came constant with the dominant character. For the determination of the proportion in which the separation of the forms with the constantly persistent character results, the two first experiments are of especial importance, since in these a larger number of plants can be compared. The ratios 1.93 to 1 and 2.13 to 1 gave together almost exactly the average ratio of 2 to 1. The sixth experiment gave a quite concordant result; in the others the ratio varies more or less, as was only to be expected in view of the smaller number of 100 trial plants. Experiment 5, which shows the greatest departure, was repeated, and then, in lieu of the ratio of 60 and 40, that of 65 and 35 resulted. *The average ratio of 2 to 1 appears, therefore, as fixed with certainty.* It is therefore demonstrated that, of those forms which possess the dominant character in the first generation, two-thirds have the hybrid-character, while one-third remains constant with the dominant character.

The ratio of 3 to 1, in accordance with which the distribution of the dominant and recessive characters results in the first generation, resolves itself therefore in all experiments into the ratio of 2:1:1 if the dominant character be differentiated according to its significance as a hybrid-character or as a parental one. Since the members of the first generation [F_2] spring directly from the seed of the hybrids [F_1], *it is now clear that the hybrids form seeds having one or other of the two differentiating characters, and of these one-half develop again the hybrid form, while the other half yield plants which remain constant and receive the dominant or the recessive characters [respectively] in equal numbers.*

THE SUBSEQUENT GENERATIONS [BRED] FROM THE HYBRIDS

The proportions in which the descendants of the hybrids develop and split up in the first and second generations presumably hold good for all subsequent progeny. Experiments 1 and 2 have already been carried through six generations, 3 and 7 through five, and 4, 5, and 6 through four, these experiments being continued from the third generation with a small number of plants, and no departure from the rule has been perceptible. The offspring of the hybrids separated in each generation in the ratio of 2:1:1 into hybrids and constant forms.

If A be taken as denoting one of the two constant characters, for instance the dominant, a, and the recessive, and Aa the hybrid form in which both are conjoined, the expression

$$A + 2Aa + a$$

shows the terms in the series for the progeny of the hybrids of two differentiating characters.

The observation made by Gärtner, Kölreuter, and others, that hybrids are inclined to revert to the parental forms, is also confirmed by the experiments described. It is seen that the number of the hybrids which arise from one fertilisation, as compared with the number of forms which become constant, and their

progeny from generation to generation, is continually diminishing, but that nevertheless they could not entirely disappear. If an average equality of fertility in all plants in all generations be assumed, and if, furthermore, each hybrid forms seed of which one-half yields hybrids again, while the other half is constant to both characters in equal proportions, the ratio of numbers for the offspring in each generation is seen by the following summary, in which A and a denote again the two parental characters, and Aa the hybrid forms. For brevity's sake it may be assumed that each plant in each generation furnishes only 4 seeds.

Generation	A	Aa	a	RATIOS A:Aa:a
1	1	2	1	1: 2 : 1
2	6	4	6	3: 2 : 3
3	28	8	28	7: 2 : 7
4	120	16	120	15: 2 :15
5	496	32	496	31: 2 :31
n				$2^n - 1$: 2 : $2^n - 1$

In the tenth generation, for instance, $2^n - 1 = 1023$. There result, therefore, in each 2,048 plants which arise in this generation 1,023 with the constant dominant character, 1,023 with the recessive character, and only two hybrids.

THE OFFSPRING OF HYBRIDS IN WHICH SEVERAL DIFFERENTIATING CHARACTERS ARE ASSOCIATED

In the experiments above described plants were used which differed only in one essential character.[15] The next task consisted in ascertaining whether the law of development discovered in these applied to each pair of differentiating characters when several diverse characters are united in the hybrid by crossing. As regards the form of the hybrids in these cases, the experiments showed throughout that this invariably more nearly approaches to that one of the two parental plants which possesses the greater number of dominant characters. If, for instance, the seed plant has a short stem, terminal white flowers, and simply inflated pods; the pollen plant, on the other hand, a long stem, violet-red flowers distributed along the stem, and constricted pods; the hybrid resembles the seed parent only in the form of the pod; in the other characters it agrees with the pollen parent. Should one of the two parental types possess only dominant characters, then the hybrid is scarcely or not at all distinguishable from it.

Two experiments were made with a considerable number of plants. In the first experiment the parental plants differed in the form of the seed and in the colour of the albumen; in the second in the form of the seed, in the colour of the albumen, and in the colour of the seed-coats. Experiments with seed characters give the result in the simplest and most certain way.

In order to facilitate study of the data in these experiments, the different characters of the seed plant will be indicated by A, B, C, those of the pollen plant by a, b, c, and the hybrid forms of the characters by Aa, Bb, and Cc.

Expt. 1.—AB, seed parents;
  A, form round;
  B, albumen yellow.
  ab, pollen parents;
  a, form wrinkled;
  b, albumen green.

The fertilised seeds appeared round and yellow like those of the seed parents. The plants raised therefrom yielded seeds of four sorts, which frequently presented themselves in one pod. In all, 556 seeds were yielded by 15 plants, and of these there were:

  315 round and yellow,
  101 wrinkled and yellow,
  108 round and green,
  32 wrinkled and green.

All were sown the following year. Eleven of the round yellow seeds did not yield plants, and three plants did not form seeds. Among the rest:

  38 had round yellow seeds . . . . . AB
  65 round yellow and green seeds ABb
  60 round yellow and wrinkled yellow seeds . . . . AaB
  138 round yellow and green, wrinkled yellow and green seeds . . . . $AaBb$

From the wrinkled yellow seeds 96 resulting plants bore seed, of which:

  28 had only wrinkled yellow seeds aB
  68 wrinkled yellow and green seeds aBb.

From 108 round green seeds 102 resulting plants fruited, of which:

  35 had only round green seeds Ab
  67 round and wrinkled green seeds Aab.

The wrinkled green seeds yielded 30 plants which bore seeds all of like character; they remained constant ab.

The offspring of the hybrids appeared therefore under nine different forms, some of them in very unequal numbers. When these are collected and co-ordinated we find:

38 plants with the sign				*AB*
35	"	"	"	" *Ab*
28	"	"	"	" *aB*
30	"	"	"	" *ab*
65	"	"	"	" *ABb*
68	"	"	"	" *aBb*
60	"	"	"	" *AaB*
67	"	"	"	" *Aab*
138	"	"	"	" *AaBb*.

The whole of the forms may be classed into three essentially different groups. The first includes those with the signs *AB, Ab, aB,* and *ab*: they possess only constant characters and do not vary again in the next generation. Each of these forms is represented on the average thirty-three times. The second group includes the signs *ABb, aBb, AaB, Aab*: these are constant in one character and hybrid in another, and vary in the next generation only as regards the hybrid-character. Each of these appears on an average sixty-five times. The form *AaBb* occurs 138 times: it is hybrid in both characters, and behaves exactly as do the hybrids from which it is derived.

If the numbers in which the forms belonging to these classes appear be compared, the ratios of 1, 2, 4 are unmistakably evident. The numbers 33, 65, 138 present very fair approximations to the ratio numbers of 33, 66, 132.

The developmental series consists, therefore, of nine classes, of which four appear therein always once and are constant in both characters; the forms *AB*, *ab*, resemble the parental forms, the two other present combinations between the conjoined characters *A, a, B, b*, which combinations are likewise possibly constant. Four classes appear always twice, and are constant in one character and hybrid in the other. One class appears four times, and is hybrid in both characters. Consequently the offspring of the hybrids, if two kinds of differentiating characters are combined therein, are represented by the expression

$$AB + Ab + aB + ab + 2ABb$$
$$+ 2aBb + 2AaB + 2Aab + 4AaBb.$$

This expression is indisputably a combination series in which the two expressions for the characters *A* and *a*, *B* and *b* are combined. We arrive at the full num-ber of the classes of the series by the combination of the expressions:

$$A + 2Aa + a$$
$$B + 2Bb + b.$$

Expt. 2.

ABC, seed parents;
  A, form round;
  B, albumen yellow;
  C, seed-coat grey-brown.
abc, pollen parents;
  a, form wrinkled;
  b, albumen green;
  c, seed-coat white.

This experiment was made in precisely the same way as the previous one. Among all the experiments it demanded the most time and trouble. From 24 hybrids 687 seeds were obtained in all: these were all either spotted, grey-brown or grey-green, round or wrinkled.[16] From these in the following year 639 plants fruited, and, as further investigation showed, there were among them:

8 plants *ABC*	22 plants *ABCc*	45 plants *ABbCc*
14 " *ABc*	17 " *AbCc*	36 " *aBbCc*
9 " *AbC*	25 " *aBCc*	38 " *AaBCc*
11 " *Abc*	20 " *abCc*	40 " *AabCc*
8 " *aBC*	15 " *ABbC*	49 " *AaBbC*
10 " *aBc*	18 " *ABbc*	48 " *AaBbc*
10 " *abC*	19 " *aBbC*	
7 " *abc*	24 " *aBbc*	
	14 " *AaBC*	78 " *AaBbCc*
	18 " *AaBc*	
	20 " *AabC*	
	16 " *Aabc*	

The whole expression contains 27 terms. Of these 8 are constant in all characters, and each appears on the average 10 times; 12 are constant in two characters, and hybrid in the third; each appears on the average 19 times; 6 are constant in one character and hybrid in the other two; each appears on the average 43 times. One form appears 78 times and is hybrid in all of the characters. The ratios 10, 19, 43, 78 agree so closely with the ratios 10, 20, 40, 80, or 1, 2, 4, 8, that this last undoubtedly represents the true value.

The development of the hybrids when the original parents differ in three characters results therefore according to the following expression:

$$ABC + ABc + AbC + Abc + aBC$$
$$+ aBc + abC + abc + 2 ABCc$$
$$+ 2 AbCc + 2 aBCc + 2 abCc$$
$$+ 2 ABbC + 2 ABbc + 2 aBbC$$
$$+ 2 aBbc + 2 AaBC + 2 AaBc$$
$$+ 2 AabC + 2 Aabc + 4 ABbCc$$
$$+ 4 aBbCc + 4 AaBCc + 4 AabCc$$
$$+ 4 AaBbC + 4 AaBbc + 8 AaBbCc.$$

Here also is involved a combination series in which the expressions for the characters *A* and *a*, *B* and *b*, *C* and *c*, are united. The expressions

$$A + 2Aa + a$$
$$B + 2Bb + b$$
$$C + 2Cc + c$$

give all the classes of the series. The constant combinations which occur therein agree with all combinations which are possible between the characters *A, B, C, a, b, c;* two thereof, *ABC* and *abc*, resemble the two original parental stocks.

In addition, further experiments were made with a smaller number of experimental plants in which the remaining characters by twos and threes were united as hybrids: all yielded approximately the same results. There is therefore no doubt that for the whole of the characters involved in the experiments the principle applies that *the offspring of the hybrids in which several essentially different characters are combined exhibit the terms of a series of combinations, in which the developmental series for each pair of differentiating characters are united.* It is demonstrated at the same time that *the relation of each pair of different characters in hybrid union is independent of the other differences in the two original parental stocks.*

If *n* represents the number of the differentiating characters in the two original stocks, 3^n gives the number of terms of the combination series, 4^n the number of individuals which belong to the series, and 2^n the number of unions which remain constant. The series therefore contains, if the original stocks differ in four characters, $3^4 = 81$ classes, $4^4 = 256$ individuals, and $2^4 = 16$ constant forms; or, which is the same, among each 256 offspring of the hybrids there are 81 different combinations, 16 of which are constant.

All constant combinations which in Peas are possible by the combination of the said seven differentiating characters were actually obtained by repeated crossing. Their number is given by $2^7 = 128$. Thereby is simultaneously given the practical proof *that the constant characters which appear in the several varieties of a group of plants may be obtained in all the associations which are possible according to the [mathematical] laws of combination, by means of repeated artificial fertilisation.*

As regards the flowering time of the hybrids, the experiments are not yet concluded. It can, however, already be stated that the time stands almost exactly between those of the seed and pollen parents, and that the constitution of the hybrids with respect to this character probably follows the rule ascertained in the case of the other characters. The forms which are selected for experiments of this class must have a difference of at least twenty days from the middle flowering period of one to that of the other; furthermore, the seeds when sown must all be placed at the same depth in the earth, so that they may germinate simultaneously. Also, during the whole flowering period, the more important variations in temperature must be taken into account, and the partial hastening or delaying of the flowering which may result therefrom. It is clear that this experiment presents many difficulties to be overcome and necessitates great attention.

If we endeavour to collate in a brief form the results arrived at, we find that those differentiating characters, which admit of easy and certain recognition in the experimental plants, all behave exactly alike in their hybrid associations. The offspring of the hybrids of each pair of differentiating characters are, one-half, hybrid again, while the other half are constant in equal proportions having the characters of the seed and pollen parents respectively. If several differentiating characters are combined by cross-fertilisation in a hybrid, the resulting offspring form the terms of a combination series in which the combination series for each pair of differentiating characters are united.

The uniformity of behaviour shown by the whole of the characters submitted to experiment permits, and fully justifies, the acceptance of the principle that a similar relation exists in the other characters which appear less sharply defined in plants, and therefore could not be included in the separate experiments. An experiment with peduncles of different lengths gave on the whole a fairly satisfactory result, although the differentiation and serial arrangement of the forms could not be effected with that certainty which is indispensable for correct experiment.

THE REPRODUCTIVE CELLS
OF THE HYBRIDS

The results of the previously described experiments led to further experiments, the results of which appear fitted to afford some conclusions as regards the composition of the egg and pollen cells of hybrids. An important clue is afforded in *Pisum* by the circumstance that among the progeny of the hybrids constant forms appear, and that this occurs, too, in respect of all combinations of the associated characters. So far as experience goes, we find it in every case confirmed that constant progeny can only be formed when the egg cells and the fertilising pollen are of like character, so that both are provided with the material for creating quite similar individuals, as is the case with the normal fertilisation of pure species. We must therefore regard it as certain that exactly similar factors must be at work also in the production of the constant forms in the hybrid plants. Since the various constant forms are produced in *one* plant, or even in *one* flower of a plant, the conclusion appears logical that in the ovaries of the hybrids there are formed as many sorts of egg cells, and in the anthers as many sorts of pollen cells, as there are possible constant combination forms, and that these egg and pollen cells agree in their internal composition with those of the separate forms.

In point of fact it is possible to demonstrate theoretically that this hypothesis would fully suffice to account for the development of the hybrids in the separate generations, if we might at the same time assume that the various kinds of egg and pollen cells were formed in the hybrids on the average in equal numbers.[17]

In order to bring these assumptions to an experimental · proof, the following experiments were designed. Two forms which were constantly different in the form of the seed and the colour of the albumen were united by fertilisation.

If the differentiating characters are again indicated as *A, B, a, b*, we have:

> *AB*, seed parent;
> *A*, form round;
> *B*, albumen yellow.
> *ab*, pollen parent;
> *a*, form wrinkled;
> *b*, albumen green.

The artificially fertilised seeds were sown together with several seeds of both original stocks, and the most vigorous examples were chosen for the reciprocal crossing. There were fertilised:

1. The hybrids with the pollen of *AB*.
2. The hybrids with the pollen of *ab*.
3. *AB* with the pollen of the hybrids.
4. *ab* with the pollen of the hybrids.

For each of these four experiments the whole of the flowers on three plants were fertilised. If the above theory be correct, there must be developed on the hybrids egg and pollen cells of the forms *AB, Ab, aB, ab*, and there would be combined:

1. The egg cells *AB, Ab, aB, ab* with the pollen cells *AB*.
2. The egg cells *AB, Ab, aB, ab* with the pollen cells *ab*.
3. The egg cells *AB* with the pollen cells *AB, Ab, aB, ab*.
4. The egg cells *ab* with the pollen cells *AB, Ab, aB, ab*.

From each of these experiments there could then result only the following forms:

1. *AB, ABb, AaB, AaBb.*
2. *AaBb, Aab, aBb, ab.*
3. *AB, ABb, AaB, AaBb.*
4. *AaBb, Aab, aBb, ab.*

If, furthermore, the several forms of the egg and pollen cells of the hybrids were produced on an average in equal numbers, then in each experiment the said four combinations should stand in the same ratio to each other. A perfect agreement in the numerical relations was, however, not to be expected, since in each fertilisation, even in normal cases, some egg cells remain undeveloped or subsequently die, and many even of the well-formed seeds fail to germinate when sown. The above assumption is also limited in so far that, while it demands the formation of an equal number of the various sorts of egg and pollen cells, it does not require that this should apply to each separate hybrid with mathematical exactness.

The first and second experiments had primarily the object of proving the compositon of the hybrid egg cells, while the third and fourth experiments were to decide that of the pollen cells.[18] As is shown by the above demonstration the first and third experiments and the second and fourth experiments should produce precisely the same combinations, and even in the second year the result should be partially visible in the form and colour of the artificially fertilised seed. In the first and third experiments the dominant characters of form and colour, *A* and *B*, appear in each union, and are also partly constant and partly in hybrid union with the recessive characters *a* and *b*, for which reason they must impress their peculiarity upon the whole of the seeds. All seeds should therefore appear round and yellow, if the theory be justified. In the second and fourth experiments, on the other hand, one union is hybrid in form and in colour, and consequently the seeds are round and yellow; another is hybrid in form, but constant in the recessive character of colour, whence the seeds are round and green; the third is constant in the recessive character of form but hybrid in colour, consequently the seeds are wrinkled and yellow; the fourth is constant in both recessive characters, so that the seeds are wrinkled and green. In both these experiments there were consequently four sorts of seed to be expected—viz, round and yellow, round and green, wrinkled and yellow, wrinkled and green.

The crop fulfilled these expectations perfectly. There were obtained in the

1st Experiment, 98 exclusively round yellow seeds;

3rd Experiment, 94 exclusively round yellow seeds.

In the 2d Experiment, 31 round and yellow, 26 round and green, 27 wrinkled and yellow, 26 wrinkled and green seeds.

In the 4th Experiment, 24 round and yellow, 25 round and green, 22 wrinkled and yellow 26 wrinkled and green seeds.

There could scarcely be now any doubt of the success of the experiment; the next generation must afford the final proof. From the seed sown there resulted for the first experiment 90 plants, and for the third 87 plants which fruited: these yielded for the

1st Exp.	3rd Exp.	
20	25	round yellow seeds . . . *AB*
23	19	round yellow and green seeds . . . . *ABb*
25	22	round and wrinkled yellow seeds . . . . *AaB*
22	21	round and wrinkled green and yellow seeds . . . . *AaBb*

In the second and fourth experiments the round and yellow seeds yielded plants with round and wrinkled yellow and green seeds, *AaBb*.

From the round green seeds, plants resulted with round and wrinkled green seeds, *Aab*.

The wrinkled yellow seeds gave plants with wrinkled yellow and green seeds, *aBb*.

From the wrinkled green seeds plants were raised which yielded again only wrinkled and green seeds, *ab*.

Although in these two experiments likewise some seeds did not germinate, the figures arrived at already in the previous year were not affected thereby, since each kind of seed gave plants which, as regards their seed, were like each other and different from the others. There resulted therefore from the

2d Exp.	4th Exp.	
31	24	plants of the form *AaBb*
26	25	" " " " *Aab*
27	22	" " " " *aBb*
26	27	" " " " *ab*

In all the experiments, therefore, there appeared all the forms which the proposed theory demands, and they came in nearly equal numbers.

In a further experiment the characters of flower-colour and length of stem were experimented upon, and selection was so made that in the third year of the experiment each character ought to appear in half of all the plants if the above theory were correct. *A*, *B*, *a*, *b* serve again as indicating the various characters.

A, violet-red flowers   *a*, white flowers
B, axis long.   *b*, axis short.

The form *Ab* was fertilised with *ab*, which produced the hybrid *Aab*. Furthermore, *aB* was also fertilised with *ab*, whence the hybrid *aBb*. In the second year, for further fertilisation, the hybrid *Aab* was used as seed parent, and hybrid *aBb* as pollen parent.

Seed parent, *Aab*.
Possible egg cells, *Ab*, *ab*.
Pollen parent, *aBb*.
Pollen cells, *aB*, *ab*.

From the fertilisation between the possible egg and pollen cells four combinations should result, viz.,

$$AaBb + aBb + Aab + ab.$$

From this it is perceived that, according to the above theory, in the third year of the experiment out of all the plants

Half should have violet-red flowers (*Aa*), Classes 1, 3.

Half should have white flowers (*a*), Classes 2, 4.

Half should have a long axis (*Bb*), Classes 1, 2.

Half should have a short axis (*b*), Classes 3, 4.

From 45 fertilisations of the second year 187 seeds resulted, of which only 166 reached the flowering stage in the third year. Among these the separate classes appeared in the numbers following:

Class	Color of flower	Stem		
1	violet-red	long	47	times
2	white	long	40	"
3	violet-red	short	38	"
4	white	short	41	"

There subsequently appeared

The violet-red flower-colour (*Aa*) in 85 plants.

The white flower-colour (*a*) in 81 plants.

The long stem (*Bb*) in 87 plants.

The short stem (*b*) in 79 plants.

The theory adduced is therefore satisfactorily confirmed in this experiment also.

For the characters of form of pod, colour of pod, and position of flowers, experiments were also made on a small scale, and results obtained in perfect agreement. All combinations which were possible through the union of the differentiating characters duly appeared and in nearly equal numbers.

Experimentally, therefore, the theory is confirmed that *the pea hybrids form egg and pollen cells which, in their constitution, represent in equal numbers all constant forms which result from the combination of the characters united in fertilisation.*

The difference of the forms among the progeny of the hybrids, as well as the respective ratios of the numbers in which they are observed, find a sufficient explanation in the principle above deduced. The simplest case is afforded by the developmental series of each pair of differentiating characters. This series is represented by the expression $A + 2Aa + a$, in which A and a signify the forms with constant differentiating characters, and Aa the hybrid form of both. It includes in three different classes four individuals. In the formation of these, pollen and egg cells of the form A and a take part on the average equally in the fertilisation; hence each form [occurs] twice, since four individuals are formed. There participate consequently in the fertilisation

The pollen cells $A + A + a + a$
The egg cells $A + A + a + a$.

It remains, therefore, purely a matter of chance which of the two sorts of pollen will become united with each separate egg cell. According, however, to the law of probability, it will always happen, on the average of many cases, that each pollen form, A and a, will unite equally often with each egg cell form, A and a, consequently one of the two pollen cells A in the fertilisation will meet with the egg cell a, and so likewise one pollen cell a will unite with an egg cell A, and the other with egg cell a.

Pollen cells A A a a

Egg cells A A a a

The result of the fertilisation may be made clear by putting the signs of the conjoined egg and pollen cells in the form of fractions, those for the pollen cells above and those for the egg cells below the line. We then have

$$\frac{A}{A} + \frac{A}{a} + \frac{a}{A} + \frac{a}{a}.$$

In the first and fourth term the egg and pollen cells are of like kind, consequently the product of their union must be constant, viz. A and a; in the second and third, on the other hand, there again results a union of the two differentiating characters of the stocks, consequently the forms resulting from these fertilisations are identical with those of the hybrid from which they sprang. *There occurs accordingly a repeated hybridisation.* This explains the striking fact that the hybrids are able to produce, besides the two parental forms, offspring which are like themselves; $\frac{A}{a}$ and $\frac{a}{A}$ both give the same union Aa, since, as already remarked above, it makes no difference in the result of fertilisation to which of the two characters the pollen or egg cells belong. We may write then

$$\frac{A}{A} + \frac{A}{a} + \frac{a}{A} + \frac{a}{a} = A + 2Aa + a.$$

This represents the average result of the self-fertilisation of the hybrids when two differentiating characters are united in them. In individual flowers and in individual plants, however, the ratios in which the forms of the series are produced may suffer not inconsiderable fluctuations.[19] Apart from the fact that the numbers in which both sorts of egg cells occur in the seed vessels can only be regarded as equal on the average, it remains purely a matter of chance which of the two sorts of pollen may fertilise each separate egg cell. For this reason the separate values must necessarily be subject to fluctuations, and there are even extreme cases possible, as were described earlier in connection with the experiments on the form of the seed and the colour of the albumen. The true ratios of the numbers can only be ascertained by an average deduced from the sum of as many single values as possible; the greater the number, the more are merely chance effects eliminated.

The developmental series for hybrids in which two kinds of differentiating characters are united contains, among sixteen individuals, nine different forms, viz.,

$$AB + Ab + aB + ab + 2ABb + 2aBb$$
$$+ 2AaB + 2Aab + 4AaBb.$$

Between the differentiating characters of the original stocks, Aa and Bb, four constant combinations are possible, and consequently the hybrids produce the corresponding four forms of egg and pollen cells AB, Ab, aB, ab and each of these will on the average figure four times in the fertilisation, since sixteen individuals are included in the series. Therefore the participators in the fertilisation are

Pollen cells $AB + AB + AB + AB$
$\qquad\qquad + Ab + Ab + Ab + Ab$
$\qquad\qquad + aB + aB + aB + aB$
$\qquad\qquad + ab + ab + ab + ab.$
Egg cells $\quad AB + AB + AB + AB$
$\qquad\qquad + Ab + Ab + Ab + Ab$
$\qquad\qquad + aB + aB + aB + aB$
$\qquad\qquad + ab + ab + ab + ab.$

In the process of fertilisation each pollen form unites on an average equally often with each egg cell form, so that each of the four pollen cells AB unites once with one of the forms of egg cell AB, Ab, aB, ab. In precisely the same way the rest of the pollen cells of the forms Ab, aB, ab unite with all the other egg cells. We obtain therefore

$$\frac{AB}{AB} + \frac{AB}{Ab} + \frac{AB}{aB} + \frac{AB}{ab} + \frac{Ab}{AB}$$
$$+ \frac{Ab}{Ab} + \frac{Ab}{aB} + \frac{Ab}{ab} + \frac{aB}{AB}$$
$$+ \frac{aB}{Ab} + \frac{aB}{aB} + \frac{aB}{ab} + \frac{ab}{AB}$$
$$+ \frac{ab}{Ab} + \frac{ab}{aB} + \frac{ab}{ab},$$

or

$$AB + ABb + AaB + AaBb + ABb$$
$$+ Ab + AaBb + Aab + AaB$$
$$+ AaBb + aB + aBb + AaBb$$
$$+ Aab + aBb + ab = AB + Ab$$
$$+ aB + ab + 2ABb + 2aBb$$
$$+ 2AaB + 2Aab + 4AaBb.[20]$$

In precisely similar fashion is the developmental series of hybrids exhibited when three kinds of differentiating characters are conjoined in them. The hybrids form eight various kinds of egg and pollen cells—ABC, ABc, AbC, Abc, aBC, aBc, abC, abc—and each pollen form unites itself again on the average once with each form of egg cell.

The law of combination of different characters, which governs the development of the hybrids, finds therefore its foundation and explanation in the principle enunciated, that the hybrids produce egg cells and pollen cells which in equal numbers represent all constant forms which result from the combinations of the characters brought together in fertilisation.

REFERENCES CITED

[1] This translation was made by the Royal Horticultural Society of London, and is reprinted, by permission of the Council of the Society, with footnotes added and minor changes suggested by Professor W. Bateson, enclosed within [ ]. The original paper was published in the *Verb. naturf. Ver. in Brunn, Abbandlungen*, iv. 1865, which appeared in 1866.

[2] [It is to the clear conception of these three primary necessities that the whole success of Mendel's work is due. So far as I know this conception was absolutely new in his day.]

[3] [Mendel uses the terms "albumen" and "endosperm" somewhat loosely to denote the cotyledons, containing food-material, within the seed.]

[4] One species possesses a beautifully brownish-red coloured pod, which when ripening turns to violet and blue. Trials with this character were only begun last year. [Of these further experiments it seems no account was published. Correns has since worked with such a variety.]

[5] [This is often called the Mummy Pea. It shows slight fasciation. The form I know has white standard and salmon-red wings.]

[6] [In my account of these experiments (*R.H.S. Journal*, vol. xxv, p. 54) I misunderstood this paragraph and took "axis" to mean the *floral* axis, instead of the main axis of the plant. The unit of measurement, being indicated in the original by a dash ('), I carelessly took to have been an *inch*, but the translation here given is evidently correct.]

[7] [It is somewhat surprising that no mention is made of Thrips, which swarm in Pea flowers. I had come to the conclusion that this is a real source of error and I see Laxton held the same opinion.]

[8] [This also happens in Sweet Peas.]

[9] [Mendel throughout speaks of his crossbred Peas as "hybrids," a term which many restrict to the offspring of two distinct *species*. He, as he explains, held this to be only a question of degree.]

[10] [Note that Mendel, with true penetration, avoids speaking of the hybrid-character as "transmitted" by either parent, thus escaping the error pervading the older views of heredity.]

[11] [Gärtner, p. 223.]

[12] [This refers to the coats of the seeds borne by F_1 plants.]

[13] [This is true also of the dwarf or "Cupid" Sweet Peas.]

[14] [This paragraph presents the view of the hybrid-character as something incidental to the hybrid, and not "transmitted" to it—a true and fundamental conception here expressed probably for the first time.]

[15] [This statement of Mendel's in the light of present knowledge is open to some misconception. Though his work makes it evident that such varieties may exist, it is very unlikely that Mendel could have had seven pairs of varieties such that the members of each pair differed from each other in *only* one considerable character (*wesentliches Merkmal*). The point is probably of little theoretical or practical consequence, but a rather heavy stress is thrown on "*wesentlich.*"]

[16] [Note that Mendel does not state the cotyledon-colour of the first crosses in this case; for as the coats were thick, it could not have been seen without opening or peeling the seeds.]

[17] [This and the preceding paragraph contain the essence of the Mendelian principles of heredity.]

[18] [To prove, namely, that both were similarly differentiated, and not one or other only.]

[19] [Whether segregation by such units is more than purely fortuitous may perhaps be determined by seriation.]

[20] [In the original the sign of equality (=) is here represented by +, evidently a misprint.]

2

Genes That Violate Mendel's Rules

by James F. Crow
February 1979

*In sexual reproduction the parental genes are
continually reshuffled and thereby exposed equally to
the stringent test of natural selection. Some genes
cheat, subverting the process to favor their own survival*

Why is sexual reproduction so ubiquitous in the living world? An elaborate biparental method of procreation is certainly not particularly efficient. It requires two to do the work of one, and there are plenty of asexual reproductive processes, such as spore formation in some fungi, that are far more sparing of time and energy. If sheer numbers were the criterion of reproductive efficacy, sex would have been abandoned a billion years ago or would never have evolved at all.

The great role of sexual reproduction is an evolutionary one: it shuffles the genes of two parents to provide the genetic endowment of their progeny, and thus exposes to the stringent test of natural selection a maximum diversity of the characters and capacities represented in a species. If the system is to work well, if the testing is to be fair, the shuffle must be an honest one. It is not always honest, however. There are genes that cheat, perpetuating themselves in the population by tricking the reproductive process to work in their own favor.

The cheating is done at the stage of sexual reproduction called meiosis, the "reduction division" whereby one male germ cell divides to form four sperm cells, each of which has half the normal complement of chromosomes, and a comparable process takes place in egg formation in the female. Meiosis is followed by fertilization, which restores the original number of chromosomes, and the two processes together constitute the physical basis of Mendelian inheritance.

Chromosomes come in pairs and therefore so do the genes, which are segments of the long thread of DNA that is the core of a chromosome. In each individual's body and germ cells one member of each homologous chromosome pair has come from each parent. In meiosis the members of each pair segregate: one or the other of them, chosen at random, is transmitted to a sperm or egg cell and so to each offspring. The various pairs of chromosomes segregate independently, so that the genes on different chromosome pairs are thoroughly scrambled at each meiosis. Moreover, there is a regular process, called crossing-over, in which the two members of a pair line up side by side, break at corresponding points and exchange partners. As a consequence even genes on the same chromosome are not constrained to stay together but can participate in the meiotic shuffle.

The evolutionary advantage of this shuffling is that it enables genes from different individuals to come together and genes that have been together to separate. By thus scrambling the genes in every sexual generation Mendelian inheritance tests all the genes in many combinations. Genes that increase the capacity for survival, fertility and the survival of progeny are retained by the sieve of natural selection; less effective genes are lost. Evolution involves the continuous testing and retesting of gene combinations and the retention of those combinations that increase reproductive success, with the result that the species becomes better adapted.

Meiosis ensures a scrupulously fair test for every gene combination by giving each gene the same chance as every other gene of being transmitted to the next generation. Genes that cheat in meiosis undermine the system generally by reducing its fairness. More directly, such genes usually affect processes other than meiosis and, like most new mutant genes, are almost always harmful; for example, one such gene found in mouse populations causes tail abnormalities in addition to meiotic disturbances. Other cheating genes have been discovered in corn, lily, tobacco, trillium, rye, mosquitoes, grasshoppers and the classical geneticist's favorite species: the fruit fly *Drosophila melanogaster*. This article will give an account of one particularly well-understood set of cheating genes, those that produce the trait *Segregation Distorter*, or *SD*, in drosophilas.

The story begins in 1956, when *SD* was discovered by Yuichiro Hiraizumi, who was then a graduate student of mine at the University of Wisconsin and is now at the University of Texas. Hiraizumi was investigating certain genes, on chromosome No. II of natural *Drosophila* populations, that affect viability. He mated males in which one member of each pair of these chromosomes was the "wild type" from nature and the other member came from a laboratory stock, and he traced the chromosomes' inheritance by noting the distribution of particular marker genes affecting eye pigment. The normal drosophila eye has two pigments, one cinnabar (a bright scarlet, like the mercury ore for which it is named) and the other brown. Together they give the eye its normal dark red color. In one mutant form no brown pigment is produced; the eye is therefore cinnabar and the mutation is called *cn*. In another mutation, *bw*, the cinnabar pigment is deleted and the eye is brown. When both mutations are present, as was the case in Hiraizumi's laboratory stock, the eye is without pigment and looks white. Each of the mutants is recessive, meaning that it has its effect only when the normal gene, which would mask it, is absent.

Hiraizumi mated hybrid red-eyed males (with one natural chromosome and one laboratory chromosome containing both mutations) to females that had two laboratory chromosomes and were therefore white-eyed. Because there is normally no crossing-over in male drosophilas each transmitted chromosome carried either both mutant eye-color genes or neither, so that according to the rules of Mendelian inheritance half of the progeny should have had white eyes and the other half should have had dark red eyes. Hiraizumi did observe the expected 50:50 distribution among the progeny of some 200 matings (except for some minor deviations caused by the viability-reducing mutant genes that were the original object of the study). Six of the matings, however, produced very strange results. Instead of half of the 100-odd progeny of each mating being red-eyed, almost all of them were. The red-eyed proportion ranged from 95 to 100 percent, and it was 99 percent or more in most cases.

What had caused six chromosomes, each descended from a different male in the wild population, to behave in this most unusual way? The possibility that there had been a selective death of white-eyed progeny early in embryonic development was quickly ruled out; far too many of the eggs that were laid developed into normal adult flies to allow such an explanation. Hiraizumi also established that whatever was skewing the eye-color ratio had its effect only in the course of sperm formation, not in egg formation; when he reversed the sexes in his mating, so that the females rather than the males carried the unusual chromosome, all the matings gave rise to the standard 50:50 ratio.

The wild flies that were the source of the unusual chromosomes had been collected in the fall from a clump of

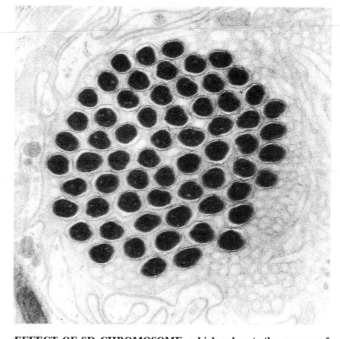 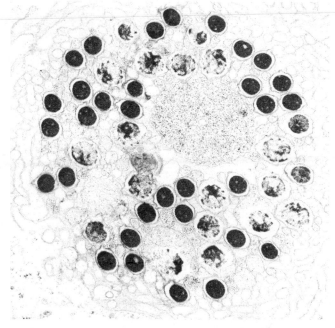

EFFECT OF SD CHROMOSOME, which subverts the process of meiosis in the fruit fly *Drosophila melanogaster*, is to prevent the proper development of half of the sperm heads. Normally the nuclei in a bundle of 64 immature sperm cells condense to form the heads of 64 sperms, as in the electron micrograph (*left*) made by Robert W. Hardy of the University of California at San Diego. In flies carrying an *SD* chromosome only 32 nuclei show the normal condensation, as in the micrograph (*right*) made by Kiyoteru T. Tokuyasu of San Diego; the others fail to develop normally. In both micrographs a section through the head region of sperm bundle is enlarged about 17,000 diameters.

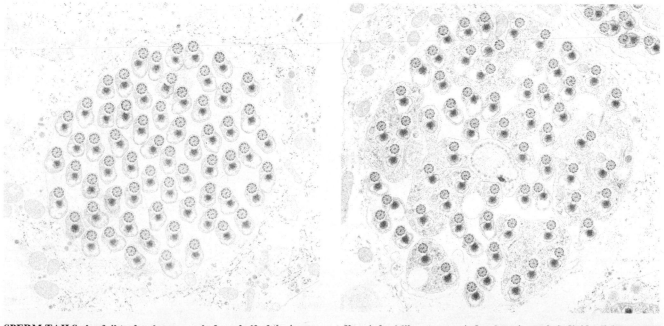

SPERM TAILS also fail to develop properly from half of the immature sperm cells in a fly carrying the *SD* chromosome, as is shown in micrographs made by Tokuyasu. Sections through a developing tail region are enlarged 10,000 diameters. In a normal fly (*left*) a tail fiber (wheel-like structure) develops in each individualizing sperm cell; in *SD* fly (*right*) half of the sperm cells do not individualize normally. They are sperms that do not carry *SD* chromosome. It "cheats" by inducing its homologous chromosome to cause sperm dysfunction.

trees near the airport in Madison; the strange results appeared during the winter. We were eager to collect more flies from the same place the next summer, and we were dismayed to find that the trees had been cut down and no drosophilas were to be found. Fortunately flies from elsewhere in nature (including the bit of nature that is my backyard) turned out to have the same peculiar chromosome; indeed, in almost every natural *Drosophila* population that has been studied, from many parts of the world, from 1 to 5 percent of the No. II chromosomes carry the *SD* trait. These chromosomes have one other peculiarity that should be mentioned. In virtually every case they carry at least one inversion, and usually more than one. Inversions (regions in which the genes run in an order that is the reverse of the normal order) are not uncommon in *Drosophila*, but it is unusual to find two or more on one chromosome. There must be a reason for the association of the *SD* trait and multiple inversions.

If natural *Drosophila* populations display this strange chromosome, why was it not discovered long ago? The main reason, I think, is simply that there is ordinarily no visible effect; the only way the phenomenon can be observed is by finding unusual inheritance ratios when crosses are made involving chromosomes marked with conspicuous mutant genes. Perhaps it actually was observed from time to time by investigators who attributed the bizarre results to some kind of experimental error.

Several years before Hiraizumi's experiment, however, Laurence M. Sandler and Edward Novitski of the Oak Ridge National Laboratory had suggested on the basis of certain experimental indications that the rules of meiosis might sometimes be violated; they called such a process "meiotic drive" and considered some of its theoretical possibilities. By a fortunate coincidence Sandler had come to Wisconsin as a postdoctoral fellow. To our delight Hiraizumi's discovery provided a perfect example of the phenomenon Sandler and Novitski had discussed theoretically. This was the beginning of a close collaboration between Hiraizumi and Sandler, who were responsible for the term *Segregation Distorter* and for most of the early understanding of the phenomenon.

By now the *SD* system has been studied in a number of laboratories in the U.S., Japan, Australia and Italy. As I have mentioned, there is no visible evidence of the *SD* chromosome on a fly that carries it; the only noticeable effect is a distortion of the ratio of progeny types. In the elaborate experiments required to analyze the system the chromosomes are therefore always marked with conspicuous mutant genes such as the eye-color genes I have described, so that the chromosomes can be followed through a complicated series of matings. I shall omit the details about how the chromosomes were labeled in various experiments and simply report the results.

One thing that makes it hard to understand how the *SD* chromosome gives rise to such a fundamental change in meiotic behavior is the fact that the chromosome must somehow inactivate precisely those sperm cells that do not contain it. How can that happen?

Sandler and Hiraizumi suggested one possibility quite early in the game: While the homologous chromosomes are still paired up during meiosis, the *SD* chromosome might do something to its normal partner (and rival) that later causes a dysfunction of the sperm receiving the normal chromosome. At first they suggested that *SD* might actually break the other chromosome. W. J. Peacock of the Commonwealth Scientific and Industrial Research Organization in Australia, working at the University of Oregon with a graduate student, John Erickson, was unable, however, to confirm the chromosome-breakage hypothesis. Microscopic observation showed

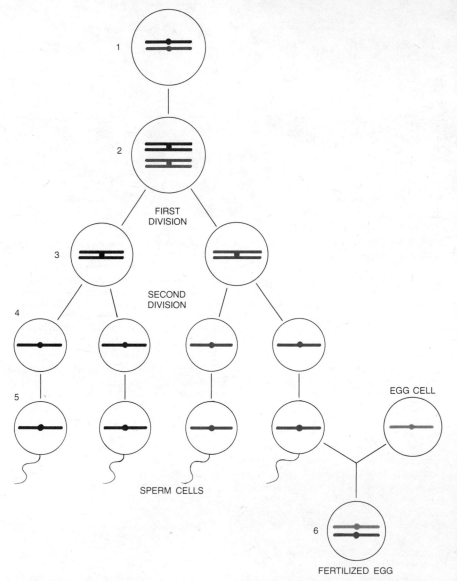

1

2

FIRST
DIVISION

3

SECOND
DIVISION

4

5

EGG CELL

SPERM CELLS

6

FERTILIZED EGG

MEIOSIS, the "reduction division" whereby a germ cell forms four sperms, is diagrammed for a hypothetical male germ cell (1) carrying only a single pair of homologous chromosomes, one from the male parent (color) and one from the female (black). The chromosomes replicate (2) and then, in the first division, one replicated chromosome goes to each of the two new germ cells (3). In the second division each strand of each replicated chromosome goes to a different spermatid (4); each resulting sperm (5) has one representative of the original chromosome pair. In a real germ cell containing many chromosomes (four pairs in *Drosophila*, 23 pairs in human beings) each pair "segregates" independently; the sperm receives a random mixture of paternal and maternal chromosomes. A similar process takes place in egg formation. When sperm and egg fuse (6), the fertilized egg has a random combination of chromosomes from each parent.

Resonance in meiosis — See Holland.

that the chromosomes in *SD* heterozygotes (cells with one *SD* chromosome and one normal chromosome) went through meiosis unharmed, so that any disabling effect of *SD* would have to be subtler than outright breakage. Peacock and Erickson put forward a clever alternative idea. Some time earlier at Oak Ridge, Novitski and Iris Sandler had suggested on the basis of circumstantial evidence that only two of the four sperm cells produced in a single meiosis are functional—even in normal males. Peacock and Erickson thought the *SD* chromosomes might take advantage of this by somehow managing to get themselves included in the sperm cells that are destined to be functional.

It would seem to be a simple matter to distinguish between the two hypotheses, induced sperm dysfunction and preferential inclusion. If *SD* caused dysfunction of half of the sperm cells, then males with an *SD* chromosome should produce only half as many functional sperm cells as normal males, and the reduced sperm production might be reflected in reduced fertility. The trouble is that ordinarily the number of sperm cells produced by a male fly is much larger than the number required to fertilize all of a female's eggs; as a result the failure of even half of the sperm cells to function might not reduce fertility. Daniel L. Hartl, who was then a graduate student at Wisconsin, Hiraizumi and I found a way to get around the problem. By mating very young males (which produce fewer sperm cells than mature males) or by mating one male with many females over several days (so that the male's sperm supply became exhausted) we made sperm production the limiting factor determining the number of progeny. Quite independently and at almost exactly the same time Benedetto Nicoletti and Gianni Trippa of the University of Rome did similar experiments. The two groups reached the same conclusions. Males carrying an *SD* chromosome did indeed produce fewer progeny than normal males and not the same number, as would be the case if preferential inclusion were the mechanism. And the number of their progeny was reduced in just the proportion expected if the *SD* chromosome was causing a dysfunction of the sperm cells that received the normal, non-*SD* chromosome.

Clinching evidence for a dysfunction induced by *SD* came from electron-microscope studies of sperm maturation. Nicoletti reported that about half of the sperm cells in *SD* males had tails with an abnormal appearance. Then Peacock, Kiyoteru T. Tokuyasu and Robert W. Hardy of the University of California at San Diego confirmed Nicoletti's finding and revealed additional detail. They showed that whereas ordinarily each nu-

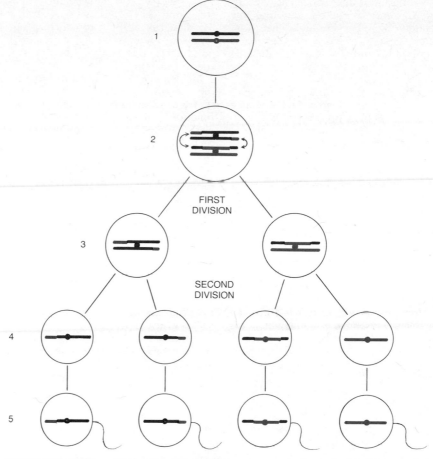

CROSSING-OVER multiplies the gene-shuffling process as is shown in this diagram of the same hypothetical meiosis as the one depicted in the illustration on the opposite page. In the course of replication the two members of a homologous pair may break at corresponding points. Homologous segments change places, so that some paternal genes (*color*) end up on maternal chromosome (*black*) and vice versa (*2*); individual genes become independent units in shuffle.

cleus in a bundle of 64 immature sperm cells (derived by two meiotic divisions from a group of 16 germ cells) becomes small and dense to form a sperm head, only half of the nuclei condensed normally in flies carrying one *SD* chromosome and one normal chromosome. The effect was even more striking in the tail region of the sperm cells, which at first are held together in a single mass of protoplasm and then normally separate to form 64 individual tails. In *SD* males only half of the sperm tails were thus individualized; the rest of them remained massed in larger groups. The sperm cells that failed to develop properly were clearly those that did not contain the *SD* chromosome. This was established by running experiments in which the males' non-*SD* chromosome was one that is resistant to the effect of *SD*. In such males all 64 sperm cells in a bundle developed normally.

The most obvious deduction from these results would be that the *SD* chromosome somehow injures its non-*SD* partner and renders it unable to carry out its usual function. That, however,

cannot be correct; the failure to develop cannot be simply the result of the failure to perform a normal function required for maturation because it has been known for a long time that a sperm cell's functioning does not depend on its chromosomal content. As long ago as 1927 H. J. Muller showed that a sperm cell can function normally even if many of its genes are missing. He produced drosophila strains in which some of the sperm cells lacked certain chromosomal material and some of the egg cells had a corresponding excess; when these sperm and egg cells were combined, "two wrongs made a right," as Muller put it, and a normal fly was hatched.

More recently Dan L. Lindsley of San Diego and Ellsworth H. Grell of Oak Ridge were able to produce sperm cells containing only the dotlike chromosome No. IV; such a sperm cell, when it was combined with an egg cell containing complementary extra chromosomes, gave rise to a normal fly. Since it was already known that the fourth chromosome is not needed for sperm-cell function, the experiment indicated that sperm function does not require any sperm chromosomes at all. The effect of the *SD* chromosome on its homologue cannot, then, be simply to inactivate some function, because no function is required. *SD* must somehow induce its partner to commit a positive act of sabotage.

Further evidence came from an analysis by Laurence Sandler and Adelaide Carpenter at the University of Washington. They reviewed the consequences of a rare error in sperm production by males with an *SD* chromosome and a normal one, as a result of which some sperm cells carried either both chromosomes or neither of them. The sperm cells with no chromosomes were functional but those with both *SD* and the normal chromosome were not, supporting the conclusion that *SD* does something to its homologue that causes the homologue in turn to produce sperm dysfunction. *SD* perpetuates itself by inducing its partner to destroy itself.

How this normal partner is instructed by *SD* to misbehave is not known, nor is the nature of the misbehavior. There is, however, one lead to the molecular details of what goes wrong in the aberrant sperm cell. During sperm-cell maturation there is normally a chemical change in the cell nucleus: lysine, one of the amino acids, is replaced by arginine. In *SD* flies this process fails, at least partly. The way is now open for a chemical

study that could yield the details. At this stage one conclusion emerges clearly: The effect of *SD* is a complicated one requiring a number of active processes, not merely a failure of some normal sperm function.

Incidentally, it is a good thing (from the organism's point of view) that sperm-cell function does not require functional genes. If functional genes were required, it would be much easier for an *SD*-like system to get started, because then all the disruptive chromosome would have to do would be to injure its partner enough to prevent its functioning. As it is the *SD* chromosome must do more: it must cause its partner to become actively harmful to sperm-cell function. It is always easier to stop something than to start something new. Furthermore, if there were genes affecting sperm-cell function, there would be competition among sperm cells, and a gene that improved the ability to fertilize would increase in the population. If such a gene happened also to cause, say, malfunction of the liver, that would be just too bad; the gene would increase anyway, since selection for good health is much less effective than selection by competition among sperm cells. Whatever the evolutionary reason for the nonfunctioning of genes in sperm cells may be, the nonfunctioning makes it more difficult for harmful *SD*-like sys-

tems to arise and also prevents harmful, or at best irrelevant, competition among sperm cells carrying different genes.

Genetic analysis of the *SD* chromosome to locate the relevant genes and learn how they interact has been difficult and time-consuming. As I have indicated, there is no way to recognize an *SD* fly by its appearance; *SD*'s effect is on the ratios of the progeny, and distorted ratios can be recognized only when the chromosome is labeled with mutant marker genes. Moreover, experiments with various strains have sometimes produced inconsistent results. Finally, gene mapping by traditional crossing-over methods has been complicated by the inverted gene sequences I mentioned above, which tend to suppress crossing-over. Nevertheless, experiments in my own laboratory and in others have established that the main segregation-distortion effect is caused by two genes that are very close to each other but that straddle the centromere, the point where the fibers that pull sister chromosomes apart during cell division are attached. Here I shall designate the genes *S*, for *segregation distorter*, and *R*, for *responder*. With a superscript plus sign to indicate the corresponding normal genes, there are then four kinds of chromosome: S-R, S-R^+, S^+-R and S^+-R^+. The first is the *SD* chromosome, with its two components; the last is the normal chromosome.

The results obtained by mating males with certain combinations of these chromosomes are particularly revealing [*see illustration on page 25*]. The first row of the table shows once again the high degree of segregation distortion caused by the *SD* chromosome. The second and third rows show that neither component has an effect by itself; both *S* and *R* are required. The fourth row brings a surprise: when *R* is only on the partner chromosome, *S* is changed from a distorter gene to a suicide gene. The fifth row shows that the S^+-R chromosome is immune to the distorting effect of S-R. The *S* gene must direct the synthesis of some product that influences the R^+ gene on the homologous chromosome (or, in the suicide case, on the same chromosome) to prevent that chromosome's sperm cell from maturing properly. As I have emphasized, the effect must be a positive act on the part of the R^+ gene.

Barry S. Ganetzky, a graduate student of Laurence Sandler's at Washington, added significant details. He prepared chromosomes in which small pieces had been deleted by X rays. When the *S* gene was deleted, the distorting effect was lost; the chromosome behaved as though the normal gene were present. And so the *S* gene appears to be doing something that was not being done by any gene on a normal chromosome; the S^+ gene does nothing (or does not exist).

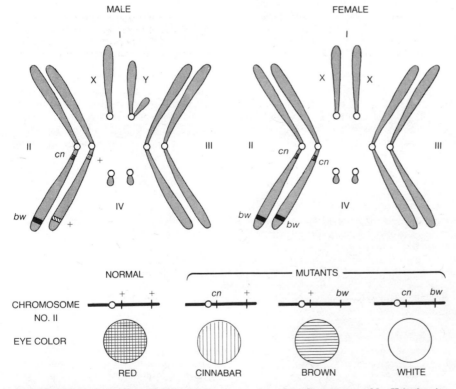

MALE   FEMALE

	NORMAL	MUTANTS		
	+ +	cn +	+ bw	cn bw
CHROMOSOME NO. II				
EYE COLOR	RED	CINNABAR	BROWN	WHITE

FRUIT FLY *D. melanogaster* has four pairs of chromosomes. Chromosome No. II is the site of eye-color mutations called *cn*, for *cinnabar* (*solid color*), and *bw*, for *brown* (*black segment*); a corresponding hatched segment indicates the corresponding normal (+) eye-color gene. The effects of the mutations are shown below. Two normal genes produce normal dark red eyes. The mutation *cinnabar* produces scarlet eyes, the mutation *brown* brown eyes; both mutations together produce white eyes. Female whose chromosomes are shown is white-eyed. Since normal genes are dominant they mask recessive mutant genes, and so male has red eyes.

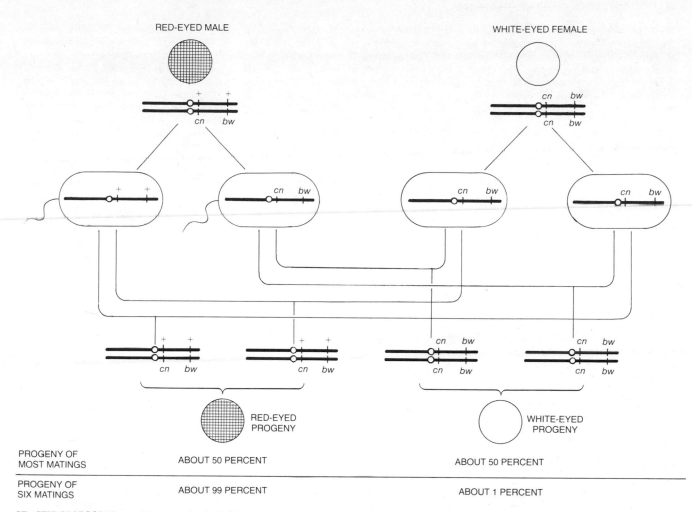

RED-EYED MALE

WHITE-EYED FEMALE

RED-EYED PROGENY

WHITE-EYED PROGENY

PROGENY OF MOST MATINGS	ABOUT 50 PERCENT	ABOUT 50 PERCENT
PROGENY OF SIX MATINGS	ABOUT 99 PERCENT	ABOUT 1 PERCENT

SD CHROMOSOME was discovered when red-eyed males having one No. II chromosome from a natural population and one from a white-eyed laboratory stock were mated to females having two laboratory chromosomes. The possible combinations of sperm cells and egg cells are shown. According to the rules of Mendelian inheritance, about half of the progeny should be red-eyed and half white-eyed. The progeny of each of some 200 matings did show the expected 50:50 ratio, but progeny of six matings were almost all red-eyed.

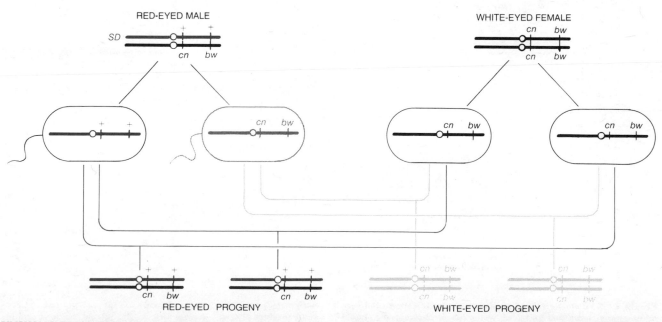

RED-EYED MALE

WHITE-EYED FEMALE

RED-EYED PROGENY

WHITE-EYED PROGENY

UNUSUAL RESULTS of the six matings are interpreted here. The eye-color ratio (almost 100 percent red-eyed) shows that the No. II chromosome from the natural population, bearing normal eye-color genes, is viable: it is combining with an egg's mutation-bearing chromosome and producing the expected red-eyed progeny. The homologous male chromosome, however, is apparently not combining successfully with a similar egg chromosome to produce white-eyed progeny. The chromosome from the natural population is distorting the results of segregation. It is a "segregation distorter" (*SD*) that favors its own survival by inactivating sperm cells containing its homologue.

When the R or R^+ gene was deleted, the chromosome behaved as if it carried an R gene; in other words, the R gene is not doing anything. To recapitulate: The S gene produces something that acts directly on the normal R^+ gene; when R^+ mutates to R or is deleted, it is no longer affected by S.

When does this effect of S on R^+ take place? The last time the homologous chromosomes are in the same nucleus and can interact easily is during early meiosis. That is eight or nine days before the time of sperm maturation, when the actual damage appears. Evidence supporting the time lag came from another experiment. If SD males are grown at a temperature of 19 degrees Celsius rather than the customary 25 degrees, the amount of distortion is substantially reduced. This temperature sensitivity provided an experimental handle for Elaine Mange, who was then a graduate student at Wisconsin. She lowered the temperature at which some flies were maintained from 25 degrees to 19 for just a brief period and mated the males at specified times after the temperature reduction. It turned out that the progeny of males mated between eight and nine days after the low-temperature pulse displayed less distorted ratios. Since it takes about that time for a cell to progress from early meiosis to the mature-sperm stage, this showed that the temperature-sensitive period is during early meiosis. These experiments were repeated carefully and extensively by Yukiko K. Hihara of Tokyo Metropolitan University, who got the same results. It appears, then, that the S gene communicates with or somehow influences the

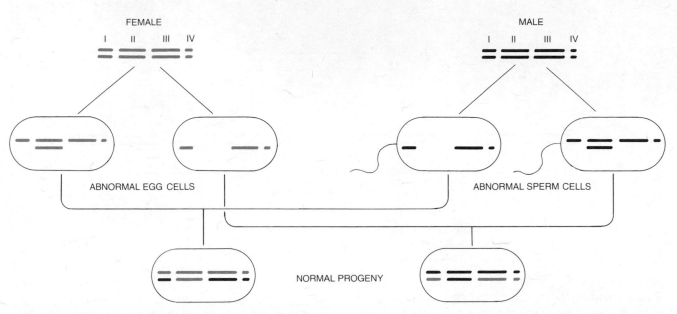

TWO WRONGS MAKE A RIGHT when abnormal egg (or sperm) cells containing excess chromosomal material happen to combine with sperm (or egg) cells having a corresponding deficiency. Here an egg has two No. II chromosomes and a sperm lacks that chromosome, or vice versa, and yet progeny are normal. If sperm cells can function without a chromosome, SD cannot merely inactivate its homologue.

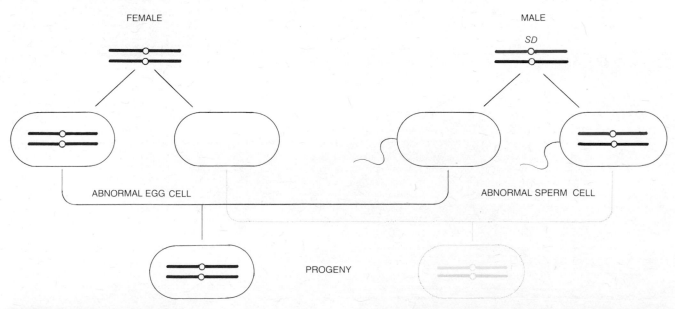

EITHER COMBINATION of an abnormal egg and sperm cell containing a deficiency and an excess in the No. II chromosome should produce progeny, according to the results shown in the upper illustration on this page. That is not the case when the two chromosomes in the male are the SD chromosome and a normal No. II chromosome. Although sperm cells with neither the SD nor the normal chromosome produce progeny, those with both the SD and the normal chromosome do not. Apparently, then, SD has some positive effect that causes its homologous chromosome to produce sperm-cell dysfunction, whether or not the SD chromosome is present in same sperm.

R^+ gene on the homologous chromosome while the two partners are paired early in meiosis.

Ganetzky also discovered a third gene close to the centromere, designated *En* for *enhancer*, that intensifies the distorting effect. The full effect (a progeny ratio of 99 percent or more) of an *SD* chromosome requires the presence of all three genes. A multiple-component system such as this must have evolved in stages over a long time. It could have a disastrous effect on the population, but it does not seem to. We have been able to learn quite a lot about its effect on a population and can make some plausible conjectures as to its evolution.

Because an *SD* chromosome is transmitted to almost all the progeny rather than to just half of them it ought to spread through a population like wildfire and quickly replace its homologue. It is clear enough why the *SD* chromosome does not in fact reach 100 percent. When a male is homozygous for *SD* (has two *SD* chromosomes), it is nearly sterile; in some strains such flies do not even survive. In a fly population the tendency of the *SD* heterozygote to increase through segregation distortion is countered by the tendency of the *SD* homozygote to be sterile or to die. One can easily compute what the frequency of *SD* chromosomes should be in such a system. The calculations predict a high frequency: more than 50 percent. And yet we actually find an *SD* frequency of less than 5 percent in natural populations. Something is holding the frequency down.

One brake on *SD* is provided by the presence of various "modifying genes" that lower the degree of distortion; such genes are scattered along all four drosophila chromosomes and are found in almost all natural populations of the flies. The most important brake, however, was revealed by Hartl's recent finding that about half of the "normal" chromosomes in one natural population are not S^+-R^+; they are S^+-R, and such chromosomes are immune to distortion by *SD*. There has not yet been any extensive analysis of other populations, but probably they too have a high proportion of S^+-R chromosomes.

I think we can reconstruct what happened. Somehow, a long time ago, the *S-R* chromosome arose. At first it increased rapidly. The increase was offset, however, by an increase of normal chromosomes that are resistant, that is, of chromosomes with the composition S^+-R. A three-way competition ensued. The *S-R* chromosome's tendency to increase was held in check by S^+-R, which nullified the former's meiotic advantage. And of course the S^+-R chromosome had an advantage over the other normal one, S^+-R^+, which is hardly

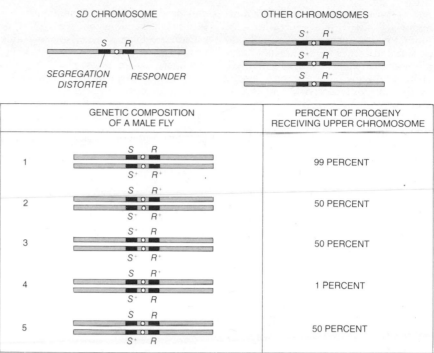

SD CHROMOSOME

SEGREGATION DISTORTER — RESPONDER

S *R*

OTHER CHROMOSOMES

S^+ R^+

S^+ R

S R^+

	GENETIC COMPOSITION OF A MALE FLY	PERCENT OF PROGENY RECEIVING UPPER CHROMOSOME
1	*S* *R* / S^+ R^+	99 PERCENT
2	*S* R^+ / S^+ R^+	50 PERCENT
3	S^+ *R* / S^+ R^+	50 PERCENT
4	*S* R^+ / S^+ *R*	1 PERCENT
5	*S* *R* / S^+ *R*	50 PERCENT

TWO MUTANT GENES, designated *segregation distorter* (*S*) and *responder* (*R*), are primarily responsible for the *SD* chromosome's effect. The four possible combinations of mutant (*color*) and normal (+) genes are shown at the top. The table gives the results of some matings involving chromosomes carrying these combinations. Genes *S* and *R* together exert the full *SD* effect (*1*), but neither gene has an effect alone (*2, 3*). When *R* is only on the homologous chromosome, *S* becomes a suicide gene (*4*). The S^+-*R* chromosome is immune to the *SD* effect (*5*). Apparently the *S* gene exerts an effect on the R^+ gene such that the R^+ gene causes sperm dysfunction.

transmitted at all when it comes up against *S-R*.

There have been a number of mathematical analyses of this system and similar ones, most recently by Hartl and Brian Charlesworth of the University of Sussex. The results of one of my own computer runs are presented in the diagram on the next page. There are three chromosome types of interest: *S-R*, S^+-R^+ and S^+-R. (The fourth, *S-R^+*, is rare, and I have ignored it. It has no competitive advantage; in fact, it commits suicide when it is combined with S^+-R.) The frequency of each of the three chromosome types at every 10th generation is plotted in triangular coordinates. The proportion of *S-R* chromosomes is indicated by the distance from the base of an equilateral triangle whose altitude is 1, the proportion of S^+-R^+ chromosomes by the perpendicular distance from the right side of the triangle and the proportion of S^+-R chromosomes by the perpendicular distance from the left side.

The original population must have been S^+-R^+. Then the *S-R* chromosome arose, presumably in two steps. Some S^+-*R* (and *S-R^+*) chromosomes must also have been present in low frequencies; they could arise by crossing-over between S^+-R^+ and *S-R* chromosomes. I assumed that the two rare types

S-R and S^+-*R* began with a frequency of 1 percent each, as indicated by the starting point near the bottom left-hand corner. What I then did was to take plausible values for the degree of distortion and the relative viability and fertility of the various chromosome combinations. These values are not known exactly, since laboratory data may not reflect the true values in nature; the diagram reflects one set of plausible values that leads eventually to a point of equilibrium very close to what is actually found in a natural population.

At first the *S-R* chromosome increases rapidly; after about 45 generations more than half of the chromosomes are of this type. Because of its resistance to distortion caused by *S-R*, the S^+-*R* chromosome then starts to increase at the expense of both *S-R* and S^+-R^+, and the trajectory moves toward the bottom right-hand corner of the diagram. When the *S-R* type becomes rare, there is no longer any advantage for S^+-*R*, and because it has a somewhat reduced fertility it decreases in frequency and the trajectory moves to the left across the bottom of the diagram. When S^+-*R* becomes rare, *S-R* again increases because of its meiotic advantage over S^+-R^+. After about 325 generations one cycle has been completed and the process is repeated through a loop with a smaller amplitude. The cycle continues indefinitely, spiraling toward the center; only about the first 1,000 generations are plotted on the diagram. Eventually the population comes to equilibrium at the point indicated by the triangle at the center of the spiral, with about 4 percent of the chromosomes *S-R* and the remainder roughly half *S+-R+* and half *S+-R*. The diagram must represent a rough history of what went on among the three chromosomes competing for meiotic advantage. Exact calculations are complicated by the many modifying genes on other chromosomes, which shift the balance one way or another.

In the course of this three-way tug-of-war the *S-R* chromosome "tries" to hold on to any modifying genes that enhance its effect. *En* is one such gene and there are many minor modifying genes on the chromosome. Now the function of the multiple inversions on the *SD* chromosome becomes clear. By preventing crossing-over they keep the enhancing modifiers locked to the *S-R* complex; without the inversions they would become separated and soon would be found just as often on the normal chromosomes as on the *SD* chromosome.

The result of this complex interaction is that the *SD* chromosome seems to be effectively held in check by the presence of the *S+-R* chromosome and the modifying genes, so that *SD's* frequency in a population is not great enough to do much harm. The *SD* system and the other meiotic-drive systems found in nature are those to which the population has become adjusted in this way and similar ones. Populations in which such an adjustment was not worked out may simply have become extinct.

What would happen if a system such as *SD* were to arise on one of the chromosomes determining sex? For example, a normal male produces an equal number of sperm cells bearing X chromosomes and Y chromosomes, so that when the sperm cells fertilize X-bearing egg cells, XX females and XY males are produced in equal numbers. If an *SD* complex were to be located on the Y chromosome, nearly 100 percent of the progeny would be expected to be males. No such Y chromosome is known in *Drosophila*, but as a graduate student at Wisconsin, Terrence W. Lyttle, now at the University of Hawaii, contrived a way to produce one. By means of radiation he promoted an exchange of chro-

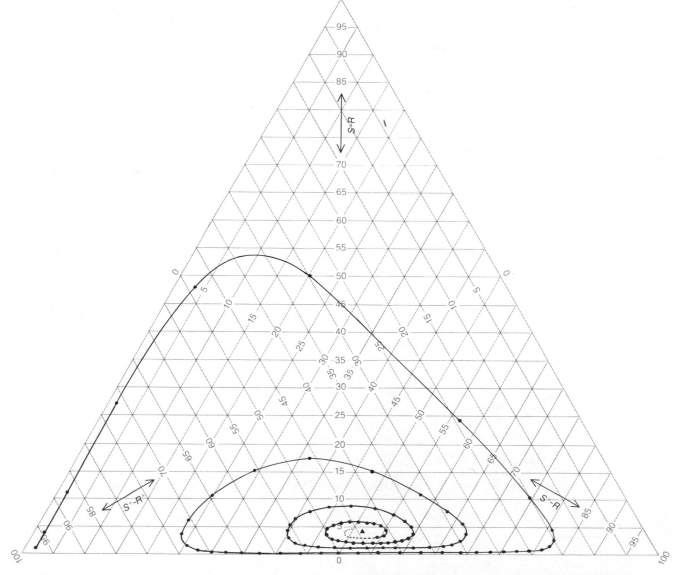

INTERPLAY OF THREE CHROMOSOMES in a fly population is traced by the spiraling curve, which gives the results of one of the author's computer runs. The proportion in the population of *S-R*, *S+-R+* and *S+-R* chromosomes is plotted, at every 10th generation (*dots*), in triangular coordinates; each chromosome's frequency in the population is measured along a different altitude of the equilateral triangle. At the beginning (*bottom left*) the population is 98 percent *S+-R+*. The distorting effect of *S-R* is countered by the resistance of *S+-R*, which in turn has reduced fertility compared with *S+-R+*. Calculations based on plausible values for distortion, fertility and viability indicate that the population goes through a succession of cycles— the loops on the graph—that eventually bring it to a point of equilibrium (*black triangle*), at which some 4 percent of the chromosomes are *S-R* and remainder of them are roughly half *S+-R+* and half *S+-R*.

mosomal segments that tied the *SD* trait to the Y chromosome. Distortion by *SD* causes males carrying this translocation to produce an excess of sperm cells containing the Y chromosome, so that almost all their progeny are males. A population afflicted by such a system should become extinct within a few generations for lack of females.

Lyttle demonstrated the effect with an artificial population in the laboratory. The drosophilas were maintained for several generations in small cages, each one holding a few thousand flies. As expected, the population soon contained no females, and it died out. Then Lyttle did an experiment in which he started with a normal population and introduced a few males carrying the translocation. The distorting chromosomes from these males gradually replaced the normal Y chromosomes, and when they became prevalent enough, the proportion of females began to decrease. After a few generations this population too was extinct.

Clearly a meiotic-drive system, which is bad enough on one of the nonsex chromosomes, can be a disaster if it is on the Y chromosome. W. D. Hamilton of Michigan State University has suggested that one reason the Y chromosome is largely devoid of genes in many species is that the lack of genes prevents meiotic-drive mutants from arising and wiping out the population.

A distorting Y chromosome would seem to provide an ideal biological control of an insect population. An example of a meiotic-drive gene has already been discovered, by George B. Craig of the University of Notre Dame, near the male-determining gene of *Aëdes aegypti*, the yellow-fever mosquito. Unfortunately for any control application, there are so many modifying genes in the *A. aegypti* population that the distorting gene is rather ineffective outside the laboratory. If meiotic-drive systems are to be adapted for practical insect control, new mutant distorting genes will presumably have to be found in the laboratory for which there is no reservoir of distortion-reducing modifiers already in the population. It will probably be difficult to develop truly new mutants that have not arisen at some time in nature. It may nonetheless be possible, and in that case meiotic-drive genes could become a major new nonchemical technique in the continuing battle with insect pests.

Among other examples of cheating genes the best-known is the one that causes a variety of tail abnormalities in various populations of house mice. These mutants are often highly damaging or even lethal in homozygous mice, and it is clear that they are much too prevalent for the good of the population; they should be eliminated by natural selection. Instead they are maintained in the population by distorted sperm ratios similar to those of *SD*. Mice are not as amenable to genetic analysis as flies, and so the details of this system are less well known; superficially, at least, the two systems appear to be very similar.

In plants a number of instances are known in which extra chromosomes that are harmful to the population are maintained by some kind of tricky behavior. One such chromosome is transmitted in excess of its rightful amount by getting itself included preferentially in the pollen-tube nucleus that fertilizes the egg. Other chromosomes take advantage of the fact that in females only one of the four products of meiosis is fertilized, the others becoming nonfunctional polar bodies; a chromosome may cheat by managing too often to get into the egg nucleus that is destined to be fertilized. In addition to these and other examples known in nature a number of cheating genes have arisen in the laboratory, sometimes by accident and sometimes in experiments designed to produce such genes.

Is it possible that meiotic-drive genes are commoner in nature than has been assumed? The examples that have been studied are all those with an extreme effect, such as *SD*. If *SD* produced a sperm ratio not of nearly 100 percent but of, say, 55 percent, it would probably never have been discovered. It is possible that mild cases of meiotic drive are rather prevalent in natural populations but have not been detected. Indeed, it has been suggested that the incidence of some human genetic diseases that are commoner than they ought to be may be explained by distorting genes, but there are other equally plausible explanations.

Mendelian inheritance is a marvelous device for making evolution by natural selection an efficient process. It would appear to be the best system that could be contrived (within the mechanical constraints imposed by the fact that genes are linked in chromosomes) for giving each gene a thorough test in combination with many other genes. The Mendelian system works with maximum efficiency only if it is scrupulously fair to all genes. It is in constant danger, however, of being upset by genes that subvert the meiotic process to their own advantage. If such genes have a harmful effect (as in the case of the sterility or lethality induced by homozygous *SD*), the population is weakened directly. Even if the cheating genes are not harmful in their own right, they inhibit the evolutionary process by reducing its efficiency. There are many refinements of meiosis and sperm formation whose purpose is apparently to render such cheating unlikely. And yet some genes have managed to beat the system.

3

The Genes of Men and Molds

by George W. Beadle
September 1948

*The study of the red fungus Neurospora crassa sheds
light on exactly how the units of heredity determine
the characteristics of all living things*

EIGHTY-FIVE years ago, in the garden of a monastery near the village of Brünn in what is now Czechoslovakia, Gregor Johann Mendel was spending his spare moments studying hybrids between varieties of the edible garden pea. Out of his penetrating analysis of the results of his studies there grew the modern theory of the gene. But like many a pioneer in science, Mendel was a generation ahead of his time; the full significance of his findings was not appreciated until 1900.

In the period following the "rediscovery" of Mendel's work biologists have developed and extended the gene theory to the point where it now seems clear that genes are the basic units of all living things. They are the master molecules that guide the development and direct the vital activities of men and amoebas.

Today the specific functions of genes in plants and animals are being isolated and studied in detail. One of the most useful genetic guinea pigs is the red bread mold *Neurospora crassa*. Its genes can conveniently be changed artificially and the part that they play in the chemical alteration and metabolism of cells can be analyzed with considerable precision. We are learning what sort of material the genes are made of, how they affect living organisms and how the genes themselves, and thereby heredity, are affected by forces in their environment. Indeed, in their study of genes biologists are coming closer to an understanding of the ultimate basis of life itself.

It seems likely that life first appeared on earth in the form of units much like the genes of present-day organisms.

Through the processes of mutation in such primitive genes, and through Darwinian natural selection, higher forms of life evolved—first as simple systems with a few genes, then as single-celled forms with

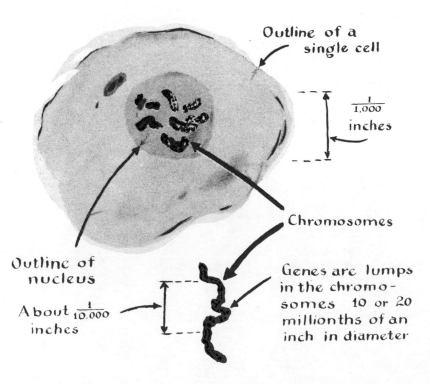

THE CELL is the site of nearly all the interactions between the gene and its environment. The genes themselves are located in the chromosomes, shown above in the stage before cell divides, duplicating each gene in the process.

THE MOLD *Neurospora* is an admirable organism for the study of genes, mainly because of its unusually simple reproductive apparatus. This may be neatly dissected to isolate a single complete set of genes. The sequence of steps in the drawing at the right shows how the tiny fruiting body of the mold is taken apart in the laboratory. With the aid of a microscope, the laboratory worker is able to spread out a set of spore sacs, each containing eight spores. One spore sac may then be separated from the others, and its spores carefully removed. The individual spores are lined up on a block of agar and finally planted in a test tube which contains all the substances that are normally required for the mold to grow.

many genes, and finally as multicellular plants and animals.

What do we know about these genes that are so all-important in the process of evolution, in the development of complex organisms, and in the direction of those vital processes which distinguish the living from the non-living worlds?

In the first place, genes are characterized by students of heredity as the units of inheritance. What is meant by this may be illustrated by examples of some inherited traits in man.

Blue-eyed people may differ by a single gene from those with brown eyes. This eye-color gene exists in two forms, which for convenience may be designated *B* and *b*.

Every person begins as a single cell a few thousandths of an inch in diameter—a cell that comes into being through the fusion of an egg cell from the mother and a sperm cell from the father. This fertilized egg carries two representatives of the eye-color gene, one from each parent. Depending on the parents, there are therefore three types of individuals possible so far as this particular gene is concerned. They start from fertilized eggs represented by the genetic formulas *BB*, *Bb* and *bb*. The first two types, *BB* and *Bb*, will develop into individuals with brown eyes. The third one, *bb*, will have blue eyes. You will note that when both forms of the gene are present the individual is brown-eyed. This is because the form of the gene for brown eyes is *dominant* over its alternative form for blue eyes. Conversely, the form for blue eyes is said to be *recessive*.

During the division of the fertilized egg cell into many daughter cells, which through growth, division and specialization give rise to a fully developed person, the genes multiply regularly with each cell division. As a result each of the millions of cells of a fully developed individual carries exact copies of the two representatives of the eye-color gene

A fruiting body is placed on a block of agar under a low power microscope.

It is pinched with tweezers until it breaks and ejects its spore sacs intact.

A drop of water disentangles the spore sacs.

With a pyrex needle, a single sac is isolated.

Individual spores are pressed out of the end of the sac and arranged in order. The spores are spaced along the edge of the agar.

Platinum-iridium knife

The agar is cut in squares.

A drop of chlorox is spread over the spores to kill bacteria and asexual spores.

The squares are lifted out of the block and placed in a labeled tube of medium to develop.

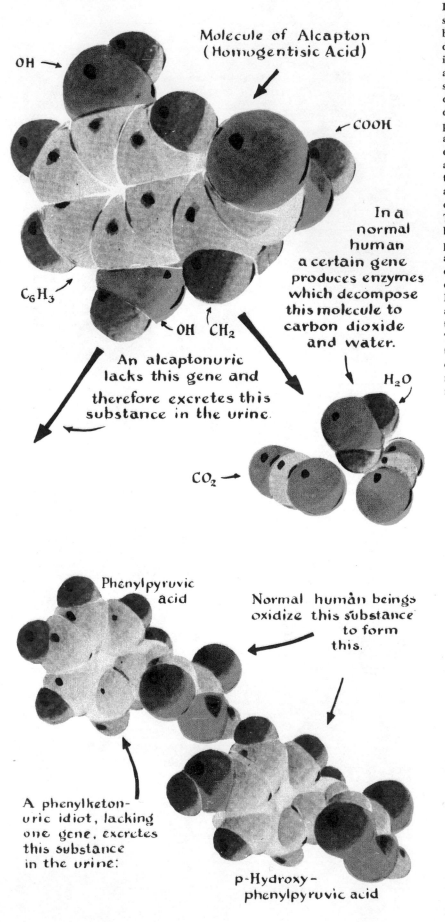

Molecule of Alcapton
(Homogentisic Acid)

OH →

← COOH

In a normal human a certain gene produces enzymes which decompose this molecule to carbon dioxide and water.

C_6H_3

OH CH_2

An alcaptonuric lacks this gene and therefore excretes this substance in the urine.

H_2O

CO_2 →

Phenylpyruvic acid

Normal human beings oxidize this substance to form this.

A phenylketon-uric idiot, lacking one gene, excretes this substance in the urine:

p-Hydroxy-phenylpyruvic acid

DEFECTIVE GENES in man can cause serious hereditary disorders. The chemical basis of two such disorders is shown in the drawings on this page. The large molecule in the drawing at the left is homogentisic acid, or alcapton. In most human beings a single gene produces an enzyme which is capable of breaking alcapton down to carbon dioxide and water. When the gene that produces the enzyme is faulty, however, alcapton is not decomposed. It must be eliminated in the urine, to which it gives a dark color. This excretion of alcapton in the urine is called alcaptonuria. The drawing at the bottom of this page shows the basis of a much more serious genetic disorder. The biochemical apparatus of most human beings, again, is able to transform phenyl-pyruvic acid into p-hydroxy phenylpyruvic acid. Those who cannot transform it are called phenylketonurics. Phenylketonuria is characterized by extreme feeblemindedness. Most phenyketonurics are imbeciles or idiots; a few are low-grade morons. The faulty genes that are responsible for both are recessive. This means that they are expressed only when two such genes are paired in the union of an egg and sperm cell. Thus most of the genes responsible for these disorders are carried by normal people without being expressed.

which has been contributed by the parents.

In the formation of egg and sperm cells, the genes are again reduced from two to one per cell. Therefore a mother of the type *BB* forms egg cells carrying only the *B* form of the gene. A type *bb* mother produces only *b* egg cells. A *Bb* mother, on the other hand, produces both *B* and *b* egg cells, in equal numbers on the average. Exactly corresponding relations hold for the formation of sperm cells.

With these facts in mind it is a simple matter to determine the types of children expected to result from various unions. Some of these are indicated in the following list:

Mother	Father	Children
BB (brown)	*BB* (brown)	All *BB* (brown)
Bb (brown)	*Bb* (brown)	¼ *BB* (brown)
		½ *Bb* (brown)
		¼ *bb* (blue)
BB (brown)	*bb* (blue)	All *Bb* (brown)
Bb (brown)	*bb* (blue)	½ *Bb* (brown)
		½ *bb* (blue)
bb (blue)	*bb* (blue)	All *bb* (blue)

This table shows that while it is expected that some families in which both parents have brown eyes will include blue-eyed children, parents who are both blue-eyed are not expected to have brown-eyed children.

LIFE CYCLE of the mold *Neurospora* is illustrated in the drawing at the right. The hyphal fusion of Sex A and Sex a at the bottom of the page is taken as a starting point. *Neurospora* enters a sexual stage rather similar to the union of sperm and egg cells in higher organisms. The union produces a fertile egg, in which two complete sets of genes are paired. The fertile egg cell then divides (*center of drawing*), and divides again. This produces four nuclei, each of which has only a single set of genes. Lined up in a spore sac, the four nuclei divide once more to produce four pairs of nuclei that are genetically identical. A group of spore sacs is gathered in a fruiting body. The sacs and the spores may then be dissected by the technique outlined on page 29. Following this, the germinating spores (*top of page*) may be planted in test tubes containing the necessary nutrients. It is at this point that genetic defects can be exposed by changing the constitution of the medium. Here also *Neurospora* may be allowed to multiply by asexual means. This makes it possible to grow large quantities of the mold without genetic change for convenient chemical analysis. The entire life cycle of the mold takes only 10 days, another reason why *Neurospora* is an exceptionally useful experimental organism.

It is important to emphasize conditions that may account for apparent exceptions to the last rule. The first is that eye-color inheritance in man is not completely worked out genetically. Probably other genes besides the one used as an example here are concerned with eye color. It may therefore be possible, when these other genes are taken into account, for parents with true blue eyes to have brown-eyed children. A second factor which accounts for some apparent exceptions is that brown-eyed persons of the *Bb* type may have eyes so light brown that an inexperienced observer may classify them as blue. Two parents of this type may, of course, have a *BB* child with dark brown eyes.

Another example of an inherited trait in man is curly hair. Ordinary curly hair, such as is found frequently in people of European descent, is dominant to straight hair. Therefore parents with curly hair may have straight-haired children but straight-haired parents do not often have children with curly hair. Again there are other genes concerned, and the simple rules based on a one-gene interpretation do not always hold.

Defective Genes

Eye-color and hair-form genes have relatively trivial effects in human beings. Other known genes are concerned with

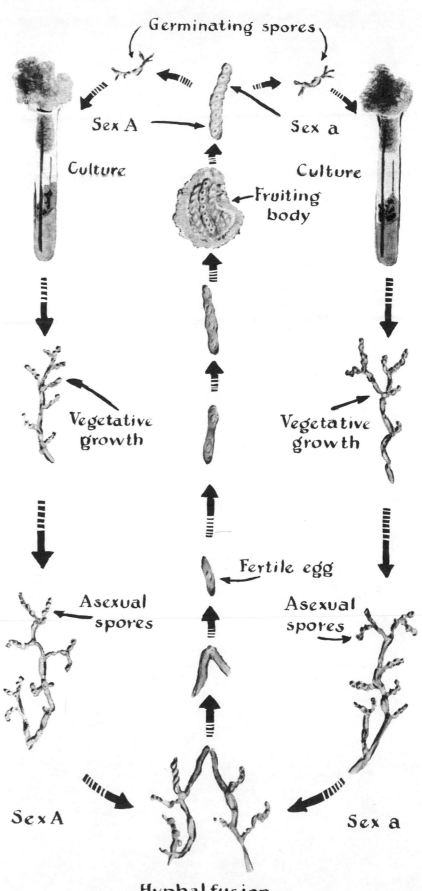

traits of deeper significance. One of these involves a rare hereditary disease in which the principal symptom is urine that turns black on exposure to air. This "inborn error of metabolism." as the English physician and biochemist Sir Archibald Garrod referred to it. has been known to medical men for probably 300 years. Its biochemical basis was established in 1859 by the German biochemist C. Bödeker, who showed that darkening of urine is due to a specific chemical substance called alcapton. later identified chemically as 2,5-dihydroxyphenylacetic acid. The disease is known as alcaptonuria, meaning "alcapton in the urine."

Alcaptonuria is known to result from a gene defect. It shows typical Mendelian inheritance, like blue eyes, but the defective form of the gene is much less frequent in the population than is the recessive form of the eye-color gene.

The excretion of alcapton is a result of the body's inability to break it down by oxidation. Normal individuals possess an enzyme (a protein-containing catalyst, often called a biocatalyst) which makes possible a reaction by which alcapton is further oxidized. This enzyme is absent in alcaptonurics. As a result alcaptonurics cannot degrade alcapton to carbon dioxide and water as normal individuals do.

Alcaptonuria is of special interest genetically and biochemically because it gives us a clue as to what genes do and how they do it. It is clear that the normal kind of gene is essential for the production of the enzyme necessary for the breakdown of alcapton. If the cells of an individual contain only the recessive or inactive form of the gene, no enzyme is formed, alcapton accumulates and is excreted in the urine. The relations between gene and chemical reaction are shown in the diagram at the top of page 30.

A hereditary error of metabolism related biochemically to alcaptonuria is phenylketonuria, a rare disease in which phenylpyruvic acid is excreted in the urine. Like alcaptonuria, this metabolic defect is inherited as a simple Mendelian recessive. It is more serious in its consequences, however, because it is invariably associated with feeble-mindedness of an extreme kind. Most phenylketonurics are imbeciles or idiots; a few are low-grade morons. It should be made clear, however, that only a small fraction of feeble-minded persons are of this particular genetic type.

Phenylketonurics excrete phenylpyruvic acid because they cannot oxidize it, as normal individuals can, to a closely related derivative differing from phenylpyruvic acid by having one more oxygen atom per molecule (see diagram at the bottom of page 30). Again it is evident that the normal form of a gene is essential for the carrying out of a specific chemical reaction.

Man, however, is far from an ideal organism in which to study genes. His life cycle is too long, his offspring are too few. his choice of a mate is not often based on a desire to contribute to the knowledge of heredity, and it is inconvenient to subject him to a complete chemical analysis. As a result, most of what we have learned about genes has come from studies of such organisms as garden peas. Indian corn plants and the fruit fly Drosophila.

In these and other plants and animals there are many instances in which genes seem. to be responsible for specific chemical reactions. It is believed that in most or all of these cases they act as pattern molecules from which enzymes are copied.

Many enzymes have been isolated in a pure crystalline state. All of them have proved to be proteins or to contain proteins as essential parts. Gene-enzyme relations such as those considered above suggest that the primary function of genes may be to serve as models from which specific kinds of enzyme proteins are copied. This hypothesis is strengthened by evidence that some genes control the presence of proteins that are not parts of enzymes.

For example, normal persons have a specific blood protein that is important in blood clotting. Bleeders, known as hemophiliacs, differ from non-bleeders by a single gene. Its normal form is presumed to be essential for the synthesis of the specific blood-clotting protein. Hemophilia, incidentally, is almost completely limited to the male because it is sex-linked; that is, it is carried in the so-called X chromosome, which is concerned with the determination of sex. As is well known, this hereditary disorder has been carried for generations by some of the royal families of Europe.

The genes that determine blood types in man and other animals direct the production of so-called antigens. These are giant molecules which apparently derive their specificity from gene models, and which are capable of inducing the formation of specific antibodies.

Neurospora

The hypothesis that genes are concerned with the elaboration of giant protein molecules has been tested by experiments with the red mold Neurospora. This fungus has many advantages in the study of what genes do. It has a short life cycle—only 10 days from one sexual spore generation to the next. It multiplies profusely by asexual spores. The result is that any strain can be multiplied a millionfold in a few days without any genetic change. Each of the cell nuclei that carry the genes of the bread mold has only a single set of genes instead of the two sets found in the cells of man and other higher organisms. This means that recessive genes are not hidden by their dominant counterparts.

During the sexual stage, in which

EXPERIMENT to determine the role of a single Neurospora gene essentially consists in disabling a gene and tracking down its missing biochemical function. Spores of the mold are first exposed to radiation that will cause mutation, i.e., change in a gene. This culture is then crossed with another. The spores resulting from this union are then planted in a medium that contains all the substances that normal Neurospora needs for growth, plus a few that the mold normally manufactures for itself. All the spores, including those which may carry a defective gene, germinate on this medium. Spores from these same cultures are then planted in a medium that contains only the bare minimum of substances required by Neurospora. Four of the cultures fail to grow, indicating that they have lost the power to manufacture one substance that Neurospora normally synthesizes. In test tubes at the bottom of opposite page, the detailed identification of exactly what synthetic power has been lost is begun by planting the defective culture in media that contain (1) all substances required by the normal mold plus vitamins, and (2) all substances plus amino acids. When mold grows on first medium, it appears it has lost the power to synthesize vitamin.

molds of opposite sex reactions come together, there is a fusion comparable to that between egg and sperm in man. The fusion nucleus then immediately undergoes two divisions in which genes are reduced again to one per cell. The four products formed from a single fusion nucleus by these divisions are lined up in a spore sac. Each divides again so as to produce pairs of nuclei that are genetically identical. The eight resulting nuclei are included in eight sexual spores, each one-thousandth of an inch long. This life cycle of Neurospora is shown in the illustration on page 31.

Using a microscope, a skilled laboratory worker can dissect the sexual spores from the spore sac in orderly sequence. Each of them can be planted separately in a culture tube (see illustration on page 29). If the two parental strains differ by a single gene, four spores always carry descendants of one form of the gene and four carry descendants of the other. Thus if a yellow and a white strain are crossed, there occur in each spore sac four spores that will give white molds and four that will give yellow.

The red bread mold is almost ideally suited for chemical studies. It can be grown in pure culture on a chemically known medium containing only nitrate, sulfate, phosphate, various other inorganic substances, sugar and biotin, a vitamin of the B group. From these relatively

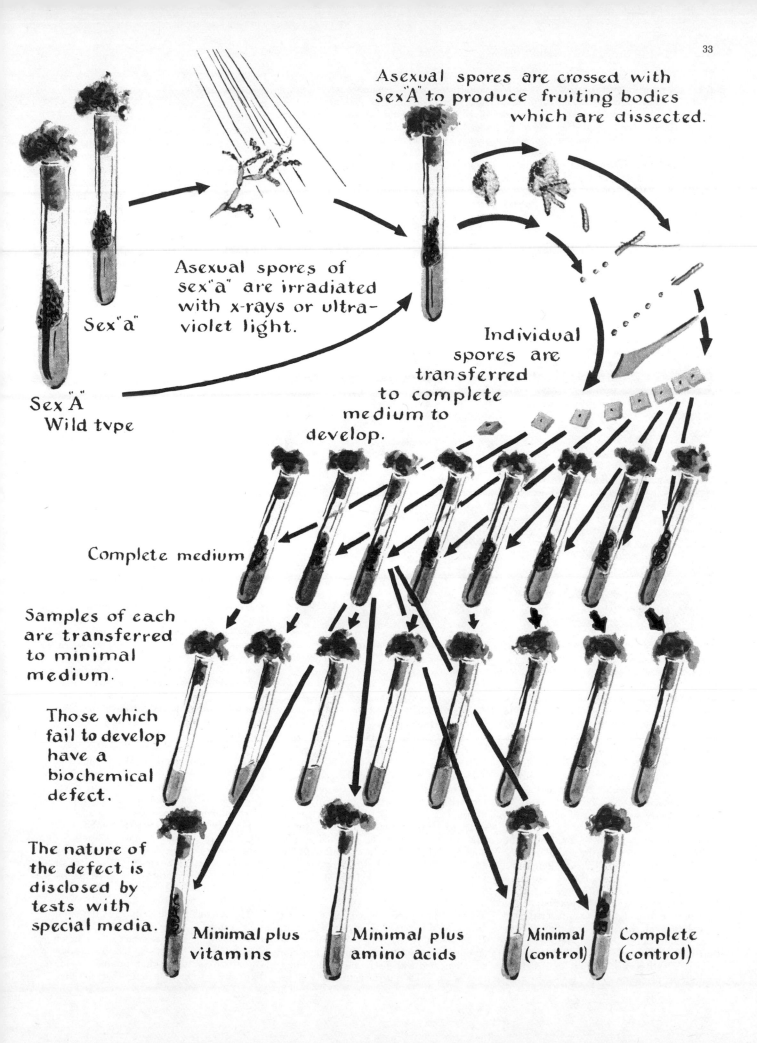

Asexual spores are crossed with sex "A" to produce fruiting bodies which are dissected.

Sex "a"

Asexual spores of sex "a" are irradiated with x-rays or ultra-violet light.

Sex "A" Wild type

Individual spores are transferred to complete medium to develop.

Complete medium

Samples of each are transferred to minimal medium.

Those which fail to develop have a biochemical defect.

The nature of the defect is disclosed by tests with special media.

Minimal plus vitamins

Minimal plus amino acids

Minimal (control)

Complete (control)

simple starting materials, the mold produces all the constituent parts of its protoplasm. These include some 20 amino acid building blocks of proteins, nine water-soluble vitamins of the B group, and many other organic molecules of vital biological significance.

To one interested in what genes do in a human being, it might at first thought seem a very large jump from a man to a mold. Actually it is not. For in its basic metabolic processes, protoplasm—Thomas Huxley's physical stuff of life—is very much the same wherever it is found.

If the many chemical reactions by which a bread mold builds its protoplasm out of the raw materials at its disposal are catalyzed by enzymes, and if the proteins of these enzymes are copied from genes, it should be possible to produce

It is known that changes in genes—mutations—occur spontaneously with a low frequency. The probability that a given gene will mutate to a defective form can be increased a hundredfold or more by so-called mutagenic (mutation producing) agents. These include X-radiation, neutrons and other ionizing radiations, ultraviolet radiation, and mustard gas. Radiations are believed to cause mutations by literally "hitting" genes in a way to cause ionization within them or by otherwise causing internal rearrangements of the chemical bonds.

A bread-mold experiment to test the hypothesis that genes control enzymes and metabolism can be set up in the manner shown in the diagrams on pages 33 and 35. Asexual spores are X-rayed or otherwise treated with mutagenic agents.

IN CONTINUATION of the experiment begun on page 33, the strain of *Neurospora* that carries a defective gene is put through another series of steps. On page 33 it had been determined that the strain in question did not grow in the absence of vitamins. This indicated that the defective gene was involved in the synthesis of a vitamin. Now the question is: exactly what vitamin? This may be found by planting the strain carrying the defective gene on a group of minimal media, each of which is supplemented by a single vitamin. The mold will then grow on the medium which contains the vitamin that it has lost the power to synthesize. In the experiment outlined on the opposite page, the missing vitamin turns out to be pantothenic acid, a vitamin of the B group. When this has been established, further experiments must be run to determine whether the deficiency of the strain involves a single gene. This is done by crossing the strain bearing the defective gene with a normal strain. All the spores from the union flourish in a medium supplemented with pantothenic acid. When they are planted in a medium that does not contain pantothenic acid, however, only four cultures grow. This is proof that one gene is involved.

PHOTOMICROGRAPH of *Neurospora* shows the sturcture of its fine red tendrils. This photograph, supplied through the courtesy of Life Magazine, was made by Herbert Gehr in the genetics laboratory of E. L. Tatum at Yale.

molds with specific metabolic errors by causing genes to mutate. Or to state the problem somewhat differently, one ought to be able to discover what genes do by making them defective.

The simplicity of this approach can be illustrated by an analogy. The manufacture of an automobile in a factory is in some respects like the development of an organism. The workmen in the factory are like genes—each has a specific job to do. If one observed the factory only from the outside and in terms of the cars that come out, it would not be easy to determine what each worker does. But if one could replace able workers with defective ones, and then observe what happened to the product, it would be a simple matter to conclude that Jones puts on the radiator grill, Smith adds the carburetor, and so forth. Deducing what genes do by making them defective is analogous and equally simple in principle.

Following a sexual phase of the life cycle, descendants of mutated genes are recovered in sexual spores. These are grown separately, and the molds that grow from them are tested for ability to produce the molecules out of which they are built.

If a gene essential for the production of vitamin B-1 by the mold is made defective, then B-1 must be supplied in the medium if a mold is to develop from a spore carrying the defective gene. But in the present state of our knowledge it is not possible to produce mutations in specific genes at will. By X-raying, for example, any one or more of several thousand genes may be mutated, or in many cases none at all will be changed. There is no known method of predicting which of the genes, if any, will be hit. It is therefore necessary to grow presumptive mutant spores on a medium supplemented with protoplasmic building blocks of which the formation could be

blocked if defective genes were present.

Molds grown on such supplemented medium may grow normally either (1) by making a particular essential part themselves or (2) by taking it ready-made from the culture medium, as they must do if the gene involved in making it is defective. The two possibilities can be distinguished by trying to grow the mold on an unsupplemented medium and on media to which single supplements are added.

Following heavy ultraviolet treatment, about two sexual spores out of every hundred tested carry defective forms of those genes which are necessary for the production of essential substances supplied in the supplemented medium. For example, strain number 5531 of the mold cannot manufacture the B-vitamin pantothenic acid. For normal growth it requires an external supply of this vitamin just as human beings do.

How do we know that the inability of the mold to produce its own pantothenic acid involves a gene defect? The only way this question can be answered at present is by seeing if inability to make pantothenic acid behaves in crosses as a single unit of inheritance.

The answer is that it does. If the mold that cannot make pantothenic acid is crossed with a normal strain of the other sex, the resulting spore sacs invariably contain four spores that produce molds like one parent and four that produce

35

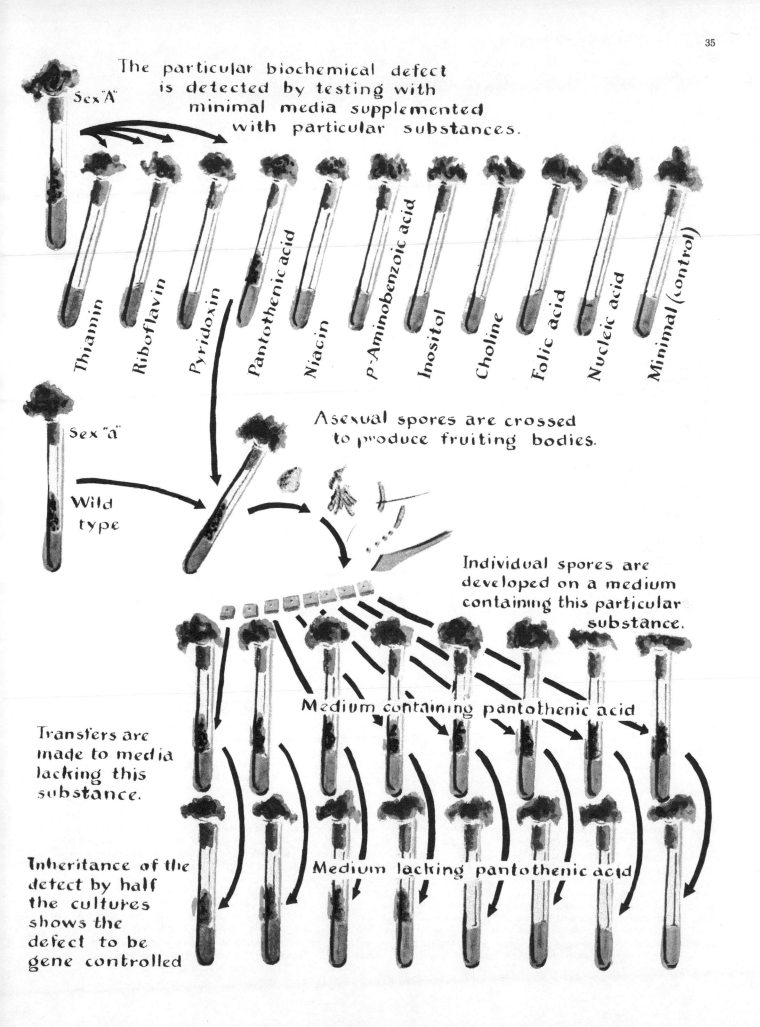

strains like the other parent. Four daughter molds out of each set of eight from a spore sac are able to make pantothenic acid, and four are not (*see page 35*).

In a similar way, genes concerned with many other specific bread-mold chemical reactions have been mutated. In each case that has been studied in sufficient detail to be sure of the relation, it has been found that single genes are directly concerned with single chemical reactions.

An example that illustrates not only that genes are concerned with specific chemical reactions but also how mutant types can be used as tools for the study of metabolic processes involves the production of the amino acid tryptophane and the vitamin niacin (also known as nicotinic acid) by bread mold. Several steps in the synthesis of tryptophane, an indispensable component of the protoplasm of all organisms, have been shown to be gene-controlled. These have been used to show that bread mold forms this component by combining indole and the amino acid serine.

It has been found that indole, in turn, is made from anthranilic acid. If the second gene in the series in the accompanying diagram is made defective, anthranilic acid cannot be converted to indole, and if the mold carrying this gene in defective form is grown on a small amount of tryptophane it accumulates anthranilic acid in much the same way as an alcaptonuric accumulates alcapton. The accumulated anthranilic acid has been chemically identified in the culture medium of such a defective strain.

A recent report that rats fed on diets rich in tryptophane did not need niacin suggested to animal biochemists that possibly niacin is made from tryptophane. Following this lead, studies were made of the strains of bread mold which require ready-made tryptophane and niacin. They gave clear evidence that the bread mold does indeed derive its niacin from tryptophane. Intermediates in the chain of reactions by which the conversion is made were then identified (*see drawing on the opposite page*).

Men and Molds

The tryptophane-niacin relation so clearly disclosed by bread mold mutants has an interesting relation to the dietary deficiency disease pellagra in man. In the past this disease has been variously attributed to poor quality of dietary proteins, to a toxic factor in Indian corn, and to lack of a vitamin. When, in 1937, C. A. Elvehjem of Wisconsin demonstrated that niacin would bring about spectacular cures of black tongue, a disease of dogs like pellagra in man, the problem seemed to be solved. It was very soon found that pellagra in man, too, is cured by small amounts of niacin in the diet. The alternative hypotheses were promptly forgotten, even though the facts that led to

them were not explained by niacin alone.

The tryptophane-niacin relation now makes it clear that the protein quality theory also is correct. Good quality proteins contain plenty of tryptophane. If this is present in sufficient amounts in the diet, niacin appears not to be needed. The corn toxin theory also has a reasonable basis. There appear to be chemical substances in this grain that interfere with the body's utilization of tryptophane and niacin in such a way as to increase the requirements of those two materials.

Another point of interest in connection with the tryptophane-niacin story is that it illustrates again that, in terms of basic protoplasmic reactions, pretty much the same things go on in men and molds. It is supposed that in much the same way as a single gene is in control of the enzyme by which alcaptonuria is broken down in man, genes of the bread mold guide chemical reactions indirectly through their control of enzyme proteins. In most instances the enzymes involved have not yet been studied directly.

Bread-mold studies have contributed strong support to the hypothesis that each gene controls a single protein. But they have not proved it to the satisfaction of all biologists. There remains a possibility that some genes possess several distinct functions and that such genes were automatically excluded by the experimental procedure followed.

What is the process by which genes direct the formation of specific proteins? This is a question to which the answer is not yet known. There is evidence that genes themselves contain proteins combined with nucleic acids to form giant nucleoprotein molecules hundreds of times larger than the relatively simple molecules pictured on the opposite page. And it has been suggested that genes direct the building of non-genic proteins in essentially the same way in which they form copies of themselves.

The general question of how proteins are synthesized by living organisms is one of the great unsolved problems of biology. Until we have made headway toward its solution, it will not be possible to understand growth, normal or abnormal, in anything but superficial terms.

Do all organisms have genes? All sexually reproducing organisms that have been investigated by geneticists demonstrably possess them. Until recently there was no simple way of determining whether bacteria and viruses also have them. As a result of very recent investigations it has been found that some bacteria and some bacterial viruses perform a kind of sexual reproduction in which hereditary units like genes can be quite clearly demonstrated.

By treatment of bacteria with mutagenic agents, mutant types can be produced that parallel in a striking manner those found in the bread mold. These

GENES DIRECT a sequence of vital chemical reactions in *Neurospora*. Each of the molecules shown in the models on the opposite page is made up of the atoms hydrogen (*white spheres*), oxygen (*light color*), carbon (*black*) and nitrogen (*dark color*). Reactions involving the genes switch these atoms around to manufacture one molecule out of another. Beginning at the upper left, a single gene is known to be involved in the synthesis of anthranilic acid. Two genes are then involved in making anthranilic acid into indole, with an unknown intermediate indicated by a question mark. Indole is combined with serine to make the amino acid tryptophane, with water left over. Tryptophane is made into kynurenine. Two genes transform kynurenine into 3-hydroxy-anthranilic acid, again with an unknown intermediate molecule. Two genes finally synthesize the last product of the chain: niacin, the B vitamin that is an essential of both plant and animal life. This sequence of events is also involved in the human nutritional disease pellagra. A diet poor in the amino acid tryptophane obviously will lead to a deficiency of niacin, which causes the symptoms of pellagra. Therefore supplying either tryptophane or niacin to patient will alleviate disease.

make it almost certain that bacterial genes are functionally like the genes of molds.

So we can sum up by asserting that genes are irreducible units of inheritance in viruses, single-celled organisms and in many-celled plants and animals. They are organized in threadlike chromosomes which in higher plants and animals are carried in organized nuclei. Genes are probably nucleoproteins that serve as patterns in a model-copy process by which new genes are copied from old ones and by which non-genic proteins are produced with configurations that correspond to those of the gene templates.

Through their control of enzyme proteins many genes show a simple one-to-one relation with chemical reactions. Other genes appear to be concerned primarily with the elaboration of antigens—giant molecules which have the property of inducing antibody formation in rabbits or other animals.

It is likely that life first arose on earth as a genelike unit capable of multiplication and mutation. Through natural selection of the fittest of these units and combinations of them, more complex forms of life gradually evolved.

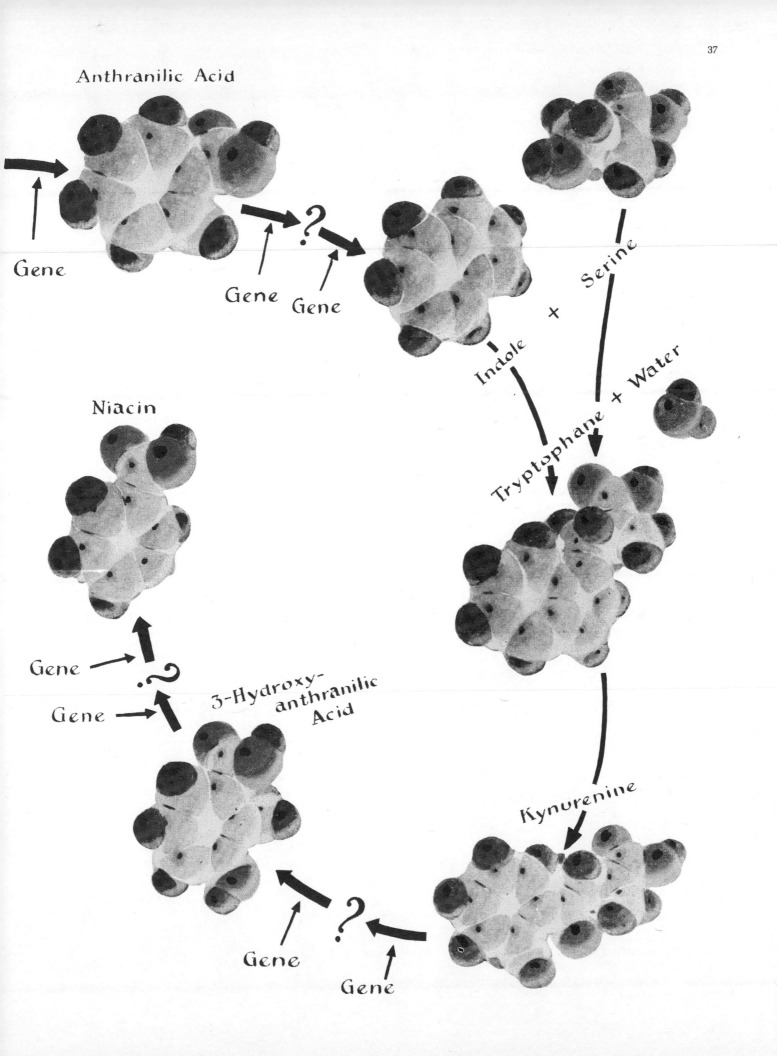

Anthranilic Acid

Gene

? Gene Gene

+

Indole

Serine

Tryptophane + Water

Niacin

Gene

Gene

?

3-Hydroxy-
anthranilic
Acid

Kynurenine

Gene

?

Gene

II

THE CHEMICAL BASIS OF HEREDITY

II THE CHEMICAL BASIS
OF HEREDITY

INTRODUCTION

Before the turn of the century, the cell nucleus and, in particular, its chromosomes had been posited as likely vehicles of heredity. Since DNA was unique to the chromosomes, that substance and not protein, the other major constituent of the chromosomes, was thought to provide the material basis for heredity. This correct view was temporarily supplanted in the mid 1920s when the analytical work of Phoebus Levene led him to postulate that DNA consisted of a monotonously repeating polymer of the four nucleotides in fixed sequence and, thus, that DNA could not be a repository of information that could specify the diversity of characteristics and properties of an organism.

This view persisted until the agent responsible for the transformation of pneumococcus bacteria from one strain type to another was identified as DNA by Oswald Avery, C. M. MacLeod, and M. J. McCarty in 1944. Even then the scientific community was reluctant to ascribe a genetic role to DNA.

Rollin Hotchkiss's painstaking purification of transforming DNA in the years immediately following had much to do with the ultimate acceptance of DNA as the genetic molecule. His and Esther Weiss's article, "Transformed Bacteria," describes the phenomenon of transformation and its utility in genetic analysis.

Then in 1952 came Alfred D. Hershey and Martha Chase's demonstration that the DNA of the bacteriophage was the carrier of its genetic specificity. When the structure of DNA was proposed in 1953 by James D. Watson and Francis H. C. Crick, they immediately recognized that the structure embodied both the capacity for self-replication and for encoding information. Any further doubts about the genetic role of DNA were abandoned by all but a few holdouts, who were wedded to protein as the carrier of genetic specificity.

The excitement and significance in what Watson and Crick proposed are recorded in Francis H. C. Crick's article "The Structure of the Hereditary Material." The relationship between the structure of the DNA molecule and the chromosome in which it is located presented a formidable problem because of the enormous length of DNA that must be organized into a single chromosome. It seemed incredible that all that DNA could be encompassed in a single gigantic DNA molecule. Thus, even though J. Herbert Taylor demonstrated that the chromosomes showed semiconservative replication and segregation (this had been demonstrated by Matthew Meselson and Franklin W. Stahl for the DNA molecule itself), Taylor felt impelled to postulate a segmented arrangement for the DNA in the chromosome. In this structure he proposed that protein linkers played a dual organizational and structural role in the chromosome. This is described in his article "The Duplication of Chromosomes."

That an entire genome could be encompassed in a single DNA molecule became apparent with John Cairns's isolation and display of the DNA of *Escherichia coli* as a circular molecule. This is described in his article "The Bacterial Chromosome."

Subsequent biophysical studies and electron microscope observation have established that the DNA in the chromosome of higher organisms exists as a single gigantic molecule with no protein linkers.

4   Transformed Bacteria

by Rollin D. Hotchkiss and Esther Weiss
November 1956

*If desoxyribonucleic acid is removed from one strain of
pneumococci and added to another strain, some of the
cells in the second strain are able to transmit
characteristics of the first to their descendants*

If man reproduced his kind the way bacteria do, a grown man at 25 would more or less abruptly become two young men in his own exact image. These two in turn would "divide" in another 25 years, so that after 50 years there would be four young men indistinguishable from the original ancestor. A rather large family could eventually be built up by this process, but all its members would be monotonously alike in appearance, abilities, temperament and vigor. The same would be true of every family. It would be entirely male or entirely female: the two sexes would be aloof from each other. There would be families of burly, competitive athletes, and others made up exclusively of gray-eyed introverts liking nothing better than to write sad poems on the haunting loveliness of subdivision.

After about 20 generations (500 years) we could expect an occasional "mutant" individual to show up, differing from the million other individuals of his family in some one trait such as eye color. In a community so ordered, a means of transferring these rare traits at will from one family to another would have dramatic value indeed. Actually such transformations can be accomplished with bacteria. It is possible to transfer mutations of a particular kind and thereby make controlled studies of heredity.

Only in the last 10 years or so have bacteria become prominent in the study of genetics. Until a little more than a decade ago it was supposed that a bacterial cell did not have a nucleus or the elaborate genetic apparatus characteristic of higher forms of life. The general feeling was that bacteria were simple enough not to need so cumbersome a system for passing along their hereditary characteristics. But in 1944 C. F. Robinow, a British bacteriologist, found signs of nuclei in bacteria. Since then bacteria have gained in complexity the more they have been studied, and their mechanisms have been found to parallel those of cells of higher organisms. The bacterial genetic system has proved to be one of the most fascinating of all to bacteriologists and geneticists alike. Undoubtedly some of the appeal of bacteria for geneticists lies in their rapid rate of multiplication. What other laboratory subject provides a new generation every half-hour? There is no problem of eyestrain involved, either. The tiny creatures take only a few hours to grow to a many-celled colony which is easily visible in the test tube or on a gelatin-like agar plate, and they are much more manageable than the classical fruit flies. When a bacterium achieves a new characteristic, it quickly produces a whole colony of the new type, which can often be readily identified.

Such a new characteristic may turn up in perhaps one in a few million cells in a growing population. The few mutant bacteria may be swamped out as they attempt to grow in the enormous competitive population. Occasionally, however, the environment may favor the rare individuals of a newly arising type, so that they outgrow the usual type. A bacterial geneticist is always on the lookout for these natural events, and he can make detection of them easier by providing a selective environment which allows only certain mutants to grow. For example, if bacteria are put upon agar plates containing penicillin, only the mutants that have resistance to the drug will grow to produce colonies. From a population of a hundred million cells, usually about five to 10 such colonies will emerge, indicating that something like one cell in 10 million of the original population became, mysteriously and suddenly, penicillin-resistant.

The technique of transformation developed in the laboratory permits us to take matters more into our own hands. We can introduce a specific hereditary characteristic into a strain of bacteria by treating the cells with an extract from killed bacteria of a related strain which possess the characteristic in question. The procedure does not create new traits but transfers traits already present in the donor bacteria. In effect we are "robbing Peter to pay Paul."

The transforming material is desoxyribonucleic acid (DNA)—the type of substance now so well known as a fundamental constituent of the chromosomes of higher plants and animals. DNA from the donor cells seems to enter into the recipient bacterial cell and, like a gene, direct part of the cell's internal mechanism. The cell even learns to make more of the directing substance itself. How this controlling substance carries its specific instructions is still a mystery: our most sensitive chemical analyses cannot distinguish any differences between DNA varieties responsible for different traits. But each variety must be chemically distinctive, because the recipient cells repeatedly respond in the same predictable way to a particular extract.

Pneumococci, the pneumonia germs, have been extensively studied by means of laboratory transformation. From about 30 million pneumococcal cells, killed and broken down by sodium desoxycholate (a substance found in bile), we can extract one microgram of purified DNA. This can be preserved for years as a white precipitate in alcohol.

The pneumococci we have used to test the transforming DNA are Type H strains which have lost the ability to produce the sugar capsule that makes the bacteria virulent. On agar plates these strains normally grow as shiny pinpoint colonies. In a broth containing a small amount of antiserum (made from the blood of an animal inoculated with pneumococci) they grow in chains and clumps which settle to the bottom as separate white colonies. By diluting samples of the culture and counting the colonies that develop, we can make quantitative studies of what happens when bacteria are transformed.

More than 25 specific characteristics have been transferred to this strain from various mutants. They include every sort of trait we can observe: acquisition by the bacterium of various types of capsule coating, development of resistance to drugs, formation of certain types of colonies, and so on. Usually only one trait is passed on to any particular cell, even if the DNA preparation is from a strain having two or three identifiable characteristics. For example, if a million pneumococci are treated with a tenth of a microgram of a DNA which carries three traits—resistance to penicillin and streptomycin and formation of a coat of Type III—some 50,000 of the million may be transformed; of these about 49,000 will acquire only one of the three traits, 800 may have two, and only four will take all three.

The DNA preparation evidently does not carry a complete package of the donor's traits into a recipient cell but only some part of the package. This part is not as small as a gene, however. Certain traits seem to travel together as if they were linked. For example, the DNA factors responsible for the ability of pneumococci to utilize mannitol (a form of sugar found in manna) as a source of energy and for the ability to resist streptomycin tend to be coupled: 20 per cent of cells transformed by a DNA carrying these two markers will show both characteristics.

Experiments in transmission of the mannitol-utilization trait illustrate an-

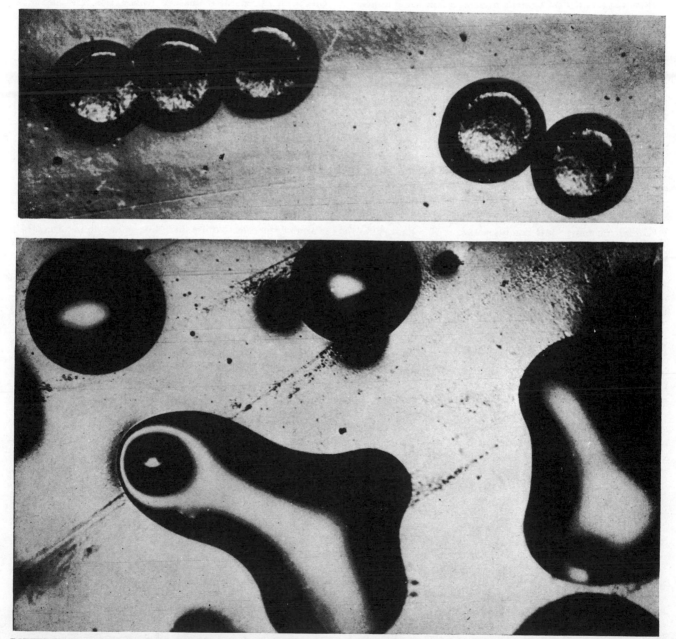

DIFFERENT STRAINS OF PNEUMOCOCCI grow on the surface of agar in colonies of characteristic form. At the top are the small colonies of pneumococci without capsules. At the bottom are the larger colonies of encapsulated Type III pneumococci. The colonies of Type III produce a sticky sugar; they are larger because they absorb moisture from the agar. If desoxyribonucleic acid (DNA) is removed from pneumococci of Type III and added to the nonencapsulated pneumococci, the latter will acquire capsules.

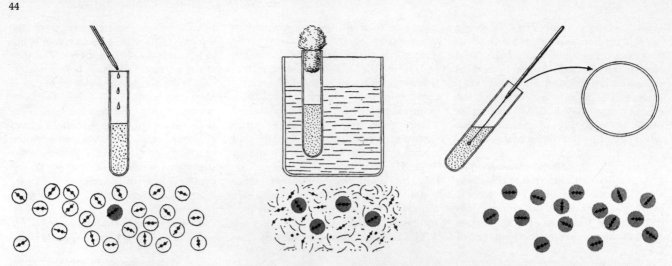

PREPARATION OF DNA from a strain of pneumococci resistant to streptomycin is traced in the drawings at the top of this and the next three pages. First, a broth containing streptomycin is inoculated with pneumococci. One cell (*color*) is a rare mutant resistant to streptomycin. Second, the broth is incubated overnight at 37 degrees centigrade. The mutant multiplies and makes the broth

other important point: namely, that DNA may carry a hereditary trait as a latent ability, regardless of whether or not the ability is developed. Like children who display their innate musical talent only after they have taken some piano lessons, the talented strains of pneumococci do not exhibit their ability to utilize mannitol until they have been exposed to this sugar for a short time. To metabolize it they have to learn to make an enzyme which oxidizes mannitol.

Under the most ideal conditions we have devised up to now, it has been possible to transmit a new trait to 17 per cent of the treated cells. Many factors influence the yield obtained. Foremost of these is the capacity of the recipient strain itself. Some strains seem to be transformed more readily than others, and many seem altogether incapable of responding to DNA. Another factor is the concentration of DNA present: one

half of a millionth of a gram per cubic centimeter is an optimal concentration, and one 10,000th of that amount will have some effect. The length of exposure to DNA also is important. We think that the bacteria are most susceptible to transformation just after cell division. After pneumococci have been cooled to a growth-arresting temperature, so that all start out "in step" in the division cycle when they are rewarmed, transformations are exceptionally numerous. About 15 minutes after the rewarming the cells abruptly lose their susceptibility, and just at this time very few cells will be dividing. Still another important factor is the composition of the medium in which the cells grow before and during DNA exposure.

After acquiring the new DNA, a cell multiplies more slowly than the others for a time and is at a disadvantage in growth until its transformation has been

completed. Some of the transformed cells are likely to survive in any event, but the percentage transformed is easier to observe if the conditions are adjusted to favor them. For example, in an experiment in which the transformation makes the cells resistant to penicillin, placing the bacteria in a penicillin broth will kill off all but the transformed cells. One must be careful, however, not to challenge the bacteria with the selective agent too soon. The transformed cells require 30 to 60 minutes to manifest their new drug resistance, and it takes still longer for the cells to set up the mechanism necessary to duplicate the new DNA. During this time the new DNA is beginning to perform its genelike functions in the cells.

A transformed pneumococcus remains susceptible to further transformations, even by the same DNA if the DNA

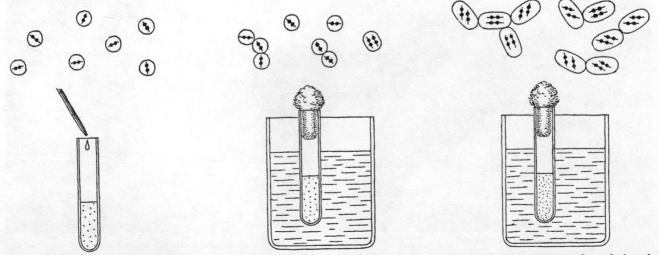

TRANSFORMATION OF PNEUMOCOCCI to a strain resistant to streptomycin from a nonresistant strain is illustrated at the bottom of this and the next three pages. First, young pneumococci are added to a rich broth containing serum albumin. Second, the tube is incubated for three hours at 37 degrees C.; the cells multiply. Third, the tube is cooled to 25 degrees for 20 minutes. This arrests

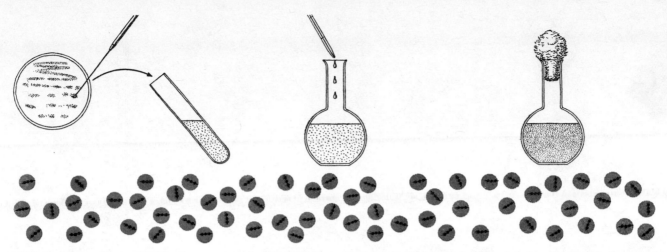

turbid; the other cells are killed by the streptomycin. Third, a bit of turbid broth is removed and spread on an agar medium. Fourth, one of the colonies on the agar plate is transferred to a tube of fresh broth. Fifth, a flask of broth is inoculated with the mutant strain. Sixth, the flask is incubated overnight until the culture is full-grown. This sequence is continued at the top of the next page.

carries more than one trait. Indeed, a particular trait, such as resistance to penicillin, may be developed by a series of stepwise mutations rather than by a single transformation. Beginning with pneumococci that survived exposure to low concentrations of penicillin, we submitted them to successively higher concentrations until we had a mutant strain which was resistant to 30 units of penicillin per 100 milliliters of culture. We then administered the DNA from this strain to pneumococci which were fully sensitive to penicillin. None of the sensitive cells acquired as much resistance to penicillin as the donor strain possessed, and most of those transformed could not resist more than five units of the drug. When these were again treated with the DNA of the highly resistant donor, some became resistant to 12 units of penicillin. It took several such steps to produce transformants able to resist 30 units. We

think these indicate the number of spontaneous mutations that must have taken place in the original evolution of the mutant strain.

Experiments with streptomycin produced the same stepwise development of resistance. However, in a large population of cells an occasional cell spontaneously acquired a high level of resistance in just one step, and the DNA from this mutant produced equally resistant cells in a single transformation. Evidently in this case one mutation of a single genetic unit modified the DNA so that it could effect the entire transformation in one step.

Another kind of phenomenon emerges when pneumococci without a capsule are treated with a DNA which confers the ability to form a capsule of Type III. Normally the DNA effects this transformation in one step, but at times it produces cells with intermediate varieties

of sugar capsules. Colonies of these cells are smaller than those of the full Type III. Harriett Ephrussi-Taylor, formerly of the Rockefeller Institute for Medical Research and now at the University of Paris, has done many experiments with two such varieties of cells. They seem to differ from each other and from normal Type III only in the quantity of capsule material they produce. If DNA from both varieties is mixed in a single culture, large, juicy colonies characteristic of the normal Type III cells will appear on the agar plates. Dr. Ephrussi-Taylor concluded that the two kinds of DNA could combine in a single recipient bacterium to yield the normal DNA of Type III cells, and this conclusion was strongly supported by other experiments.

The more the action of DNA is studied, the clearer it becomes that each organism's DNA is biologically distinc-

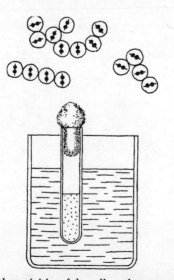
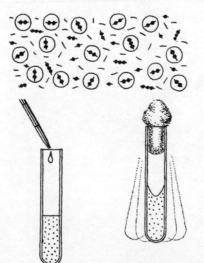
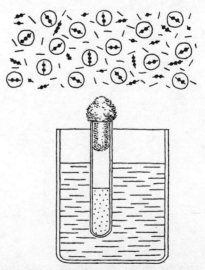

the activities of the cells at the same point in their cycle of division. Fourth, the culture is rewarmed to 37 degrees. Now all the cells start dividing at the same time. Fifth, DNA removed from a re-sistant strain is added to the culture. The DNA spreads through the broth, and some of the cells react with it. Sixth, the culture is re-incubated for five minutes; the cells continue to react with DNA.

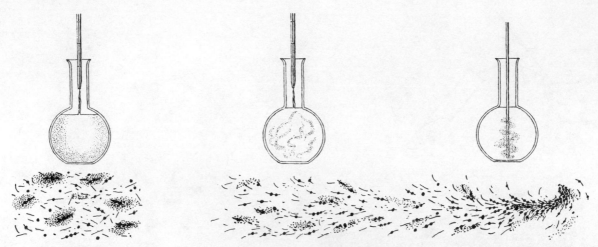

PREPARATION OF DNA IS CONTINUED from the drawings at the top of the preceding two pages. First, sodium desoxycholate is added to the flask containing the full-grown culture of pneumococ- ci. This kills and breaks up the cells; the broth clears. Second, alcohol is added to the cleared flask. This causes the DNA to pre- cipitate in threads. Third, the threads are collected from the flask

tive, even though we cannot detect any chemical difference. Attempts have been made to bring about transformations in pneumococci by injecting DNA from species of bacteria distantly related to them. The attempts have not succeeded. The foreign DNA may enter the cell, but it does not produce any detectable change in the cell's traits. The machinery of the cell apparently recognizes some- thing unusual about the foreign DNA and makes no genetic response to it, so far as we can determine. However, the cell's incorporation of the bogus DNA prevents it from reacting freely with a suitable DNA. (In this respect it behaves something like a fertilized egg: the egg, having accepted one spermatozoon, re- pels all others.) It seems that the various kinds of DNA are sufficiently alike to penetrate the outer defenses of the bac- terial cell. Even DNA from thymus- gland cells of the calf will react with

pneumococci, inhibiting their transfor- mation by an appropriate DNA. Indeed, a foreign DNA can compete on about equal terms with pneumococcal DNA for entry into a susceptible pneumococ- cus. But only the native DNA seems capable of producing a genetic effect on the bacterium.

The experiments in transforming bac- teria go back to a discovery made in 1928 by Fred Griffith, an English bac- teriologist. Pneumonia was then one of the most challenging problems in medi- cal research (the "miracle" drugs having not yet been discovered), and the pneu- mococcus was as popular a subject of study as the viruses are now. Griffith in- jected into mice pneumococci without capsules, which are not virulent, to- gether with killed pneumococci of a viru- lent type (Type III capsules). To his surprise, the tissues of the mice were

soon teeming with live, virulent pneu- mococci of Type III. Since the dead cells could not have come to life, it be- came evident that their material must somehow have transformed the nonen- capsulated bacteria into virulent germs which had the ability to make Type III capsules.

Griffith's discovery was followed up by workers at the Rockefeller Institute, under the great and inspiring leader- ship of the late Oswald T. Avery. One cannot fail to note that the study of pneumococcal transformation since the initial discovery has been carried out es- sentially by a single "school," consisting of Avery's students and their followers. This school, now in its second generation and widely spread, can trace its lines of descent as accurately as those of the bac- teria it studies.

By 1944 Avery, with Colin M. Mac- Leod and Maclyn McCarty, had identi-

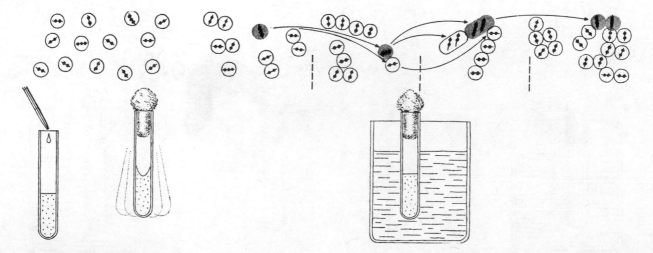

TRANSFORMATION OF PNEUMOCOCCI IS CONTINUED from the preceding two pages. First, desoxyribonuclease is shaken in the culture, which destroys the DNA that is not inside the cells. Second, the culture is reincubated for two hours. After about 45 minutes some cells show the new trait. These cells are slow to divide; for several generations the new DNA is passed on to only one daughter

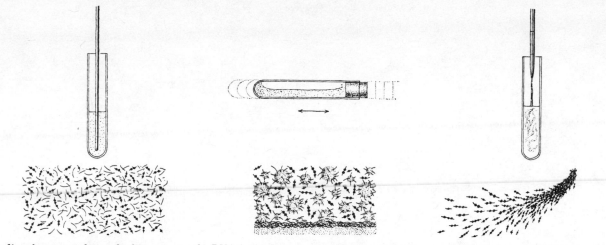

by winding them on a glass rod; this separates the DNA from most of the other substances in the debris of the broken cells. Fourth, the DNA is dissolved in a salt solution. Fifth, impurities are removed

by adding various solvents to the solution and shaking it; the solvents tend to precipitate the impurities in particles. Sixth, alcohol is added to the purified DNA, causing it to form threads again.

fied the substance responsible for the transformation of pneumococci as DNA. There followed a long series of transformation experiments, not only on pneumococci but also on other species of bacteria. The germ once thought to be the cause of influenza (*Hemophilus influenzae*) has been studied extensively by Hattie Alexander and her associates at Columbia University, and their findings largely parallel those on pneumococci. Other investigators have reported success in transforming strains of *Escherichia coli, Shigella paradysenteriae* and meningococci with DNA.

Over the last 30 years there has been an impressive accumulation of evidence that nucleic acids play a central role in the hereditary mechanism of all living creatures [see "The Chemistry of Heredity," by A. E. Mirsky; Scientific American, Offprint No. 28]. The trans-

formation work has had a decisive part in that vast investigation. This lead has generated a great number of exciting genetic experiments with animal and plant cells, bacteria and viruses, as the many recent articles on the subject in Scientific American have made plain. In the nucleic acids biologists at last have definite chemical substances which embody the properties of the somewhat hypothetical units long known as genes. Biochemists have not been slow to take up the challenge to explore the structure of the nucleic acids for the key to the machinery of heredity [see "The Structure of the Hereditary Material," by F. H. C. Crick, *beginning on page 49;* and "The Gene," by Norman H. Horowitz, Scientific American Offprint 17].

Happily, in the mid-20th century we can feel that we are on the threshold of still more exciting discoveries. We can expect to learn new kinds of facts about

heredity in the coming years, and also to find theories which will unify the facts. The transformation of bacteria is one of our most promising laboratory tools for further discovery. It means that we can interbreed organisms by transferring a comparatively small and simple genetic unit—much simpler than the intricate apparatus of chromosomes involved in other genetic systems. The simplicity of this process gives many possibilities for controlled manipulation and variation. Transforming agents may be added or withheld at will, used in various concentrations or in combination with other materials, pretreated with chemicals, modified or damaged by exposure to acid, heat or radiant energy. The outcome of the transformations with DNA so treated should do much to throw light upon the still mysterious processes set in motion by the fascinating entities that we call genes.

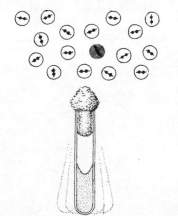
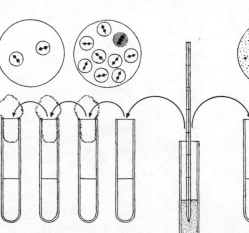
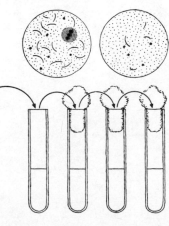

cell. Then one cell duplicates the DNA and passes it on to both daughter cells, which continue the duplication. Third, the tube is shaken to break up clumps of cells. Fourth, samples of the culture

are placed in tubes, some of which (*right*) contain streptomycin. The streptomycin kills the untransformed cells. Later, colonies are visible in the tubes. In higher dilutions they can be counted.

The Structure of the Hereditary Material

by F. H. C. Crick
October 1954

*An account of the investigations which have led to
the formulation of an understandable structure for
DNA. The chemical reactions of this material within
the nucleus govern the process of reproduction*

Viewed under a microscope, the process of mitosis, by which one cell divides and becomes two, is one of the most fascinating spectacles in the whole of biology. No one who watches the event unfold in speeded-up motion pictures can fail to be excited and awed. As a demonstration of the powers of dynamic organization possessed by living matter, the act of division is impressive enough, but even more stirring is the appearance of two identical sets of chromosomes where only one existed before. Here lies biology's greatest challenge: How are these fundamental bodies duplicated? Unhappily the copying process is beyond the resolving power of microscopes, but much is being learned about it in other ways.

One approach is the study of the nature and behavior of whole living cells; another is the investigation of substances extracted from them. This article will discuss only the second approach, but both are indispensable if we are ever to solve the problem; indeed some of the most exciting results are being obtained by what might loosely be described as a combination of the two methods.

Chromosomes consist mainly of three kinds of chemical: protein, desoxyribonucleic acid (DNA) and ribonucleic acid (RNA). (Since RNA is only a minor component, we shall not consider it in detail here.) The nucleic acids and the proteins have several features in common. They are all giant molecules, and each type has the general structure of a main backbone with side groups attached. The proteins have about 20 different kinds of side groups; the nucleic acids usually only four (and of a different type). The smallness of these numbers itself is striking, for there is no obvious chemical reason why many more types of side groups should not occur. Another interesting feature is that no protein or nucleic acid occurs in more than one optical form; there is never an optical isomer, or mirror-image molecule. This shows that the shape of the molecules must be important.

These generalizations (with minor exceptions) hold over the entire range of living organisms, from viruses and bacteria to plants and animals. The impression is inescapable that we are dealing with a very basic aspect of living matter, and one having far more simplicity than we would have dared to hope. It encourages us to look for simple explanations for the formation of these giant molecules.

The most important role of proteins is that of the enzymes—the machine tools of the living cell. An enzyme is specific, often highly specific, for the reaction which it catalyzes. Moreover, chemical and X-ray studies suggest that the structure of each enzyme is itself rigidly determined. The side groups of a given enzyme are probably arranged in a fixed order along the polypeptide backbone. If we could discover how a cell produces the appropriate enzymes, in particular how it assembles the side groups of each enzyme in the correct order, we should have gone a long way toward explaining the simpler forms of life in terms of physics and chemistry.

We believe that this order is controlled by the chromosomes. In recent years suspicion has been growing that the key to the specificity of the chromosomes lies not in their protein but in their DNA. DNA is found in all chromosomes —and only in the chromosomes (with minor exceptions). The amount of DNA per chromosome set is in many cases a fixed quantity for a given species. The sperm, having half the chromosomes of the normal cell, has about half the amount of DNA, and tetraploid cells in the liver, having twice the normal chromosome complement, seem to have twice the amount of DNA. This constancy of the amount of DNA is what one might expect if it is truly the material that determines the hereditary pattern.

Then there is suggestive evidence in two cases that DNA alone, free of protein, may be able to carry genetic information. The first of these is the discovery that the "transforming principles" of bacteria, which can produce an inherited change when added to the cell, appear to consist only of DNA. The second is the fact that during the infection of a bacterium by a bacteriophage the DNA of the phage penetrates into the bacterial cell while most of the protein, perhaps all of it, is left outside.

The Chemical Formula

DNA can be extracted from cells by mild chemical methods, and much experimental work has been carried out to discover its chemical nature. This work has been conspicuously successful. It is now known that DNA consists of a very long chain made up of alternate sugar and phosphate groups [see diagram below]. The sugar is always the same sugar, known as desoxyribose. And it is always joined onto the phosphate in the same way, so that the long chain is perfectly regular, repeating the same phosphate-sugar sequence over and over again.

But while the phosphate-sugar chain is perfectly regular, the molecule as a whole is not, because each sugar has a "base" attached to it and the base is not always the same. Four different types of base are commonly found: two of them are purines, called adenine and guanine, and two are pyrimidines, known as thymine and cytosine. So far as is known the order in which they follow one another along the chain is irregular, and probably varies from one piece of DNA to another. In fact, we suspect that the order of the bases is what confers specificity on a given DNA. Because the sequence of the bases is not known, one can only say that the general formula for DNA is established. Nevertheless this formula should be reckoned one of the major achievements of biochemistry, and it is the foundation for all the ideas described in the rest of this article.

At one time it was thought that the four bases occurred in equal amounts, but in recent years this idea has been shown to be incorrect. E. Chargaff and his colleagues at Columbia University, A. E. Mirsky and his group at the Rockefeller Institute for Medical Research and G. R. Wyatt of Canada have accurately measured the amounts of the bases in many instances and have shown that the relative amounts appear to be fixed for any given species, irrespective of the individual or the organ from which the DNA was taken. The proportions usually differ for DNA from different species, but species related to one another may not differ very much.

Although we know from the chemical formula of DNA that it is a chain, this does not in itself tell us the shape of the molecule, for the chain, having many single bonds around which it may rotate, might coil up in all sorts of shapes. However, we know from physical-chemical measurements and electron-microscope pictures that the molecule usually is long, thin and fairly straight, rather like a stiff bit of cord. It is only about 20 Angstroms thick (one Angstrom = one 100-millionth of a centimeter). This is very small indeed, in fact not much more than a dozen atoms thick.

The length of the DNA seems to depend somewhat on the method of preparation. A good sample may reach a length of 30,000 Angstroms, so that the structure is more than 1,000 times as long as it is thick. The length inside the cell may be much greater than this, because there is always the chance that the extraction process may break it up somewhat.

Pictures of the Molecule

None of these methods tells us anything about the detailed arrangement in space of the atoms inside the molecule. For this it is necessary to use X-ray diffraction. The average distance between bonded atoms in an organic molecule is about 1½ Angstroms; between unbonded atoms, three to four Angstroms. X-rays have a small enough wavelength (1½ Angstroms) to resolve the atoms, but unfortunately an X-ray diffraction photograph is not a picture in the ordinary sense of the word. We cannot focus X-rays as we can ordinary light; hence a picture can be obtained only by roundabout methods. Moreover, it can show clearly only the periodic, or regularly repeated, parts of the structure.

With patience and skill several English workers have obtained good diffraction pictures of DNA extracted from cells and drawn into long fibers. The first studies, even before details emerged, produced two surprises. First, they revealed that the DNA structure could take two forms. In relatively low hu-

midity, when the water content of the fibers was about 40 per cent, the DNA molecules gave a crystalline pattern, showing that they were aligned regularly in all three dimensions. When the humidity was raised and the fibers took up more water, they increased in length by about 30 per cent and the pattern tended to become "paracrystalline," which means that the molecules were packed side by side in a less regular manner, as if the long molecules could slide over one another somewhat. The second surprising result was that DNA from different species appeared to give identical X-ray patterns, despite the fact that the amounts of the four bases present varied. This was particularly odd because of the existence of the crystalline form just mentioned. How could the structure appear so regular when the bases varied? It seemed that the broad arrangement of the molecule must be independent of the exact sequence of the bases, and it was therefore thought that the bases play no part in holding the structure together. As we shall see, this turned out to be wrong.

The early X-ray pictures showed a third intriguing fact: namely, that the repeats in the crystallographic pattern came at much longer intervals than the chemical repeat units in the molecule. The distance from one phosphate to the next cannot be more than about seven Angstroms, yet the crystallographic repeat came at intervals of 28 Angstroms in the crystalline form and 34 Angstroms

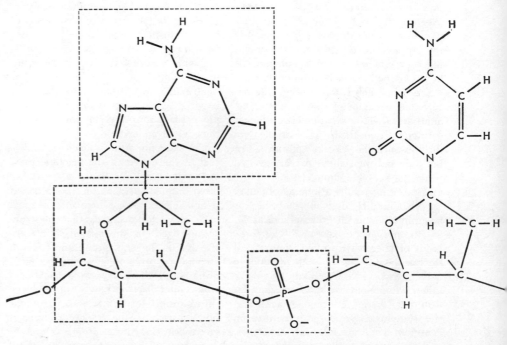

FRAGMENT OF CHAIN of deoxyribonucleic acid shows the three basic units that make up the molecule. Repeated over and over in a long chain, they make it 1,000 times as long

in the paracrystalline form; that is, the chemical unit repeated several times before the structure repeated crystallographically.

J. D. Watson and I, working in the Medical Research Council Unit in the Cavendish Laboratory at Cambridge, were convinced that we could get somewhere near the DNA structure by building scale models based on the X-ray patterns obtained by M. H. F. Wilkins, Rosalind Franklin and their co-workers at Kings' College, London. A great deal is known about the exact distances between bonded atoms in molecules, about the angles between the bonds and about the size of atoms—the so-called van der Waals' distance between adjacent nonbonded atoms. This information is easy to embody in scale models. The problem is rather like a three-dimensional jig saw puzzle with curious pieces joined together by rotatable joints (single bonds between atoms).

The Helix

To get anywhere at all we had to make some assumptions. The most important one had to do with the fact that the crystallographic repeat did not coincide with the repetition of chemical units in the chain but came at much longer intervals. A possible explanation was that all the links in the chain were the same but the X-rays were seeing every tenth link, say, from the same angle and the others from different angles. What sort of chain might produce this pattern? The answer was easy: the chain might be coiled in a helix. (A helix is often loosely called a spiral; the distinction is that a helix winds not around a cone but around a cylinder, as a winding staircase usually does.) The distance between crystallographic repeats would then correspond to the distance in the chain between one turn of the helix and the next.

We had some difficulty at first because we ignored the bases and tried to work only with the phosphate-sugar backbone. Eventually we realized that we had to take the bases into account, and this led us quickly to a structure which we now believe to be correct in its broad outlines.

This particular model contains a pair of DNA chains wound around a common axis. The two chains are linked together by their bases. A base on one chain is joined by very weak bonds to a base at the same level on the other chain, and all the bases are paired off in this way right along the structure. In the diagram on page 52, the two ribbons represent the phosphate-sugar chains, and the pairs of bases holding them together are symbolized as horizontal rods. Paradoxically, in order to make the structure as symmetrical as possible we had to have the two chains run in opposite directions; that is, the sequence of the atoms goes one way in one chain and the opposite way in the other. Thus the figure looks exactly the same whichever end is turned up.

Now we found that we could not arrange the bases any way we pleased; the four bases would fit into the structure only in certain pairs. In any pair there must always be one big one (purine) and one little one (pyrimidine). A pair of pyrimidines is too short to bridge the gap between the two chains, and a pair of purines is too big to fit into the space.

At this point we made an additional assumption. The bases can theoretically exist in a number of forms depending upon where the hydrogen atoms are attached. We assumed that for each base one form was much more probable than all the others. The hydrogen atoms can be thought of as little knobs attached to the bases, and the way the bases fit together depends crucially upon where these knobs are. With this assumption the only possible pairs that will fit in are: adenine with thymine and guanine with cytosine.

The way these pairs are formed is shown in the diagrams on page 54. The dotted lines show the hydrogen bonds, which hold the two bases of a pair together. They are very weak bonds; their energy is not many times greater than the energy of thermal vibration at room temperature. (Hydrogen bonds are the main forces holding different water molecules together, and it is because of them that water is a liquid at room temperatures and not a gas.)

Adenine must always be paired with

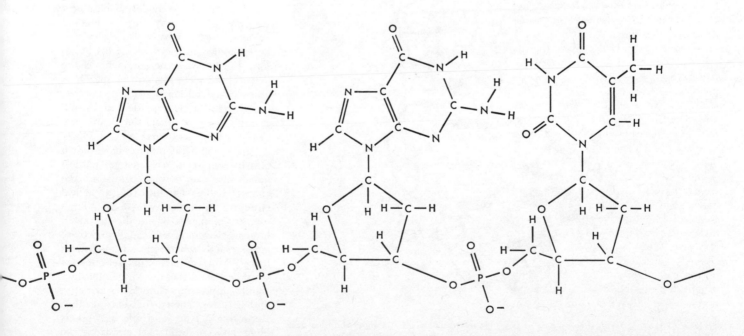

as it is thick. The backbone is made up of pentose sugar molecules (marked by the middle colored square), linked by phosphate groups (bottom square). The bases (top square), adenine, cytosine, guanine and thymine protrude off each sugar in irregular order.

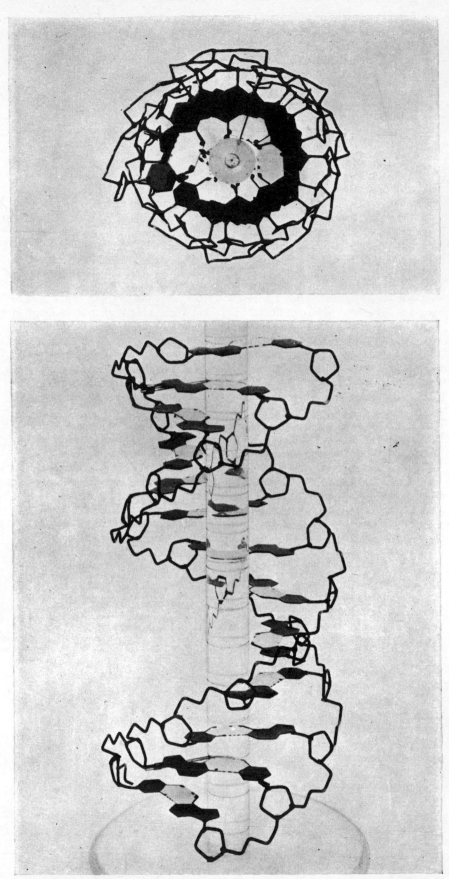

STRUCTURAL MODEL shows a pair of DNA chains wound as a helix about the fiber axis. The pentose sugars can be plainly seen. From every one on each chain protrudes a base, linked to an opposing one at the same level by a hydrogen bond. These base-to-base links act as horizontal supports, holding the chains together. Upper photograph is a top view.

thymine, and guanine with cytosine; it is impossible to fit the bases together in any other combination in our model. (This pairing is likely to be so fundamental for biology that I cannot help wondering whether some day an enthusiastic scientist will christen his newborn twins Adenine and Thymine!) The model places no restriction, however, on the sequence of pairs along the structure. Any specified pair can follow any other. This is because a pair of bases is flat, and since in this model they are stacked roughly like a pile of coins, it does not matter which pair goes above which.

It is important to realize that the specific pairing of the bases is the direct result of the assumption that both phosphate-sugar chains are helical. This regularity implies that the distance from a sugar group on one chain to that on the other at the same level is always the same, no matter where one is along the chain. It follows that the bases linked to the sugars always have the same amount of space in which to fit. It is the regularity of the phosphate-sugar chains, therefore, that is at the root of the specific pairing.

The Picture Clears

At the moment of writing, detailed interpretation of the X-ray photographs by Wilkins' group at Kings' College has not been completed, and until this has been done no structure can be considered proved. Nevertheless there are certain features of the model which are so strongly supported by the experimental evidence that it is very likely they will be embodied in the final correct structure. For instance, measurements of the density and water content of the DNA fibers, taken with evidence showing that the fibers can be extended in length, strongly suggest that there are two chains in the structural unit of DNA. Again, recent X-ray pictures have shown clearly a most striking general pattern which we can now recognize as the characteristic signature of a helical structure. In particular there are a large number of places where the diffracted intensity is zero or very small, and these occur exactly where one expects from a helix of this sort. Another feature one would expect is that the X-ray intensities should approach cylindrical symmetry, and it is now known that they do this. Recently Wilkins and his co-workers have given a brilliant analysis of the details of the X-ray pattern of the crystalline form, and have shown that they

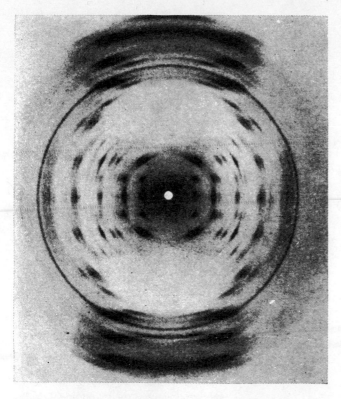

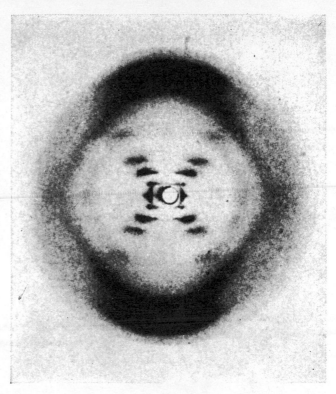

STRUCTURE A is the crystalline form of DNA found at relatively low humidity. This X-ray photograph is by H. R. Wilson.

STRUCTURE B is the paracrystalline form of DNA. The molecules are less regularly arranged. Picture is by R. E. Franklin.

are consistent with a structure of this type, though in the crystalline form the bases are tilted away from the fiber axis instead of perpendicular, as in our model. Our construction was based on the paracrystalline form.

Many of the physical and chemical properties of DNA can now be understood in terms of this model. For example, the comparative stiffness of the structure explains rather naturally why DNA keeps a long, fiber-like shape in solution. The hydrogen bonds of the bases account for the behavior of DNA in response to changes in pH. Most striking of all is the fact that in every kind of DNA so far examined—and over 40 have been analyzed—the amount of adenine is about equal to the amount of thymine and the guanine equal to the cytosine, while the cross-ratios (between, say, adenine and guanine) can vary considerably from species to species. This remarkable fact, first pointed out by Chargaff, is exactly what one would expect according to our model, which requires that every adenine be paired with a thymine and every guanine with a cytosine.

It may legitimately be asked whether the artificially prepared fibers of extracted DNA, on which our model is based, are really representative of intact DNA in the cell. There is every indication that they are. It is difficult to see

how the very characteristic features of the model could be produced as artefacts by the extraction process. Moreover, Wilkins has shown that intact biological material, such as sperm heads and bacteriophage, gives X-ray patterns very similar to those of the extracted fibers.

The present position, therefore, is that in all likelihood this statement about DNA can safely be made: its structure consists of two helical chains wound around a common axis and held together by hydrogen bonds between specific pairs of bases.

The Mold

Now the exciting thing about a model of this type is that it immediately suggests how the DNA might produce an exact copy of itself. The model consists of two parts, each of which is the complement of the other. Thus either chain may act as a sort of mold on which a complementary chain can be synthesized. The two chains of a DNA, let us say, unwind and separate. Each begins to build a new complement onto itself. When the process is completed, there are two pairs of chains where we had only one. Moreover, because of the specific pairing of the bases the sequence of the pairs of bases will have been duplicated exactly; in other words, the mold has not only assembled the build-

ing blocks but has put them together in just the right order.

Let us imagine that we have a single helical chain of DNA, and that floating around it inside the cell is a supply of precursors of the four sorts of building blocks needed to make a new chain. Unfortunately we do not know the makeup of these precursor units; they may be, but probably are not, nucleotides, consisting of one phosphate, one sugar and one base. In any case, from time to time a loose unit will attach itself by its base to one of the bases of the single DNA chain. Another loose unit may attach itself to an adjoining base on the chain. Now if one or both of the two newly attached units is not the correct mate for the one it has joined on the chain, the two newcomers will be unable to link together, because they are not the right distance apart. One or both will soon drift away, to be replaced by other units. When, however, two adjacent newcomers are the correct partners for their opposite numbers on the chain, they will be in just the right position to be linked together and begin to form a new chain. Thus only the unit with the proper base will gain a permanent hold at any given position, and eventually the right partners will fill in the vacancies all along the forming chain. While this is going on, the other single chain of the original pair also will

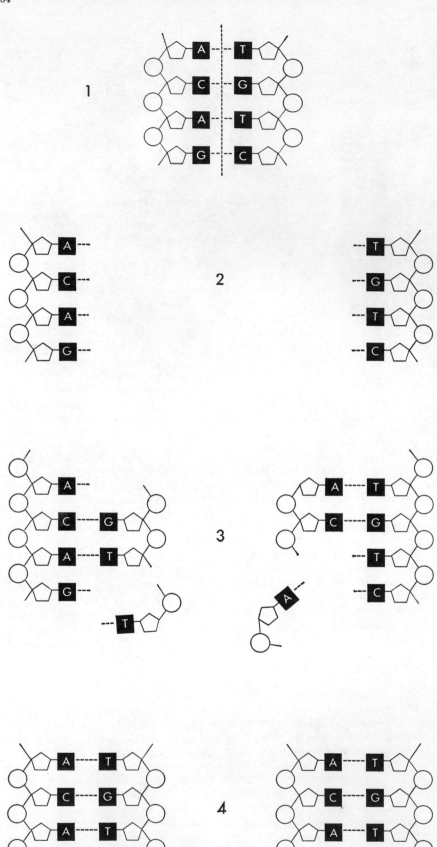

REPLICATION mechanism by which DNA might duplicate itself is shown in diagram. A helix of two DNA chains unwinds and separates (1). Two complementary chains of DNA (2) within the cell begin to attach DNA precursor units floating loosely (3). When the proper bases are joined, two new helixes will build up (4). Letters represent the bases.

be forming a new chain complementary to itself.

At the moment this idea must be regarded simply as a working hypothesis. Not only is there little direct evidence for it, but there are a number of obvious difficulties. For example, certain organisms contain small amounts of a fifth base, 5-methyl cytosine. So far as the model is concerned, 5-methyl cytosine fits just as well as cytosine and it may turn out that it does not matter to the organism which is used, but this has yet to be shown.

A more fundamental difficulty is to explain how the two chains of DNA are unwound in the first place. There would have to be a lot of untwisting, for the total length of all the DNA in a single chromosome is something like four centimeters (400 million Angstroms). This means that there must be more than 10 million turns in all, though the DNA may not be all in one piece.

The duplicating process can be made to appear more plausible by assuming that the synthesis of the two new chains begins as soon as the two original chains start to unwind, so that only a short stretch of the chain is ever really single. In fact, we may postulate that it is the growth of the two new chains that unwinds the original pair. This is likely in terms of energy because, for every hydrogen bond that has to be broken, two new ones will be forming. Moreover, plausibility is added to the idea by the fact that the paired chain forms a rather stiff structure, so that the growing chain would tend to unwind the old pair.

The difficulty of untwisting the two chains is a topological one, and is due to the fact that they are intertwined. There would be no difficulty in "unwinding" a single helical chain, because there are so many single bonds in the chain about which rotation is possible. If in the twin structure one chain should break, the other one could easily spin around. This might relieve accumulated strain, and then the two ends of the broken chain, still being in close proximity, might be joined together again. There is even some evidence suggesting that in the process of extraction the chains of DNA may be broken in quite a number of places and that the structure nevertheless holds together by means of the hydrogen bonding, because there is never a break in both chains at the same level. Nevertheless, in spite of these tentative suggestions, the difficulty of untwisting remains a formidable one.

There remains the fundamental puzzle as to how DNA exerts its hereditary

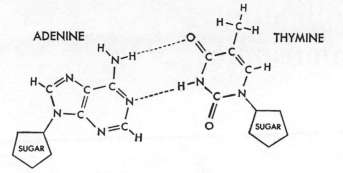

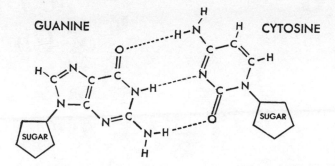

ONE LINKAGE of base to base across the pair of DNA chains is between adenine and thymine. For the structure proposed, the link of a large base with a small one is required to fit chains together.

ANOTHER LINKAGE is comprised of guanine with cytosine. Assuming the existence of hydrogen bonds between the bases, these two pairings, and only these, will explain the actual configuration.

influence. A genetic material must carry out two jobs: duplicate itself and control the development of the rest of the cell in a specific way. We have seen how it might do the first of these, but the structure gives no obvious clue concerning how it may carry out the second. We suspect that the sequence of the bases acts as a kind of genetic code. Such an arrangement can carry an enormous amount of information. If we imagine that the pairs of bases correspond to the dots and dashes of the Morse code, there is enough DNA in a single cell of the human body to encode about 1,000 large textbooks. What we want to know, however, is just how this is done in terms of atoms and molecules. In particular, what precisely is it a code for? As we have seen, the three key components of living matter—protein, RNA and DNA—are probably all based on the same general plan. Their backbones are regular, and the variety comes from the sequence of the side groups. It is therefore very natural to suggest that the sequence of the bases of the DNA is in some way a code for the sequence of the

amino acids in the polypeptide chains of the proteins which the cell must produce. The physicist George Gamow has recently suggested in a rather abstract way how this information might be transmitted, but there are some difficulties with the actual scheme he has proposed, and so far he has not shown how the idea can be translated into precise molecular configurations.

What then, one may reasonably ask, are the virtues of the proposed model, if any? The prime virtue is that the configuration suggested is not vague but can be described in terms acceptable to a chemist. The pairing of the bases can be described rather exactly. The precise positions of the atoms of the backbone is less certain, but they can be fixed within limits, and detailed studies of the X-ray data, now in progress at Kings' College, may narrow these limits considerably. Then the structure brings together two striking pieces of evidence which at first sight seem to be unrelated—the analytical data, showing the one-to-one ratios for adenine-thymine and guanine-cytosine, and the helical

nature of the X-ray pattern. These can now be seen to be two facets of the same thing. Finally, is it not perhaps a remarkable coincidence, to say the least, to find in this key material a structure of exactly the type one would need to carry out a specific replication process; namely, one showing both variety and complementarity?

The model is also attractive in its simplicity. While it is obvious that whole chromosomes have a fairly complicated structure, it is not unreasonable to hope that the molecular basis underlying them may be rather simple. If this is so, it may not prove too difficult to devise experiments to unravel it. It would, of course, help enormously if biochemists could discover the immediate precursors of DNA. If we knew the monomers from which nature makes DNA, RNA and protein, we might be able to carry out very spectacular experiments in the test tube. Be that as it may, we now have for the first time a well-defined model for DNA and for a possible replication process, and this in itself should make it easier to devise crucial experiments.

6

The Duplication
of Chromosomes

by J. Herbert Taylor
June 1958

*It is generally assumed that the basic genetic material
is DNA (deoxyribonucleic acid). How are the molecules
of DNA arranged in the chromosome, and how do they
make replicas of themselves?*

In the search for the secret of how living things reproduce themselves, geneticists have recently focused on the substance called DNA (deoxyribonucleic acid). DNA seems pretty clearly to be the basic hereditary material—the molecule that carries the blueprints for reproduction. F. H. C. Crick and J. D. Watson have found that the molecule consists of two complementary strands, and they have developed the theory that it duplicates itself by a template process, each strand acting as a mold to form a new partner [see "Nucleic Acids," by F. H. C. Crick; SCIENTIFIC AMERICAN Offprint 54].

The problem now is: How does this model fit into the larger picture of chromosomes? For half a century we have known that chromosomes, the rod-like bodies found in the nucleus of every cell, are the carriers of heredity. They contain the genes that pass on the hereditary traits to offspring. As each cell divides, the chromosomes duplicate themselves, so that every daughter cell has copies of the originals. How is the reproduction of chromosomes related to the reproduction of DNA? The question is being pursued by two approaches: from DNA up toward chromosomes and from chromosomes down toward the molecular level. This article will report some experimental studies of the behavior of chromosomes which have suggested a general model of the mechanism of reproduction.

Chromosomes take their name from the fact that they readily absorb dyes and stand out in strong color when cells are stained [*see photographs on the opposite page*]. They become visible under the microscope shortly before a cell is ready to divide. At that time each chromosome consists of a pair of rods side by side. When the cell divides, the two members of the pair (called chromatids) separate, and one chromatid goes to each daughter cell.

In the new cell the chromatid disappears. Then as this cell approaches division, each chromatid reappears, now twinned with a new partner. It has made a copy of itself for the destined daughter cell. There are two possible ways it may have done this: (1) by staying intact (even though invisible in the microscope) and acting as a template, or (2) by breaking down and generating small

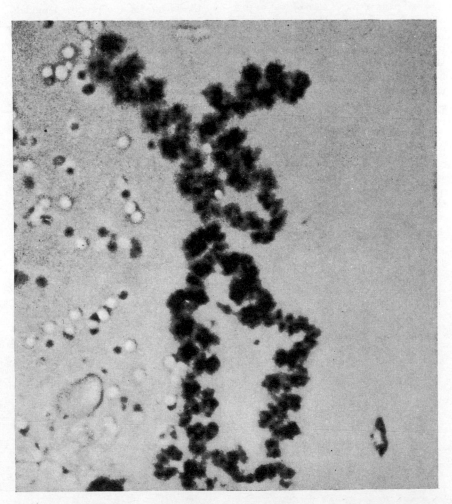

DOUBLE COILED STRUCTURE of a chromosome can be seen in this photomicrograph, the magnification of which is about 4,000 diameters. Chromosome was partly unwound by treatment with dilute potassium cyanide. The chromosome is from a cell of the Easter lily.

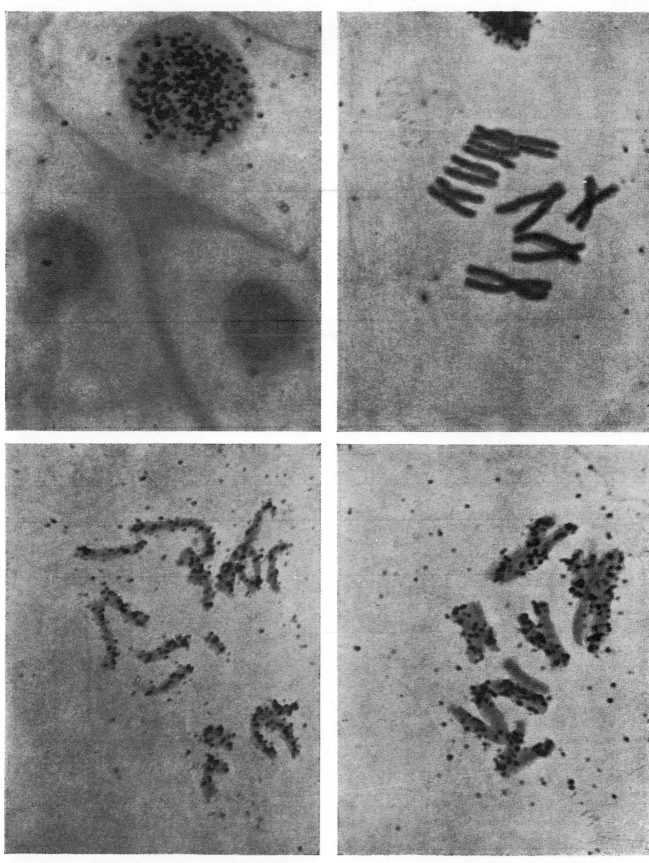

CHROMOSOMES of Bellevalia, a plant of the lily family, are tagged with radioactive thymidine in experiment on duplication. Radiation from the thymidine strikes photographic film placed over the cells, producing black specks. The upper nucleus in the photomicrograph at top left has taken up the tracer material but its chromosomes have not yet become visible. Chromosomes at top right completed their duplication before the cells were placed in radioactive solution and are not tagged. Those at bottom left duplicated once in radioactive solution. Both members of each pair are labeled. The chromosomes at bottom right duplicated once in radioactive solution and once after cells were removed. Only one member of each pair is tagged, except where segments crossed.

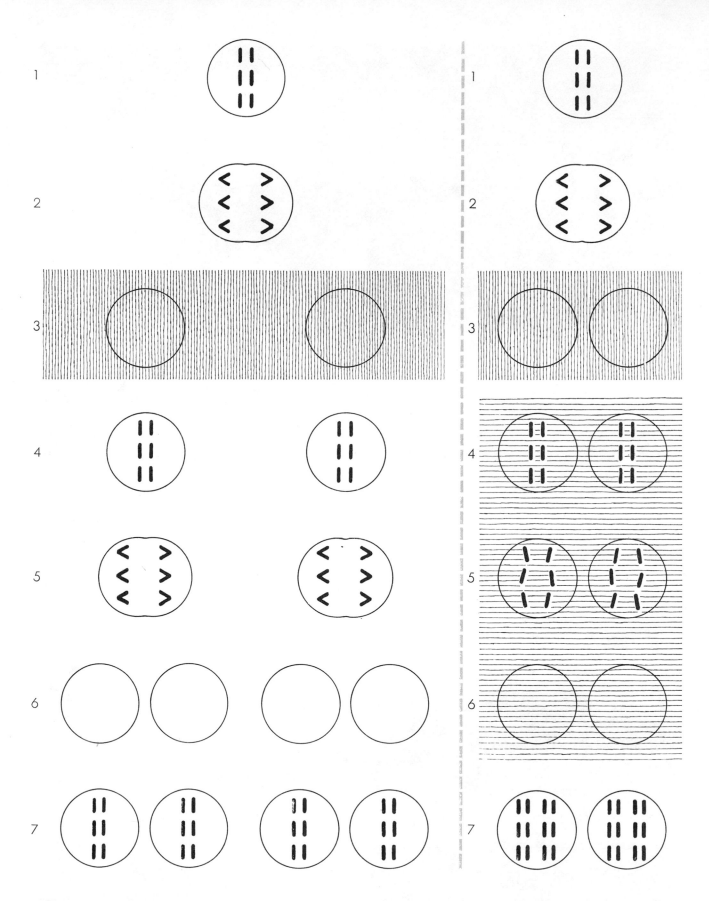

DUPLICATION CYCLES in tracer experiment are diagrammed schematically. Cells to left of the vertical broken line are allowed to divide normally. Those to the right are prevented from dividing when they are placed in colchicine (*black shading*), but their chromosomes continue to duplicate. Black rods represent unlabeled chromatids; colored rods, labeled chromatids. Colored shading indicates radioactive thymidine solution. The empty circles represent stage when chromatids are invisible and duplicating themselves.

units which then reassemble themselves in the form of the original chromatid.

It has recently become possible to resolve this question by means of radioactive tracers. When cells grow in a medium containing thymidine, a component of DNA, all of the thymidine is taken up by the chromosomes; none of it is built into any other part of the cell. Thus if we label thymidine with radioactive atoms (the radioisotopes of hydrogen or carbon), we can follow the transmission of the material through successive replications of the chromosomes. For our own experiments, which I conducted in collaboration with Walter L. Hughes and Philip S. Woods of the Brookhaven National Laboratory, we chose radiohydrogen (tritium) as the tracer. This substance makes it possible to distinguish a radioactive chromatid from a nonradioactive one lying next to it. To localize the radioactivity we use the technique of autoradiography. The cells are squashed flat on a glass slide and covered with a thin sheet of photographic film. Radioactive emanations from the cells produce darkened spots on the film. The emissions from radioactive carbon are fairly penetrating and therefore darken a comparatively wide area of the film; we selected tritium instead because its emissions travel only a short distance—so short that we can narrow down the source to a single chromosome or part of a chromosome.

Our first experiment followed the fate of the thymidine through one duplication of labeled chromosomes. In order to control the situation so that we could identify newly formed chromosomes we treated the cells with colchicine—a drug which prevents cells from dividing but allows chromosomes to go on duplicating themselves. This enabled us to sequester the new chromosomes within the original cells and to tell how many generations had been produced. The cells we studied were those in the growing roots of plants, cultured in a solution containing tritium-labeled thymidine.

We found, to begin with, that in cells that had taken up this thymidine (*i.e.*, produced a new generation of chromosomes preparatory to division), all the chromosomes were labeled, and radioactivity was distributed equally between the two chromatids of each chromosome. This might suggest that the new chromosomes had been formed from a mixture of materials generated by a breakdown of the original chromatids. But when we

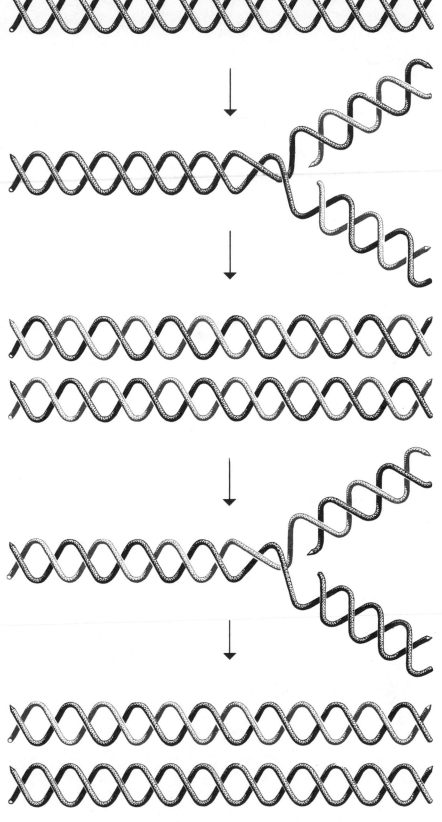

DNA MOLECULES consist of two complementary chains wound around each other in a double helix. When they duplicate, they unwind and each chain builds itself a new partner. Shown here are two cycles of duplication. The first cycle takes place in radioactive solution, producing two labeled chains (*colored helixes*). When a labeled molecule duplicates itself again in nonradioactive solution, only one of its descendants contains a labeled chain.

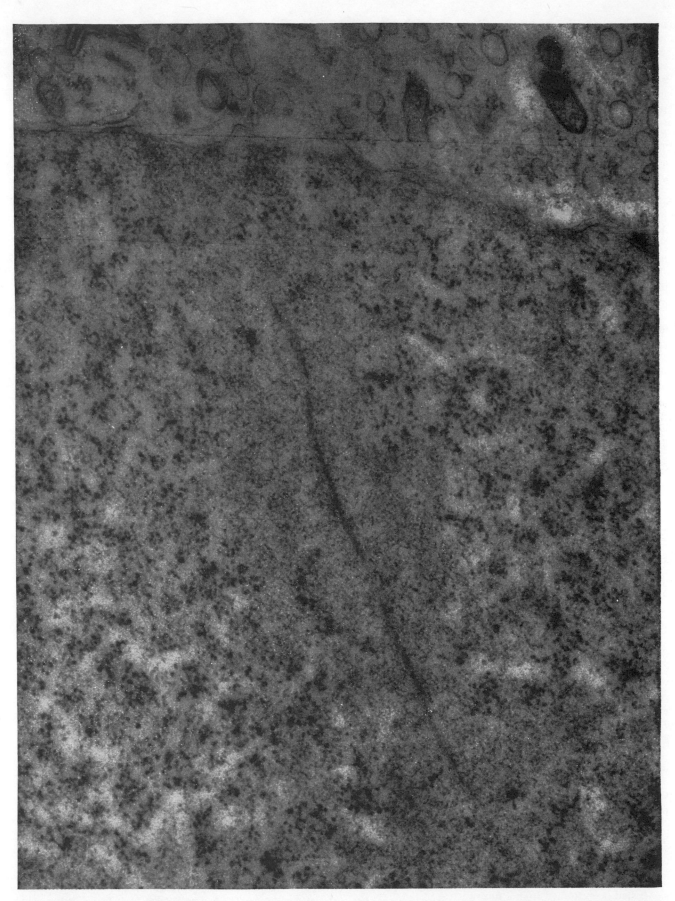

CHROMATID appears as a linear structure running diagonally down the middle of this electron micrograph, made by Montrose J. Moses of the Rockefeller Institute for Medical Research. The mag- nification is some 50,000 diameters. The line may represent the central column in the author's models and the fuzzy material surrounding the line may be DNA strands running perpendicularly outward.

followed root cells through a second generation of chromosome reproduction, where the second generation was synthesized in a medium containing nonradioactive thymidine, we found to our delight that in each new doubled chromosome one chromatid was labeled and the other was not!

What might this mean? The simplest and most likely answer was that a chromatid itself consists of two parts, each of which remains intact and acts as a template. In the radioactive medium each of the original chromatids, after splitting in two, builds itself a radioactive partner. Therefore all the new chromosomes are labeled. Now when the labeled chromatids split again to produce a second generation, half of the strands are labeled and half are not. In a nonradioactive medium all of them will build unlabeled partners. As a result, half of the newly formed chromatids will be partly labeled, half will have no label at all [see diagrams on page 58].

Our picture of the chromatid as a two-part structure fits very well with what we know about the DNA molecule and with the Crick-Watson theory. DNA too is a double structure, consisting of two complementary helical chains wound around each other. And some of our recent experiments indicate that the two strands of a chromatid are complementary structures. It is tempting, therefore, to suppose that a chromatid is simply a chain of DNA. But when we consider the question of scale, we realize that the matter cannot be so simple. If all the DNA in a chromatid formed a single linear chain, the chain would be more than a yard long, and its two strands would be twisted around each other more than 300 million times! It seems unthinkable that so long a chain could untwist itself completely, as the chromatid must each time it generates a new chromosome. Furthermore, the chromatid has the wrong dimensions to be a single DNA chain. When fully extended, it is about 100 times thicker and only one 10,000th as long as the linear DNA chain would be.

Under a high-power microscope we can see that the chromatid is a strand of material tightly wound in a helical coil—in fact, so strongly wound that the coil itself often winds up helically, like a coiled telephone cord which is twisted into a series of secondary kinks [see photograph on page 56]. But beyond this the optical microscope cannot resolve details of the chromatid's structure. Assuming that it is made up of pieces of DNA as its

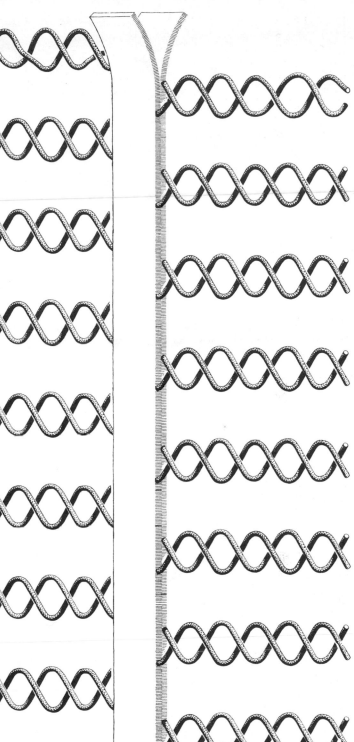

RIBBON MODEL of a chromatid consists of a two-layered central column to which DNA molecules are attached. One chain of each molecule is anchored to the front layer and the other to the back layer. When the chromatid duplicates, the central ribbon peels apart, unwinding the DNA molecules. Each half of the structure then builds itself a new partner.

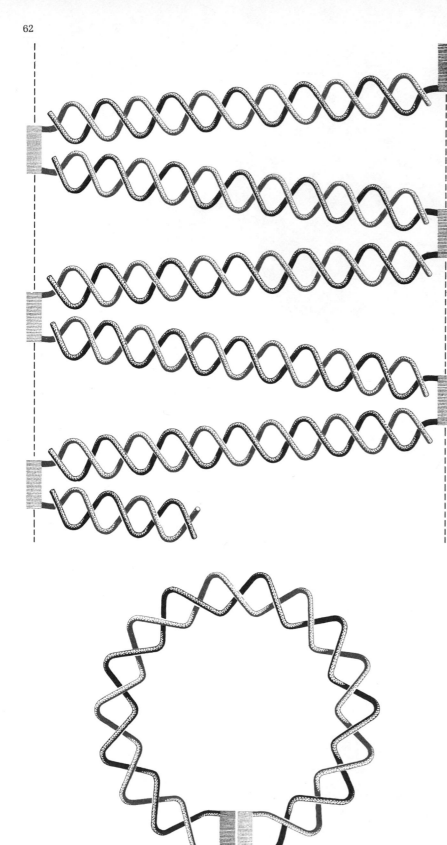

ANOTHER MODEL of the chromatid places its molecules of DNA between two columns, with one chain of each molecule attached to the left-hand column and the other to the right-hand column. Shaded rectangles represent structural material of the columns; broken lines indicate bonds which may include calcium. When the chromatid becomes visible, the columns come together so that the structure appears as in lower drawing when viewed end on.

basic replicating units, what sort of model can we imagine to explain its construction?

We know that chromosomes contain protein. So as a start we may picture the chromatid as a long protein backbone with DNA molecules branching out to the sides like ribs [see diagram on page 61]. Because the chromatid splits in two, we visualize the backbone as a two-layered affair whose layers can separate. The ends of the two strands of a DNA molecule are attached to these layers: one strand to one layer, the complementary strand to the other [see diagram]. As the layers peel apart, they unwind the strands. The unwinding strands promptly begin to build matching new strands for themselves. Eventually the new strands also assemble a new backbone, and the original chromatid is thus fully duplicated.

This model seems to have all the necessary mechanical specifications except one. The genes in a chromosome are arranged in a fixed linear order. Here the DNA segments, with one end waving freely about, are not so arranged. To meet this objection Ernst Freese of Harvard University has suggested a slightly different model which joins the free ends of the DNA molecules so that they form a definite sequence. Instead of one spine there are two, with the DNA segments crossing between them somewhat like the rungs of a ladder [see diagram at the left]. The spines may consist of blocks of protein joined by flexible bonds involving calcium atoms. The DNA rungs zigzag so that they march up the ladder, and the points on the rungs thus have a sequential order.

Now we can suppose that the calcium bonds give the structure considerable flexibility, allowing it to fold and coil on itself. The two spines may come together and so form a long tube [see lower drawing at left]; the tube may then coil into a tight helix. Replication in this model is accomplished by a stretching of the chain and unwinding of the DNA strands, each of which has one free end, as the diagrams show.

Recently Montrose J. Moses of the Rockefeller Institute for Medical Research and Don W. Fawcett of Cornell Medical College, using the electron microscope, have obtained pictures of chromatids which do indeed show a spine structure with DNA branches [see photograph on page 60]. It appears that we are beginning to penetrate down to the detailed mechanisms of the duplication of life.

7

The Bacterial Chromosome

by John Cairns
January 1966

*When bacterial DNA is labeled with radioactive atoms,
it takes its own picture. Autoradiographs reveal that
the bacterial chromosome is a single very long DNA
molecule and show how it is duplicated*

The information inherited by living things from their forebears is inscribed in their deoxyribonucleic acid (DNA). It is written there in a decipherable code in which the "letters" are the four subunits of DNA, the nucleotide bases. It is ordered in functional units—the genes—and thence translated by way of ribonucleic acid (RNA) into sequences of amino acids that determine the properties of proteins. The proteins are, in the final analysis, the executors of each organism's inheritance.

The central event in the passage of genetic information from one generation to the next is the duplication of DNA. This cannot be a casual process. The complement of DNA in a single bacterium, for example, amounts to some six million nucleotide bases; this is the bacterium's "inheritance." Clearly life's security of tenure derives in large measure from the precision with which DNA can be duplicated, and the manner of this duplication is therefore a matter of surpassing interest. This article deals with a single set of experiments on the duplication of DNA, the antecedents to them and some of the speculations they have provoked.

When James D. Watson and Francis H. C. Crick developed their two-strand model for the structure of DNA, they saw that it contained within it the seeds of a system for self-duplication. The two strands, or polynucleotide chains, were apparently related physically to each other by a strict system of *complementary* base pairing. Wherever the nucleotide base adenine occurred in one chain, thymine was present in the other; similarly, guanine was always paired with cytosine. These rules meant that the sequence of bases in each chain inexorably stipulated the sequence in

the other; each chain, on its own, could generate the entire sequence of base pairs. Watson and Crick therefore suggested that accurate duplication of DNA could occur if the chains separated and each then acted as a template on which a new complementary chain was laid down. This form of duplication was later called "semiconservative" because it supposed that although the individual parental chains were conserved during duplication (in that they were not thrown away), their association ended as part of the act of duplication.

The prediction of semiconservative replication soon received precise experimental support. Matthew S. Meselson and Franklin W. Stahl, working at the California Institute of Technology, were able to show that each molecule of DNA in the bacterium *Escherichia coli* is composed of equal parts of newly synthesized DNA and of old DNA that was present in the previous generation [*see top illustration on page 65*]. They realized they had not proved that the two parts of each molecule were in fact two chains of the DNA duplex, because they had not established that the molecules they were working with consisted of only two chains. Later experiments, including some to be described in this article, showed that what they were observing was indeed the separation of the two chains during duplication.

The Meselson-Stahl experiment dealt with the end result of DNA duplication. It gave no hint about the mechanism that separates the chains and then supervises the synthesis of the new chains. Soon, however, Arthur Kornberg and his colleagues at Washington University isolated an enzyme from *E. coli* that, if all the necessary precursors were provided, could synthesize in the test tube

chains that were complementary in base sequence to any DNA offered as a template. It was clear, then, that polynucleotide chains could indeed act as templates for the production of complementary chains and that this kind of reaction could be the normal process of duplication, since the enzymes for carrying it out were present in the living cell.

Such, then, was the general background of the experiments I undertook beginning in 1962 at the Australian National University. My object was simply (and literally) to look at molecules of DNA that had been caught in the act of duplication, in order to find out which of the possible forms of semiconservative replication takes place in the living cell: how the chains of parent DNA are arranged and how the new chains are laid down [*see bottom illustration on page 65*].

Various factors dictated that the experiments should be conducted with *E. coli*. For one thing, this bacterium was known from genetic studies to have only one chromosome; that is, its DNA is contained in a single functional unit in which all the genetic markers are arrayed in sequence. For another thing, the duplication of its chromosome was known to occupy virtually the entire cycle of cell division, so that one could be sure that every cell in a rapidly multiplying culture would contain replicating DNA.

Although nothing was known about the number of DNA molecules in the *E. coli* chromosome (or in any other complex chromosome, for that matter), the dispersal of the bacterium's DNA among its descendants had been shown to be semiconservative. For this and other reasons it seemed likely that the

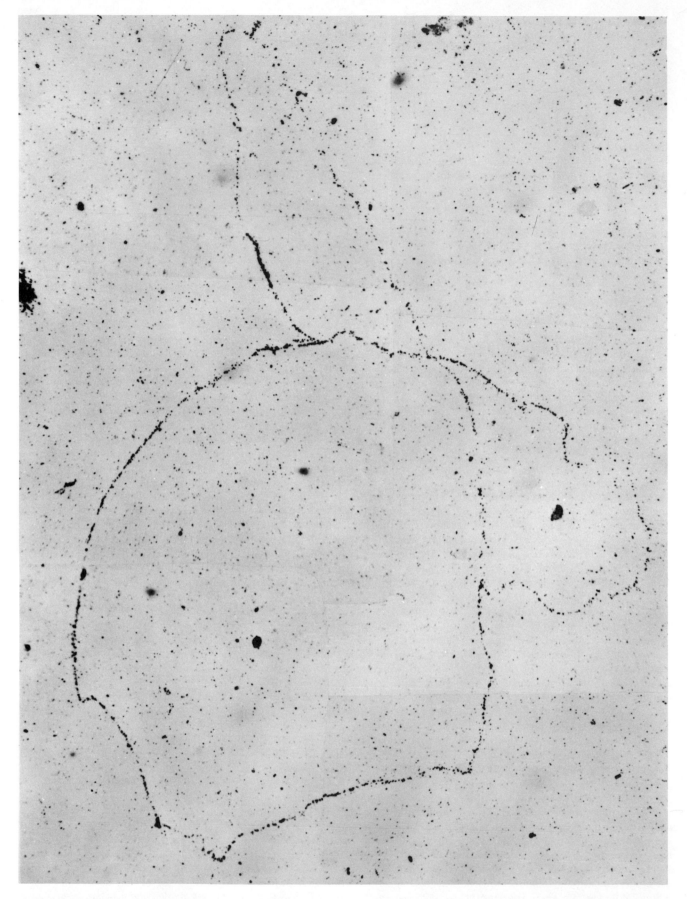

AUTORADIOGRAPH shows a duplicating chromosome from the bacterium *Escherichia coli* enlarged about 480 diameters. The DNA of the chromosome is visible because for two generations it incorporated a radioactive precursor, tritiated thymine. The thy-mine reveals its presence as a line of dark grains in the photographic emulsion. (Scattered grains are from background radiation.) The diagram on the opposite page shows how the picture is interpreted as demonstrating the manner of DNA duplication.

bacterial chromosome would turn out to be a single very large molecule. All the DNA previously isolated from bacteria had, to be sure, proved to be in molecules much smaller than the total chromosome, but a reason for this was suggested by studies by A. D. Hershey of the Carnegie Institution Department of Genetics at Cold Spring Harbor, N.Y. He had pointed out that the giant molecules of DNA that make up the genetic complement of certain bacterial viruses had been missed by earlier workers simply because they are so large that they are exceedingly fragile. Perhaps the same thing was true of the bacterial chromosome.

If so, the procedure for inspecting the replicating DNA of bacteria would have to be designed to cater for an exceptionally fragile molecule, since the bacterial chromosome contains some 20 times more DNA than the largest bacterial virus. It would have to be a case of looking but not touching. This was not as onerous a restriction as it may sound. The problem was, after all, a topographical one, involving delineation of strands of parent DNA and newly synthesized DNA. There was no need for manipulation, only for visualization.

Although electron microscopy is the

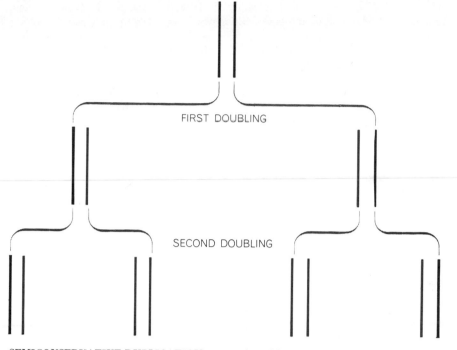

SEMICONSERVATIVE DUPLICATION was confirmed by the Meselson-Stahl experiment, which showed that each DNA molecule is composed of two parts: one that is present in the parent molecule, the other comprising new material synthesized when the parent molecule is duplicated. If radioactive labeling begins with the first doubling, the unlabeled (*black*) and labeled (*colored*) nucleotide chains of DNA form two-chain duplexes as shown here.

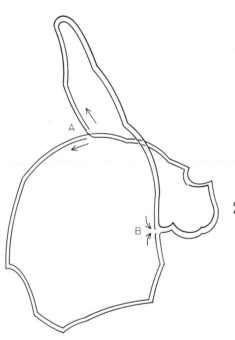

INTERPRETATION of autoradiograph on opposite page is based on the varying density of the line of grains. Excluding artifacts, dense segments represent doubly labeled DNA duplexes (*two colored lines*), faint segments singly labeled DNA (*color and black*). The parent chromosome, labeled in one strand and part of another, began to duplicate at *A*; new labeled strands have been laid down in two loops as far as *B*.

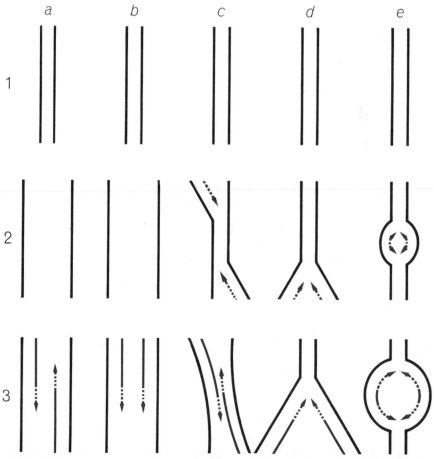

DUPLICATION could proceed in various ways (*a–e*). In these examples parental chains are shown as black lines and new chains as colored lines. The arrows show the direction of growth of the new chains, the newest parts of which are denoted by broken-line segments.

obvious way to get a look at a large molecule, I chose autoradiography in this instance because it offered certain peculiar advantages (which will become apparent) and because it had already proved to be the easier, albeit less accurate, technique for displaying large DNA molecules. Autoradiography capitalizes on the fact that electrons emitted by the decay of a radioactive isotope produce images on certain kinds of photographic emulsion. It is possible, for example, to locate the destination within a cell of a particular species of molecule by labeling such molecules with a radioactive atom, feeding them to the cell and then placing the cell in contact with an emulsion; a developed grain in the emulsion reveals the presence of a labeled molecule [see "Autobiographies of Cells," by Renato Baserga and Walter E. Kisieleski; SCIENTIFIC AMERICAN Offprint 165].

It happens that the base thymine, which is solely a precursor of DNA, is susceptible to very heavy labeling with tritium, the radioactive isotope of hydrogen. Replicating DNA incorporates the labeled thymine and thus becomes visible in autoradiographs. I had been able to extend the technique to demonstrating the form of individual DNA molecules extracted from bacterial viruses. This was possible because, in spite of the poor resolving power of autoradiography (compared with electron microscopy), molecules of DNA are so extremely long in relation to the resolving power that they appear as a linear array of grains. The method grossly exaggerates the apparent width of the DNA, but this is not a serious fault in the kind of study I was undertaking.

The general design of the experiments called for extracting labeled DNA from bacteria as gently as possible and then

mounting it—without breaking the DNA molecules—for autoradiography. What I did was kill bacteria that had been fed tritiated thymine for various periods and put them, along with the enzyme lysozyme and an excess of unlabeled DNA, into a small capsule closed on one side by a semipermeable membrane. The enzyme, together with a detergent diffused into the chamber, induced the bacteria to break open and discharge their DNA. After the detergent, the enzyme and low-molecular-weight cellular debris had been diffused out of the chamber, the chamber was drained, leaving some of the DNA deposited on the membrane [see illustration below]. Once dry, the membrane was coated with a photographic emulsion sensitive to electrons emitted by the tritium and was left for two months. I hoped by this procedure to avoid subjecting the DNA to appreciable turbulence and so to find

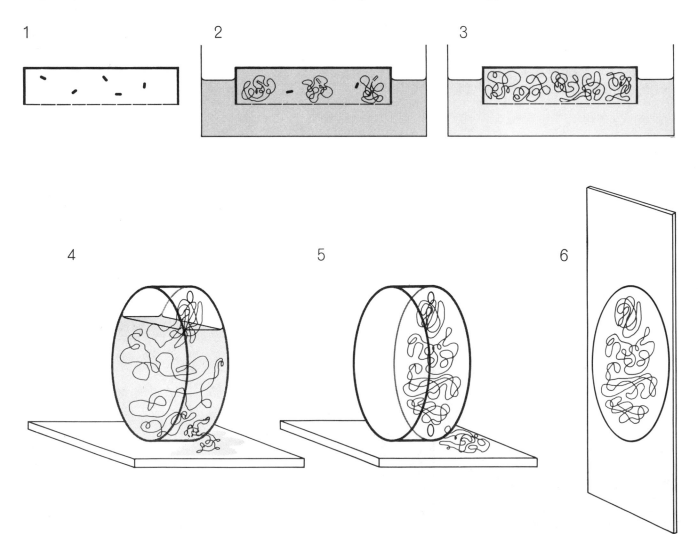

AUTORADIOGRAPHY EXPERIMENT begins with bacteria whose DNA has been labeled with radioactive thymine. The bacteria and an enzyme are placed in a small chamber closed by a semipermeable membrane (1). Detergent diffused into the chamber causes the bacteria to discharge their contents (2). The detergent and cellular debris are washed away by saline solution diffused through the chamber (3). The membrane is then punctured. The saline drains out slowly (4), leaving some unbroken DNA molecules (color) clinging to the membrane (5). The membrane, with DNA, is placed on a microscope slide and coated with emulsion (6).

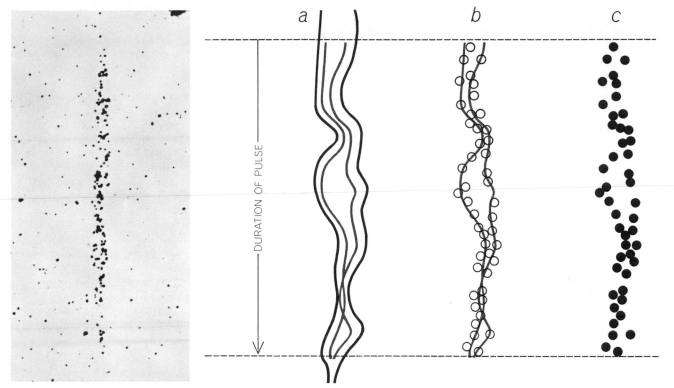

DNA synthesized in *E. coli* fed radioactive thymine for three minutes is visible in an autoradiograph, enlarged 1,200 diameters, as an array of heavy black grains (*left*). The events leading to the autoradiograph are shown at right. The region of the DNA chains synthesized during the "pulse-labeling" is radioactive and is shown in color (*a*). The radioactivity affects silver grains in the photographic emulsion (*b*). The developed grains appear in the autoradiograph (*c*), approximately delineating the new chains of DNA.

some molecules that—however big—had not been broken and see their form. Inasmuch as *E. coli* synthesizes DNA during its entire division cycle, some of the extracted DNA should be caught in the act of replication. (Since there was an excess of unlabeled DNA present, any tendency for DNA to produce artificial aggregates would not produce a spurious increase in the size of the labeled molecules or an alteration in their form.)

It is the peculiar virtue of autoradiography that one sees only what has been labeled; for this reason the technique can yield information on the history as well as the form of a labeled structure. The easiest way to determine which of the schemes of replication was correct was to look at bacterial DNA that had been allowed to duplicate for only a short time in the presence of labeled thymine. Only the most recently made DNA would be visible (corresponding to the broken-line segments in the bottom illustration on page 65), and so it should be possible to determine if the two daughter molecules were being made at the same point or in different regions of the parent molecule. A picture obtained after labeling bacteria for three minutes, or a tenth of a generation-time [*at left in illustration above*], makes it clear that two labeled structures are being made in the same place. This place is presumably a particular region of a larger (unseen) parent molecule [*see diagrams at right in illustration above*].

The autoradiograph also shows that at least 80 microns (80 thousandths of a millimeter) of the DNA has been duplicated in three minutes. Since duplication occupies the entire generation-time (which was about 30 minutes in these experiments), it follows that the process seen in the autoradiograph could traverse at least 10×80 microns, or about a millimeter, of DNA between one cell division and the next. This is roughly the total length of the DNA in the bacterial chromosome. The autoradiograph therefore suggests that the entire chromosome may be duplicated at a single locus that can move fast enough to traverse the total length of the DNA in each generation.

Finally, the autoradiograph gives evidence on the semiconservative aspect of duplication. Two structures are being synthesized. It is possible to estimate how heavily each structure is labeled (in terms of grains produced per micron of length) by counting the number of exposed grains and dividing by the length. Then the density of labeling can be compared with that of virus DNA labeled similarly but uniformly, that is, in both of its polynucleotide chains. It turns out that each of the two new structures seen in the picture must be a single polynucleotide chain. If, therefore, the picture is showing the synthesis of two daughter molecules from one parent molecule, it follows that each daughter molecule must be made up of one new (labeled) chain and one old (unlabeled) chain—just as Watson and Crick predicted.

The "pulse-labeling" experiment just described yielded information on the isolated regions of bacterial DNA actually engaged in duplication. To learn if the entire chromosome is a single molecule and how the process of duplication proceeds it was necessary to look at DNA that had been labeled with tritiated thymine for several generations. Moreover, it was necessary to find, in the jumble of chromosomes extracted from *E. coli*, autoradiographs of unbroken chromosomes that were disen-

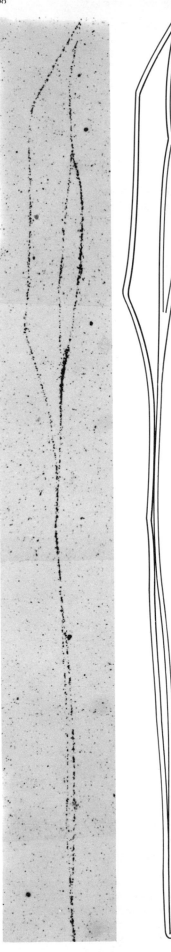

tangled enough to be seen as a whole. Rather than retrace all the steps that led, after many months, to satisfactory pictures of the entire bacterial chromosome in one piece, it is simpler to present two sample autoradiographs and explain how they can be interpreted and what they reveal.

The autoradiographs on page 64 and at the left show bacterial chromosomes in the process of duplication. All that is visible is labeled, or "hot," DNA; any unlabeled, or "cold," chain is unseen. A stretch of DNA duplex labeled in only one chain ("hot-cold") makes a faint trace of black grains. A duplex that is doubly labeled ("hot-hot") shows as a heavier trace. The autoradiographs therefore indicate, as shown in the diagrams that accompany them, the extent to which new, labeled polynucleotide chains have been laid down along labeled or unlabeled parent chains. Such data make it possible to construct a bacterial family history showing the process of duplication over several generations [see illustration on opposite page].

The significant conclusions are these:

1. The chromosome of E. coli apparently contains a single molecule of DNA roughly a millimeter in length and with a calculated molecular weight of about two billion. This is by far the largest molecule known to occur in a biological system.

2. The molecule contains two polynucleotide chains, which separate at the time of duplication.

3. The molecule is duplicated at a single locus that traverses the entire length of the molecule. At this point both new chains are being made: two chains are becoming four. This locus has come to be called the replicating "fork" because that is what it looks like.

4. Replicating chromosomes are not Y-shaped, as would be the case for a linear structure [see "d" in bottom illustration on page 65]. Instead the three ends of the Y are joined: the ends of the daughter molecules are joined to each other and to the far end of the parent molecule. In other words, the chromo-

COMPLETE CHROMOSOME is seen in this autoradiograph, enlarged about 370 diameters. Like the chromosome represented on pages 64 and 65, this one is circular, although it happens to have landed on the membrane in a more compressed shape and some segments are tangled. Whereas the first chromosome was more than halfway through the duplication process, this one is only about one-sixth duplicated (from A to B).

some is circular while it is being duplicated.

It is hard to conceive of the behavior of a molecule that is about 1,000 times larger than the largest protein and that exists, moreover, coiled inside a cell several hundred times shorter than itself. Apart from this general problem of comprehension, there are two special difficulties inherent in the process of DNA duplication outlined here. Both have their origin in details of the structure of DNA that I have not yet discussed.

The first difficulty arises from the opposite polarities of the two polynucleotide chains [see illustration on page 70]. The deoxyribose-phosphate backbone of one chain of the DNA duplex has the sequence $-O-C_3-C_4-C_5-O-P-O-C_3-C_4-C_5-O-P-...$ (The C_3, C_4 and C_5 are the three carbon atoms of the deoxyribose that contribute to the backbone.) The other chain has the sequence $-P-O-C_5-C_4-C_3-O-P-O-C_5-C_4-C_3-O-...$

If both chains are having their complements laid down at a single locus moving in one particular direction, it follows that one of these new chains must grow by repeated addition to the C_3 of the preceding nucleotide's deoxyribose and the other must grow by addition to a C_5. One would expect that two different enzymes should be needed for these two quite different kinds of polymerization. As yet, however, only the reaction that adds to chains ending in C_3 has been demonstrated in such experiments as Kornberg's. This fact had seemed to support a mode of replication in which the two strands grew in opposite directions [see "a" and "c" in bottom illustration on page 65]. If the single-locus scheme is correct, the problem of opposite polarities remains to be explained.

The second difficulty, like the first, is related to the structure of DNA. For the sake of simplicity I have been representing the DNA duplex as a pair of chains lying parallel to each other. In actuality the two chains are wound helically around a common axis, with one complete turn for every 10 base pairs, or 34 angstrom units of length (34 tenmillionths of a millimeter). It would seem, therefore, that separation of the chains at the time of duplication, like separation of the strands of an ordinary rope, must involve rotation of the parent molecule with respect to the two daughter molecules. Moreover, this rotation must be very rapid. A fast-multiplying bacterium can divide every 20 minutes;

during this time it has to duplicate—and consequently to unwind—about a millimeter of DNA, or some 300,000 turns. This implies an average unwinding rate of 15,000 revolutions per minute.

At first sight it merely adds to the difficulty to find that the chromosome is circular while all of this is going on. Obviously a firmly closed circle—whether a molecule or a rope—cannot be unwound. This complication is worth worrying about because there is increasing evidence that the chromosome of *E. coli* is not exceptional in its circularity. The DNA of numerous viruses has been shown either to be circular or to become circular just before replication begins. For all we know, circularity may therefore be the rule rather than the exception.

There are several possible explanations for this apparent impasse, only one of which strikes me as plausible.

First, one should consider the possibility that there is no impasse—that in the living cell the DNA is two-stranded but not helical, perhaps being kept that way precisely by being in the form of a circle. (If a double helix cannot be unwound when it is firmly linked into a circle, neither can relational coils ever be introduced into a pair of uncoiled circles.) This hypothesis, however, requires a most improbable structure for two-strand DNA, one that has not been observed. And it does not really avoid the unwinding problem because there would still have to be some mechanism for making nonhelical circles out of the helical rods of DNA found in certain virus particles.

Second, one could avoid the unwinding problem by postulating that at least one of the parental chains is repeatedly broken and reunited during replication, so that the two chains can be separated over short sections without rotation of the entire molecule. One rather sentimental objection to this hypothesis (which was proposed some time ago) is that it is hard to imagine such cavalier and hazardous treatment being meted out to such an important molecule, and one so conspicuous for its stability. A second objection is that it does not explain circularity.

The most satisfactory solution to the unwinding problem would be to find some reason why the ends of the chromosome actually *must* be joined together. This is the case if one postulates that there is an active mechanism for unwinding the DNA, distinct from the mechanism that copies the unwound

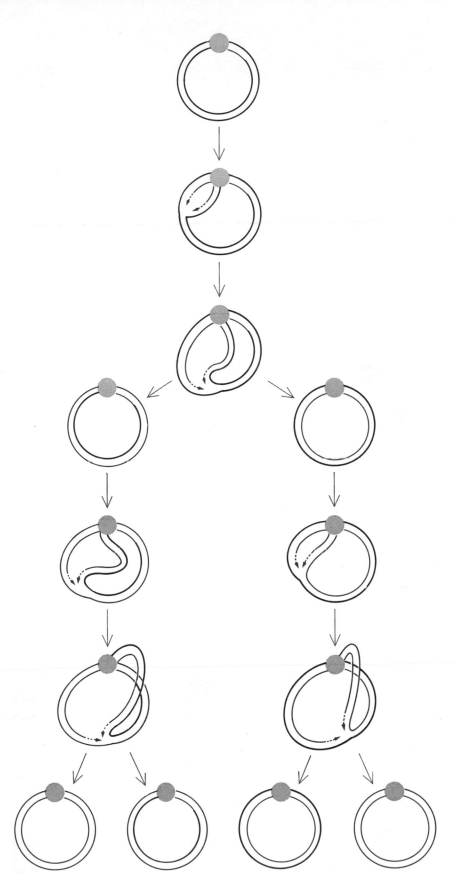

BACTERIAL DNA MOLECULE apparently replicates as in this schematic diagram. The two chains of the circular molecule are represented as concentric circles, joined at a "swivel" (*gray spot*). Labeled DNA is shown in color; part of one chain of the parent molecule is labeled, as are two generations of newly synthesized DNA. Duplication starts at the swivel and, in these drawings, proceeds counterclockwise. The arrowheads mark the replicating "fork": the point at which DNA is being synthesized in each chromosome. The drawing marked A is a schematic rendering of the chromosome in the autoradiograph on page 64.

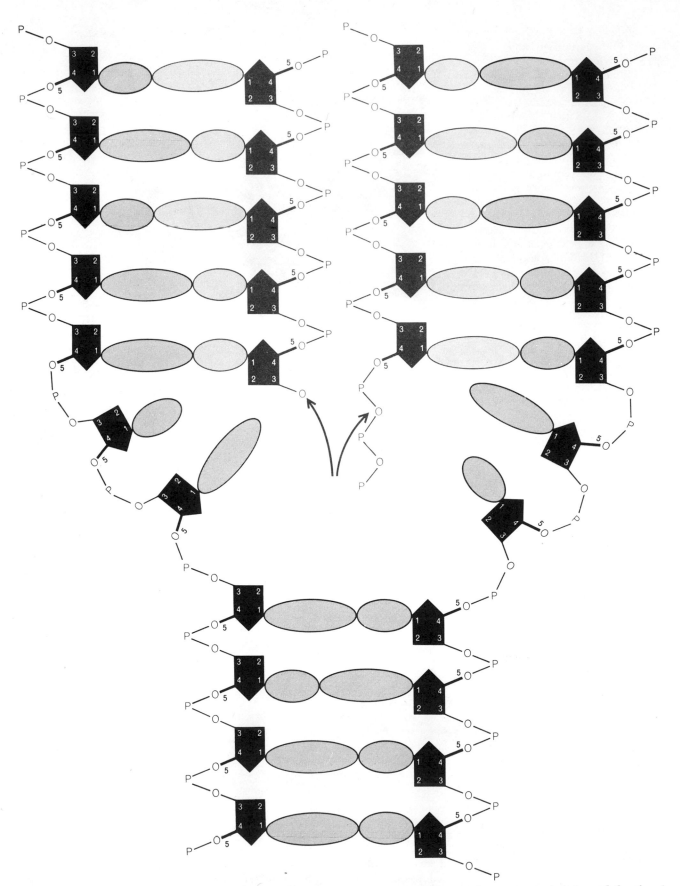

OPPOSITE POLARITIES of the two parental chains of the DNA duplex result in opposite polarities and different directions of growth in the two new chains (*color*) being laid down as complements of the old ones during duplication. Note that the numbered carbon atoms (*1 to 5*) in the deoxyribose rings (*solid black*) are in different positions in the two parental chains and therefore in the two new chains. As the replicating fork moves downward, the new chain that is complementary to the left parental chain must grow by addition to a C$_3$, the other new chain by addition to a C$_5$, as shown by the arrows. The elliptical shapes are the four bases.

chains. Now, any active unwinding mechanism must rotate the parent molecule with respect to the two new molecules—must hold the latter fast, in other words, just as the far end of a rope must be held if it is to be unwound. A little thought will show that this can be most surely accomplished by a machine attached, directly or through some common "ground," to the parent molecule and to the two daughters [*see illustration below*]. Every turn taken by such a machine would inevitably unwind the parent molecule one turn.

Although other kinds of unwinding machine can be imagined (one could be situated, for example, at the replicating fork), a practical advantage of this particular hypothesis is that it accounts for circularity. It also makes the surprising —and testable—prediction that any irreparable break in the parent molecule will instantly stop DNA synthesis, no matter how far the break is from the replicating fork. If this prediction is fulfilled, and the unwinding machine acquires the respectability that at present it lacks, we may find ourselves dealing with the first example in nature of something equivalent to a wheel.

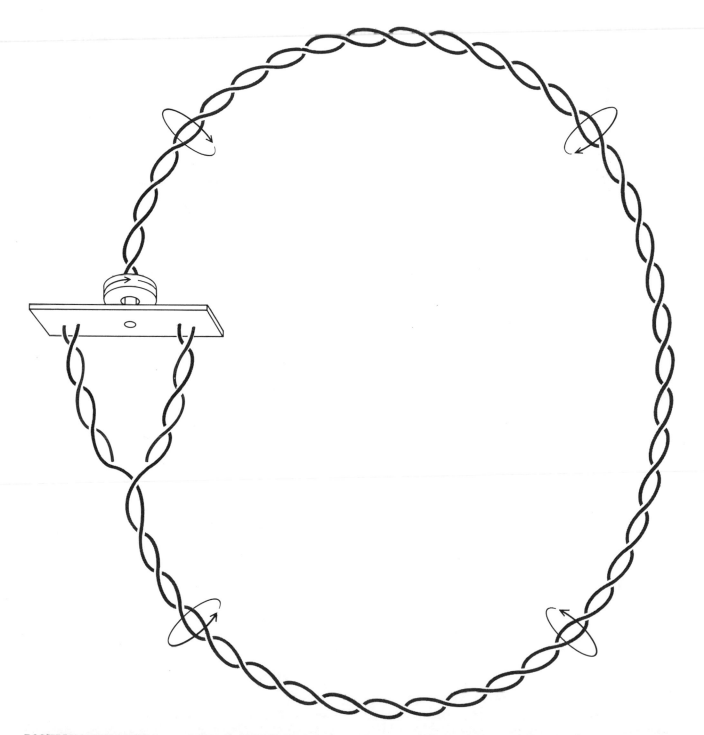

POSSIBLE MECHANISM for unwinding the DNA double helix is a swivel-like machine to which the end of the parent molecule and also the ends of the two daughter molecules are joined. The torque imparted by this machine is considered to be transmitted along the parent molecule, producing unwinding at the replicating fork. If this is correct, chromosome breakage should halt duplication.

III

GENETIC ANALYSIS

GENETIC ANALYSIS

INTRODUCTION

The essential condition for genetic analysis is the cross where the genetic material of two (or more) separate lineages is combined within a common cell. Two outcomes are consequent upon such a condition: one provides the basis for a functional analysis in the form of a complementation test, and the other—providing there is a resolution of the genetic mixture —allows a mapping of the genetic differences that differentiate the parental lineages entering into the cross.

Classically, the combination of lineages is achieved by sexual interaction, which leads to the fusion of haploid gametes. Such crosses characteristically require the combination of the entire genomes from each parent. The data for the functional analysis is derived from observation of the characteristics of the resulting F_1 hybrid. The mapping data are obtained from the progeny of a cross between the F_1 hybrid and a true breeding line that is homozygous recessive for the characters differing in the cross. The various proportions of the different classes of backcross progeny so resulting allow the assignment of genetic loci to linkage groups, and, for those that belong to the same linkage group, their relative ordering and spacing on a genetic map.

The comparison of the data obtained from extensive genetic analysis with those derived from direct cytological observation on the organism's chromosome constitution has amply demonstrated that linkage groups correspond to chromosomes, while the ordering and spacing of the loci of a linkage group map correspond to their physical ordering and approximate spacing on the chromosome.

For the purposes of functional analysis, one examines the F_1 hybrid or its formal genetic equivalent, where two sets of genetic information differing with respect to one or more functions are combined within a single cell or organism. This situation comprises an assay known as the complementation test. For example, if a cross of two recessive mutant strains of similar phenotype leads to a wild-type F_1, or a joint infection of a cell by two mutant strains of a virus leads to a normal infection, then the two mutants involved in each case are said to "complement" one another. That is, the two defective parental genomes in combination mutually compensate each other's defects. From such a result it is concluded that in each case the two mutants affect different genes. If two recessive mutants fail to complement in combination, then it is concluded that the mutants in question affect the same gene.

The functional or complementation analysis of an extensive set of mutants of a given phenotypic class provides an estimate of the number of genes that contribute to that phenotype function. For very simple biological entities such as the viruses, the complementation analysis can provide an estimate of the number of genes in a viral genome. We shall return to this problem presently.

With the discovery of the phenomena of transformation and transduction (see Rollin D. Hotchkiss and Esther Weiss's article "Transformed Bacteria" and Norton D. Zinder's article " 'Transduction' in Bacteria") it became clear that conditions for genetic analysis are not restricted to those involving sexual interaction. In these instances small fragments of DNA are introduced, either directly or indirectly, into the recipient cells, by transformation or transduction, respectively. While the primary significance of the phenomenon of transformation was that it demonstrated unambiguously that DNA was indeed a carrier of genetic specificity, transformation has also proved to be a very valuable tool for high resolution genetic analysis in those microorganisms that are susceptible to transformation. Similarly, while the historic significance of genetic transduction lay in its revealing that unexpected transactions can be mediated by virus parasites between different host cells, it too has proved immensely useful as a means of achieving a very high resolution of genetic analysis.

The discoveries of genetic recombinants in bacteria in 1946, which resulted from bacterial crosses made by Joshua Lederberg and Edward Tatum and from bacteriophage crosses by Alfred Hershey, helped to usher in the new phase of high resolution genetic analysis. This advance in genetic analysis and the circumstances allowing it are described in two articles, "Viruses and Genes" by François Jacob and Elie L. Wollman and "The Genetics of a Bacterial Virus" by R. S. Edgar and R. H. Epstein.

The opportunity for high resolution analysis was afforded by the availability of very large numbers of progeny issuing from a particular cross; these allowed geneticists to screen for rare recombinants with very little effort and thus to perform a fine structure analysis of a particular gene down to the nucleotide level.

The classical instance of "running the genetic map into the ground" is presented in Seymour Benzer's article, "The Fine Structure of the Gene." However, it is not sufficient to have just the means for achieving a high resolution analysis, especially if one of the goals of analysis is to catalog all the genes that specify a virus or bacterial species, unless one has a way of both recognizing the consequence of mutation in a function and being able to work with such mutations in a genetic cross. This leads us to a discussion of the "marker" problem in genetic analysis, as described in Edgar and Epstein's article.

Ordinarily, the occurrence of mutation in an organism is recognized by its phenotypic consequence, either immediately (when the mutation is dominant) or subsequently, after inbreeding to obtain some progeny homozygous for the mutation in question (when it is recessive).

Mendel, for example, was limited to using mutations that led to morphological but noncritical changes in his pea plants.

George Beadle and Edward Tatum were able to use mutations that affected genes coding for critical enzymes involved in the production of essential vitamins and amino acids, by being able to remedy these defects with appropriate nutritional supplements added to the culture medium. However, they too were limited in the scope of the kinds of mutations they analyzed—namely, to those affecting functions that could be remedied by an external supplement.

While lethal mutations can be recognized and, if manifested late in development, can be the subject of limited analysis, they are understandably quite difficult to work with in diploid organisms and impossible to work with in the lower life forms such as fungi, bacteria, and viruses, where the cell or virus possesses but one copy of each gene or is functionally haploid. Yet, mutations in functions critical to the viability of an organism or virus being studied are of the utmost interest if one is to achieve a significant genetic description of an organism or virus in terms of the ensemble of structural and catalytic proteins that specify it.

In their article, Edgar and Epstein describe how they overcame these limitations by discovering a class of mutation that inactivates *any* essential protein-specified function—irrespective of the particular nature of that function—under one particular condition of growth but not under another. Their article describes two such conditional lethal systems that they developed for this purpose.

The final article in this section, "Hybrid Cells and Human Genes" by Frank H. Ruddle and Raju S. Kucherlapati, reports the opportunities for genetic analysis afforded by the rather bizarre and unexpected consequences of cell fusion between the diploid somatic tissue culture cells of two mammalian species—in this instance, cells of mice and man. This situation allows a crude level of analysis limited to the assignment of human mutations to a particular chromosome, and in some instances to one or other arm of a chromosome. Despite this lack of resolution, the system has important applications in medical genetics in that it provides opportunities for more refined prognostication in counseling people who are at risk with respect to giving birth to children with known genetic anomalies.

8

"Transduction" in Bacteria

by Norton D. Zinder
November 1958

A virus can transfer genetic material, and thereby hereditary traits, from one bacterium to another. The phenomenon sheds light on the behavior of viruses and the mechanism of heredity

Viruses first made their existence known as especially tiny germs that cause disease in animals and plants. Then it was discovered that a virus can multiply itself only inside the living cell; the new viruses are released, and the cell dies. During the time it is inside, the virus vanishes in the biochemical system of its host, so that its activities there have been largely obscure. Bit by bit, however, the life story of the virus is being pieced together. As the parts fall into place, the virus is assuming a new identity. From the point of view of the biologist the germ has become a valuable ally in the exploration of the life processes of the cell.

Geneticists have found viruses particularly useful as a means of getting at the mechanism of heredity. The typical virus is a bit of genetic material encapsulated in a protein coat. As such it may affect the genetics of its host in important ways. This was first observed in the case of viruses that infect bacteria. At times these viruses cause a latent infection; they do not kill their host but become a part of its genetic apparatus. The latent virus then acts like a bit of the bacteria's genetic material and induces new traits in its host. In this role it may be reproduced through many generations along with the bacteria's own genes before it resumes its existence as a separate virus. It is such a latent virus infection which causes the normally innocuous diphtheria bacillus to make its lethal toxin.

This article is concerned with a discovery which implicates the virus even more deeply in the genetic processes of its host. It now appears that a bacterial virus can carry the bacteria's own genes from cell to cell. Like a disease carrier, it infects one bacterium with hereditary material picked up from another.

This "transduction" of bacterial heredity was discovered by accident and good luck in an investigation that was at first not concerned with viruses at all. The discovery occurred during an attempt to induce sexual mating in the bacteria which cause a disease in mice resembling typhoid fever in man. Mating is a rare process in bacteria. A bacterium ordinarily multiplies simply by dividing into two cells, each of which usually has the same genetic constitution as the other. In 1946, however, Joshua Lederberg and Edward L. Tatum (then at Yale University) found that a bacterium in one strain of the bacillus *Escherichia*

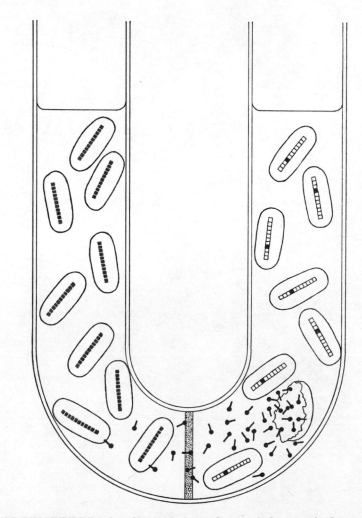

U-TUBE EXPERIMENT led to discovery of transduction. At bottom of tube was a filter. On left side of filter was one strain of the bacterium *Salmonella typhimurium*; on right side, another strain. The strain at right harbored "latent" virus (*small black square on*

coli could mate under certain conditions with a bacterium in another strain of the same species, and thus give rise to bacteria with some characteristics of both strains. There seemed to be no reason why *E. coli* alone among bacteria should have the ability to mate, so in 1949 Lederberg and I (then at the University of Wisconsin) undertook to induce mating in another species.

For our first experiment we chose two strains of the mouse-typhoid bacterium, *Salmonella typhimurium*. Each strain lacked the capacity to synthesize a particular amino acid needed for its growth, but was able to make the amino acid which the other strain could not produce. If mating occurred, some of the offspring should be able to synthesize both factors; they could then be isolated by transfer to an agar medium that did not contain either of the two amino acids. On such a medium the parent strains would not be able to proliferate, but the new cells would form visible colonies in a few hours. Accordingly we mixed cells of the two parent strains and spread them on the selective agar. Colonies of the new type of cells appeared; apparently we had succeeded in mating *Salmonella*.

To be sure that the new cells were the product of mating, we tried to mate strains that were distinguished from each other by more than one trait. The offspring of bacterial mating may combine genes from their two parents in any proportion. We could therefore expect to find a variety of new cells, some more like one parent and some more like the other. We were surprised to find, however, that the offspring of these matings all resembled one parent, except for the trait by which they were isolated and which was supplied by the other parent. The transfer of only one trait at a time was not consistent with the idea of a mating process. Furthermore, one of the two strains always acted as the donor, and the other as the recipient, of the genetic trait.

We considered first the simplest alternative: the change was merely a random mutation. But this possibility had to be rejected because the change appeared much more frequently in mixed cultures of the parent strains than in unmixed cultures. Nor could it be an increase in the mutation rate in the mixed cultures, for the new trait of the recipient strain was always related to traits of the donor strain. We concluded that we had stumbled upon an instance of bacterial "transformation." In this process the genetic traits of one strain of bacteria are transformed by contact with the genetic material of another strain. No contact between the cells is necessary. In familiar instances of transformation, in fact, the cells of the donor strain are dead, and their genetic material—the deoxyribonucleic acid, or DNA, contained in their chromosomes—is released into the culture medium. The recipient strain incorporates some of this free DNA into its chromosomes, and thus acquires traits of the dead strain [see "Transformed

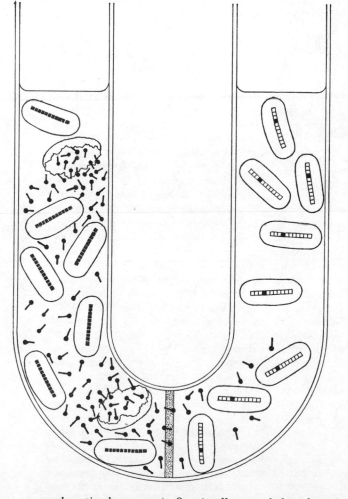

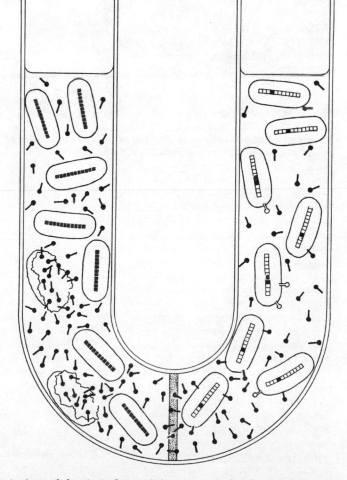

schematic chromosome). Occasionally one of these bacteria released live viruses (*tadpole-shaped objects*) which passed through filter (*first drawing*). In strain at left the viruses multiplied rapidly (*second drawing*). Some of the viruses (*colored*) carried bits of genetic material of the strain at left back through the filter to become part of genetic material of the strain at right (*third drawing*).

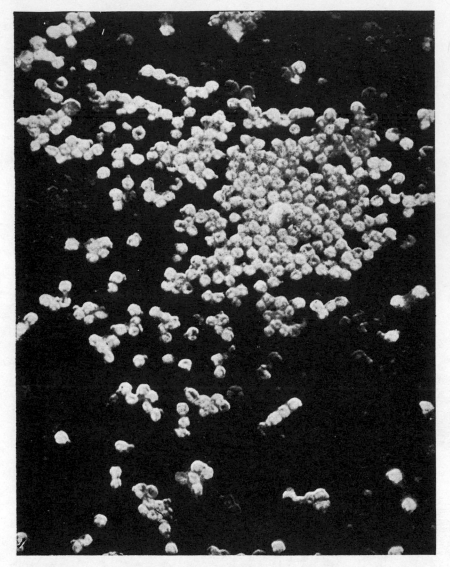

VIRUS PARTICLES used for transduction of *Salmonella* are enlarged some 60,000 diameters in this electron micrograph made by Keith R. Porter of the Rockefeller Institute.

Bacteria," by Rollin D. Hotchkiss and Esther A. Weiss, *beginning on page 42*]. Though well-established in certain species of bacteria, transformation had never been observed in *Salmonella.*

To test the transformation hypothesis, we now grew our two *Salmonella* strains in a specially constructed U-shaped tube. One strain was grown in one arm of the tube; the other strain, in the other arm. Between the two arms was a filter which prevented mating contact between the two strains [*see illustration on pages 78 and 79*]. To promote the exchange of substances secreted by the two strains, we gently flushed the nutrient broth from one side of the tube to the other. When the two strains were transferred to a selective medium, some offspring of the recipient strain showed a new trait picked up from the donor strain. Transformation seemed to be the answer. We then performed a more conventional transformation experiment, treating the recipient culture with pure DNA extracted from the donor strain. This, unexpectedly, had no effect at all. With our hypothesis now shaken, we went back to the U-tube. This time, with the idea of eliminating the possibility of transformation, we added an enzyme which destroys free DNA. In spite of the presence of the enzyme, genetic changes appeared just as before. We were obliged to discard transformation as well as mating, and to look for another explanation.

What could be happening in the U-tube? The donor bacterium passed its genetic material to the recipient in pieces small enough to pass through the filter, but the pieces were not damaged by the DNA-destroying enzyme. Both parents had to be present for the transfer to take place, but they did not have to be in direct contact. We wondered whether the donor perhaps gave off something other than DNA which could affect the heredity of the recipient. To test this idea, we added broth filtered from a culture of donor cells to a culture of recipient cells. No heritable changes resulted. However, when we grew the two strains together and filtered the broth, this fluid did produce changes in a fresh culture of recipient cells. At this point we realized there were two steps in the process: first the recipient had to produce something to stimulate the donor, and then the donor could send genetically active material back to the recipient.

Now we made a further discovery: once a culture of donor cells had been stimulated by exposure to fluids from the recipient strain, the fluid from this culture could stimulate other donor cells. The stimulating material was somehow reproduced in the donor bacteria. This reproductive capacity made us think of bacterial viruses. Viruses are small enough to pass with ease through the filter in the U-tube, and the protein coat of the virus protects its DNA from the enzyme which destroys free DNA.

The most familiar bacterial viruses are so destructive that an infected culture virtually disappears before your eyes. Obviously we would have noticed a virus of this type immediately. The "temperate" virus that causes a latent infection is harder to detect. Killing off only a small fraction of the cells, a latent infection may scarcely change the density of a culture. But occasionally one of the latent viruses regains its original form, multiplies in the cell and bursts forth to invade others. As a result a little free virus is always present in a culture of bacteria harboring a latent infection.

Sure enough, when we looked for viruses in cultures of donor *Salmonella* which had been treated with stimulating fluids, we found large numbers of virus particles. Further experiments indicated that virus activity and stimulating activity went hand in hand, and confirmed the fact that the virus was the stimulating agent.

It was but one step further to the theory that the viruses also acted as carriers of genetic material from donor to recipient bacteria. We hesitated to take this step, for the theory implied too much. It suggested, for example, that the

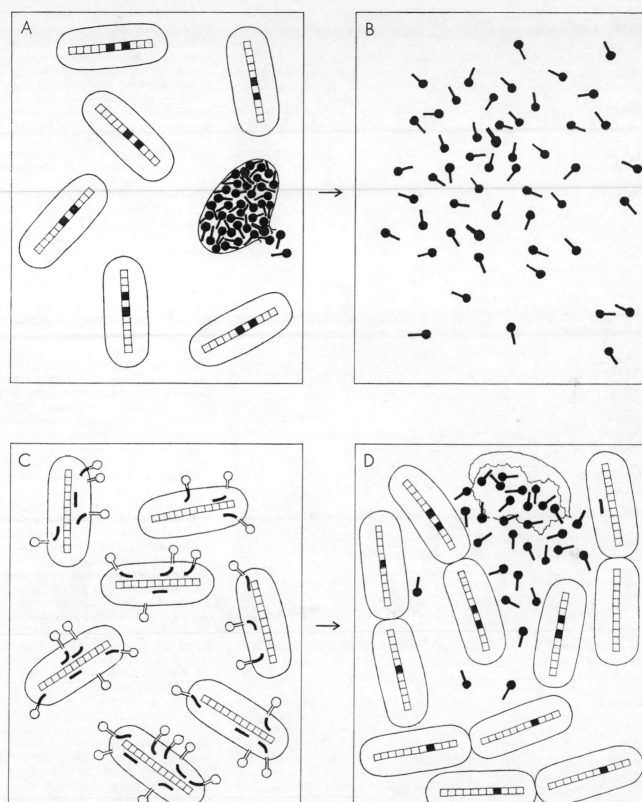

TRANSDUCTION requires first (A) a culture of donor cells which is infected with virus. Liquid filtered from this culture contains viruses (B), some of which carry genetic material from the bacteria (*colored*). Another strain of bacteria can be infected with this virus (C). Some of them (D) then produce more virus (*top*), some acquire genes of the donor bacteria (*colored square*), some acquire the genes of latent virus (*black square*), and some receive both. Occasionally a cell takes in a bacterial gene but does not incorporate it into its chromosome (*upper right*); gene is not duplicated when cell divides, so only one daughter cell receives it.

viruses of human diseases may carry genetic material from one host cell to another. But no other theory could explain the compelling evidence that the bacterial virus and the carrier of the genetic material from the donor to the recipient bacteria were identical in physical, chemical and biological properties.

The *Salmonella* mystery was now easily resolved. One of our two strains (the recipient) carried a temperate virus as a latent infection. The second strain (the donor) was especially susceptible to this virus. When the two were mixed, an occasional infective virus developed in the recipient strain and invaded a cell of the donor strain. When the offspring viruses erupted from the dying cell, most of them had genetic cores of virus DNA synthesized from the substance of the bacterial cell. But some of them incorporated particles of *Salmonella* DNA unchanged in their genetic cores. The

normal viruses went on to cause infections in other bacterial cells, active or latent depending upon the strain of the bacteria. Those that carried bacterial DNA and invaded the recipient strain of bacteria brought about entirely different consequences. Instead of killing their host, they simply disappeared. The DNA they brought with them took its place in the chromosomes of the recipient cell and modified the nutritional or other characteristics of the cell's offspring.

It was pure chance that one of the strains chosen for our studies contained a latent virus, and that another was susceptible to this virus. We had certainly not expected to encounter a new genetic mechanism, and, considering the many factors that had to be in harmony, the discovery was extremely fortuitous.

In some ways the transduction of genetic traits by a virus closely resembles latent infection by a virus. The difference between the two processes lies in the nature of the genetic material carried by the virus, whether it is DNA picked up from a bacterial chromosome or true virus DNA. The DNA in a virus of a given strain has the same composition as the DNA in other viruses of the same strain. In latent infection this DNA will always take the same station in the host cell's chromosome, and will induce the same change in the characteristics of the host cell. The traits associated with latent virus—a new synthetic capacity or a change in the cell wall—appear in every cell harboring the virus. If the bacterial cells lose the virus genetic material, as evidenced by the disappearance of free virus from the culture, they simultaneously lose the trait associated with the virus.

On the other hand, the bacterial DNA carried by the virus may vary in composition, and each kind of DNA may be capable of producing a different trait. The properties of this DNA do not depend on the virus at all, but only on the bacterium from which it came. In the virus's new host, the bacterial DNA takes a station in the chromosome corresponding to the position it had occupied in the chromosome of the previous host. Only a very few of the cells in a culture will gain traits by transduction, but once incorporated such traits remain even after the strain loses the virus DNA.

Actually the transduction of a new trait to a bacterium is not always accompanied by a latent infection. Conversely a virus can take the latent form in a bacterium without establishing a transduction. Whether one virus particle can produce both effects is uncertain; a given cell is usually invaded by more than one virus. It is quite possible that a virus must lose some of its own DNA in order to acquire bacterial DNA, and as a result it may no longer be able to produce an active infection.

The piece of DNA that is picked up by the virus must be very small indeed. It was thought for a time that this fragment might correspond to a single gene unit, since a virus particle appeared to transfer just one bacterial trait at a time. In the course of our work, however, we discovered several traits which the virus regularly transfers in pairs. The piece of DNA carried by a virus must therefore be large enough for at least two genes, the two presumably lying adjacent or closely linked in the chromosome.

Studies of the swimming ability of

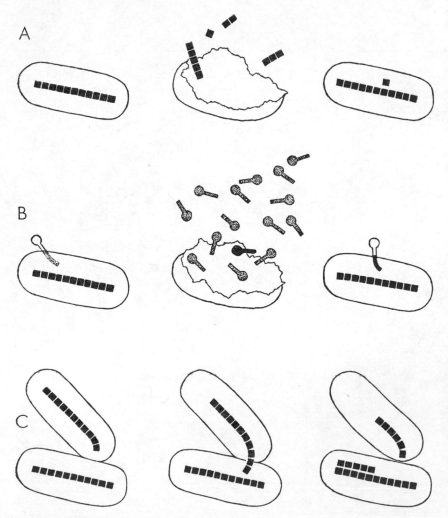

A

B

C

THREE METHODS of transferring genetic material from one bacterial cell to another are known. In "transformation" (A) the cells are disrupted to free the genetic material (*colored*), a small bit of which may then enter another cell. In transduction (B) viruses which can infect the cells carry genetic material from one to another. In mating (C) the material is passed by direct contact of cells, and a greater quantity can be transferred at one time.

Salmonella led us to the first pair of linked genes. Some *Salmonella* strains have whiplike tails (flagella) with which they can swim through liquids or semi-solid gelatins; others have no tails and cannot move. The tailless bacteria stay put and grow into colonies wherever they are placed, but the others swim off and spread in a cloudy swarm throughout the culture medium. The swimming strains are of different types, distinguished by the proteins of which their tails are made. We found that the ability to swim depends on two genes—one to determine whether or not the tail is made, and the other to determine the specific protein of which it is made. A tailless cell may already have a dormant gene for a kind of tail protein; if it acquires a gene for tail-making by transduction, it will make a tail of the protein type determined by its own gene. Almost as often, we found, the tailless cell will pick up a new gene for tail protein along with the tail-making gene; as a result it produces a tail of the donor's type instead of its own. The new genes introduced by transduction push out the corresponding genes in the bacterial chromosome and take their place.

The linkage between genetic traits revealed by transduction offers a clue to the linkage of the genes in the structure of the chromosomes. Transduction thus promises to be a useful tool in the important task of "mapping" chromosomes.

Bruce Stocker, a British bacteriologist, has demonstrated an abortive mode of transduction. The transducing DNA does not, in this case, replace a gene in the chromosome of the recipient cell, but by its very presence it induces a new trait. When the cell divides, however, the new gene is not duplicated, and only one of the daughter cells exhibits the trait. Stocker made this interesting discovery in an investigation of the swimming trait, when the recipient cells produced a trail of colonies rather than a spreading swarm. The colonies along the trail consisted entirely of tailless cells. But the trail had been laid by the swimming daughter cell which moved on to the next site after each division.

It seems possible to move any heritable trait from one cell to another by transduction. We have succeeded in transducing almost every trait we can reliably detect by experiment, including drug resistance, motility factors and antigenic factors. Transduction has been demonstrated in many kinds of *Salmonella*, in related species and in mating strains of *E. coli*. But ironically no one has succeeded in mating *Salmonella*.

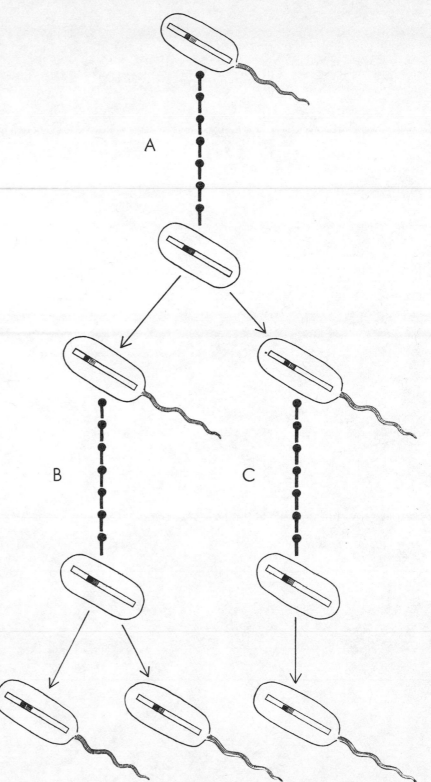

TRANSDUCTION TO MOTILITY revealed that two genes, presumably linked in the bacterial chromosome, can be carried together in a virus. Virus (A) from a culture of tailed cells (*top*) was used for transduction of untailed cells. Two types of motile cells resulted: some (*right center*) had received only the gene for tail-making and made tails of the protein type for which they already had a latent gene; an equal number (*left center*) had tails like the donor because they had also received a tail-protein gene. Transductions of untailed cells were repeated with virus from each of the new types of cells. One (B) gave rise to two types of tailed cells; the other (C) produced only one type of tailed cell.

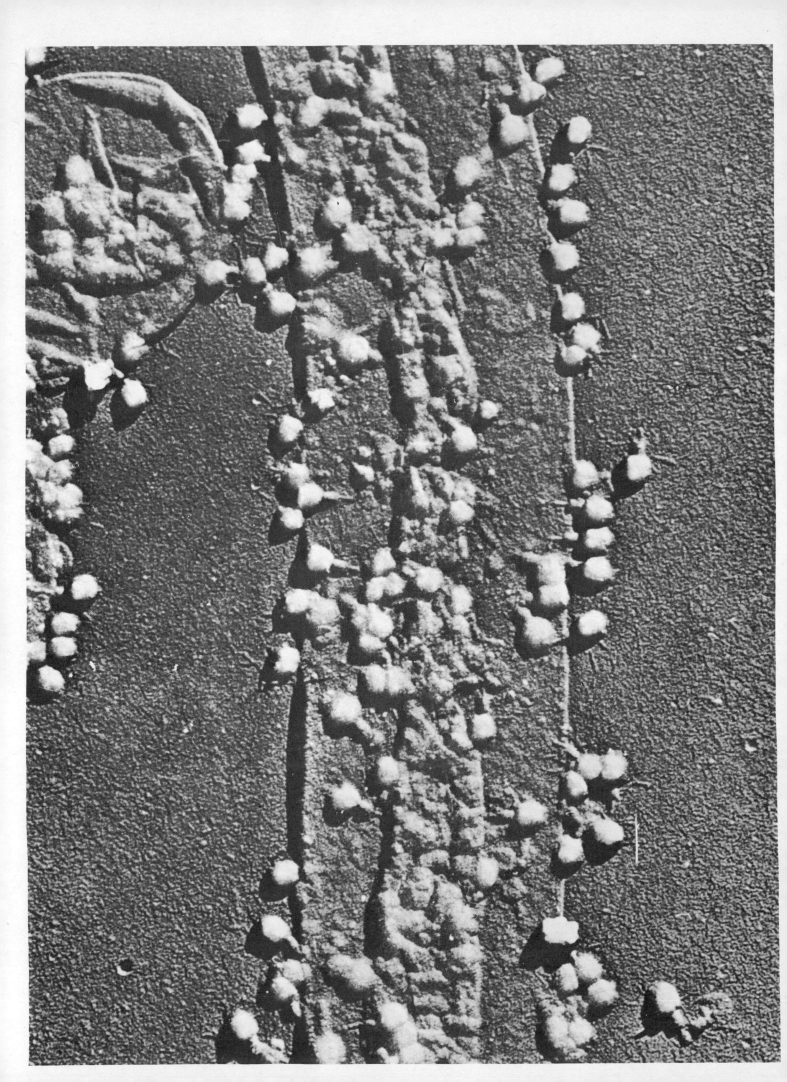

Viruses and Genes

by François Jacob and Elie L. Wollman
June 1961

*When a virus infects a bacterium, the genes of the
virus sometimes act as genes of their host. The
phenomenon has illuminated the mechanism of both
heredity and infection*

Almost everyone now accepts the unity of the inanimate physical world. Physicists do not hesitate to extrapolate laboratory results obtained with a small number of atoms to explain the source of the energy produced by stars. In the world of living things a comparable unity is more difficult to demonstrate; in fact, it is not altogether conceded by biologists. Nevertheless, most students of bacteria and viruses are inclined to believe that what is true for a simple bacillus is probably true for larger organisms, be they mice, men or elephants.

Accordingly we shall be concerned here with seeking lessons in the genetic behavior of the colon bacillus (*Escherichia coli*) and of the still simpler viruses that are able to infect the bacillus and destroy it. Viruses are the simplest things that exhibit the fundamental properties of living systems. They have the capacity to produce copies of themselves (although they require the help of a living cell) and they are able to undergo changes in their hereditary properties. Heredity and variation are the subject matter of genetics. Viruses, therefore, possess for biologists the elemental qualities that atoms possess for

SCORES OF VIRUSES of the strain designated T$_2$ are attached to the wall of a colon bacillus in this electron micrograph. The viruses are fastened to the bacterial wall by their tails, through which they inject their infectious genetic material. (Walls of the cell collapsed when the specimen was dried by freezing. "Shadowing" with uranium oxide makes objects stand out in relief.) The electron micrograph was made by Edouard Kellenberger of the University of Geneva. The magnification is 70,000 diameters.

physicists. When a virus penetrates a cell, it introduces into the cell a new genetic structure that interferes with the genetic information already contained within the cell. The study of viruses has thus become a branch of cellular genetics, a view that has upset many old notions, including the traditional distinction between heredity and infection.

For a long time geneticists have worked with such organisms as maize and the fruit fly *Drosophila*. They have learned how hereditary traits are transmitted from parents to progeny, they have discovered the role of the chromosomes as carriers of heredity and they have charted the results of mutations—the events that modify genes. Complex organisms, however, multiply too slowly and in insufficient numbers for the high-resolution analyses needed to clarify such problems as the chemical nature of genes and the processes by which a gene makes an exact copy of itself and influences cellular activity. These detailed problems are most readily studied in bacteria and in viruses. Within the space of a day or two the student of bacteria or bacterial viruses can grow and study more specimens than the fruit-fly geneticist could study in a lifetime. An operation as simple as the mixing of two bacterial cultures on a few agar plates can provide information on a billion or more genetic interactions in which genes recombine to form those of a new generation.

It is the events of recombination, together with mutations, that model and remodel the chromosomes, the structures that contain in some kind of code the entire pattern of every organism. In recent years geneticists and biologists have clarified the nature of the hereditary message and have gained some clues as to what the letters of the code

are. The primary, and perhaps the unique, bearers of genetic information in all forms of life appear to be molecules of nucleic acid. In living organisms, with the exception of some of the viruses, these long-chain molecules are composed of deoxyribonucleic acid (DNA). In all plant viruses and in some animal viruses the genetic substance is not DNA but its close chemical relative ribonucleic acid (RNA). DNA molecules are built up of hundreds of thousands or even millions of simple molecular subunits: the nucleotides of the four bases adenine, thymine, guanine and cytosine. These subunits, in an almost infinite variety of combinations, seem capable of encoding all the characteristics that all organisms transmit from one generation to the next. RNA molecules, which are somewhat shorter in length and not so well understood, act similarly for the viruses in which RNA is the genetic material.

Ultimately the role of the genes—the words of the hereditary message—is to specify the molecular organization of proteins. Proteins are long-chain molecules built up of hundreds of molecular subunits: the 20 amino acids. The sequence of nucleotides in the nucleic acid that contains the hereditary message is thought to determine the sequence of amino acids in the protein it manufactures. This process involves a "translation" from the nucleic-acid code into the protein code through a mechanism that is not yet understood.

The Bacterial Chromosome

Before considering viruses as cellular genetic elements, we shall summarize the present knowledge of the genetics of the bacterial cell. In bacteria the hereditary message appears to be written in a single linear structure, the bacterial chromosome. For the study of this chromosome an excellent tool was discovered in 1946 by Joshua Lederberg and Edward L. Tatum, who were then working at Yale University. They used the colon bacillus, which is able to synthesize all the building blocks required for the manufacture of its nucleic acids and proteins and therefore to grow on a minimal nutrient medium containing glucose and inorganic salts. Mutant strains, with defective or altered genes, can be produced that lack the ability to synthesize one or more of the building blocks and therefore cannot grow in the absence of the building block they cannot make. If, however, two different mutant strains are mixed, bacteria like the original strain reappear and are able to grow on a minimal medium.

Lederberg and Tatum were able to demonstrate that such bacteria are the result of genetic recombination occurring when a bacterium of one mutant strain conjugates with a bacterium of another mutant strain. Further work by Lederberg, and by William Hayes in London, has shown that the colon bacillus also has sex: some individuals act as males and transmit genetic material by direct contact to other individuals that act as recipients, or females. The difference between the two mating types may be ascribed to the fertility factor (or sex factor) *F*, present only in males. Curiously, females can easily be converted into males; during conjugation certain types of male, called F^+, transmit their sex factor to the females, which then become males.

The Chromosome "Essay"

Our own work at the Pasteur Institute in Paris has shed light on the different steps involved in bacterial conjugation and on the mechanism ensuring the transfer of the chromosome from certain strains of male, called *Hfr*, to females. When cultures of such males and of females are mixed, pairings take place between male and female cells through random collisions. A bridge forms between the two mating bacteria; one of the chromosomes of the male (bacteria have generally two to four identical chromosomes during growth) begins to migrate across the bridge and to enter the female. In the female, portions of the male chromosome have the ability to recombine with suitable portions of one of the female chromosomes. The chromosomes may be compared to written essays that differ only by a few letters, or a few words, corresponding to the

mutations. Portions of the two essays may become paired, word for word and letter for letter. Through the process known as genetic recombination, which is still very mysterious and challenging, fragments of the male chromosome, which can be anything from a word or a phrase up to several sentences, may be exactly substituted for the corresponding part of the female chromosome. This process gives rise to a complete new chromosome that contains a full bacterial essay in which some words from the male have replaced corresponding words from the female. The new chromosome is then replicated and transmitted to the daughter cell.

Perhaps the most remarkable feature of bacterial conjugation is the way in which the male chromosome migrates across the conjugation bridge. For a given type of male the migration always starts at the same end of the chromosome, which, if we represent the bacterial chromosome by the letters of the alphabet, we can call A. Then, with the chromosome proceeding at constant speed, it takes two hours before the other end, Z, has penetrated the female. After the mating has begun, conjugation can be interrupted at will by violently stirring the mating mixture for a minute or so in a blender. The mechanical agitation does not kill the cells but it disrupts the bridge and breaks the male chromosome during its migration. The fragment of the male chromosome that has entered the female before the interruption is still functional and has the ability to provide words or sentences for a chromosome [*see illustration on pages 88 and 89*]. If conjugation is mechanically interrupted at various intervals after the onset of mating, it is found that any gene carried by the male chromosome, from A to Z, enters the female at a precise time. We have therefore been able to draw two kinds of detailed chromosome map showing the location of genes. One map, the conventional kind, is based on the observed frequency of different sorts of genetic recombination; the second is a new kind of map reflecting the time at which any gene penetrates the female cell. The latter can be compared to a road map drawn by measuring the times at which a car proceeding at a constant speed passes through various cities.

Finally, the mode of the male chromosome's migration has provided a unique opportunity for correlating genetic measurements with chemical measurements of the chromosome. In collaboration with Clarence Fuerst, who is now working at the University of Toronto, we have grown male bacteria in a medium con-

taining the radioactive isotope phosphorus 32, which is incorporated into the DNA of the bacterial chromosome. The labeled bacteria are then frozen and kept in liquid nitrogen to allow some of the radioactive atoms to disintegrate. At various times samples are thawed and the labeled males are then mated with unlabeled females. The experiments show that the radioactive disintegrations sometimes break the chromosomes. If the break occurs between two markers, say E and F, the head part, *ABCDE*, is transferred to the female, but the tail part, *FGHIJKLMNOPQRSTUVWXYZ*, is not. Therefore the greater the number of phosphorus atoms between the A extremity of the chromosome and a given gene, the greater the chance that a break will prevent this gene from being transferred to the female. It is thus possible to draw a chromosomal map showing the location of the genes in terms of numbers of phosphorus atoms contained in the chromosome between the known genes. When we compare this map with those obtained by genetic analysis or by mechanical interruption, we find that for a given type of male all three maps are consistent.

In some types of male mutant the genetic characters have the same sequence along the chromosome but the character injected first differs from one mutant to another. The characters can also be injected either in the forward direction or in the backward direction, that is, from A to Z or from Z to A, with the alphabet capable of being broken at any point. These observations can be explained most simply by assuming that all the genetic "letters" of the colon bacillus are arranged linearly in a ring and that the ring can be opened at various points by mutation. It seems, furthermore, that the opening of the ring is a consequence of the attachment of the sex factor to the chromosome. The ring opens at precisely the point where the factor F, which is free to move, happens to affix itself. A cell with the F factor affixed to the chromosome is called an *Hfr* male, or "supermale," because it enhances the transmission of chromosomal markers. *Hfr* stands for "high frequency of recombination." When the chromosome is opened by the F factor, one of the free ends initiates the penetration of the chromosome into the female, carrying the sequence of characters after it. The other end carries the sex factor itself and is the last to enter the female. The sex factor has other remarkable properties and we shall bring it back into our story later.

The long-range objective of such stud-

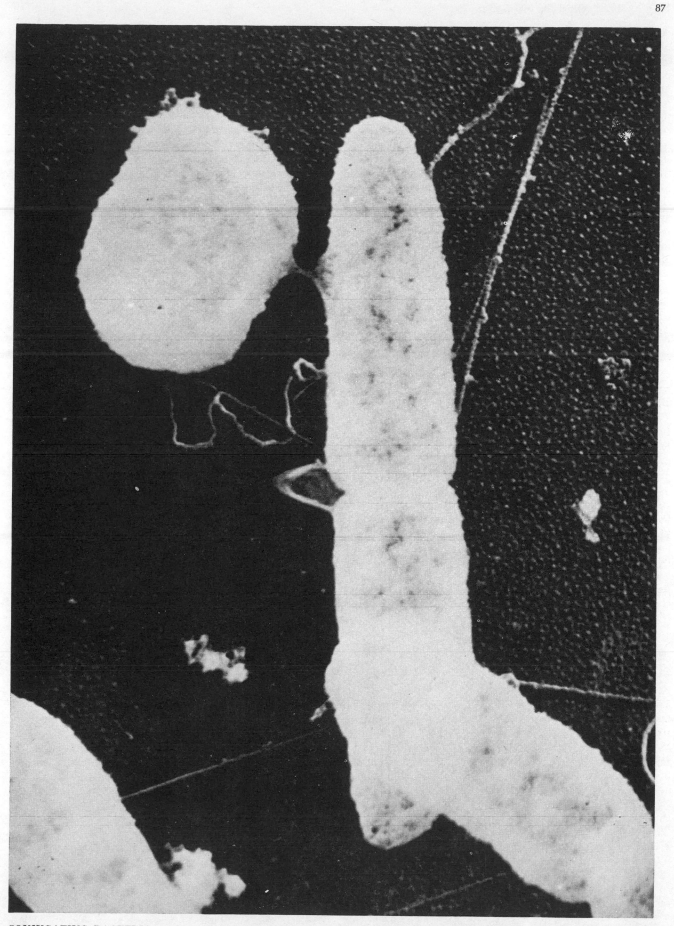

CONJUGATING BACTERIA conduct a transfer of genetic material. Long cell (*right*) is an *Hfr* "supermale" colon bacillus, which is attached by a short temporary bridge to a female colon bacillus (*see illustration on next two pages*). This electron micrograph, shown at a magnification of 100,000 diameters, was made by Thomas F. Anderson of the Institute for Cancer Research in Philadelphia.

ies is to learn how the thousands of genes strung along the chromosome control the molecular pattern of the bacterial cell: its metabolism, growth and division. These processes imply precise regulatory mechanisms that maintain a harmonious equilibrium between the cellular constituents. At any time the bacterial cell "knows" which components to make and how much of each is needed for it to grow in the most economical way. It is able to recognize which kind of food is available in a culture medium and to manufacture only those protein enzymes that are required to get energy and suitable building blocks from the available food.

At the Pasteur Institute, in collaboration with Jacques Monod, we have recently found new types of gene that determine specific systems of regulation. Mutants have been isolated that have become "unintelligent" in the sense that they cannot adjust their syntheses to their actual requirements. They make, for example, a certain protein in large amounts when they need only a little of it or even none at all. This waste of energy decreases the cells' growth rate. It seems that the production of a particular protein is controlled by two kinds of gene. One, which may be called the structural gene, contains the blueprint for determining the molecular organiza-

tion of the protein—its particular sequence of amino acid subunits. Other genes, which may be called control genes, determine the rate at which the information contained in the structural gene is decoded and translated into protein. This control is exercised by a signal embodied in a repressor molecule, probably a nucleic acid, that migrates from the chromosome to the cytoplasm of the cell. One of the control genes, called the regulator gene, manufactures the repressor molecule; thus it acts as a transmitter of signals. These are picked up by the operator gene, a specific receiver able to switch on or off the activity of the adjacent structural genes. Metabolic

PARENT CELLS   START OF TRANSFER   END OF TRANSFER

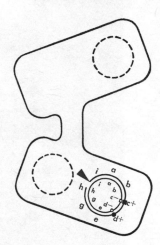

TRANSFER INTERRUPTED

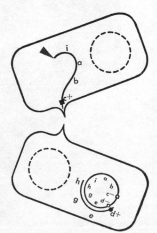

CHROMOSOMAL TRANSFER provides a primitive sexuality for colon bacillus. The bacterial chromosome, which appears to be ring-shaped, carries genetic markers (*designated by letters*), the presence or absence of which can be determined by studying cell's nutritional requirements. When the sex, or *F*, agent is attached to the chromosome, opening the ring, the cell is called an *Hfr* supermale. Two markers, labeled *c*+ and *d*+ when present and *c*− and *d*− when absent, can be traced from parents to daughter cells. When male and female cells conjugate, one of the male chromosomes (there are usually several, all identical) travels through the bridge.

products can interfere with the signals, either activating or inactivating the proper repressor molecules and thereby initiating or inhibiting the production of proteins.

Within the bacterial cell, then, there exists a complex system of transmitters and receivers of specific signals, by means of which the cell is kept informed of its metabolic requirements and enabled to regulate its syntheses. The bacterial chromosome contains not only a series of blueprints for the manufacture of individual molecular components but also a plan for the co-ordinated production of these components.

Let us now turn to the events that

DAUGHTER CELLS

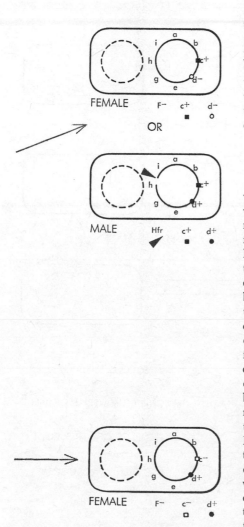

If transfer is complete, daughter cells may be male or female and carry any marker of the male. If transfer is interrupted, daughters are all female and can carry only those markers passed before bridge was broken.

take place when a bacterial virus of the strain designated T_2 infects the colon bacillus. A T_2 virus is a structure shaped like a tadpole; by weight it is about half protein and half DNA. The DNA is enclosed in the head, the outside of which is protein; the tail is also composed of protein. The roles of the DNA and the protein in the infective process were clarified in 1952 by the beautiful experiments of Alfred D. Hershey and Martha Chase of the Carnegie Institution of Washington's Department of Genetics in Cold Spring Harbor, N. Y. By labeling the DNA fraction of the virus with one radioactive isotope and the protein fraction with another, Hershey and Chase were able to follow the fate of the two fractions. They found that the DNA is injected into the bacterium, whereas the protein head and tail parts of the virus remain outside and play no further role. Electron micrographs reveal that the tail provides the method of attachment to the bacterium and that the DNA is injected through the tail. The Hershey-Chase experiment was a landmark in virology because it demonstrated that the nucleic acid carries into the cell all the information necessary for the production of complete virus particles.

How Viruses Destroy Bacteria

A bacterium that has been infected by virus DNA will break open, or lyse, within about 20 minutes and release a new crop of perhaps 100 particles of infectious virus, complete with protein head and tail parts. In this brief period the virus DNA subverts the cell's chemical facilities for its own purposes. It brings into the cell a plan for the synthesis of new molecular patterns and the cell faithfully carries it out. The infected cell creates new protein subunits needed for the virus head and tail, and filaments of nucleic acid identical to the DNA of the invading particle. These pools of building blocks pile up more or less at random, and in excess amounts, inside the cell. Then the long filaments of virus DNA suddenly condense and the protein subunits assemble around them, creating the complete virus particle. The whole process can be compared to the occupation of one country by another; the genetic material of the virus overthrows the lawful rule of the cell's own genetic material and establishes itself in power.

A virus can therefore be considered a genetic element enclosed in a protein coat. The protein coat protects the genetic material, gives it rigidity and stability

and ensures the specific attachment of the virus to the surface of the cell. As André Lwoff of the Pasteur Institute has pointed out, viruses can be uniquely defined as entities that reproduce from their own genetic material and that possess an apparatus specialized for the process of infection. The definition excludes both the cell and the specialized particles within the cell that serve its normal functions.

Another important criterion of viral growth is that of unrestricted synthesis. Infection with a virus is a sort of molecular cancer. The replication of the genetic material of the virus and the synthesis of the viral building blocks do not appear to be subject to any control system at all.

Lysogenic Bacteria

When a T_2 virus infects a bacterium, it forces the host to make copies of it and ultimately to destroy itself. Such a virus is said to be virulent, and when it is inside the cell, reproducing itself, it is said to be in the vegetative state.

There are, however, other bacterial viruses, called temperate viruses, which behave differently. After entering a cell the genetic material of a temperate virus can take two distinct paths, depending on the conditions of infection. It can enter the vegetative state, replicate itself and kill the host, just as a virulent virus does. Under other circumstances it does not replicate freely and does not kill the host. Instead it finds its way to the bacterial chromosome, anchors itself there and behaves like an integrated constituent of the host cell. Thereafter it will be transmitted for years to the progeny of the bacterium like a bacterial gene. We know that the bacterial host has not destroyed the invading particle, because from time to time one of the daughter cells in the infected line will break open and yield a crop of virus particles, as it would if it had been freshly attacked by a virulent virus. When the virus is in the subdued and integrated state, it is called a provirus. Bacteria carrying a provirus are called lysogenic, meaning that they carry a property that can lead to lysis and death.

Lysogeny was discovered in the early 1920's, soon after the discovery of the bacterial virus itself, and it remained a profound mystery for some 25 years. The mystery was explained by the fine detective work of Lwoff and his colleagues [see "The Life Cycle of a Virus," by André Lwoff; SCIENTIFIC AMERICAN, March, 1954]. Lwoff found that when he exposed certain types of lysogenic bac-

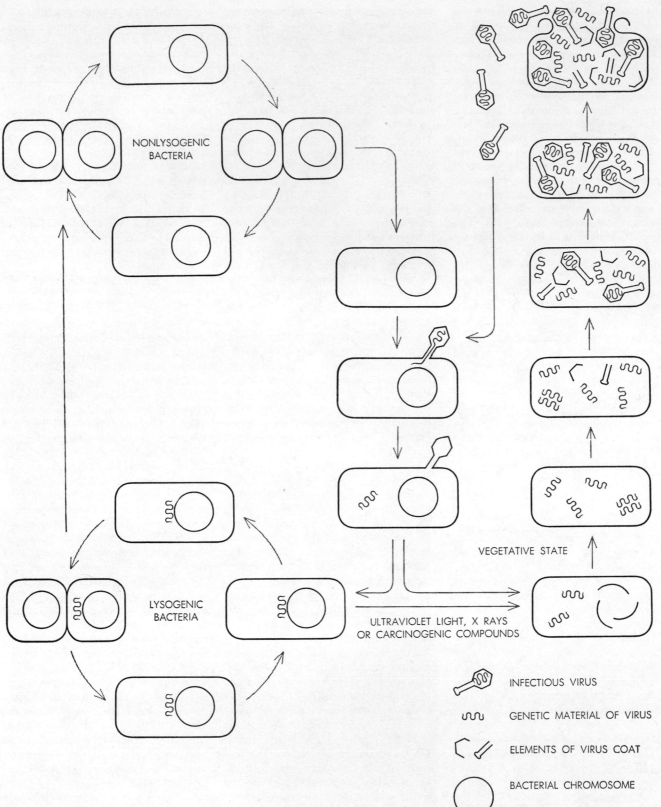

NONLYSOGENIC
BACTERIA

LYSOGENIC
BACTERIA

ULTRAVIOLET LIGHT, X RAYS
OR CARCINOGENIC COMPOUNDS

VEGETATIVE STATE

INFECTIOUS VIRUS

GENETIC MATERIAL OF VIRUS

ELEMENTS OF VIRUS COAT

BACTERIAL CHROMOSOME

LIFE CYCLE OF BACTERIAL VIRUS shows that, for the bacterium attacked, infection and death are not inevitable. After the genes of the virus (*color*) enter a cell descended from a completely healthy line (*top left*), the cell may take either of two paths. One (*far right*) leads to destruction as the virus enters the vegetative state, makes complete copies of its infective self and bursts open the cell, a process called lysis. The other path leads to the so-called lysogenic state, in which the viral genes attach themselves to the bacterial chromosome and become a provirus; the cell lives. Exposure to ultraviolet light, however, can dislodge the provirus and induce the vegetative state. The provirus is sometimes lost during cell division, returning the cell to the nonlysogenic state.

teria to ultraviolet light, X rays or active chemicals such as nitrogen mustard or organic peroxides, the whole bacterial population would lyse within an hour, releasing a multitude of infectious virus particles. When a provirus is thus activated, or "induced," it leaves the integrated state and enters the vegetative state, eventually destroying the cell [*see illustration on opposite page*].

To determine the position of the provirus inside the host cell, we can apply the method of interrupting the sexual conjugation of bacteria that carry a provirus and are therefore lysogenic. In this way we can correlate the location of the provirus with that of known characters on the bacterial chromosome. Each of 15 different types of provirus takes a particular position at a specific site on the bacterial chromosome. Only one is an exception; it seems free to take a position anywhere. In the proviral state the genetic material of the virus has not become an integral part of the bacterial chromosome; instead it appears to be added to the chromosome in an unknown but specific way. However it may be hooked on, the genetic material of the virus is replicated together with the genetic material of the host. It behaves like a gene, or rather as a group of genes, of the host.

Nonviral Effects of Provirus

The presence of this apparently innocuous genetic element, the provirus, can confer on the lysogenic bacteria that harbor it some new and striking properties. It is not at all obvious why some of these properties should be related to the presence of a provirus. As one example, diphtheria bacilli are able to produce diphtheria toxin only if the bacilli carry certain specific types of provirus. The disease diphtheria is caused solely by this toxin.

In other instances the presence of a provirus is responsible for a particular type of substance coating the surface of a bacterium. The substance can be identified by various immunological tests (typically by noting if a precipitate forms when a certain serum is added). The nonlysogenic strain, carrying no provirus, will bear a different substance. In such cases the genes of the virus are responsible for hereditary properties of the host. They can scarcely be distinguished from the genes of the bacterium.

The most striking property the provirus confers on its bacterial host is immunity from infection by external viruses of the same type as the provirus. When

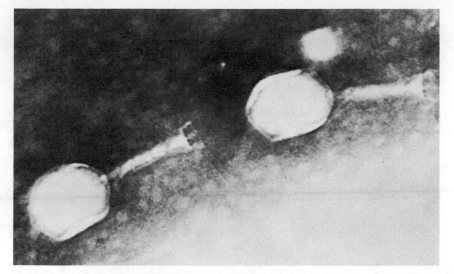

INTACT T$_2$ VIRUS has polyhedral head membrane and a curious pronged device at the end of its tail. The magnification is 200,000 diameters. This electron micrograph and the two below were made by S. Brenner and R. W. Horne at the University of Cambridge.

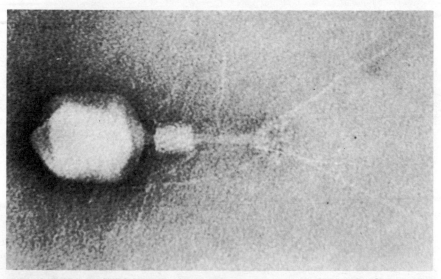

"TRIGGERED" T$_2$ VIRUS results from exposure to a specific bacterial substance that causes contraction of the tail sheath (*stubby cylinder*) and discharge of viral genes.

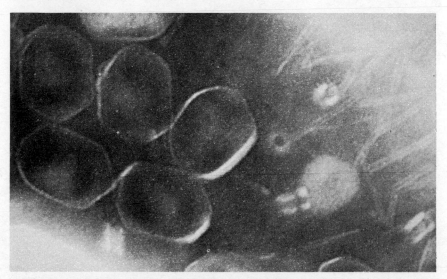

ISOLATED T$_2$ PARTS can be found still unassembled if host cell is forced to burst open before synthesis of virus particles is complete. Parts include head membranes and tails.

92

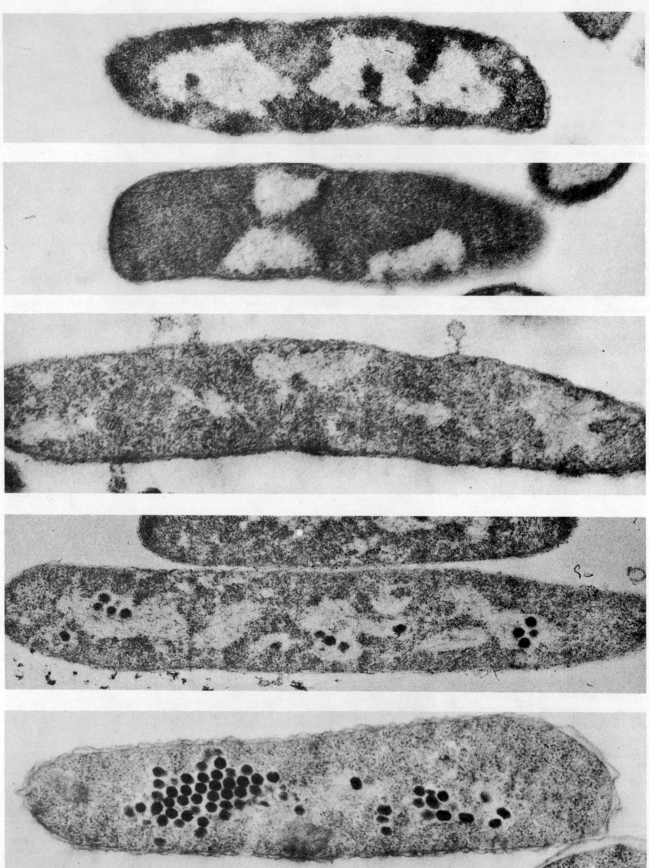

GROWTH OF T₂ VIRUS inside bacterial host is revealed in a striking series of electron micrographs by Kellenberger. Top picture shows the colon bacillus before infection. Four minutes after infection (*second from top*) characteristic vacuoles form along the cell wall. Ten minutes after infection (*third from top*) the virus has reorganized the entire cell interior and has created pools of new viral components. Twelve minutes after infection (*fourth from top*) new virus particles have started to condense. Thirty minutes after infection (*bottom*) more than 50 fully developed T₂ viruses have been produced and the cell is about ready to burst open.

lysogenic cells are mixed with such viruses, the virus particles adsorb on the cell and inject their genetic material into the cell, but the cell survives. The injected material is somehow prevented from multiplying vegetatively and is diluted out in the course of normal bacterial multiplication.

In the past two years we have attempted to learn more about the mechanism of this immunity. It seems clear that the mere attachment of the provirus to the host chromosome cannot account for the immunity of the host. The provirus must do something or produce something. We have evidence that the immunity is expressed by a substance or factor not tied to the chromosome. Remarkably enough, the system of immunity appears to be similar to the cellular systems already described that regulate the synthesis of protein in growing bacteria. It seems that the provirus produces a chemical repressor capable of inhibiting one or several reactions leading to the vegetative state. Thus immunity can be visualized as a specific system of regulation, involving the transmission of signals (repressors), which are received by an invading virus particle carrying the appropriate receptor.

Transduction

The close association that may take place between the genetic material of the virus and that of the host becomes even more striking in the phenomenon of transduction, discovered in 1952 by Norton D. Zinder and Lederberg at the University of Wisconsin [see "'Transduction' in Bacteria," by Norton D. Zinder; *beginning on page 78*]. They found that when certain proviruses turn into infective viruses, thereby killing their hosts, they may carry away with them pieces of genetic material from their dead hosts. When the viruses infect a host that is genetically different, the genes from the old host— the transduced genes—may be recombined with the genes of the new host. The transduction process seems able to move any sort of gene from one bacterial host to another.

Lysogeny and transduction therefore represent two complementary processes. In lysogeny the genes of the virus become an integral part of the genetic apparatus of the host and replicate at the pace of the host's chromosome. In transduction genes of the host become linked to the genes of the virus and can replicate at the unrestricted viral pace when the virus enters the vegetative state.

Viruses, like all other genetic ele-

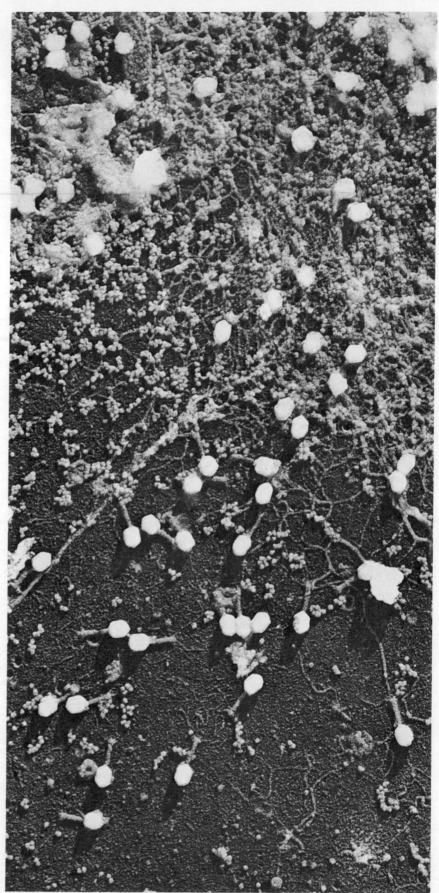

DEATH OF A BACTERIUM occurs when T_2 virus particles, having multiplied inside their host (*see sequence on opposite page*), dissolve the walls of the bacterial cell and spill out— a phenomenon called lysis. Viruses are the large white objects; the other matter is cellular debris. The electron micrograph (magnification: 50,000 diameters) is by Kellenberger.

ments, can undergo mutations, and these produce a variety of stable, heritable changes. The mutations of particular interest are those that prevent the formation of mature, infectious virus particles. Lysogenic bacteria in which such mutations have taken place are called defective lysogenic bacteria. These bacteria hereditarily perpetuate a mutated provirus, which is perfectly able to replicate together with the host's chromosome. If these cells are exposed to ultraviolet radiation, which activates the provirus, we observe that the defective lysogenic cells die without releasing any infectious viruses. Examination of such bacteria usually shows that virus subunits have started to appear inside the cell but have failed to reach maturity [*see illustrations on pages 96 and 97*]. Evidently some essential step in the formation of

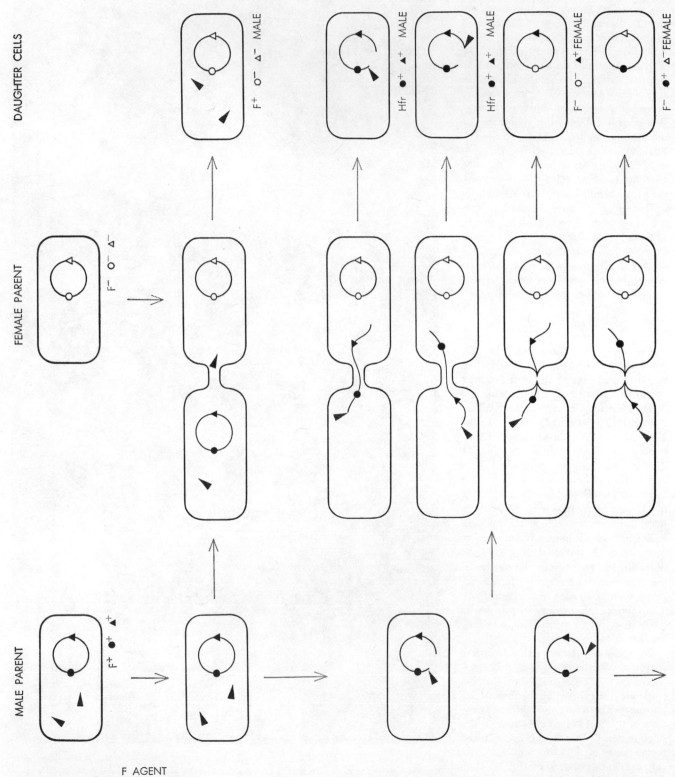

DAUGHTER CELLS

FEMALE PARENT

MALE PARENT

F AGENT
NONINTEGRATED

F AGENT INTEGRATED (Hfr); CHROMOSOMAL TRANSFER

an infectious virus has been blocked by the mutation.

Just as we can study how other kinds of mutation block biochemical pathways associated with cell nutrition, we can try to identify the biochemical blockages that keep the provirus from multi-plying normally. When a defective provirus turns to the vegetative state, some viral components begin to appear but the process halts. By using various biological tests, together with electron microscopy, we try to establish how far the process has gone. We have been able to identify two ways in which the process is halted and to relate the blockage to two main groups of viral genes.

One group of genes is concerned with the autonomous reproduction of the genetic material of the virus. The DNA of the provirus, which was able to repli-

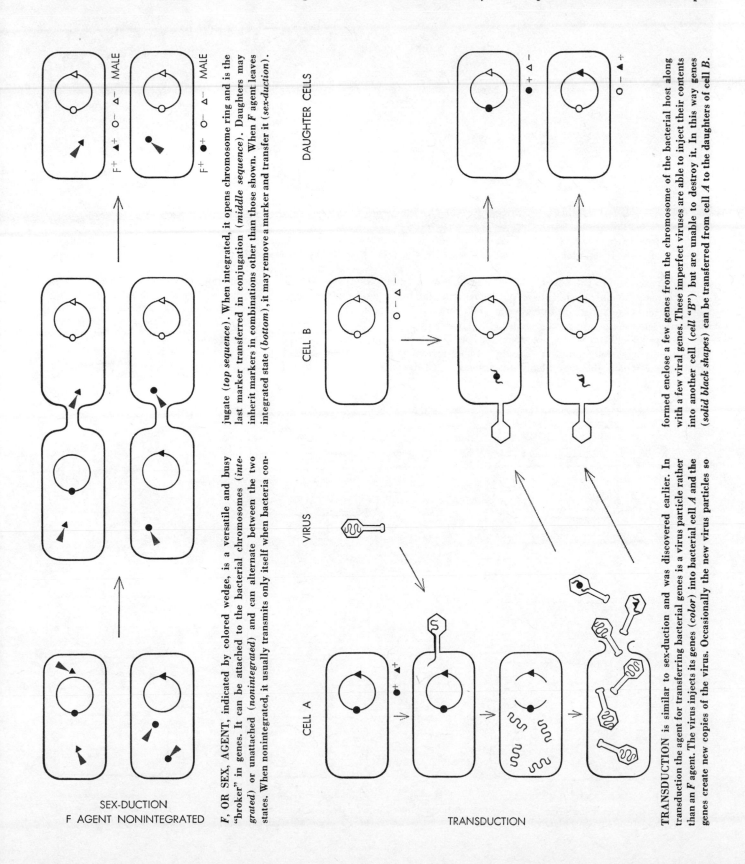

SEX-DUCTION
F AGENT NONINTEGRATED

F, OR SEX, AGENT, indicated by colored wedge, is a versatile and busy "broker" in genes. It can be attached to the bacterial chromosomes (integrated) or unattached (nonintegrated) and can alternate between the two states. When nonintegrated, it usually transmits only itself when bacteria conjugate (top sequence). When integrated, it opens chromosome ring and is the last marker transferred in conjugation (middle sequence). Daughters may inherit markers in combinations other than those shown. When F agent leaves integrated state (bottom), it may remove a marker and transfer it (sex-duction).

TRANSDUCTION

TRANSDUCTION is similar to sex-duction and was discovered earlier. In transduction the agent for transferring bacterial genes is a virus particle rather than an F agent. The virus injects its genes (color) into bacterial cell A and the genes create new copies of the virus. Occasionally the new virus particles so formed enclose a few genes from the chromosome of the bacterial host along with a few viral genes. These imperfect viruses are able to inject their contents into another cell (cell "B") but are unable to destroy it. In this way genes (solid black shapes) can be transferred from cell A to the daughters of cell B.

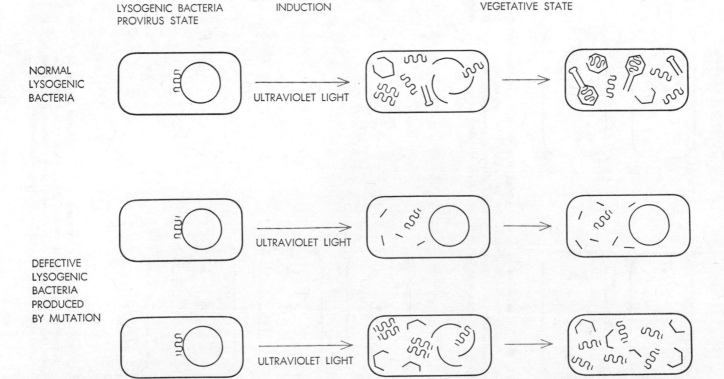

INCOMPLETE VIRUS PARTICLES are created by defective proviruses (*see illustration below*). The electron micrograph at left shows virus heads and tails that remain unassembled because of some defect. Occasionally (*right*) only heads can be found. Electron

cate when attached to the host chromosome, becomes unable to replicate on its own. A second group of genes is involved in the manufacture of the protein molecules that provide the coat and infectious apparatus of a normal virus. We have examples in which there is plenty of viral DNA, and many components of the coat material, but one or another essential protein is missing.

This study leads us to conclude that what distinguishes the genetic material of a virus from genetic elements of other types is that the virus carries two sets of information, one of which is necessary for the unrestricted multiplication of the viral genes and the other for the manufacture of an infectious envelope and traveling case.

The concept of a virus as it has emerged from the study of bacterial vi-

LYSOGENIC BACTERIA PROVIRUS STATE   INDUCTION   VEGETATIVE STATE

NORMAL LYSOGENIC BACTERIA

ULTRAVIOLET LIGHT

DEFECTIVE LYSOGENIC BACTERIA PRODUCED BY MUTATION

ULTRAVIOLET LIGHT

ULTRAVIOLET LIGHT

DEFECTIVE LYSOGENIC BACTERIA appear as mutations among normal lysogenic bacteria. Upon induction with ultraviolet light a normal provirus (*color, top left*) leaves the bacterial chromosome, replicates, produces infectious virus particles and kills its host. When defective proviruses are induced, the host cell may also be killed, but no infectious viruses appear at lysis. In

micrographs (magnification: 57,000) were
made by Kellenberger and W. Arber.

ruses is far more complex and more fascinating than the concept that prevailed
only a decade ago. As we have seen, a
virus may exist in three states; the only
thing common to the virus in the three
states is that it carries at all times much
the same genetic information encoded in

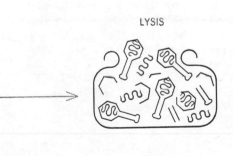

LYSIS

MAY OR MAY NOT LYSE

some cases (*middle*) the viral genes fail to
replicate. In others (*bottom*) they replicate
but the jacketing components are defective.

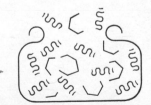

DNA. In the extracellular infectious state
the nucleic acid is enclosed in a protective, resistant shell. The virus then
remains inert like the spore of a bacterium, the seed of a plant or the pupa
of an insect. In the vegetative state of
autonomous replication the genetic material is free of its shell, overrides the
regulatory mechanism of the host and
imposes its own commands on the synthetic machinery of the cell. The viral
genes are fully active. Finally, in the
proviral state the genetic material of the
virus has become subject to the regulatory system of the host and replicates
as if it were part of the bacterial chromosome. A specific system of signals
prevents the genes of the virus from ex-
pressing themselves; complete virus particles are therefore not manufactured.

The Concept of the "Episome"

Less than a decade ago there was no
reason to doubt that virus genetics and
cell genetics were two different subjects
and could be kept cleanly apart. Now
we see that the distinction between viral
and nonviral genetics is extremely difficult to draw, to the point where even
the meaning of such a distinction may be
questionable.

As a matter of fact there appear to
be all kinds of intermediates between
the "normal" genetic structure of a bacterium and that of typical bacterial
viruses. Recent findings in our laboratory have shown that phenomena that
once seemed unrelated may share a deep
identity. We note, for example, that certain genetic elements of bacteria, which
we have no reason to class as viral, actually behave very much like the genetic
material of temperate viruses. One of
these is the fertility, or *F*, factor in colon
bacilli; in the so-called *Hfr* strains of
males the *F* agent is attached to one of
various possible sites on the host chromosome. In the males bearing the *F* agent
designated F$^+$ the agent is not fixed to
the chromosome and so it replicates as
an autonomous unit. It bears one other
striking resemblance to provirus. The
integrated state of the *F* factor excludes
the nonintegrated replicating state, just
as a provirus immunizes against the
vegetative replication of a like virus.

Another genetic agent resembling provirus is the factor that controls the production of colicines. These are extremely
potent protein substances that are released by some strains of colon bacillus;
the proteins are able to kill bacteria of
other strains of the same or related species. The colicinogenic factors also seem

to exist in two alternative states: integrated and nonintegrated. In the latter
state they seem able to replicate freely
and eventually at a faster rate than does
the bacterial chromosome. Bacteria that
lack these genetic elements—*F* agents
and colicinogenic factors—cannot, so far
as we know, gain them by mutation but
can only receive them (by sexual conjugation, for example) from an organism
that already possesses them. They may
replicate either along with the chromosome or autonomously. Such genetic elements, which may be present or absent,
integrated or autonomous, we have proposed to call "episomes," meaning "added bodies" [*see illustration on page 98*].

The concept of episomes brings together a variety of genetic elements that
differ in their origin and in their behavior. Some are viruses; others are not.
Some are harmful to the host cell;
others are not. The important lesson,
learned from the study of mutant temperate bacterial viruses, is that the tran-
sition from viral to nonviral, or from
pathogenic to nonpathogenic, can be
brought about by single mutations. We
also have impressive evidence that any
chromosomal gene of the host may be
incorporated in an episome through
some process of genetic recombination.
During the past year, in collaboration
with Edward A. Adelberg of the University of California, we have shown that the
sex factor, when integrated, is able to
pick up the adjacent genes of the bacterial chromosome. Then this new unit
formed by the sex factor and a few bacterial genes is able to return to the autonomous state and to be transmitted by
conjugation as a single unit. This process, in many respects similar to transduction, has been called sex-duction [*see
illustration at left on pages 94 and 95*].

Do episomes exist in organisms higher
than bacteria? We do not know; but if
we accept the basic unity of all cellular
biology, we should be confident that the
answer is yes and that mice, men and
elephants must harbor episomes. So far
the great precision and resolution that
can be achieved in the study of bacterial
viruses cannot be duplicated for more
complex organisms. There is, nevertheless, evidence for episome-like factors in
the fruit fly and in maize. There have
been reports of two viruses in the fruit
fly, transmitted through the egg to the
offspring, which may exist either as
nonintegrated or as integrated elements.
Although it does not seem that the virus
is actually located on the chromosome in
the latter state, the resemblance to provirus is striking. Barbara McClintock, of

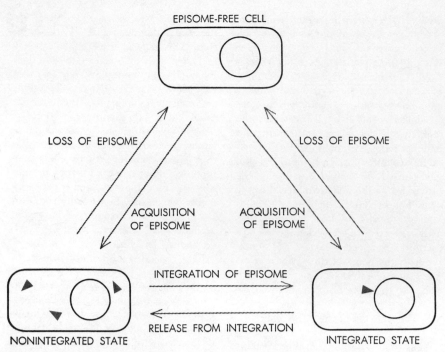

EPISOME-FREE CELL

LOSS OF EPISOME

LOSS OF EPISOME

ACQUISITION OF EPISOME

ACQUISITION OF EPISOME

INTEGRATION OF EPISOME

RELEASE FROM INTEGRATION

NONINTEGRATED STATE

INTEGRATED STATE

CONCEPT OF THE "EPISOME," as put forward by the authors, describes a genetic element, such as the *F* agent, that may be either attached to the chromosome or unattached. When integrated, it replicates at host's pace; nonintegrated, it replicates autonomously.

the Carnegie Institution of Washington's laboratory at Cold Spring Harbor, has discovered in maize "controlling elements" that are able to switch a gene off or on. (A gene responsible for a reddish color in corn may be switched on and off so fast that a single kernel may turn out speckled.) The controlling elements in maize are not always present, but when they are, they are added to specific chromosomal sites and can move from one site to another or even from one chromosome to another. These elements, therefore, act like episomes.

The discovery of proviruses and episomes has brought to light a phenomenon that biologists would scarcely have considered possible a few years ago: the addition to the cell's chromosome of pieces of genetic material arising outside the cell. The bacterial episomes provide new models to explain how two cells that otherwise possess an identical heredity can differ from each other. The episome brings into the cell a supplementary set of instructions governing additional biochemical reactions that can be superimposed on the basic metabolism of the cell.

The episome concept has implications for many problems in biology. For example, two main hypotheses have been advanced for the origin of cancer. One assumes that a mutation occurs in some cell of the body, enabling the cell to escape the normal growth-regulating mechanism of the organism. The other suggests that cancers are due to the presence in the environment of viruses that can invade healthy cells and make them malignant [see "The Polyoma Virus," by Sarah E. Stewart; SCIENTIFIC AMERICAN Offprint 77]. In the light of the episome concept the two hypotheses no longer appear mutually exclusive. We have seen that proviruses, living peacefully with their hosts, can be induced to turn to the vegetative, replicating state by radiation or by certain strong chemicals—the very agents that can be used to produce cancer experimentally in mice. If defective, the provirus will not even make viral particles. Malignant transformation involves a heritable change that allows a cell to escape the growth control of the organism of which it is a part. We can easily conceive that such a heritable change may result from a mutation of the cell, from an infection with some external virus or from the action of an episome, viral or not. Thus in the no man's land between heredity and infection, between physiology and pathology at the cellular level, episomes provide a new link and a new way of thinking about cellular genetics in bacteria and perhaps in mice, men and elephants.

The Genetics of a Bacterial Virus

by R. S. Edgar and R. H. Epstein
February 1965

*The T4 virus is a simple form of life with a precise
architecture dictated by genes in its DNA molecule.
By mapping the genes and learning their function one
learns how the virus is put together*

Viruses, the simplest living things known to man, have two fundamental attributes in common with higher forms of life: a definite architecture and the ability to replicate that architecture according to the genetic instructions encoded in molecules of nucleic acid. Yet in viruses life is trimmed to its bare essentials. A virus particle consists of one large molecule of nucleic acid wrapped in a protective coat of protein. The virus particle can do nothing for itself; it is able to reproduce only by parasitizing, or infecting, a living host cell that can supply the machinery and materials for translating the viral genetic message into the substance and structure of new virus particles. Since a virus is an isolated packet of genetic information unencumbered by the complex supporting systems characteristic of living cells, it is a peculiarly suitable subject for genetic investigation. One can study the molecular basis of life by identifying the individual genes in viral nucleic acid and learning what part each plays in the formation of virus progeny. That is what we have been doing for the past four years, working with the T4 bacteriophage, a virus that infects the colon bacterium *Escherichia coli*.

The T4 virus is one of the most complex viral structures. About .0002 millimeter long, the T4 particle consists of a head in the shape of a bipyramidal hexagonal prism and a tail assembly with several components. The head is a protein membrane stuffed with a long, tightly coiled molecule of deoxyribonucleic acid (DNA). The protein tail plays a role in attaching the virus to the host bacterial cell and injecting the viral DNA through the cell wall. Six tail fibers resembling tentacles bring the virus to the surface of the cell; a flat end plate fitted with prongs anchors the virus

there as the muscle-like sheath of the tail contracts to extrude the viral DNA through a hollow core into the cell.

Within a few minutes after the DNA enters the bacterium the metabolism of the infected bacterial cell undergoes a profound change. The cell's own DNA is degraded and its normal business —the synthesis of bacterial protein— ceases; synthetic activity has come under the control of the viral DNA, which takes over the synthesizing apparatus of the cell to direct the synthesis of new types of protein required for the production of new virus particles. The first proteins to appear include enzymes needed for the replication of the viral DNA, which has components not present in bacterial DNA and for the synthesis of which there are therefore no bacterial enzymes. Once these "early enzymes" are available the replication of viral DNA begins. Soon thereafter a new class of proteins appears in the cell: the proteins that will be required for the head membrane and tail parts.

About 15 minutes after the viral DNA was first injected new viral DNA begins to condense in the form of heads; protein components assemble around these condensates and soon whole virus particles are completed. For perhaps 10 minutes the synthesis and assembly of DNA and protein components continue and mature virus particles accumulate. The lysis, or dissolution, of the infected cell brings this process to an abrupt halt. Some 200 new virus particles are liberated to find new host cells to infect and so repeat the cycle of reproduction.

The remarkable sequence of synthesis, assembly and lysis is directed by the message borne by the genes of the viral DNA. Each gene is a segment of the DNA molecule, a twisted molecular ladder in which the rungs are pairs of nitrogenous bases: either adenine paired with thymine or guanine paired with cytosine. (In T4 DNA the cytosine is hydroxymethyl cytosine.) The sequence of base pairs in the DNA molecule, like

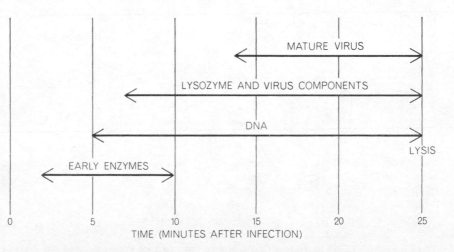

VIRUS INFECTION of a colon bacterium (at 37 degrees centigrade) proceeds on schedule, with the sequence of syntheses leading up to the lysis, or dissolution, of the host cell.

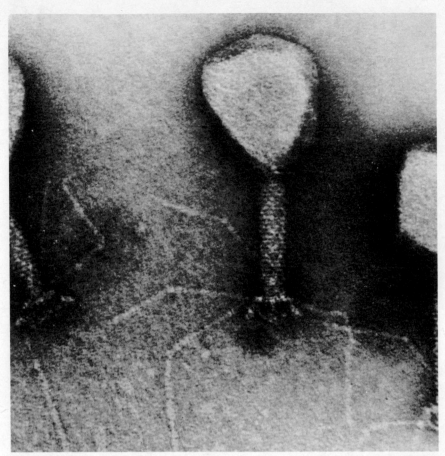

T4 BACTERIOPHAGE is enlarged about 300,000 diameters in an electron micrograph made by Michael Moody of the California Institute of Technology. The preparation was negatively stained with electron-dense uranyl acetate, which makes the background dark.

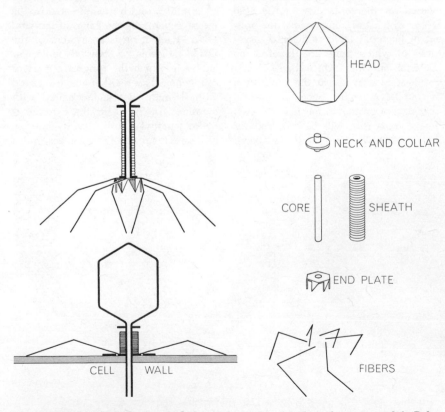

T4 COMPONENTS are diagrammed. A complete virus particle is shown at top left. Below it is a particle attached to a bacterial cell wall, with its sheath contracted and its hollow core penetrating the cell wall. The various components are shown separately at the right.

the sequence of letters in a word, spells out the information for the assembly of amino acids into protein molecules; a gene is defined as a segment of DNA sufficient to encode a single protein molecule. Since the average protein molecule consists of about 200 amino acid units and the code of DNA requires three base pairs per amino acid, the average length of a gene should be about 600 base pairs. Since there are about 200,000 base pairs in a molecule of T4 DNA, we began by assuming that the molecule contains several hundred genes and initiates the production of several hundred proteins in the host cell. Our task was first to map the location in the T4 DNA molecule of as many genes as possible and then to associate these genes with specific functions.

In order to identify a gene, map its location and learn its function one must find a gene that has undergone mutation: a molecular mistake that occurs like a typographical error in the sequence of base pairs and results either in genetic nonsense, meaning the inability to form protein, or in "missense," meaning the formation of faulty protein. Once a mutation occurs it is copied in successive replications of the DNA and reveals itself by its malfunction in protein synthesis. A mutation therefore serves as a marker for a gene. Moreover, by comparing the growth of a mutant strain of an organism with the growth of a "wild type," or normal, strain one can often infer the normal function of the gene under examination.

The trouble is that most mutations important enough to be recognized and studied are lethal; that is, they result in offspring that cannot survive, or at least cannot reproduce. How, then, can one study lethal mutations? In advanced plants and animals there are two copies of every gene, and it is possible to study "recessive" mutations that are lethal only when they happen to occur in both copies. Less advanced forms of life such as molds, bacteria and viruses, however, have only one copy of each gene, so some other method of studying lethal mutations must be found.

One such method was developed by George W. Beadle and Edward L. Tatum for the study of mutations in the genes of molds and bacteria. The genes that can be investigated by this method are those that direct the synthesis of enzymes required for the formation of nutrients, such as amino acids and vitamins, that are essential to the mold or bacterial cell. In these cases a mutation, although inherently lethal, will not prevent cell growth if the missing nutrient is supplied by the experimenter: it is a "conditional" lethal mutation. Such mutations are restricted to genes whose function can be supplanted by the experimenter. Our aim is to study mutations that affect the synthesis and assembly of virus components, and we had no way of supplying proteins or pieces of virus to infected cells. We needed other kinds of conditional lethal mutations.

One of us (Edgar), working at the California Institute of Technology, has dealt primarily with a class of mutations that are temperature-sensitive: they render the gene inactive at one temperature but not at temperatures a few degrees lower. An example of such a gene in a higher animal is the gene that controls the hair pigment in Siamese cats. The gene is inactive at body temperature, with the result that most of the cat's coat is white. On the cooler parts of the body—the paws, the tip of the tail, the nose and the ears—the gene becomes functional and the hair is pigmented. Of course, this defect is not lethal to the cat, but similar mutations in genes with functions essential to an organism are conditional lethal mutations if one can control the temperature. A strain of T4 bacteriophage with a temperature-sensitive lethal mutation, for example, grows perfectly well if it is incubated on bacteria at 25 degrees cen-

tigrade but not if it is incubated at 42 degrees. Temperature-sensitive mutations can occur in many different genes, since what they do is simply render a protein—regardless of its particular function—more readily inactivated by heat. They apparently do so by substituting one amino acid for another at some sensitive point in the structure of the protein molecule; in other words, they are "missense" mutations.

Epstein has worked with another class of conditional lethal mutations: the "amber" mutations, which he developed at Cal Tech and has studied primarily in the laboratory of Edouard Kellenberger at the University of Geneva. (We call them the amber mutations because they were discovered with the help of a graduate student named Bernstein, and *bernstein* is the German word for "amber"; it is often safer to give a new discovery a silly name than a speculatively descriptive one!) In these mutations the conditional property is not temperature-sensitivity but the ability of a virus to grow in certain host cells. Whereas the wild-type T4 virus grows equally well in colon bacteria of strains *B* and *CR*, amber mutants grow only in *CR*. Apparently only *CR* bacteria are able to translate the mutant message into protein properly; in strain *B* the mutant gene is translated into protein only up to the point of mutation and the resulting protein fragment is inactive. In other words, amber mutations are trans-

lated as "nonsense" in strain *B* but as "sense," or at worst as "missense," in strain *CR*. Again we could expect the amber mutations to occur in many different genes, since these mutations affect the overall translatability of any affected gene rather than the ability of specific genes to direct the synthesis of specific proteins.

Mutations arise at random in the normal course of virus infection and reproduction; we amplify the process by treating virus particles with one of a variety of chemical mutagens. We then plate the virus on cultures of colon bacteria. Any amber or temperature-sensitive mutant reveals itself by its failure to grow under "restrictive" conditions, that is, on strain *B* in the case of an amber mutation or at 42 degrees in the case of a temperature-sensitive mutation. In this manner we have isolated more than 1,000 amber and temperature-sensitive mutant strains. The mutations, however, occur at random at various sites in the many genes of the viral DNA. Since we are trying to identify genes, not merely mutations, we need to determine which mutant strains contain mutations affecting the same gene.

We do this by performing complementation tests [see illustration on next two pages]. The test consists in infecting bacteria simultaneously with two mutant viruses under restrictive conditions in which each mutant alone would be unable to grow in the bacterial cells.

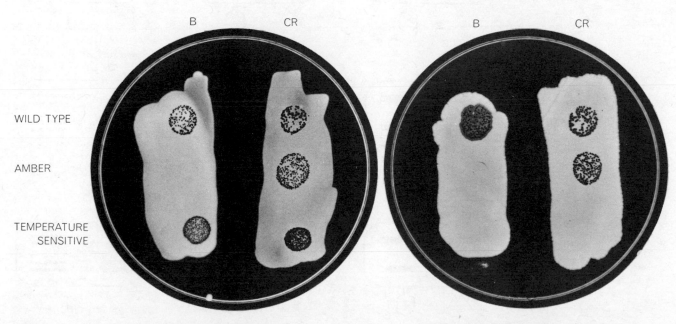

INCUBATED AT 25 DEGREES C.   INCUBATED AT 42 DEGREES C.

GROWTH CHARACTERISTICS of "wild type" virus and "amber" and temperature-sensitive mutants are compared. The photographic prints were made by exposing actual Petri dishes in an enlarger. On each dish bacterial strains *B* and *CR* had been streaked, with drops of virus suspensions placed on each streak, and the plates had been incubated at two temperatures, as shown. The amber mutants grew only on strain *CR*, the temperature-sensitive mutants grew only at 25 degrees C. and the wild-type virus grew under all conditions.

Infection of strain *B* bacteria at high temperature is restrictive for both amber and temperature-sensitive mutants. If, under these restrictive conditions, a yield of progeny virus is produced from cells infected by two mutants, the mutations must be complementary defects. Each mutant can perform the function the other mutant is unable to perform, and we can conclude that the two mutations are in different genes. If, on the other hand, the doubly infected bacteria produce no progeny virus, the two mutant strains must be unable to complement each other. Their mutations must affect a common function, and we conclude that they are in the same gene.

When complementation tests are applied to amber mutants, the results are clear-cut. These mutants, when tested against one another, fall into mutually exclusive classes: mutations in different genes result in full complementation no matter how they are paired, whereas mutations within the same gene fail to complement each other no matter how they are paired. In the case of temperature-sensitive mutants, however, the results are equivocal: some of the mutants display "intragenic" complementation and yield virus progeny even under restrictive conditions. Apparently two different "missense" mutations can give rise to "hybrid" proteins that, although

altered, are nevertheless complete and functional. The amber mutants, as we have mentioned, involve "nonsense" mutations and therefore would not be expected to show intragenic complementation. Since both amber and temperature-sensitive mutations occur in many genes, the ambers provide a check on the equivocal temperature-sensitive results.

By means of complementation tests we subdivided our many hundreds of amber and temperature-sensitive mutants into separate groups, each of which identifies one gene of the virus; our mutations turned out to be located in 56 different genes. The next step was

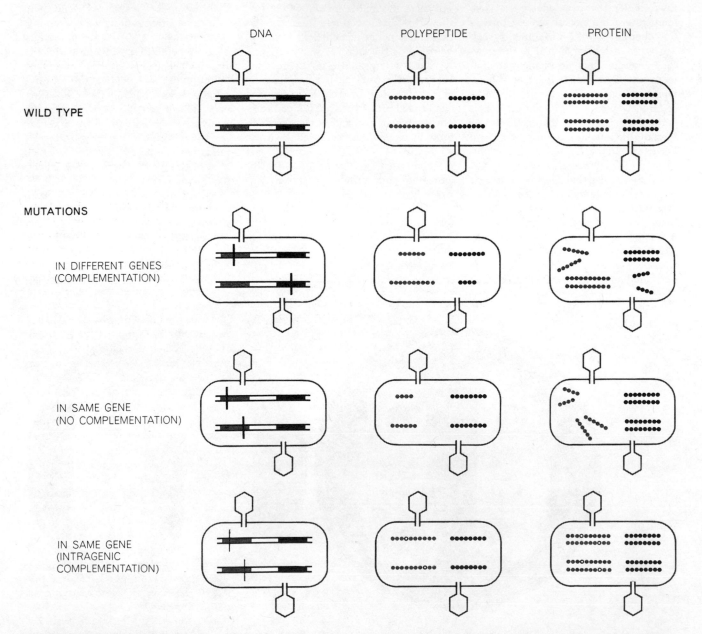

COMPLEMENTATION TEST identifies individual genes. The top row shows how, in wild-type virus, two genes of the deoxyribonucleic acid (DNA) molecule (*color and black*) might direct the synthesis of two polypeptide chains that form proteins and end up as virus components. An infection with wild-type virus results in a large number of plaques on a bacterial culture (*right*). If two mutations being tested occur in different genes, one gene makes the protein the other cannot make; they complement each other and virus particles are produced (*second row*). Two mutations in the same gene will ordinarily not complement each other, as

to locate those genes, and four that had been identified earlier by other investigators, on a genetic map—a representation of the position of the genes in relation to one another.

Such a map is constructed on the basis of recombination, the process by which the genetic material from two parents is mixed in the progeny. In viruses recombination can occur when viruses of two different strains infect the same cell. The mechanism of recombination is still poorly understood, but it probably involves the breakage of DNA molecules and the reassociation of pieces derived from both strains to form a new "hybrid" DNA molecule. Recom-

bination between two different mutants can result in some virus progeny that carry both mutations and in some wild-type viruses with no mutations. The wild-type recombinations can be recognized by their ability to multiply under restrictive conditions. The closer together two genes are on the DNA molecule, the less likely it is that breaks and reunions will occur between them, so the frequency of recombination is a measure of the distance between the two genes. We infect a bacterial culture with two strains that are mutant, say, in genes *a* and *x* respectively, and incubate it under "permissive" conditions in which both mutants can grow. Among millions of virus progeny of such a cross there will be some wild-type recombinants. By plating measured amounts of the progeny under permissive and under restrictive conditions we can determine what fraction of the progeny are wild-type. From this we calculate the frequency of recombination between genes *a* and *x* and thus the distance between them.

By plotting the results of hundreds of crosses we constructed a genetic map of the T4 DNA molecule [*see illustration on page 105*]. A remarkable feature of the map is that it has no "ends" and must be drawn as a linear array that closes on itself—a circle. This is rather surprising, since it has been established by electron microscopy and other means that the actual form of the T4 DNA molecule is that of a strand with two ends. (Just to confuse matters, some other viruses do have circular molecules!) Why the map should be circular is not yet known with certainty. It is probably because different viral DNA molecules have different sequences of genes, all of them circular permutations of the same basic sequence. In alphabetical terms, it is as if one DNA were *a*, *b*...*y*, *z* while another were *n*, *o*... *z*, *a*...*l*, *m*. In the second case *z* and *a* would be "closely linked" and would map close together.

Recombination occurs between mutation sites within genes as well as between genes, so we have been able to make a number of "intragenic" maps. These show that the genes are not uniform in size. Although most of them are quite small, each accounting for about half of 1 percent of the length of the map, gene No. 34 is about 20 times larger, and genes No. 35 and No. 43 are also outsized. Average gene size is therefore not a precise indicator of the number of genes in the virus. It looks as if the mutations discovered to date cover about half of the map, so we con-

clude that roughly half of the genes remain to be discovered. Unfortunately a kind of law of diminishing returns seems to be taking effect: for every 100 new mutants we isolate and test we are lucky to discover one new gene. Apparently amber and temperature-sensitive mutations are rare in the genes that are as yet undiscovered. We are devising new techniques with which to seek them out, but there will probably be a number of genes that are simply not susceptible to the conditional-lethal procedure. This could be because neither amber nor temperature-sensitive mutations occur in them or because, if they do occur, the loss of gene function is not lethal and the mutation therefore goes unnoticed.

While attempting to uncover the remaining genes, we have begun to determine the functions performed by the genes already identified. The mutants were originally detected because of their inability to produce progeny virus under restrictive conditions. In order to investigate the nature of the abortive infections more closely in an attempt to find out just what step in the growth cycle goes awry, we have employed a large number of mutants involving several different defects in each of the 60 genes. We chose just a few aspects of bacteriophage growth to examine, largely because they are easy to observe or measure and because they provide information on the major events of the cycle.

1. Can the infecting mutant virus accomplish the disruption of the bacterial DNA molecule? With the phase microscope one can observe whether or not the bacterial nucleoid, or DNA-containing body, disintegrates. So far every mutant we have tested has been able to disrupt the host DNA, so it is clear that in every case the infective process is at least initiated.

2. Does DNA synthesis occur in the infected cell? After the disruption of the bacterial nucleoid all host functions cease. Any new DNA that is revealed in chemical tests is viral DNA and an indication that the genes responsible for DNA synthesis are operative.

3. Do the infected cells lyse at the normal time? During the last half of the growth cycle an enzyme, lysozyme, is synthesized that is responsible for disrupting the cell wall. Normal lysis indicates that this enzyme is synthesized and does its work.

4. Are complete virus particles or components such as heads and tails produced in the infected cells? Electron

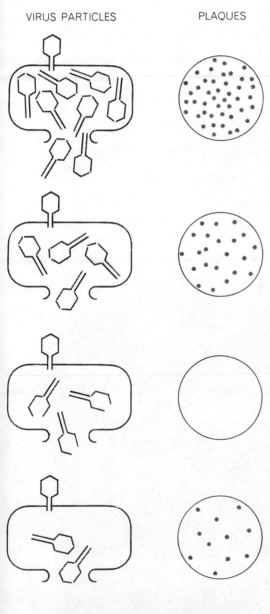

VIRUS PARTICLES   PLAQUES

seen in the third row. In some cases involving the temperature-sensitive mutants, however, "intragenic complementation" occurs: some virus is produced in spite of errors in polypeptide synthesis (*bottom row*).

ROLE OF INDIVIDUAL GENES of the T4 bacterial virus was investigated by the authors. This electron micrograph made by E. Boy de la Tour of the University of Geneva shows a complete virus particle with its hexagonal head and springlike tail assembly (*upper center*) and a number of "polyheads": cylindrical tubes of hexagonally arranged protein subunits that were not assembled into virus heads. The failure in assembly is due to a mutation in gene No. 20. The enlargement is about 270,000 diameters.

microscopy tells us the extent to which protein virus components have been synthesized and assembled in an infected cell.

Our data indicate that the various genes can be assigned to two groups. There are genes that appear to govern early steps in the infective process, as indicated by the fact that they affect DNA synthesis, and genes that appear to govern later steps, as indicated by their role in the maturation of new phage particles.

The major class of "early" genes includes those that are essential if any DNA synthesis is to occur. Mutations in these genes must cause the loss of some enzyme function necessary for DNA synthesis. Seven genes of this type have been identified, the precise function of one of which has been determined: John M. Buchanan and his co-workers at the Massachusetts Institute of Technology have found that gene No. 42 controls the synthesis of an enzyme necessary for the manufacture of hydroxymethyl cyto-

sine, one of the four bases in the T4 DNA molecule.

The "no DNA" mutants reveal an interesting regulatory feature of gene action. Not only is there no DNA synthesis in cells infected by these mutants, but also the cells do not lyse and no virus components are made. It appears that the decoding of the late-functioning genes depends somehow on the prior synthesis of viral DNA. Buchanan's group has found, moreover, that in these cells any of the early enzymes that are

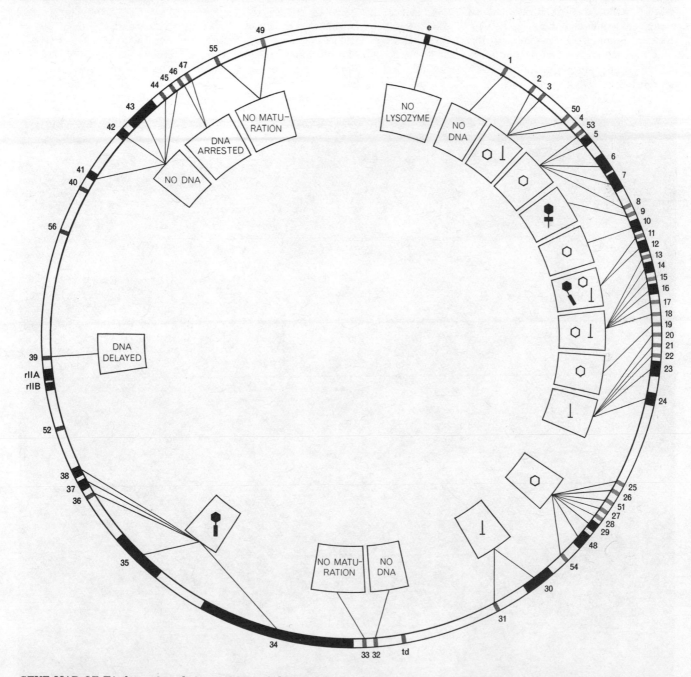

GENE MAP OF T4 shows the relative positions of the 60 genes identified to date and the major physiological properties of mutants defective in various genes. Minimum length is shown for some of the genes (*black segments*) but is not yet known for others (*gray*). The boxes indicate deficiencies in synthesis associated with mutations in some genes or, in the case of other genes, the components

that are present in defective lysates of corresponding mutants. There may be no DNA synthesis or it may be delayed or arrested. There may be no virus maturation at all. Synthesis and lysis may proceed normally but, as shown by the symbols, incomplete viruses may be produced, ranging from heads or tails only to complete particles lacking tail fibers (*genes No. 34 through No. 38*).

not eliminated by the particular mutation continue to be synthesized well beyond the normal shutoff time of 10 minutes. It appears, then, that in the absence of normal DNA synthesis some timing mechanism for switching early genes off and turning late ones on fails to function.

Among the early genes some others have been found that appear to delay or modify DNA synthesis or to block the activity of late genes without disturbing DNA synthesis, but the manner in which they function is still obscure.

Most of the genes—about 40 of those we have identified so far—clearly play roles in forming and assembling the virus components. Mutations in these morphogenetic genes seem not to affect the synthesis of DNA or the lysis of the

cell. What happens is that no infective progeny virus particles are produced, only bits and pieces of virus. For example, mutations in genes No. 20 through No. 24 result in the production of normal numbers of virus tails but no heads; mutations in the segment from gene No. 25 through No. 54 produce heads but no tails; mutations in genes No. 34 through No. 38 produce particles that are complete except for the tail fibers. Presumably the defective gene in each case is concerned with synthesis or assembly of the missing component.

A glance at the map [page 105] shows that the arrangement of the genes in the DNA molecule is far from random: genes with like functions tend to fall into clusters. Similar clusters of certain genes in bacteria are called "oper-

ons," and all the genes within an operon function together as a unit under the control of separate regulatory genes. There is no indication that the clusters in viral DNA act as operons; the available evidence suggests, indeed, that each gene acts independently. Still, it is difficult to believe the clustering does not reflect in some meaningful way a high degree of coordination in the activities of the genes.

The large number of genes associated with morphogenesis is of particular interest. What do all these genes do? There is evidence that only a few of them are concerned with the actual synthesis of protein components. For example, the head of the virus particle is made up of about 300 identical protein

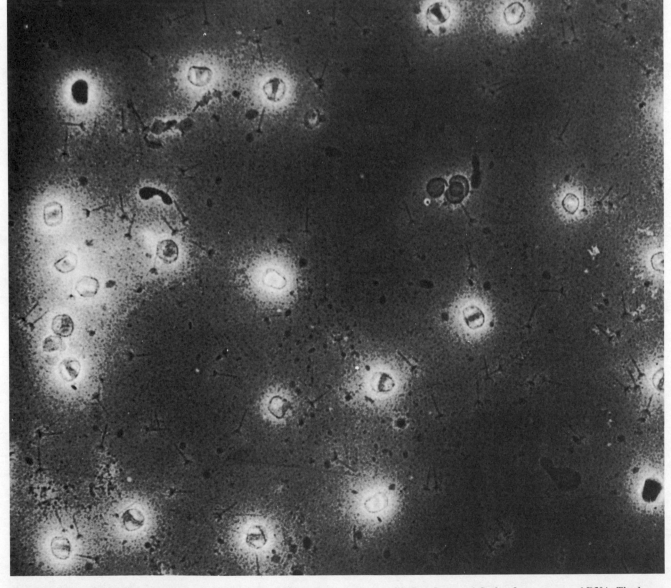

DEFECTIVE LYSATE of a temperature-sensitive strain mutant in gene No. 18 is enlarged about 60,000 diameters in this electron micrograph made by Edgar. Heads and tails have been formed but not assembled, and most of the heads are empty of DNA. The lysate was negatively stained with phosphotungstic acid, which filled the empty virus heads, and the exposed plate was printed as a negative.

subunits aggregated in a precise pattern; if there are any other protein molecules in the head membrane, they must be present in very small amounts. Yet at least seven genes and probably more are involved in the production of virus heads. Sydney Brenner and his associates at the University of Cambridge have found that just one of these genes, No. 23, is responsible for the actual synthesis of the protein subunits; cells infected with mutants defective in any other genes contain normal numbers of the subunits. The other genes must therefore be concerned with the assembly of the units rather than with their synthesis. When gene No. 20 is defective, for instance, the subunits assemble in the form of long cylindrical tubes instead of forming hexagonal heads [see illustration on page 104].

At this time we can only speculate as to the precise roles of the many morphogenetic genes. One possibility is that the proteins made by all of them are incorporated into the virus but in minor amounts that have escaped detection. Such minor components might be necessary to serve as the hinges, joints, nuts and bolts of the virus. Another possibility is that the proteins made by some of these late genes do not appear in the completed virus at all but instead play accessory roles in the assembly process —perhaps "gluing" subunits together in the specific configurations necessary for the proper construction of the virus. This notion of accessory morphogenetic genes is somewhat novel to many students of virus structure, who have generally believed that the assembly of viruses comes about through a spontaneous "crystallization" of subunits. In other words, it has been assumed that the form of a virus is inherent in its structural components. Although this may be true of viruses with simple spherical or cylindrical forms, it may not be true of viruses with more complex forms. The study of the effects of mutations on the assembly of viruses should serve as a powerful tool with which to explore this problem.

The relation between genes and form should be of general interest. Life is characterized by the complexity of its architecture. This complexity is manifested at all levels of organization, from molecules to the assemblages of specialized cells that make up higher animals and plants. The building blocks of all living things are, like virus particles, intricate molecular aggregations. Knowing how a bacteriophage such as T4 is put together may help us to understand the origins of form in all living systems.

11

The Fine Structure
of the Gene

by Seymour Benzer
January 1962

*The question "What is a gene?" has bothered
geneticists for fifty years. Recent work with a small
bacterial virus has shown how to split the gene and
make detailed maps of its internal structure*

Much of the work of science takes the form of map making. As the tools of exploration become sharper, they reveal finer and finer details of the region under observation. In the December, 1961 issue of *Scientific American* John C. Kendrew of the University of Cambridge described the mapping of the molecule of the protein myoglobin, revealing a fantastically detailed architecture. A living organism manufactures thousands of different proteins, each to precise specifications. The "blueprints" for all this detail are stored in coded form within the genes. In this article we shall see how it is possible to map the internal structure of a single gene, with the revelation of detail comparable to that in a protein.

It has been known since about 1913 that the individual active units of heredity—the genes—are strung together in one-dimensional array along the chromosomes, the threadlike bodies in the nucleus of the cell. By crossing such organisms as the fruit fly *Drosophila*, geneticists were able to draw maps showing the linear order of various genes that had been marked by the occurrence of mutations in the organism. Most geneticists regarded the gene as a more or less indivisible unit. There seemed to be no way to attack the questions "Exactly what is a gene? Does it have an internal structure?"

In recent years it has become apparent that the information-containing part of the chromosomal chain is in most cases a giant molecule of deoxyribonucleic acid, or DNA. (In some viruses the hereditary material is ribonucleic acid, or RNA.) Indeed, the threadlike molecule of DNA can be seen in the electron microscope [*see bottom illustration on opposite page*]. For obtaining information about the fine structure of DNA, however, modern methods of genetic analysis are a more powerful tool than even the electron microscope.

It is important to understand why this fine structure is not revealed by conventional genetic mapping, as is done with fruit flies. Genetic mapping is possible because the chromosomes sometimes undergo a recombination of parts called crossing over. By this process, for example, two mutations that are on different chromosomes in a parent will sometimes emerge on the same chromosome in the progeny. In other cases the progeny will inherit a "standard" chromosome lacking the mutations seen in the parent. It is as if two chromosomes lying side by side could break apart at any point and recombine to form two new chromosomes, each made up of parts derived from the original two. As a matter of chance two points far apart will recombine frequently; two points close together will recombine rarely. By carrying out many crosses in a large population of fruit flies one can measure the frequency—meaning the ease—with which different genes will recombine, and from this one can draw a map showing the parts in correct linear sequence. This technique has been used to map the chromosomes of many organisms. Why not, then, use the technique to map mutations inside the gene? The answer is that points within the same gene are so close together that the chance of detecting recombination between them would be exceedingly small.

In the study of genetics, however, everything hinges on the choice of a suitable organism. When one works with fruit flies, one deals with at most a few thousand individuals, and each generation takes roughly 20 days. If one works with a microorganism, such as a bacterium or, better still, a bacterial virus (bacteriophage), one can deal with billions of individuals, and a generation takes only minutes. One can therefore perform in a test tube in 20 minutes an experiment yielding a quantity of genetic data that would require, if humans were used, the entire population of the earth. Moreover, with microorganisms special tricks enable one to select just those individuals of interest from a population of a billion. By exploiting these advantages it becomes possible not only to split the gene but also to map it in the utmost detail, down to the molecular limits of its structure.

Replication of a Virus

An extremely useful organism for this fine-structure mapping is the T4 bacteriophage, which infects the colon bacillus. T4 is one of a family of viruses that has been most fruitfully exploited by an entire school of molecular biologists founded by Max Delbrück of the California Institute of Technology. The T4 virus and its relatives each consist of a head, which looks hexagonal in electron micrographs, and a complex tail by which the virus attaches itself to the bacillus wall [*see top illustration on opposite page*]. Crammed within the head of the virus is a single long-chain molecule of DNA having a weight about 100 million times that of the hydrogen atom. After a T4 virus has attached itself to a bacillus, the DNA molecule enters the cell and dictates a reorganization of the cell machinery to manufacture 100 or so copies of the complete virus. Each copy consists of the DNA and at least six distinct protein components. To make these components the invading DNA specifies the formation of a series of special enzymes, which themselves are proteins. The entire process is controlled by the battery of genes that constitutes the DNA molecule.

According to the model for DNA de-

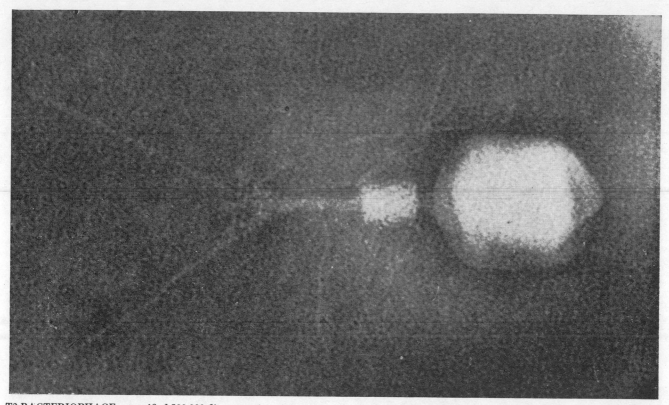

T2 BACTERIOPHAGE, magnified 500,000 diameters, is a virus that
contains in its head complete instructions for it own replication.
To replicate, however, it must find a cell of the colon bacillus
into which it can inject a giant molecule of deoxyribonucleic acid
(DNA). This molecule, comprising the genes of the phage, subverts the machinery of the cell to make about 100 copies of the
complete phage. The mutations that occasionally arise in the DNA
molecule during replication enable the geneticist to map the detailed structure of individual genes. The electron micrograph was
made by S. Brenner and R. W. Horne at the University of Cambridge.

MOLECULE OF DNA is the fundamental carrier of genetic information. This electron micrograph shows a short section of DNA
from calf thymus; its length is roughly that of the rII region in
the DNA of T4 phage studied by the author. The DNA molecule in
the phage would be about 30 feet long at this magnification of 150,-
000 diameters. The white sphere, a polystyrene "measuring stick," is
880 angstrom units in diameter. The electron micrograph was made
by Cecil E. Hall of the Massachusetts Institute of Technology.

vised by James D. Watson and F. H. C. Crick, the DNA molecule resembles a ladder that has been twisted into a helix. The sides of the ladder are formed by alternating units of deoxyribose sugar groups and phosphate groups. The rungs, which join two sugar units, are composed of pairs of nitrogenous bases: either adenine paired with thymine or guanine paired with cytosine. The particular sequence of bases provides the genetic code of the DNA in a given organism.

The DNA in the T4 virus contains some 200,000 base pairs, which, in amount of information, corresponds to much more than that contained in this article. Each base pair can be regarded as a letter in a word. One word (of the DNA code) may specify which of 20-odd amino acids is to be linked into a polypeptide chain. An entire paragraph might be needed to specify the sequence of amino acids for a polypeptide chain that has functional activity. Several polypeptide

units may be needed to form a complex protein.

One can imagine that "typographical" errors may occur when DNA molecules are being replicated. Letters, words or sentences may be transposed, deleted or even inverted. When this occurs in a daily newspaper, the result is often humorous. In the DNA of living organisms typographical errors are never funny and are often fatal. We shall see how these errors, or mutations, can be used to analyze a small portion of the genetic information carried by the T4 bacteriophage.

Genetic Mapping with Phage

Before examining the interior of a gene let us see how genetic experiments are performed with bacteriophage. One starts with a single phage particle. This provides an important advantage over higher organisms, where two different individuals are required and the male and female may differ in any number of respects besides their sex. Another simplification is that phage is haploid, meaning that it contains only a single copy of its hereditary information, so that none of its genes are hidden by dominance effects. When a population is grown from a single phage particle, using a culture of sensitive bacteria as fodder, almost all the descendants are identical, but an occasional mutant form arises through some error in copying the genetic information. It is precisely these errors in reproduction that provide the key to the genetic analysis of the structure [*see upper illustration on pages 112 and 113*].

Suppose that two recognizably different kinds of mutant have been picked up; the next step is to grow a large population of each. This can be done in two test tubes in a couple of hours. It is now easy to perform a recombination experiment. A liquid sample of each phage population is added to a culture of bacterial cells in a test tube. It is arranged that the phage particles outnumber the bacterial cells at least three to one, so that each cell stands a good chance of being infected by both mutant forms of phage DNA. Within 20 minutes about 100 new phage particles are formed within each cell and are released when the cell bursts. Most of the progeny will resemble one or the other parent. In a few of them, however, the genetic information from the two parents may have been recombined to form a DNA molecule that is not an exact copy of the molecule possessed by either parent but a combination of the two. This new recombinant phage particle can carry

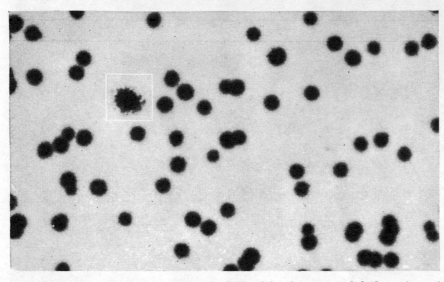

SPONTANEOUS MUTATIONAL EVENT is disclosed by the one mottled plaque (*square*) among dozens of normal plaques produced when standard T4 phage is "plated" on a layer of colon bacilli of strain B. Each plaque contains some 10 million progrny descended from a single phage particle. The plaque itself represents a region in which cells have been destroyed. Mutants found in abnormal plaques provide the raw material for genetic mapping.

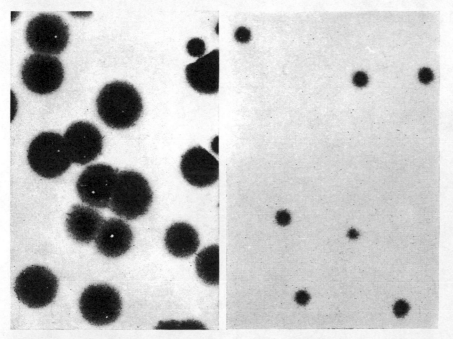

DUPLICATE REPLATINGS of mixed phage population obtained from a mottled plaque, like that shown at top of page, give contrasting results, depending on the host. Replated on colon bacilli of strain B (*left*), rII mutants produce large plaques. If the same mixed population is plated on strain K (*right*), only standard type of phage produce plaques.

both mutations or neither of them [*see lower illustration on next two pages*].

When this experiment is done with various kinds of mutant, some of the mutant genes tend to recombine almost independently, whereas others tend to be tightly linked to each other. From such experiments Alfred D. Hershey and Raquel Rotman, working at Washington University in St. Louis, were able to construct a genetic map for phage showing an ordered relationship among the various kinds of mutation, as had been done earlier with the fruit fly *Drosophila* and other higher organisms. It thus appears that the phage has a kind of chromosome —a string of genes that controls its hereditary characteristics.

One would like to do more, however, than just "drosophilize" phage. One would like to study the internal structure of a single gene in the phage chromosome. This too can be done by recombination experiments, but instead of choosing mutants of different kinds one chooses mutants that look alike (that is, have modifications of what is apparently the same characteristic), so that they are likely to contain errors in one or another part of the same gene.

Again the problem is to find an experimental method. When looking for mutations in fruit flies, say a white eye or a bent wing, one has to examine visually every fruit fly produced in the experiment. When working with phage, which reproduce by the billions and are invisible except by electron microscopy, the trick is to find a macroscopic method for identifying just those individuals in which recombination has occurred.

Fortunately in the T4 phage there is a class of mutants called *r*II mutants that can be identified rather easily by the appearance of the plaques they form on a given bacterial culture. A plaque is a clear region produced on the surface of a culture in a glass dish where phage particles have multiplied and destroyed the bacterial cells. This makes it possible to count individual phage particles without ever seeing them. Moreover, the shape and size of the plaques are hereditary characteristics of the phage that can be easily scored. A plaque produced in several hours will contain about 10 million phage particles representing the progeny of a single particle. T4 phage of the standard type can produce plaques on either of two bacterial host strains, B or K. The standard form of T4 occasionally gives rise to *r*II mutants that are easily noticed because they produce a distinctive plaque on B cultures. The key to the whole mapping technique is that

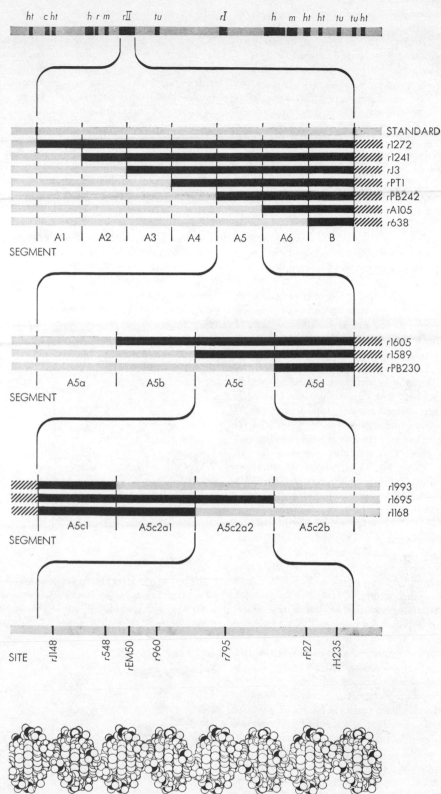

MAPPING TECHNIQUE localizes the position of a given mutation in progressively smaller segments of the DNA molecule contained in the T4 phage. The *r*II region represents to start with only a few percent of the entire molecule. The mapping is done by crossing an unknown mutant with reference mutants having deletions (*dark gray tone*) of known extent in the *r*II region (*see illustration of method on page 114*). The order and spacing of the seven mutational sites in the bottom row are still tentative. Each site probably represents the smallest mutable unit in the DNA molecule, a single base pair. The molecular segment (*extreme bottom*), estimated to be roughly in proper scale, contains a total of about 40 base pairs.

these mutants do not produce plaques on K cultures.

Nevertheless, an *r*II mutant can grow normally on bacterial strain K if the cell is simultaneously infected with a particle of standard type. Evidently the standard DNA molecule can perform some function required in K that the mutants cannot. This functional structure has been traced to a small portion of the DNA molecule, which in genetic maps of the T4 phage is designated the *r*II region.

To map this region one isolates a number of independently arising *r*II mutants (by removing them from mutant plaques visible on B) and crosses them against one another. To perform a cross, the two mutants are added to a liquid culture of B cells, thereby providing an opportunity for the progeny to recombine portions of genetic information from either parent. If the two mutant versions are due to typographical errors in different parts of the DNA molecule, some individuals of standard type may be regenerated. The standards will produce plaques on the K culture, whereas the mutants cannot. In this way one can easily detect a single recombinant among a billion progeny. As a consequence one can "resolve" two *r*II mutations that are extremely close together. This resolving power is enough to distinguish two mutations that are only one base pair apart in the DNA molecular chain.

What actually happens in the recombination of phage DNA is still a matter of conjecture. Two defective DNA molecules may actually break apart and rejoin to form one nondefective molecule, which is then replicated. Some recent evidence strongly favors this hypothesis. Another possibility is that in the course of replication a new DNA molecule arises from a process that happens to copy only the good portions of the two mutant molecules. The second process is called copy choice. An analogy for the two different processes can be found in the methods available for making a good tape recording of a musical performance from two tapes having defects in different places. One method is to cut the defects out of the two tapes and splice the good sections together. The second method (copy choice) is to play the two tapes and record the good sections on a third tape.

Mapping the *r*II Mutants

A further analogy with tape recording will help to explain how it has been established that the *r*II region is a simple linear structure. Given three tapes, each with a blemish or deletion in a different place, labeled *A*, *B* and *C*, one can imagine the deletions so located that deletion *B* overlaps deletion *A* and deletion *C*, but that *A* and *C* do not overlap each other. In such a case a good performance can be re-created only by recombining *A* and *C*. In mutant forms of phage DNA containing comparable deletions the existence of overlapping can be established by recombination experiments of just the same sort.

To obtain such deletions in phage one looks for mutants that show no tendency to revert to the standard type when they reproduce. The class of nonreverting mutants automatically includes those in which large alterations or deletions have occurred. (By contrast, *r*II mutants that revert spontaneously behave as if their alterations were localized at single points). The result of an exhaustive study covering hundreds of nonreverting *r*II mutants shows that all can be represented as containing deletions of one size or another in a single linear structure. If the structure were more complex, containing, for example, loops or branches, some mutations would have been expected to overlap in such a way as to make it impossible to represent them in a linear map. Although greater complexity cannot be absolutely excluded, all observations to date are satisfied by the postulate of simple linearity.

Now let us consider the *r*II mutants that do, on occasion, revert spontaneously when they reproduce. Conceivably they arise when the DNA molecule of the phage undergoes an alteration of a single base pair. Such "point" mutants are those that must be mapped if one is to probe the fine details of genetic structure. However, to test thousands of point mutants against one another for recombination in all possible pairs would

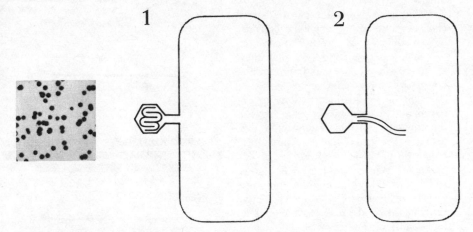

REPLICATION AND MUTATION occur when a phage particle infects a bacillus cell. The experiment begins by isolating a few standard particles from a normal plaque (*photograph at far left*) and growing billions of progeny in a broth culture of strain B colon bacilli. A sample of the broth is then spread on a Petri dish containing the same strain, on which the

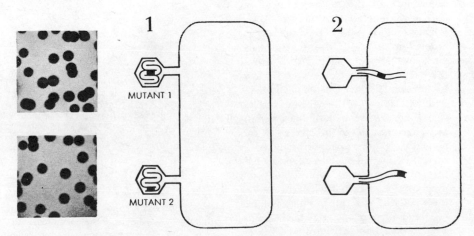

MUTANT 1

MUTANT 2

PROCESS OF RECOMBINATION permits parts of the DNA of two different phage mutants to be reassembled in a new DNA molecule that may contain both mutations or neither of them. Mutants obtained from two different cultures (*photographs at far left*) are introduced into a broth of strain B colon bacilli. Crossing occurs (*1*) when DNA from each mutant type

require millions of crosses. Mapping of point mutations by such a procedure would be totally impracticable.

The way out of this difficulty is to make use of mutants of the nonreverting type, whose deletions divide up the *r*II region into segments. Each point mutant is tested against these reference deletions. The recombination test gives a negative result if the deletion overlaps the point mutation and a positive result (over and above the "noise" level due to spontaneous reversion of the point mutant) if it does not overlap. In this way a mutation is quickly located within a particular segment of the map. The point mutation is then tested against a second group of reference mutants that divide this segment into smaller segments, and so on [*see illustration on pages 116 and 117*]. A point mutation can be assigned by this method to any of 80-odd ordered segments.

The final step in mapping is to test against one another only the group of mutants having mutations within each segment. Those that show recombination are concluded to be at different sites, and each site is then named after the mutant indicating it. (The mutants themselves have been assigned numbers according to their origin and order of discovery.) Finally, the order of the sites within a segment can be established by making quantitative measurements of recombination frequencies with respect to one another and neighbors outside the segment.

The Functional Unit

Thus we have found that the hereditary structure needed by the phage to multiply in colon bacilli of strain K consists of many parts distinguishable by mutation and recombination. Is this region to be thought of as one gene (because it controls one characteristic) or as hundreds of genes? Although mutation at any one of the sites leads to the same observed physiological defect, it does not necessarily follow that the entire structure is a single functional unit. For instance, growth in strain K could require a series of biochemical reactions, each controlled by a different portion of the region, and the absence of any one of the steps would suffice to block the final result. It is therefore of interest to see whether or not the *r*II region can be subdivided into parts that function independently.

This can be done by an experiment known as the *cis-trans* comparison. It will be recalled that the needed function can be supplied to a mutant by simultaneous infection of the cell with standard phage; the standard type supplies an intact copy of the genetic structure, so that

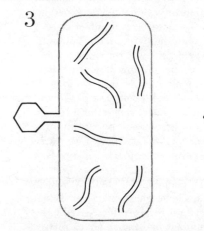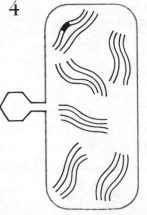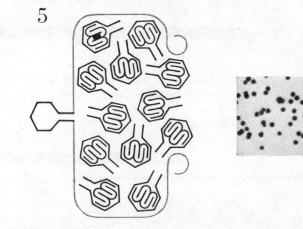

mutants and standard phage produce different plaque types. The diagrams show a bacillus infected by a single standard phage. The DNA molecule from the phage enters the cell (2) and is replicated (3 and 4). Among scores of perfect replicas, one may contain a mutation (*dark patch*). Encased in protein jackets, the phage particles finally burst out of the cell (5). When a mutant arises during development of a plaque, the mixture of its mutant progeny and standard types makes plaque look mottled (*photograph at right*).

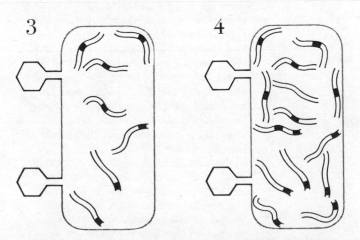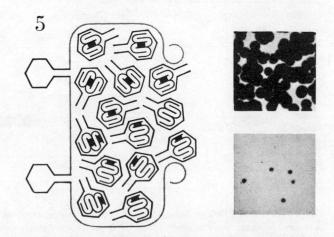

infects a single bacillus. Most of the DNA replicas are of one type or the other, but occasionally recombination will produce either a double mutant or a standard recombinant containing neither mutation. When the progeny of the cross are plated on strain B (*top* *photograph at far right*), all grow successfully, producing many plaques. Plated on strain K, only the standard recombinants are able to grow (*bottom photograph at right*). A single standard recombinant can be detected among as many as 100 million progeny.

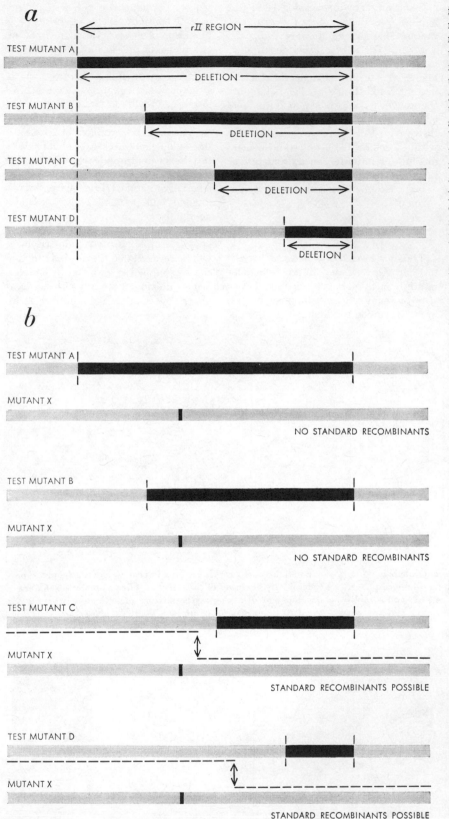

a

rII REGION

TEST MUTANT A

DELETION

TEST MUTANT B

DELETION

TEST MUTANT C

DELETION

TEST MUTANT D

DELETION

b

TEST MUTANT A

MUTANT X

NO STANDARD RECOMBINANTS

TEST MUTANT B

MUTANT X

NO STANDARD RECOMBINANTS

TEST MUTANT C

MUTANT X

STANDARD RECOMBINANTS POSSIBLE

TEST MUTANT D

MUTANT X

STANDARD RECOMBINANTS POSSIBLE

DELETION MAPPING is done by crossing an unknown mutant with a selected group of reference mutants (*four at top*) whose DNA molecules contain deletions—or what appear to be deletions—of known length in the *r*II region. Thus when mutant X is crossed with test mutants *A* and *B*, no standard recombinants are observed because both copies of the DNA molecule are defective at the same place. When X is crossed with *C* and *D*, however, standard recombinants can be formed, as indicated by broken lines and arrows. By using other reference mutants with appropriate deletions the location of X can be further narrowed.

it does not matter what defect the *r*II mutant has and both types are enabled to reproduce. Now suppose the intact structure of the standard type could be split into two parts. If this were to destroy the activity, the two parts could be regarded as belonging to a single functional unit. Although the experiment as such is not feasible, one can do the next best thing. That is to supply piece *A* intact by means of a mutant having a defect in piece *B*, and to use a mutant with a defect in piece *A* to supply an intact piece *B*. If the two pieces *A* and *B* can function independently, the system should be active, since each mutant supplies the function lacking in the other. If, however, both pieces must be together to be functional, the split combination should be inactive.

The actual experimental procedure is as follows. Let us imagine that one has identified two mutational sites in the *r*II region, *X* and *Y*, and that one wishes to know if they lie within the same functional unit. The first step is to infect cells of strain K with the two different mutants, *X* and *Y*, this is called the *trans* test because the mutations are borne by different DNA molecules. Now in K the decision as to whether or not the phage will function occurs very soon after infection and *before* there is any opportunity for recombination to take place. To carry out a control experiment one needs a double mutant (obtainable by recombination) that contains both *X* and *Y* within a single phage particle. When cells of strain K are infected with the double mutant and the standard phage, the experiment is called the *cis* test since one of the infecting particles contains both mutations in a single DNA molecule. In this case, because of the presence of the standard phage, normal replication is expected and provides the control against which to measure the activity observed in the *trans* test. If, in the *trans* test, the phage fails to function or shows only slight activity, one can conclude that *X* and *Y* fall within the same functional unit. If, on the other hand, the phage develops actively, it is probable (but not certain) that the sites lie in different functional units. (Certainty in this experiment is elusive because the products of two defective versions of the same functional unit, tested in a *trans* experiment, will sometimes produce a partial activity, which may be indistinguishable from that produced by a *cis* experiment.)

As applied to *r*II mutants, the test divides the structure into two clear-cut parts, each of which can function independently of the other. The functional units have been called cistrons, and we say that the rII region is composed of an A cistron and a B cistron.

We have, then, genetic units of various sizes: the small units of mutation and recombination, much larger cistrons and finally the rII region, which includes both cistrons. Which one of these shall we call the gene? It is not surprising to find geneticists in disagreement, since in classical genetics the term "gene" could apply to any one of these. The term "gene" is perfectly acceptable so long as one is working at a higher level of integration, at which it makes no difference which unit is being referred to. In describing data on the fine level, however, it becomes essential to state unambiguously which operationally defined unit one is talking about. Thus in describing experiments with rII mutants one can speak of the rII "region," two rII "cistrons" and many rII "sites."

Some workers have proposed using the word "gene" to refer to the genetic unit that provides the information for one enzyme. But this would imply that one should not use the word "gene" at all, short of demonstrating that a specific enzyme is involved. One would be quite hard pressed to provide this evidence in the great majority of cases in which the term has been used, as, for example, in almost all the mutations in *Drosophila*. A genetic unit should be defined by a genetic experiment. The absurdity of doing otherwise can be seen by imagining a biochemist describing an enzyme as that which is made by a gene.

We have seen that the topology of the rII region is simple and linear. What can be said about its topography? Are there local differences in the properties of the various parts? Specifically, are all the subelements equally mutable? If so, mutations should occur at random throughout the structure and the topography would then be trivial. On the other hand, sites or regions of unusually high or low mutability would be interesting topographic features. To answer this question one isolates many independently arising rII mutants and maps each one to see if mutations tend to occur more frequently at certain points than at others. Each mutation is first localized into a main segment, then into a smaller segment, and finally mutants of the same small segment are tested against each other. Any that show recombination are said to define different sites. If two or more reverting mutants are found to show no detectable recombination with each other, they are considered to be

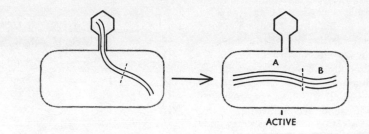

PHAGE ACTIVITY requires that the coded information inside functional units of the DNA molecule be available intact. The rII region consists of two functional units called A cistron and B cistron. When both are present intact (*right*), the phage actively replicates inside colon bacillus of strain K. Colored lines indicate effective removal of coded information.

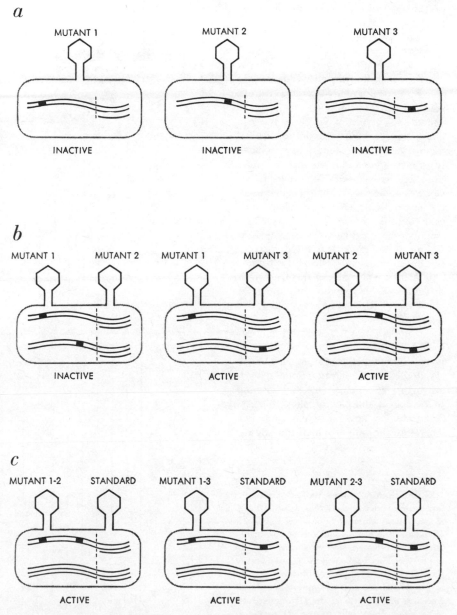

CIS-TRANS TEST determines the size of functional units. In bacillus of strain K, T4 phage is active only if both A and B cistrons are provided intact; hence mutants *1*, *2* and *3* are inactive. (The sites of mutation have been previously established.) Tests with the three mutants taken two at a time (*b*) show that sites *1* and *2* must be in the same cistron. A test of each mutant with standard phage (*c*) provides a control; in this case all are active.

repeats, and one of them is chosen to represent the site in further tests. A set of distinct sites is thereby obtained, each with its own group of repeats. The designation of a mutant as a repeat is, of course, tentative, since in principle it remains possible that a more sensitive test could show some recombination.

The illustration on pages 118 and 119 shows a map of the *r*II region with each occurrence of a spontaneous mutation indicated by a square. These mutations, as well as other data from induced mutations, subdivide the map into more than 300 distinct sites, and the distribution of repeats is indeed far from random. The topography for spontaneous mutation is evidently quite complex, the structure consisting of elements with widely different mutation rates.

Spontaneous mutation is a chronic disease; a spontaneous mutant is simply one for which the cause is unknown. By using chemical mutagens such as nitrous acid or hydroxylamine, or physical agents such as ultraviolet light, one can alter the DNA in a more controlled manner and induce mutations specifically. A method of inducing specific mutations has long been the philosophers' stone of genetics. What the genetic alchemist desired, however, was an effect that could be directed at the gene controlling a particular characteristic. Chemical mutagenesis is highly specific but not in this way. When Rose Litman and Arthur B. Pardee at the University of California discovered the mutagenic effect of 5-bromouracil on phage, they regarded it as a nonspecific mutagen because mutations were induced that affected a wide assortment of different phage characteristics. This nonspecificity resulted because each functional gene is a structure with many parts and is bound to contain a number of sites that are responsive to any particular mutagen. Therefore the rate at which mutation is

DELETION MAP shows the reference mutants that divided the *r*II region into 80 segments. These mutants behave as if various sections of the DNA molecule had been deleted or inactivated, and as a class they do not revert, or back-mutate, spontaneously to produce standard phage. Mutants that do revert usually act as if the mutation is localized at a single point on the DNA molecule. Where this point falls in the *r*II region is determined by systematically crossing the revertible mutant with these reference deletion mutants, as illustrated on page 114. The net result is to assign the point mutation to smaller and smaller segments of the map.

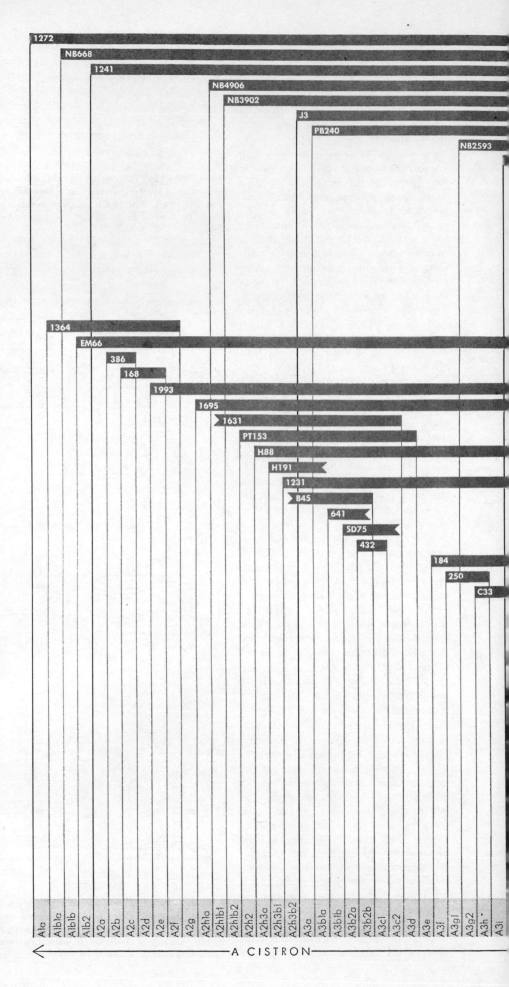

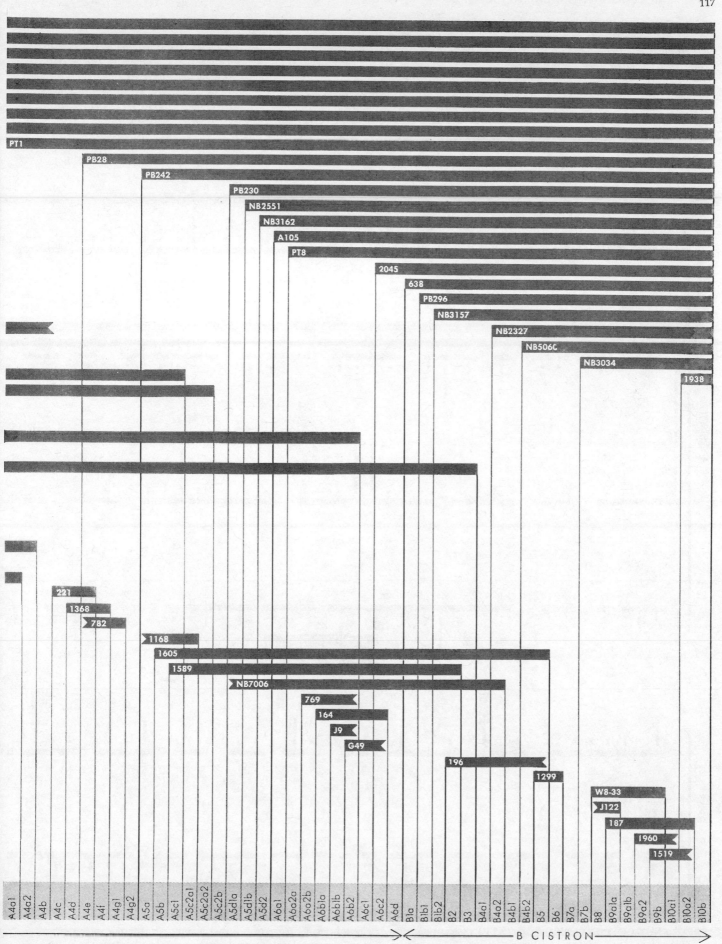

induced in various genes is more or less the same. By fine-structure genetic analysis, however, Ernst Freese and I, working in our laboratory at Purdue University, have found that 5-bromouracil increases the mutation rate at certain sites by a factor of 10,000 or more over the spontaneous rate, while producing no noticeable change at some other sites. This indicates a high degree of specificity indeed, but at the level within the cis-

tron. Furthermore, other mutagens specifically alter other sites. The response of part of the B cistron to a variety of mutagens is shown in the illustration on the following two pages.

Each site in the genetic map can, then, be characterized by its spontaneous mutability and by its response to various mutagens. By this means many different kinds of site have been found. Some response patterns are represented at only

a single site in the entire structure; for example, the prominent spontaneous hot spot in segment B4. This is at first surprising, because according to the Watson-Crick model for DNA the structure should consist of only two types of element, adenine-thymine (AT) pairs and guanine-hydroxymethylcytosine (GC) pairs. One possible explanation for the uneven reactivity among various sites is that the response may depend not

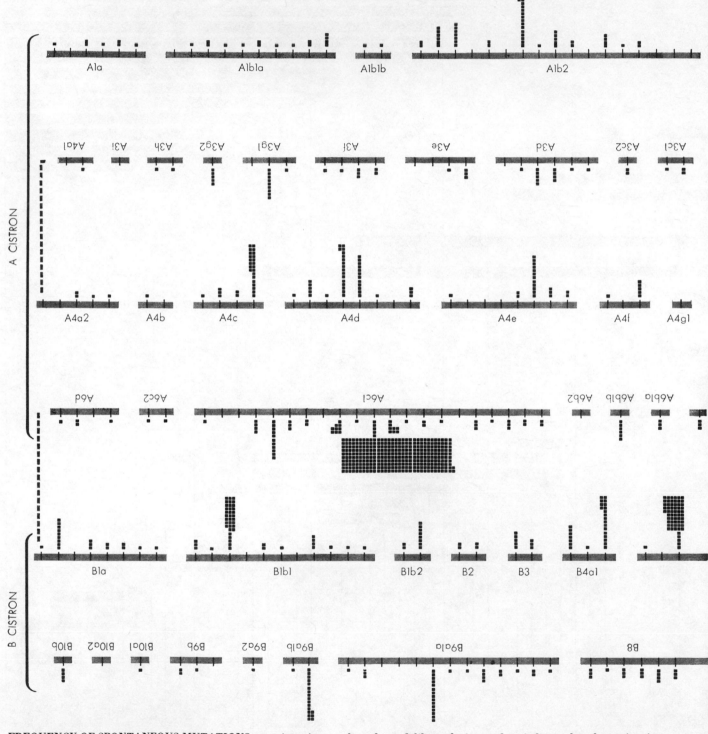

FREQUENCY OF SPONTANEOUS MUTATIONS at various sites is shown in this complete map of the rII region. Alternate rows have been deliberately inverted to indicate that the region is a continuous molecular thread. Each spontaneous mutation at a site

only on the particular base pair at a site but also very much on the type and arrangement of neighboring base pairs.

Once a site is identified it can be further characterized by the ease with which a particular mutagen makes reverse mutations produce phage of standard type. Combining such studies with studies of the chemical mechanism of mutagenesis, it may be possible eventually to translate the genetic map, bit by bit, into the actual base sequence.

Saturation of the Map

How far is the map from being run into the ground? Since many of the sites are represented by only one occurrence of a mutation, it is clear that there must still exist some sites with zero occurrences, so that the map as it stands is not saturated. From the statistics of the distribution it can be estimated that there must exist, in addition to some 350 sites now known, at least 100 sites not yet discovered. The calculation provides only a minimum estimate; the true number is probably larger. Therefore the map at the present time cannot be more than 78 per cent saturated.

Everything that we have learned about the genetic fine structure of T4 phage is compatible with the Watson-

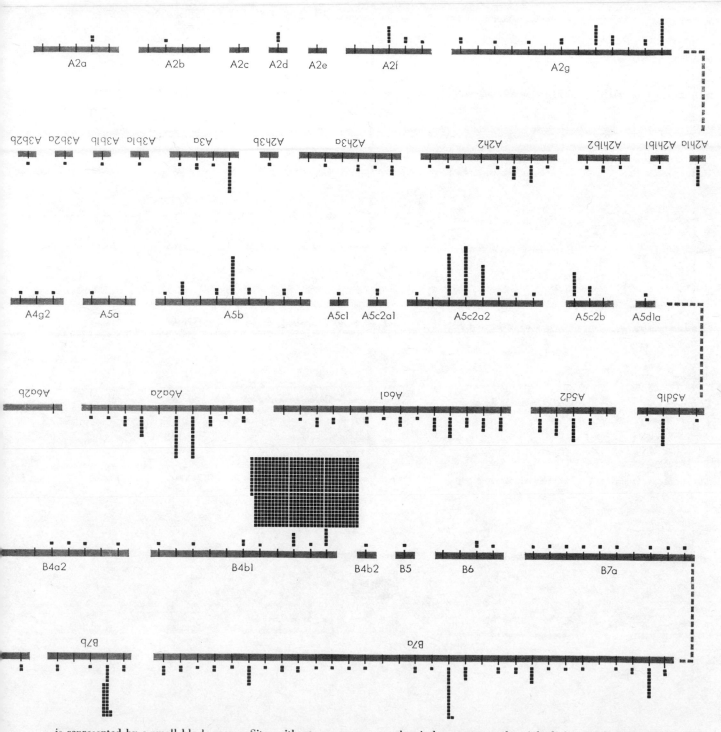

is represented by a small black square. Sites without squares are known to exist because they can be induced to mutate by use of chemical mutagens or ultraviolet light (see *illustration on next two pages*), but they have not been observed to mutate spontaneously.

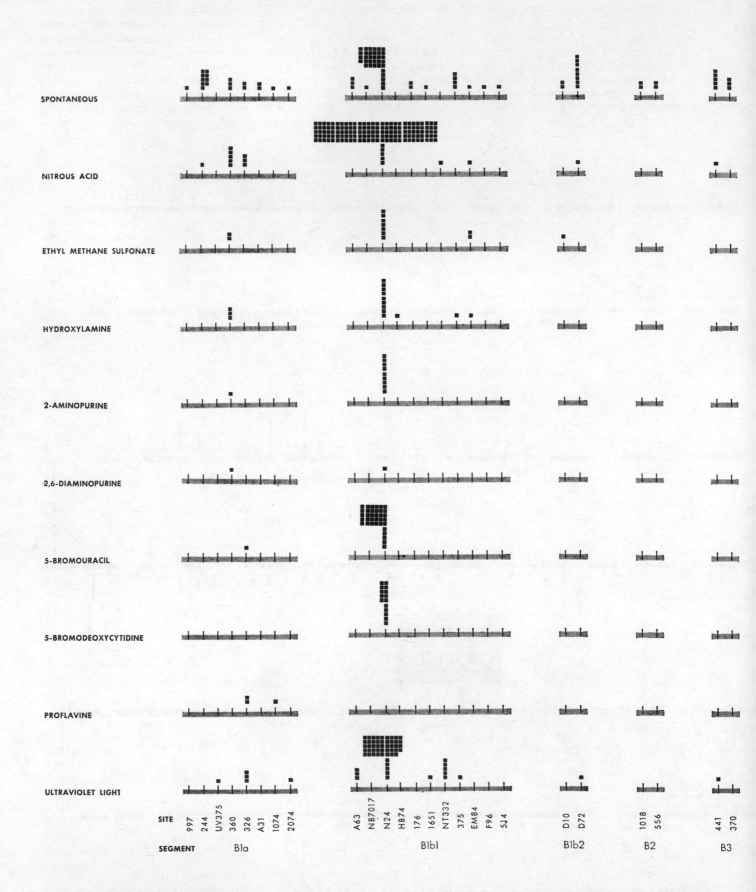

| | SPONTANEOUS | NITROUS ACID | ETHYL METHANE SULFONATE | HYDROXYLAMINE | 2-AMINOPURINE | 2,6-DIAMINOPURINE | 5-BROMOURACIL | 5-BROMODEOXYCYTIDINE | PROFLAVINE | ULTRAVIOLET LIGHT |

| SITE | 997 | 244 | UV375 | 360 | 326 | A31 | 1074 | 2074 | A63 | NB7017 | N24 | HB74 | 176 | 1651 | NT332 | 375 | EM84 | F96 | 5J4 | D10 | D72 | 1018 | 556 | 441 | 370 |

| SEGMENT | B1a | | | | | | | | B1b1 | | | | | | | | | | | B1b2 | | B2 | | B3 | |

RESPONSE OF PHAGE TO MUTAGENS is shown for a portion of the B cistron. The total number of mutations studied is not the same for each mutagen. It is clear, nevertheless, that mutagenic action is highly specific at certain sites. For example, site EM26,

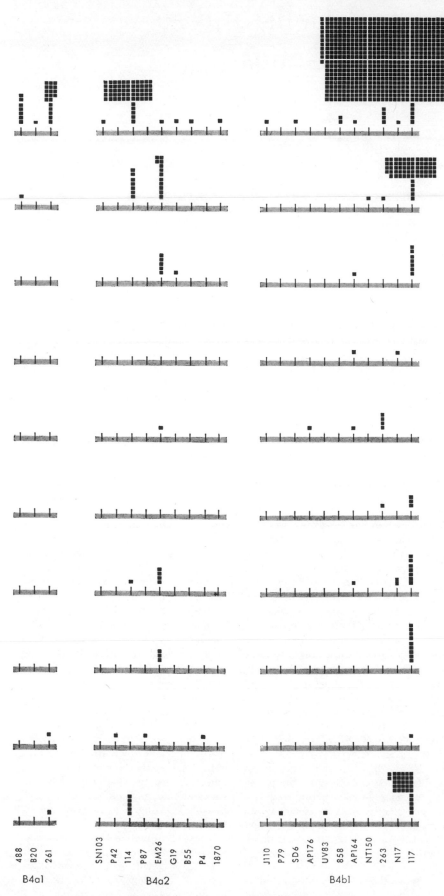

Crick model of the DNA molecule. In this model the genetic information is contained in the specific order of bases arranged in a linear sequence. The four-letter language of the bases must somehow be translated into the 20-letter language of the amino acids, so that at least several base pairs must be required to specify one amino acid, and an entire polypeptide chain should be defined by a longer segment of DNA. Since the activity of the resulting enzyme, or other protein, depends on its precise structure, this activity should be impaired by any of a large number of changes in the DNA base sequence leading to amino acid substitutions.

One can also imagine that certain changes in base sequence can lead to a "nonsense" sequence that does not specify any amino acid, and that as a result the polypeptide chain cannot be completed. Thus the genetic unit of function should be vulnerable at many different points within a segment of the DNA structure. Considering the monotonous structure of the molecule, there is no obvious reason why recombination should not be possible at every link in the molecular chain, although not necessarily with the same probability. In short, the Watson-Crick model leads one to expect that the functional units—the genes of traditional genetics—should consist of linear segments that can be finely dissected by mutation and recombination.

Mapping Other Genes

The genetic results fully confirm these expectations. All mutations can in fact be represented in a strictly linear map, the functional units correspond to sharply defined segments, and each functional unit is divisible by mutation and recombination into hundreds of sites. Mutations are induced specifically at certain sites by agents that interact with the DNA bases. Although the data on mutation rates are complex, it is quite probable that they can be explained by interactions between groups of base pairs.

In confining this investigation to rII mutants of T4, attention has been focused on a tiny bit of hereditary material constituting only a few per cent of the genetic structure of a virus, enabling the exploration to be carried almost to the limits of the molecular structure. Similar results are being obtained in many other microorganisms and even in higher organisms such as corn. Given techniques for handling cells in culture in the appropriate way, man too may soon be a subject for genetic fine-structure analysis.

which resists spontaneous mutation, responds readily to certain mutagens. However, site 117 in segment B4b1 is more apt to mutate spontaneously than in response to a mutagen.

12 Hybrid Cells and Human Genes

by Frank H. Ruddle and Raju S. Kucherlapati
July 1974

The mapping of human genes and the study of how they are regulated is facilitated by an experimental substitute for sexual breeding: the fusion of human somatic cells with the cells of other mammals

In order to recognize, understand and eventually treat human diseases that arise from the inheritance of faulty genes one must learn what genes are responsible for each disorder, where they are situated on the 24 different chromosomes in the human cell and how they function in the complex milieu of cell and organism. In the past decade the increased medical interest in human genetic analysis has been matched by the development of a major new experimental approach that makes such analysis feasible in the laboratory: the hybridization of somatic cells, which is to say the fusion of body cells (as opposed to male and female germ cells) from other mammals with those of humans in such a way that the resulting hybrid cells contain different assortments of only a few human chromosomes each. Somatic-cell hybridization experiments, facilitated by new chromosome-staining techniques and an enhanced ability to distinguish homologous gene products from different species, are beginning to fill in the human genetic map. Moreover, human genetic analysis is providing information on not only the location of genes on chromosomes but also the regulation of their expression in cells. It should therefore contribute to our understanding both of such medical problems as hereditary disease and cancer and of such fundamental biological problems as differentiation, development and evolution.

The classical method of mapping human genes has been to trace the pattern of inheritance of individual characteristics through many generations of a family and thus learn what traits are linked, or associated with one another on a single chromosome, and in some cases what that chromosome is. Such studies were extremely difficult because human families are relatively small, the generation time is long and it is impractical to carry out the controlled sexual breeding on which the classical genetic analysis of organisms higher than bacteria depends. Human geneticists developed statistical methods that eased some of these difficulties [see "The Mapping of Human Chromosomes," by Victor A. McKusick; SCIENTIFIC AMERICAN; Offprint 1220], but the procedures are at best laborious, time-consuming and expensive. They are also limited in scope: the low frequency of rare traits in the total population makes many interesting disorders hard to study, and some genes cannot be investigated by familial analysis because they have no variant forms.

As long as 20 years ago geneticists such as Guido Pontecorvo, Curt Stern and Joshua Lederberg were suggesting that parasexual experimental systems should be developed for human genetic analysis. In 1960 George Barski and his colleagues in Paris discovered that two different mouse-cell lines would fuse in a laboratory culture, forming hybrid cells. The techniques of hybridization were improved by a number of workers, notably Boris Ephrussi, first in France and later at Case Western Reserve University, and Henry Harris of the University of Oxford. In 1967 Mary C. Weiss and Howard Green, then at the New York University School of Medicine, first demonstrated the potential of mouse-human hybrid cells for human genetic analysis [see "Hybrid Somatic Cells," by Boris Ephrussi and Mary C. Weiss; SCIENTIFIC AMERICAN; Offprint 1137].

In essence the hybridization procedure is simple. Mammalian cells can be treated in ways that adapt them to survival and multiplication in a suitable medium. Cells from two species—human fibroblasts (connective-tissue cells), say, and mouse cells adapted to tissue culture—are mixed in a laboratory dish and allowed to remain in contact for several hours. Spontaneous fusion between cells of the two different species occurs at a low frequency, but in practice fusion is usually enhanced by the addition of agents such as inactivated Sendai virus, which forms intercellular bridges between closely adjacent cells. When a suspension of highly dispersed parental and fused cells is plated on a culture medium, clones, or colonies, of cells arise, each stemming from the repeated division of an individual parental or fused cell and its progeny.

To give the hybrid-cell clones (and the investigator trying to isolate them) an advantage, the cells are usually grown in a "selective" medium. The one most commonly used is based on a selective strategy developed by Waclaw Szybalski, then at Rutgers University, and John W. Littlefield of the Harvard Medical School. It depends on the drug aminopterin, which blocks the production of folic acid, which is necessary for the normal synthesis of nucleotides, the building blocks of DNA. The nucleotides can be synthesized in the presence of aminopterin if their precursors, hypoxanthine and thymidine, are added to the medium, but only if the cells contain the "salvage pathway" enzymes hypoxanthine guanine phosphoribosyltransferase (HGPRT) and thymidine kinase (TK).

The usual practice is to fuse human fibroblasts that are capable of making both enzymes with rodent cells that are deficient in either HGPRT or TK, in a medium containing hypoxanthine, aminopterin and thymidine (HAT medium). The enzyme-deficient rodent parental cells die whereas the human parental

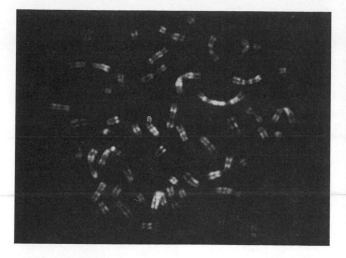

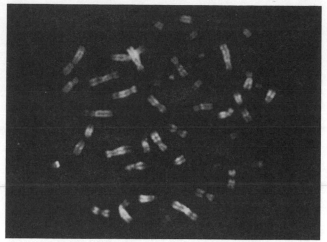

CHROMOSOMES of a mouse tumor cell (*left*) and a human fibroblast, a connective-tissue cell (*right*), were stained with the fluorescent dye quinacrine for these photomicrographs. The chromosomes are at metaphase, the stage in the cell cycle in which they are highly condensed and have replicated preparatory to cell division; the twin strands of each chromosome are joined to form *V*-shaped or *X*-shaped double chromosomes. There are 40 chromosomes in a normal mouse cell and varying larger numbers in tumor cells; there are 46 in a human cell. The fluorescent dye delineates characteristic pattern of bright and dark bands on each chromosome.

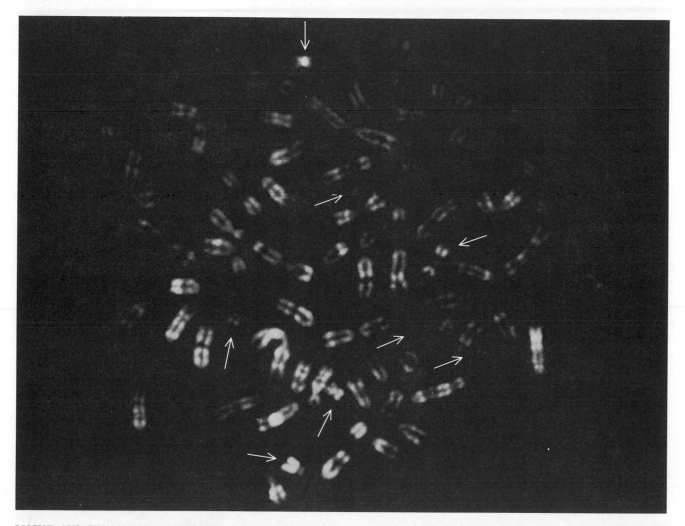

MOUSE AND HUMAN CHROMOSOMES are both present in a metaphase preparation of a hybrid cell that was formed by the fusion of two cells like those whose chromosomes are pictured at the top of the page. There are 73 chromosomes in this aberrant fused cell, only eight of them human chromosomes (*arrows*). This situation is typical of mouse-human hybrid cells, in which most of the human chromosomes are eliminated. Because different human chromosomes remain in different hybrid clones, or cell lines, it is possible to match them with the specific products of their genes and thus locate genes on chromosomes. This process is facilitated by stains, such as the quinacrine stain used here, that delineate the characteristic banding patterns. The bands tell an investigator, for example, that the arrowed chromosome at the top is human chromosome No. 3, distinguished by a particularly bright centromere.

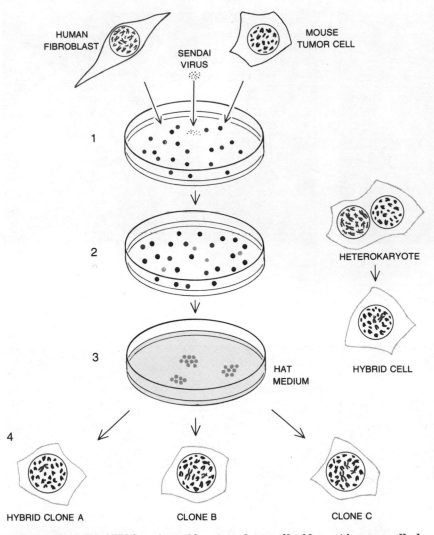

HYBRID SOMATIC CELLS are formed by mixing human fibroblasts with mouse cells that are deficient in either of two enzymes, TK or HGPRT, and adding a fusion-enhancing agent such as Sendai virus (1). Some of the cells fuse (2), forming first heterokaryons with two nuclei and then new hybrid cells (gray). When cells are plated on HAT, a "selective" medium, fused cells proliferate and form colonies (3); the mouse parental cells cannot for lack of an enzyme and the fibroblasts do not multiply rapidly. The fused cells retain all the mouse chromosomes but only a few of 46 human chromosomes; human-chromosome loss is random, so that each hybrid clone has a different human chromosome complement (4).

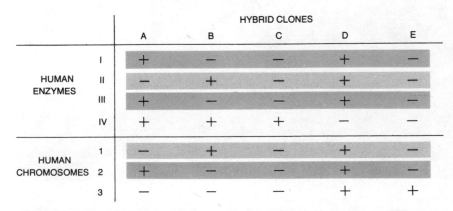

| | | HYBRID CLONES | | | | |
		A	B	C	D	E
HUMAN ENZYMES	I	+	−	−	+	−
	II	−	+	−	+	−
	III	+	−	−	+	−
	IV	+	+	+	−	−
HUMAN CHROMOSOMES	1	−	+	−	+	−
	2	+	−	−	+	−
	3	−	−	−	+	+

LINKAGE AND LOCATION of genes is studied in hybrid cells by noting which gene products, usually enzymes, consistently appear together in a clone and then noting which chromosome is consistently present along with each product. In this table of hypothetical results of testing for enzymes and chromosomes, for example, enzymes I and III are seen to be linked on the same chromosome, that is, they are "syntenic." Moreover, they are assignable to chromosome No. 2. Enzyme II is seen to be assignable to chromosome No. 1.

cells and the fused cells survive. If the human cells are either fibroblasts that have been cultured repeatedly and therefore grow slowly or white blood cells, which do not normally proliferate, there will not be many human parental clones, and it is easy to isolate pure clonal populations of hybrid cells, each of which is derived from a single fusion event.

There are two important things about such hybrids. One is that the mouse genome, or genetic material, is retained essentially intact whereas the human genome is preferentially, partially and randomly lost. Normal human cells have 46 chromosomes: an X and a Y sex chromosome (in male cells) or two X chromosomes (in female cells) and 22 pairs of autosomes, or nonsex chromosomes. Normal mouse cells have 40 chromosomes. The hybrids always have fewer than the expected combined total of 86, with the usual range being from 41 to 55. All the mouse chromosomes are present; it is most of the human ones that are lost, and that loss is random: hybrids that are the result of different fusions retain different sets of human chromosomes. The other important thing about the hybrids is that, as Weiss and Green discovered, both mouse and human genomes are functional: both sets of genes are expressed at the same time, each coding for its appropriate proteins. The mouse chromosomes and the human chromosomes can be distinguished from one another, and the mouse and the human enzymes or other protein products can also be distinguished.

The chromosomes of the two species and some of the 24 different human chromosomes differ fairly obviously in shape, in size and in the location of the centromere: the point at which the two strands of a metaphase chromosome converge. The problem of distinguishing certain very similar human chromosomes from one another has been largely solved in the past few years by the development, primarily by Torbjörn Caspersson of the Royal Caroline Institute in Sweden, of a fluorescent staining method. When the chromosomes are stained with quinacrine, the dye binds and fluoresces in such a way that different chromosomes have different patterns of dark and bright bands. The banding patterns serve to identify the chromosomes and also provide landmarks along the chromosomal strand. As for the gene products, the evolutionary distance between mouse and man is large enough so that homologous proteins generally differ in their amino acid constitution, and these differences can be detected by electrophoresis [*see illustration at right*].

The task in mapping is first to learn whether two or more genes are on the same chromosome and then if possible to learn which chromosome they are on and just where on the chromosome individual genes are located. The human chromosomes are usually lost from hybrid cells as discrete units. Genes that are on the same chromosome will therefore usually be expressed together. In other words, two genes that are consistently either present together or lost together can be considered to be syntenic, or on the same chromosomal strand. Assaying a number of clones for various human enzymes therefore provides information on the synteny of genes. It requires only one more step to assign genes to chromosomes by essentially the same method: one simply compares the pattern of gene expression with the presence or absence of specific chromosomes [*see bottom illustration on opposite page*].

With these methods it has been possible in the past several years to assign more than 50 genes to 18 human chromosomes [*see illustration on next two pages*]. A number of the genes already mapped are implicated in human genetic diseases. For example, a deficiency in the enzyme hexosaminidase A is associated with Tay-Sachs disease; a deficiency in the enzyme galactose-1-phosphate uridyl transferase is associated with galactosemia. Increased knowledge of the map positions of such genes and their syntenic relations with other genes increases the ability to predict abnormal offspring by prenatal examination of the fetus. Progress in mapping has been so rapid that we can expect that in a few years at least one gene will have been assigned to each of the human chromosomes.

Although the basic method is logical and straightforward, various refinements have been developed in order to get results more quickly and with more precision. Theoretically if one had 24 rodent-human hybrid-cell lines, each retaining only one human chromosome, and if each hybrid retained a different chromosome, the gene for any testable enzyme could be assigned to the proper chromosome. In practice it is hard to get such one-chromosome hybrid lines, and so we have developed a scheme that should enable us to obtain similar information with as few as five hybrid lines. Each line has been selected for its possession of a unique subset of chromosomes. For example, consider three different lines, each of which has a different

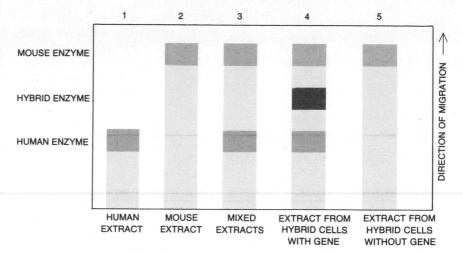

HOMOLOGOUS ENZYMES are distinguished by electrophoresis. The diagram shows how the process works for glucose-6-phosphate dehydrogenase (G6PD). Cell extracts from each of five sources are placed in channels at base of a starch gel. G6PD is a dimer: a polymer with two subunits. Human cells make a human homopolymer and mouse cells a mouse homopolymer, each of which migrates at a different speed in an electric field, so that the human enzyme (*1*) does not move as far as the mouse enzyme (*2*). A mixture of extracts from both kinds of cell shows the two components (*3*). In a hybrid cell human and mouse subunits are synthesized and interact randomly, giving rise to the two homopolymers and to mouse-human heteropolymers that migrate at an intermediate speed. The enzyme from a hybrid cell that includes the gene for human G6PD therefore separates into three bands (*4*). Hybrids that do not retain the human G6PD gene produce only the homopolymer from the mouse (*5*).

subset of the human chromosomes Nos. 1, 2, 3, 4, 5, 6, 7 and 8 [*see illustration below*]. Each of the three clones has a unique binary combination of chromosome presence or absence; to put it another way, the pattern of presence or absence in each clone is unique for each chromosome: chromosome No. 2 is present or absent in the panel of three clones in the sequence +, +, −; chromosome No. 6 has the pattern −, +, −.

The panel can be tested for any human phenotype (which is to say for any enzyme or other gene product) and will yield a specific pattern of expression that depends on the location of the responsible gene. That is, if testing the panel for the presence of human enzyme A yields the result −, −, +, one can say that the gene responsible for

the expression of enzyme A must be on chromosome No. 7. Given five clones (with 2^5, or 32, combinatorial possibilities), one can cover the entire set of human chromosomes. In ·practice we work with from eight to 10 clones because increasing the level of redundancy helps us to avoid false interpretations based on errors that are encountered in testing for the presence or absence of an enzyme.

The methods we have described so far can pin a gene down to an individual chromosome but cannot go further; they cannot locate a gene at a precise point on a chromosome or establish the order in which several genes are arrayed along a chromosome. These things can often be done if one makes use of chromosomal aberrations such as translocations or de-

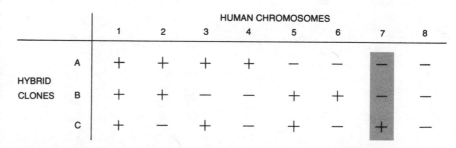

		HUMAN CHROMOSOMES							
		1	2	3	4	5	6	7	8
HYBRID CLONES	A	+	+	+	+	−	−	−	−
	B	+	+	−	−	+	+	−	−
	C	+	−	+	−	+	−	+	−

PANEL OF THREE HYBRID-CELL CLONES, each containing a unique subset of eight human chromosomes, can be tested for the presence or absence of any gene product. The pattern of presence or absence will match a pattern associated with one of the eight chromosomes. For example, if an enzyme is present only in clone C, its gene must be on chromosome No. 7. A larger panel of five clones, each with a unique subset of the 24 chromosomes, offers 32 combinatorial possibilities, enough to cover the 24 different human chromosomes.

letions, which disturb the normal syntenic relations of genes.

Florence C. Ricciuti of our laboratory at Yale University capitalized on a translocation to localize three genes that were known to be on the human X chromosome: the genes for the enzymes HGPRT, phosphoglycerate kinase (PGK) and glucose-6-phosphate dehydrogenase (G6PD). The translocation, which appeared in one family, is such that most of the long arm of the X chromosome is translocated to an almost intact chromosome No. 14. The two translocation products are a short one bearing the centromere and the short arm of the X and a long one made up of most of the No. 14 with the long arm of the X attached to it; they are easily distinguished from each other and from the rest of the human chromosome complement. As for the location of the three genes, there were three possibilities: they could all be on the long arm of the X, they could all be on the short arm or they could be distributed on both arms.

Ricciuti hybridized human cells that carried the translocation with a mouse-cell line that was deficient in HGPRT. The hybrids were selected in HAT medium and then were analyzed for the presence of a number of human enzymes. The three human enzymes in question were present in all the hybrids. Since the ability of such hybrids to grow in HAT depended on their retention of human HGPRT, the results suggested that all three genes must be on a single translocation product, that is, all three were on the short arm of the X or all three were on the long arm. The enzyme screening showed, moreover, that the gene for nucleoside phosphorylase (NP), which was known to be on an autosome (a nonsex chromosome), was segregating along with the X-linked markers. That was strong evidence that the NP gene is on No. 14 and therefore also that the three X-linked enzymes are most probably on the translocated long arm of X. Detailed analysis of the chromosome pattern confirmed this by demonstrating a positive correlation between the long translocation product and the four enzymes: NP, PGK, HGPRT and G6PD.

Similar experiments were done by Park S. Gerald of the Children's Hospital Medical Center in Boston and by Dirk Bootsma at Erasmus University in Rotterdam. They worked with translocations that had different break points, and so their results not only confirmed Ricciuti's but also provided a unique sequence for the three genes on the X chromosome [see illustration on page 128]. At the present time several workers, including John Hamerton of the University of Manitoba and Bootsma, are doing similar analyses in an effort to localize genes on chromosome No. 1.

Another translocation has helped us, in collaboration with James K. McDougall of the University of Birmingham, to localize the gene for human thymidine kinase (TK) rather precisely. TK, the first gene mapped through somatic-cell hybridization, is on chromosome No. 17. Charlotte Boone and T. R. Chen of our laboratory examined several mouse-human cell lines in which the mouse cells had been deficient in TK. One of the hybrid lines was found to have a spontaneous translocation between a mouse chromosome and the long arm of human chromosome No. 17; the human TK was synthesized by that line, and so

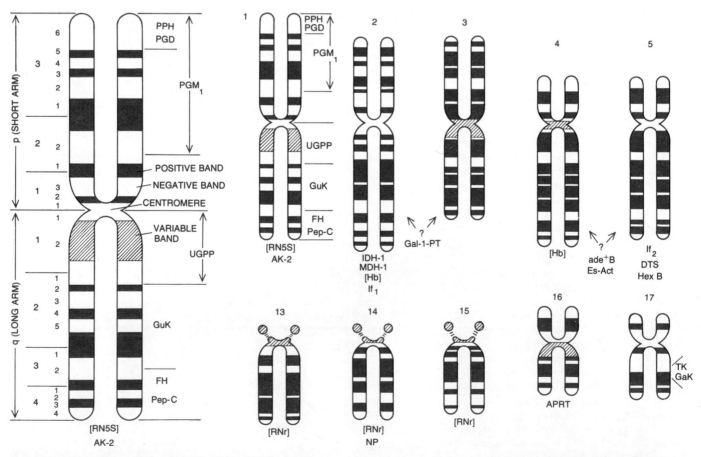

BANDING PATTERNS AND KNOWN GENE LOCI are shown for the 24 different human chromosomes. Chromosome No. 1 is drawn large (*left*) to illustrate the standard nomenclature and the numbered designations of the various segments. The colored bands are the positive ones, those that are stained brightly by a fluorescent dye such as quinacrine; the white bands are the negative ones; hatched regions are variable bands that stain differently in the chromosomes of different individuals. Genes are designated by the initials of the enzymes or other products they code for: *PPH* is phosphopyruvate hydratase, for example; *RNr* is ribosomal RNA.

we could conclude that the TK gene is on the long arm of No. 17. (The translocation, incidentally, provided the first indication that chromosomes from different species could exchange segments.)

In order to localize the TK gene more precisely we capitalized on McDougall's discovery that human cells infected with adenovirus 12 exhibit a high frequency of breakage in chromosome No. 17. The breakage is preceded by the formation of an "uncoiler" region in the chromosome: the chromosome strand, which is ordinarily tightly coiled in these metaphase chromosomes, is uncoiled here, as it is during the stage of the cell cycle when the DNA is being transcribed into RNA and the RNA is being translated into protein. As a matter of fact, it was known that the level of host-specific TK in cells increases after adenovirus-12 infection; apparently extra TK is being synthesized for a time. We postulated therefore that the primary effect of the virus was to induce the uncoiling or to interfere with normal compaction of the chromosome and that the chromosome breaks were subsequent, secondary results of the infection. If that was the

case, we expected the chromosomes might break at different points, so that some broken chromosomes would retain the TK gene and others would not. That would provide a unique opportunity for determining the exact location of the gene.

We chose hybrid cells in which the mouse-human translocation represented the only detectable human chromosome segment, treated the cells with adenovirus 12 in various concentrations, cultured the infected cells and then tested different colonies for their ability to grow in HAT medium, which would reflect their retention or loss of the human TK gene. We found that cells without the ability to make TK had lost a portion of the long arm of No. 17. There were two kinds of cells that retained TK activity: cells that had not lost any part of No. 17 and cells in which the deletion was somewhat smaller than the one that resulted in the loss of TK activity. Comparison of the breaking points for the two classes of cells gave us the location of the TK gene [see illustration on page 129]. The three cell lines, incidentally, constitute a small panel for the long arm

of chromosome No. 17. Working with this panel, we could map an unrelated gene, for galactokinase, to a region close to TK. The concept can be extended to map genes in all other subregions of chromosomes.

The methods we have described so far involve the mapping of "constitutive" genes: genes that are expressed during the entire life of most or all types of cells. There are two other classes of genes: those that are expressed only when the cell is exposed to a specific external challenge and those that are expressed continuously but only in specialized cell types. Mapping such genes calls for somewhat different techniques, and it can also provide some particularly rewarding insights into mechanisms of gene regulation.

Mammalian cells protect themselves against attack by a virus by producing a protein, interferon, that somehow inhibits virus replication. Just what interferon does and how it works are not yet clear, although the matter has been under vigorous investigation because of interferon's obvious potential for treat-

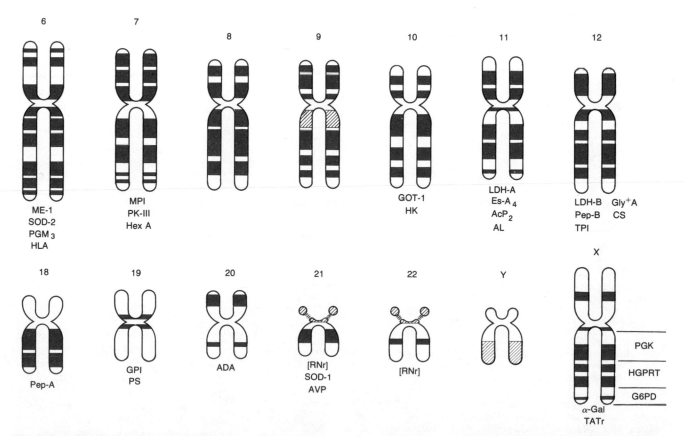

The genes are listed under the chromosome with which they have been associated or, in cases in which a location on the chromosome has been determined, are listed opposite that location. Genes that are listed with question marks may be on either of the chromosomes to which they are referred. Genes that are listed in brackets were mapped by annealing techniques involving the preparation of complementary nucleic acid; all the other genes on the map were located through the technique of somatic-cell hybridization. Still other genes have been mapped over the years through studies of genetic anomalies in families, but they are not included here.

128

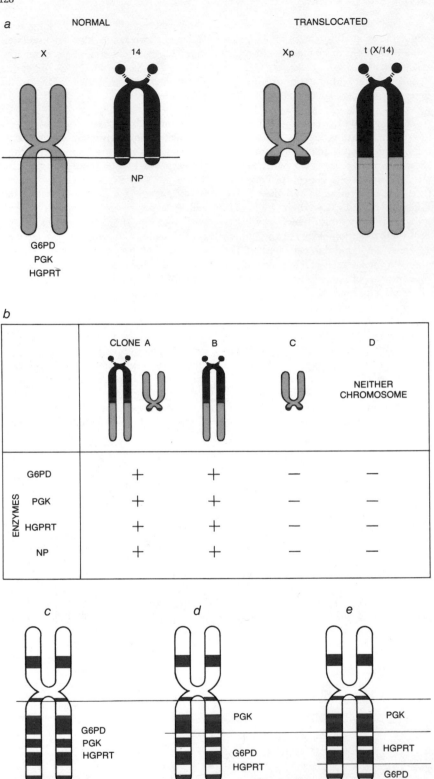

a NORMAL — TRANSLOCATED

X   14   Xp   t (X/14)

NP

G6PD
PGK
HGPRT

b

ENZYMES	CLONE A	B	C	D
				NEITHER CHROMOSOME
G6PD	+	+	—	—
PGK	+	+	—	—
HGPRT	+	+	—	—
NP	+	+	—	—

c

G6PD
PGK
HGPRT

d

PGK

G6PD
HGPRT

e

PGK

HGPRT

G6PD

TRANSLOCATIONS can be capitalized on to locate genes in specific regions of chromosomes. Genes for three enzymes, HGPRT, PGK and G6PD, were known to be on the human X chromosome; another gene, for the enzyme NP, was known to be on some autosome (non-sex chromosome). The translocation is such that most of the long arm of the X is exchanged with the tip of chromosome No. 14 (*a*). Cells bearing the translocation were fused with mouse cells deficient in HGPRT and hybrid cells were selected in HAT medium. Examination of clones yielded the results shown (*b*), indicating that all three of the X-chromosome genes must be somewhere on the translocated long arm (and that the NP gene must be on No. 14). Comparison of the X-chromosome results from the authors' laboratory at Yale University (*c*) with results of similar experiments, in which the translocations had different break points, in two other laboratories (*d*, *e*) mapped the sequence and the approximate loci of the three genes along the arm, as shown in illustration on the preceding two pages.

ing virus diseases [see "The Induction of Interferon," by Maurice R. Hilleman and Alfred A. Tytell; SCIENTIFIC AMERICAN; Offprint 1226]. An important fact about interferon is that in many instances it is species-specific: mouse interferon is made by mouse cells and protects only mouse cells, not human ones; human cells make human interferon.

In our laboratory Y. H. Tan and Richard P. Creagan treated a number of different mouse-human hybrid cells with viruses and with chemical agents that mimic the virus in inducing cells to produce interferon. They found that two chromosomes, No. 2 and No. 5, had to be present together if interferon was to be induced. Even then, however, the antiviral response was not forthcoming. If chromosome No. 21 was present, on the other hand, and if interferon was then supplied, the cells did exhibit the antiviral response; cells with No. 21 alone, however, did not respond to the inducer by making interferon or by inhibiting virus replication.

Our interpretation is that two different genes, one on chromosome No. 2 and one on No. 5, are required for the expression of human interferon. We do not know which of the genes codes for interferon itself and which is required for some other, supportive function, and so studies to determine the nature and action of the two genes are in progress. These results constitute, among other things, the first reported case in which two genes appear to be required for the expression of a specific function in human cells.

As for chromosome No. 21, we believe it codes for still another protein, whose identity and function have not yet been determined, that may interfere directly with viral gene replication or may possibly code for a species-specific receptor site on the cell membrane that accepts interferon. The hybridization system is well suited for the genetic characterization of just such discrete membrane functions. Recently Dorothy Miller and her associates at Columbia University have shown that the poliomyelitis virus receptor site is coded by a gene on chromosome No. 19. In our laboratory Creagan and Susie Chen have shown that the diphtheria-toxin receptor site is coded by a gene on chromosome No. 5. It may soon be possible to identify gene products, coded by each of the human chromosomes, that are expressed at the cell surface. This may provide investigators with additional opportunities to regulate the human chromosomal composition of hybrid cells and will also contribute to our knowledge of the structure and function of the cell membrane.

We shall describe two examples of genes that are expressed in specialized cell types: the gene for the blood-plasma protein albumin, which is expressed in the liver, and the gene for globin (the protein of hemoglobin), which is ordinarily expressed in red blood cells and their immediate precursors. Characterizing such specialized functions as these and understanding the mechanisms whereby they are expressed to the exclusion of other functions in specialized cells should provide important information bearing on the regulation of gene activity and thus on the processes of differentiation and development.

Gretchen Darlington and Hans Peter Bernhard of our laboratory set out to study the regulation of albumin production by hybridizing liver cells with cells that do not express this phenotype, and to map the human albumin gene. Since normal liver cells do not survive well in culture, they worked with mouse hepatoma (liver cancer) cells, which are easily cultured and which synthesize and secrete albumin. They fused the hepatoma cells with human white blood cells, which do not make albumin, and examined the resulting hybrid cells for their ability to secrete albumin and for their human-enzyme and chromosome composition.

All the hybrid lines produced mouse albumin, and two of them also produced human albumin (which can be distinguished from the mouse version of the protein by immunological methods). What this meant was that white blood cells contain the hereditary information for the synthesis of albumin: the albumin gene is there but it is repressed, or inactivated, and it is somehow activated by factors present in the albumin-making mouse-liver cell. A similar result was attained by Mary Weiss and Jerry Petersen and by Stephen Malawista of the National Center for Scientific Research in France, who crossed rat hepatoma cells with mouse cells that do not normally make albumin and found that the hybrid cells produced some mouse albumin. These results provide additional evidence to support the assumption that all cells of an organism contain all its genes, that cells are genetically equivalent even if they have become differentiated to the point where they are functionally quite distinctive.

So far not enough hybrid-cell lines have produced human albumin for us to establish the syntenic relations of the

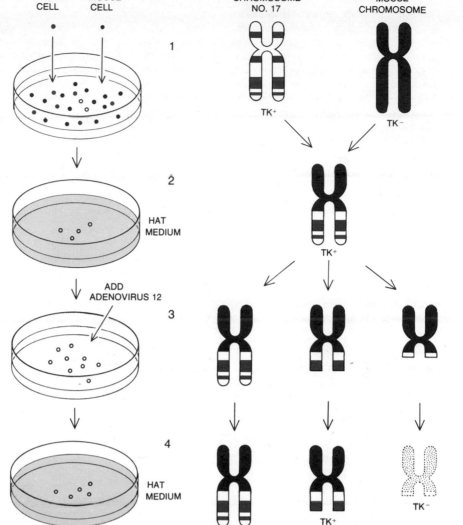

INDUCED CHROMOSOME BREAKAGE defined the location of the gene for the enzyme TK, which had previously been mapped to chromosome No. 17. The investigators fused mouse cells deficient in TK with human cells (1). A hybrid line in which there was a translocation of the long arm of a human No. 17 to a mouse chromosome was viable in HAT medium, showing that the TK gene was on the long arm of No. 17 (2). Adenovirus 12 was added to induce chromosome breakage at several different points, and cells with different translocation products were grown in nonselective mediums (3). When they were tested in HAT medium (4), it was found that the cells with translocation products retaining the whole long arm of No. 17 or most of the arm still had the TK gene; the cells that had lost more of the arm did not have the gene. Comparison of the three break points located the TK gene.

human albumin gene with any other human markers or to associate the gene with a chromosome; we hope that will be possible after study of a larger series of hybrid cells. Quite apart from the mapping, however, the techniques developed in the albumin experiments provide a new way to get at the complex and fascinating subject of gene repression and derepression, which is currently under intensive study. And they should also be applicable to the prenatal detection of hereditary diseases affecting such specialized tissues as the brain, the liver and the kidneys.

Prenatal detection is now confined to

diseases involving gross chromosomal abnormalities or enzymes that happen to be produced and expressed in fibroblasts, the only cells that are easily recovered from the amniotic fluid. Perhaps fibroblasts (like the white blood cells in the albumin experiment) could be made to reveal the properties of their entire genetic complement (as the white blood cells were made to reveal their albumin gene) through fusion with established cell lines that do express specific genes that are medically significant. Then many more defects in the fetal genome could presumably be diagnosed.

Rather than activating a repressed

gene, hybridization can have the opposite result: it can extinguish a tissue-specific, differentiated function that was expressed by one of the parental cells. Arthur Skoultchi, now at the Albert Einstein College of Medicine in New York, developed a system in our laboratory that makes it possible to investigate the mechanisms of this extinction process, which are still not understood.

When certain strains of mice are treated with a complex of viruses known collectively as Friend leukemia virus, they develop a cancerous condition called erythroleukemia. Infected cells (precursors of red blood cells) from the spleen of such mice have the property of growing indefinitely in a culture. Charlotte Friend of the Mount Sinai School of Medicine in New York, who first isolated the cells, showed that they can also be made to produce hemoglobin if they are treated with the chemical dimethyl sulfoxide. Skoultchi found that if the erythroleukemic cells are fused with mouse fibroblasts, the hybrid cells cannot synthesize hemoglobin even after treatment with dimethyl sulfoxide.

One obvious explanation for their loss of this capability would be that they have lost the chromosome or chromosomes that carry the genes for globin, the hemoglobin protein. That explanation can be ruled out, however. The reasoning is as follows: the gene for the enzyme glucose phosphate isomerase (GPI) is known to be linked to the gene for one of the two kinds of globin; all the hybrid-cell lines made GPI in the two distinguishable forms characteristic of each type of parental cell but did not make hemoglobin; therefore the globin gene must be present even though its product is not. Such a result is consistent with gene modulation, that is, with such processes as repression and derepression.

The globin system is particularly adaptable to the study of gene modulation because one can look into the cell and learn not only whether or not globin is being produced but also, if it is not, at what stage its synthesis was blocked. Globin messenger RNA, the nucleic acid that is transcribed from the DNA of the gene and is subsequently translated into the protein globin, is relatively easy to purify because it constitutes a large fraction of the RNA of any cell that makes it and because it can be separated from other RNA's by sedimentation. Once globin messenger RNA has been purified it is possible, with the help of the enzyme reverse transcriptase (also called RNA-directed DNA polymerase), to prepare DNA that is complementary to the RNA. This is DNA whose subunit sequence forms a template that in effect fits the sequence of the RNA subunits and that therefore tends, when exposed to the globin RNA, to become "annealed" to the RNA.

The complementary DNA can be used as a probe with which to measure the amount of globin RNA present in a cell. When the probing technique is applied to the hybrid cells that result from fusion of mouse fibroblasts and erythroleukemic cells, the data indicate that there is very little if any globin messenger RNA in their cytoplasm. The extinction of gene activity, at least in this instance, is apparently at the level of transcription from DNA to RNA, not at the level of translation from RNA into protein. (It is also possible that it is the transport, processing or stability of the messenger RNA that is affected.)

Somatic-cell hybridization has proved to be an efficient tool for the mapping of human genes. It has opened new doors to the understanding of human hereditary disorders and of the regulatory mechanisms that underlie both the malignant transformation of cells and the mutually interdependent functions of that most complex group of cells, the human organism.

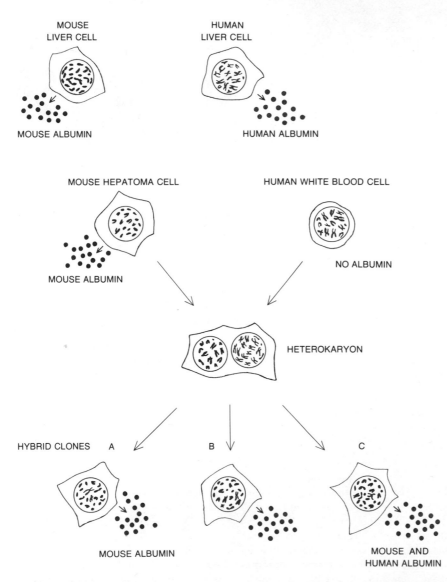

ACTIVATION of a repressed gene coding for the protein albumin was demonstrated in hybrid cells. The albumin gene is ordinarily expressed in mouse liver or hepatoma (liver cancer) cells, as it is in human liver cells. White blood cells do not make albumin. Investigators fused mouse hepatoma cells with human white blood cells. The resulting hybrid cells all produced mouse albumin. Some of them also produced human albumin, which can be distinguished from the mouse protein by immunochemical methods. The result shows that the gene for human albumin is present in white blood cells but is not expressed. It is activated in those hybrid cells that contain its chromosome (which has not yet been surely identified), presumably by some regulatory factor that is supplied by the mouse liver cell.

IV

GENE EXPRESSION
AND REGULATION

GENE EXPRESSION AND REGULATION   IV

INTRODUCTION

A. GENETIC FINE STRUCTURE AND COLINEARITY

Classical genetic analysis had revealed that genes were on chromosomes and that the Mendelian principles of segregation and independent assortment could be understood in terms of the mechanism of meiosis. With the discoveries that both the genetic material of the chromosomes and the immediate functional product of the genes are polymers of nucleotides and amino acids, respectively, the possibility for analysis of genetic changes within genes was realized.

That genes and their functional products were linear polymers was well known before there was any explicit evidence that mutations, with their striking functional consequences for the organism carrying them, could be attributed to the alteration of a single amino acid in the polypeptide chain.

It was Vernon Ingram who first demonstrated that sickle-cell anemia, a genetic disease, was due to such a single amino acid substitution—namely, that of valine for glutamic acid in hemoglobin. This discovery, which Ingram describes in his article "How Do Genes Act?," implied that the mutation itself may only involve one or very few nucleotides in the gene specifying hemoglobin. (The reader should note that Ingram's analysis was performed on unfractionated hemoglobin, which is a tetramer consisting of two alpha and two beta polypeptide chains. Further characterization of the sickle-cell hemoglobin revealed that this substitution took place at the sixth amino acid from the amino terminus of the beta polypeptide chain.)

You will recall from reading Seymour Benzer's article in the previous section that the gene has a fine structure, as would be expected if a gene were a segment of the DNA molecule whose nucleotide sequence specifies the amino acid sequence of the protein, and that so-called point mutations may involve as little as a single base pair in the nucleotide sequence specifying a gene. Unfortunately, Benzer was denied the opportunity to demonstrate the corollary of the existence of a genetic fine structure in the form of localized changes in the amino acid sequence of the rII gene product, because the rII protein proved, at the time, refractory to isolation. Before we can put the two halves of the story together, it is necessary to review the nature of the genetic code. Two articles by Francis H. C. Crick—"The Genetic Code" and "The Genetic Code: III"—tell of the elucidation and the nature of the genetic code and how it provides the material basis for Beadle and Tatum's correspondence theorem of "one gene, one enzyme."

It fell to Sydney Brenner and to Charles Yanofsky, as the latter describes in his article on "Gene Structure and Protein Structure," to complete the story with their fine structure analyses of both gene and product, confirming the colinearity of cause and consequence between the location of the mutations on the genetic map and the corresponding changes of amino acids in the polypeptide sequence.

Yanofsky's and Brenner's separate demonstrations of colinearity, together with the understanding of how the genetic information encoded in a nucleotide sequence of a gene is transcribed and translated according to the dictates of the genetic code, refined and generalized Beadle and Tatum's dictum of "one gene, one enzyme" to "one gene, one polypeptide chain."

The ultimate resolution achievable by genetic analysis is demonstrated in John C. Fiddes's article, "The Nucleotide Sequence of a Viral DNA," in which he presents the nucleotide sequence of the entire genome of the simple singlestranded DNA virus ϕX174.

You will see from Fiddes's article that a tract of nucleotides can contribute its information to more than one polypeptide product. This circumstance does not, however, require any basic change in the seminal statement of "one gene, one polypeptide chain" because it is theoretically possible that the double strands of sequences of nucleotides, given the three reading frame registers and the two polarities for transcription, can encode as many as six distinct polypeptide products. Since the sequences are not independent of one another, however, it must be conjectured that there is a considerable tradeoff in evolutionary flexibility in trying to pack so much information in such a restricted space.

Other recent discoveries would appear at first glance to contravene the principle of "one gene, one polypeptide chain." This situation arises out of the fact that the RNA transcript that is ultimately used for translation is, in some instances, an edited version of the original RNA transcript. Specific intercalary sequences have, by some mechanism, been excised from the primary RNA transcript to give a shorter version. Moreover, there is evidence that a given primary sequence of messenger RNA may be processed in different ways, generating a set of final messenger RNAs and their corresponding translation products. Even so, it is semantically appropriate to conserve the concept "one gene, one polypeptide sequence" statement with the understanding that a gene not be defined at the DNA level but as a specific posttranscriptionally edited version of an RNA sequence that is to be translated in a particular phase.

B. REGULATION OF GENE EXPRESSION

It is always difficult when dealing with genetics not to stray from matters strictly genetic into other fields. Differentiation is one such field, but since the genetic approach has made such formidable contributions to the understanding of what one should be looking for to explain the phenomenon of differentiation, an article on the regulation of gene expression is included in this section. "A DNA Operator-Repressor System," by Tom Maniatis and Mark Ptashne, is concerned with the analysis, at the molecular level, of the regulation of gene expression in procaryotes. The study of genetic regulation in these organisms, in which "differentiation" is restricted to temporal responses to changing nutritional circumstances, has provided models to guide the investigation of differentiation in higher organisms, in which differentiation takes place in space as well as in time.

Since this article was written in 1976, another mode of gene-expression regulation has come to light. It is quite distinct from the repression and activation regulatory systems represented by the lactose activation system and arabinose operons, respectively. The newer mode is the "attenuator" control system for the tryptophan and histidine operons. In each instance, the operon sequence of genes concerned with specifying the enzymes of the amino acid synthesis pathway is preceded by a "leader" sequence coding for a leader polypeptide rich in the amino acid that is the product of the pathway. The availability of the amino acid governs the spacing between the RNA polymerase and the first ribosome translating the nascent messenger. Thus, when the amino acid is present in sufficient concentration, the nascent messenger RNA transcript assumes a secondary structure, which in some way aborts further transcription. If the concentration of the required amino acid is limiting, a larger portion of the transcript is available, and it forms an alternative secondary structure allowing transcription to proceed.

13 How Do Genes Act?

by Vernon M. Ingram
January 1958

In which the effect of a human mutation that causes a disease of the blood is traced to a change in one of the 300 amino acid units that make up the structure of the protein hemoglobin

During the last few years geneticists and biochemists have made exciting progress toward learning how heredity works. They have found more and more evidence that the substance of the genes is deoxyribonucleic acid (DNA). These studies have led to an extremely interesting hypothesis about how DNA controls the making of living material—that is, proteins. In this article I shall describe some experiments, carried out in recent months at the University of Cambridge, which not only provide strong support for this hypothesis but have actually pinpointed the chemical effects of certain genes, thus giving us our first detailed picture of gene action.

The meaning of these experiments will be clearer if we first review very briefly the main outlines of the present view of the chemistry of heredity, which has been discussed in many articles in SCIENTIFIC AMERICAN. The blueprints of heredity reside in the threadlike chromosomes, of which there are about 24 pairs in every human cell. Each chromosome contains several hundred genes, strung together in a row. The gene responsible for any particular trait has a specific position, or "locus," in the string. The present concept is that the backbone of the chromosome consists of long DNA molecules and that the genes are segments of these molecules. The DNA molecule is made up of two chains twined around each other in a helical structure and cross-linked by pairs of bases—adenine and thymine or guanine and cytosine. It is like an enormously long winding staircase, with the stair treads corresponding to these cross-links.

The DNA, according to the simple and attractive hypothesis that has excited so much recent attention, holds the key to the manufacture of proteins by the cell. Its structure controls the construction of the protein molecules for which it is responsible. It carries a kind of Morse code: that is to say, a given DNA has its bases arranged in a particular order, and this order determines the order in which amino acids fall into place in the corresponding protein molecule [see "Nucleic Acids," by F. H. C. Crick; SCIENTIFIC AMERICAN; Offprint 54]. There are some 20 or more different amino acids. A protein molecule is made of hundreds or thousands of amino acid units, and all the evidence indicates that each protein owes its uniqueness to the specific sequence of its amino acids.

To sum up the theory: A segment of a DNA molecule (*i.e.*, a gene) carries the code for a protein or section of a protein molecule. Putting the hypothesis this way suggests a possible test for a gene's action. Suppose we have an abnormal protein arising from mutation of a gene. By chemical analysis of this deviant form and the normal protein we might discover how they differ. If we can determine this, we may learn what the site of the gene's action is and how large or how small a chemical change it has produced.

Such a protein is, in fact, conveniently available for study. It is the human hemoglobin molecule, the oxygen-carrying part of the red blood-cell. As readers of this magazine well know, several mutant forms of hemoglobin have been discovered in recent years. The best-known is the sickle cell form, responsible for sickle cell anemia. The discovery of this defect has been enormously fruitful to biological science in several fields [see "Sickle Cells and Evolution," by Anthony C. Allison; SCIENTIFIC AMERICAN; Offprint 1065]. Various researches, particularly those of J. V. Neel at the University of Michigan, have proved that the defect is hereditary, that it is traceable to the action of a single mutant gene and that the gene causes the synthesis of a hemoglobin chemically different from the normal form.

We undertook to track down this difference. To start with, the only definite sign of the chemical difference was the finding by Linus Pauling and his co-workers at the California Institute of Technology that the normal hemoglobin molecule and sickle cell hemoglobin have different electric charges, as shown by the technique called electrophoresis. Our problem was to find out precisely what part of the molecule was altered—this in a huge molecule consisting of some 8,000 atoms! The situation was not, however, quite as desperate as may appear. The hemoglobin molecule is made up of two identical halves, so we can confine ourselves to one half of it (only 4,000 atoms!). We are further favored by the fact that we can consider the atoms in groups rather than individually, the groups being amino acids. The half-molecule contains about 300 amino acid units, of 19 different kinds.

The problem thus resolved itself into trying to learn just how the arrangement of these 300 units in a sickle cell hemoglobin molecule differed from that in a normal molecule. It was a sufficiently formidable task. To work out the complete structure of the insulin molecule, with 51 amino acids, took a group of workers at Cambridge nearly 10 years! We did not venture to analyze the entire structure of hemoglobin. There were indications that the change to the abnormal molecule lay in some small section of it, and I attacked the problem by breaking the molecule down to small fragments, in the hope of locating the change in a single fragment. Whereas the chances of finding a small change are pretty remote in a molecule made of

300 amino acid units, they should be much better in fragments consisting of, say, only 10 amino acids.

To break down the hemoglobin molecules we used trypsin, a digestive enzyme. Trypsin breaks an amino acid chain only at points where lysine or arginine occurs. Since the half-molecule of hemoglobin contains about 26 units of these two amino acids, we could expect to get about the same number of fragments. This proved to be the case: the molecule broke down to 28 fragments and a resistant "core" comprising about a quarter of the molecule. We removed the core in a centrifuge and proceeded to analyze the fragments—each a small group of amino acids, known as a peptide.

For this I employed a combination of electrophoresis and chromatography. First a drop of the solution containing the 28 peptides was deposited as a spot near the edge of a large sheet of moist filter paper. An electric current, passed along the edge of the paper, separated the fragments, for various peptides migrate at different speeds according to the electric charges they carry. There was now a line of spots, each consisting of several peptides. These were next separated by chromatography. The sheet was dried and hung so that the edge with the line of spots touched a liquid; as the liquid moved up the paper, the peptides of each spot migrated upward, again at different rates according to their constitution, and eventually each peptide came to rest as a distinct spot. The end result was a network of 28 spots on the paper, each representing a fragment of the broken molecule of hemoglobin. I call this map a "fingerprint" of the hemoglobin.

The next step, of course, was to compare the fingerprints of normal hemoglobin and the errant sickle cell form. They proved to be exactly alike in all respects except one: in the sickle cell fingerprint, one peptide (which I call the No. 4 peptide) was displaced slightly from the position it occupied in the normal fingerprint. The amount of displacement showed a difference in electrical charge similar to the difference between the whole sickle cell hemoglobin molecule and the normal one. We were therefore encouraged to believe that we were on the right track: that the entire difference between the molecules lay in this small fraction.

At this point I should interject that we have carefully analyzed the rest of the molecule—all the other fragments and the core—and we have been unable to find any difference between the normal and abnormal molecules in these other fractions. Whether these pieces are identical in structure cannot yet be decided, but they are the same in composition and it is reasonable to believe that they are identical in every way.

So peptide No. 4 became the center of our search for the chemical peculiarity in sickle cell hemoglobin. We now had to analyze the peptide, both in the normal and in the sickle cell molecule, to see what amino acids it was made of in each case. This was a very tedious

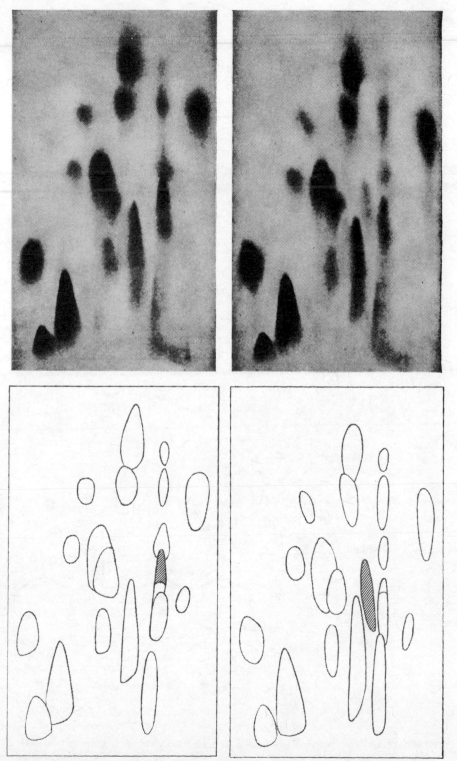

"FINGERPRINTS" of fragmented hemoglobin were made by a combination of electrophoresis and chromatography. At top left is the fingerprint of normal hemoglobin; at top right, of hemoglobin S. The hatched spots in the diagrams show where they differ significantly.

138

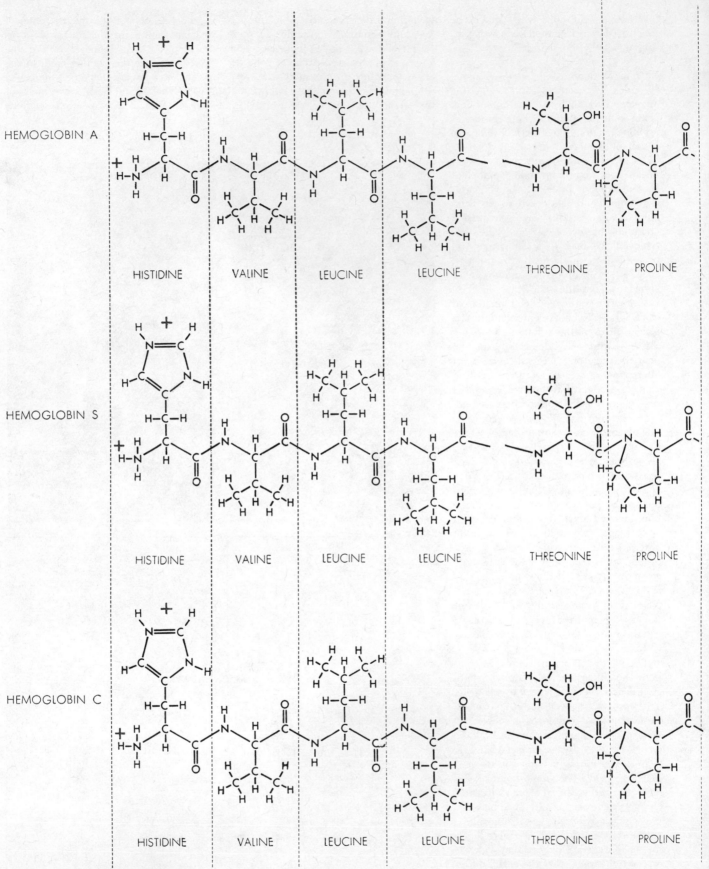

HEMOGLOBIN A

HISTIDINE VALINE LEUCINE LEUCINE THREONINE PROLINE

HEMOGLOBIN S

HISTIDINE VALINE LEUCINE LEUCINE THREONINE PROLINE

HEMOGLOBIN C

HISTIDINE VALINE LEUCINE LEUCINE THREONINE PROLINE

SHORT SECTION of the normal hemoglobin molecule (hemoglobin A) is depicted in the molecular diagram at the top of these two pages. The amino acid units of the molecule are set off by the vertical colored lines. The name of each amino acid is given below the diagram. The corresponding section of hemoglobin S, which is found in individuals suffering from sickle cell anemia, is second from the top. It differs from the normal molecule only in that a glutamic acid unit has been replaced by a valine unit. The same section of hemoglobin C, another abnormal form of the molecule, is at the bottom. It differs from the normal molecule only in that the same glutamic acid unit has been

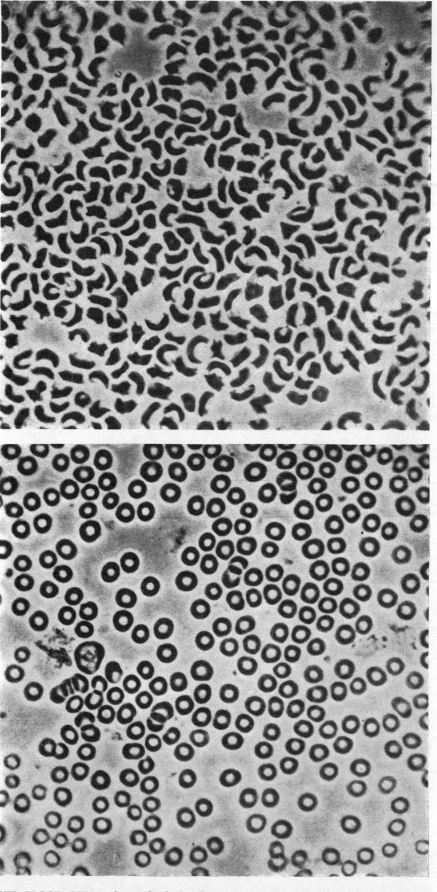

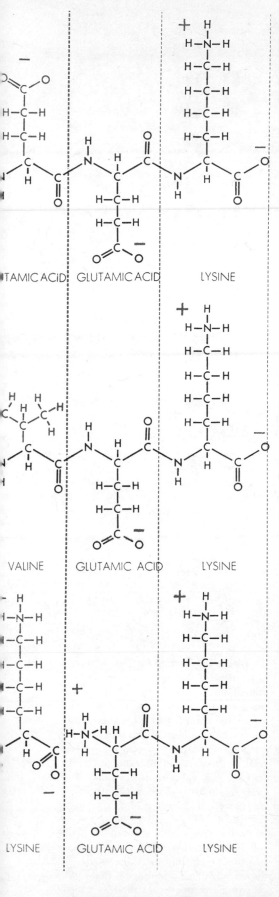

TAMIC ACiD   GLUTAMIC ACID   LYSINE

VALINE   GLUTAMIC ACID   LYSINE

LYSINE   GLUTAMIC ACID   LYSINE

replaced by a lysine unit. The characteristic electric charge of certain groups of atoms in the chain is indicated by the colored plus and minus signs. The colored column here occupies the colored row on page 140.

RED BLOOD CELLS of an individual suffering from sickle cell anemia are enlarged 600 diameters in the phase-contrast photomicrograph at the top. Normal red cells are shown in the photomicrograph at the bottom. The distorted shape of the sickle cell is due to the fact that, when its defective hemoglobin molecules lose oxygen, they clump together in rods.

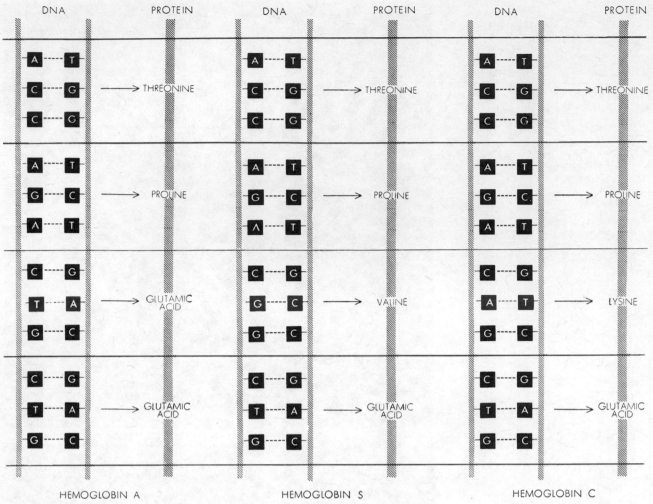

| DNA | PROTEIN | DNA | PROTEIN | DNA | PROTEIN |

HEMOGLOBIN A HEMOGLOBIN S HEMOGLOBIN C

HYPOTHETICAL SCHEME of how deoxyribonucleic acid (DNA) is related to hemoglobin A (normal hemoglobin), hemoglobin S and hemoglobin C is outlined. At the far left is a diagram of a short segment of DNA. The two chains of the DNA are joined by pairs of bases (*squares*). There are four bases: adenine (A), thymine (T), cytosine (C) and guanine (G). To the right of this diagram is a parallel diagram of the protein of hemoglobin A. The amino acid units of the protein are presumably related to segments of the DNA. In the mutated DNA which controls the synthesis of hemoglobin S and hemoglobin C the pairing of bases may be altered at one point (*shown by colored squares*). This might account for the difference in one amino acid of the three forms of hemoglobin.

business. We first had to purify the peptide, which involved cutting out the No. 4 spot on the paper, washing it off and running it again through the separation process of paper electrophoresis or chromatography. To get enough material for analysis we had to fingerprint dozens of batches of broken-down hemoglobin and purify their No. 4 peptides. Fortunately, with the very delicate methods of chemical analysis available nowadays a few thousandths of a gram of purified material was sufficient.

What did the analysis finally show? Both the normal and the sickle cell peptides turned out to contain the same types of amino acids: glutamic acid, valine, histidine, leucine, threonine, pro-

line and lysine. But there was a difference in amount. The normal peptide had two glutamic acid units and a single valine, whereas the abnormal version had a single glutamic acid and a double dose of valine. In other words, in the sickle cell peptide a valine unit replaces a glutamic acid.

The next step was to find the order of arrangement of the amino acids in the peptide—a short chain of nine units. We broke the peptide down (this time with hydrochloric acid) into fragments consisting of from two to five amino acids. Step by step, sometimes peeling off one amino acid at a time, we determined the order of the amino acids in each fragment, and then we were able to fit

the pieces together like a jigsaw puzzle to learn the sequence in the whole peptide. Thus we established that in the sickle cell peptide a valine unit occupies the place of the usual glutamic acid in the seventh position of the nine-unit sequence [*see diagrams on pages 138 and 139*]. A glutamic acid unit has an electrical charge; a valine unit has none. This explains the difference in the electrical charges on the two peptides.

Thus it appears that we have tracked down the difference between the sickle cell hemoglobin molecule and the normal one. According to all our evidence, the sole chemical difference is that in the abnormal molecule a valine is substituted for glutamic acid at one point. A change of one amino acid in nearly 300 is certainly a very small change indeed, and yet this slight alteration can be fatal to the unfortunate possessor of the errant hemoglobin. Equally remarkable is the fact that the sickle cell gene operates so delicately on the synthesis of a protein, changing just one amino acid and leaving the rest of the molecule's structure unaltered.

After I had worked through the sickle cell problem, my colleague John Hunt tackled another abnormal hemoglobin, known as hemoglobin C. This deviant, rarer than the sickle cell form, was known to have parallel properties, and indeed there was genetic evidence that it derived from a mutation in the same part of the chromosome that is responsible for the sickle cell defect. Was it possible that the C gene altered the very same site in the hemoglobin molecule as the sickle cell gene? This seemed almost too pat to be true, but Hunt's chemical investigation fulfilled the expectation beyond all hope.

The fingerprint of hemoglobin C again showed a change only in the No. 4 peptide. But this time the abnormal peptide spot was not merely displaced but broke down into two separate spots. The chemical analysis showed why. It developed that in the hemoglobin C molecule the very same glutamic acid unit that gives way to valine in the sickle cell case is again replaced, this time by another amino acid—namely, lysine. Now lysine, as I have mentioned, offers a vulnerable point for splitting by trypsin. Thus the peptide I call No. 4 breaks into two at this point. There is a further significant fact. Whereas valine has no electrical charge, lysine does have one: it is positive. Since glutamic acid carries a negative charge, the substitution by positively charged lysine produces a greater electrical change in the molecule than occurs in the case of sickle cell hemoglobin.

The exciting point of these findings is that they demonstrate that we are apparently on the right track in our ideas about the mechanism of heredity. We can go on with more confidence to explore the possibilities raised by this work and the hypothesis it supports.

One question certainly calling for further study is why the change of a single amino acid alters the behavior of hemoglobin so drastically as to cause the severe sickle cell disease. We should also like to know more about changes brought about by other mutations of the hemoglobin gene or genes. Some elegant recent researches by Seymour Benzer of Purdue University have brought to light that a gene can mutate at 100 or more spots along its length. If we could correlate mutations at these spots with changes of the bases along the DNA molecule, and those changes in turn with a series of chemical alterations along the length of a protein molecule, we would be a long way toward understanding how heredity works. There is no way to make a systematic study of these matters with human genes, but it might be possible through experiments on microorganisms.

14 The Genetic Code

by F. H. C. Crick
October 1962

*How does the order of bases in a nucleic acid
determine the order of amino acids in a protein? It
seems that each amino acid is specified by a triplet
of bases, and that triplets are read in simple sequence*

Within the past year important progress has been made in solving the "coding problem." To the biologist this is the problem of how the information carried in the genes of an organism determines the structure of proteins.

Proteins are made from 20 different kinds of small molecule—the amino acids—strung together into long polypeptide chains. Proteins often contain several hundred amino acid units linked together, and in each protein the links are arranged in a specific order that is genetically determined. A protein is therefore like a long sentence in a written language that has 20 letters.

Genes are made of quite different long-chain molecules: the nucleic acids DNA (deoxyribonucleic acid) and, in some small viruses, the closely related RNA (ribonucleic acid). It has recently been found that a special form of RNA, called messenger RNA, carries the genetic message from the gene, which is located in the nucleus of the cell, to the surrounding cytoplasm, where many of the proteins are synthesized [see "Messenger RNA," by Jerard Hurwitz and J. J. Furth; SCIENTIFIC AMERICAN Offprint 119].

The nucleic acids are made by joining up four kinds of nucleotide to form a polynucleotide chain. The chain provides a backbone from which four kinds of side group, known as bases, jut at regular intervals. The order of the bases, however, is not regular, and it is their precise sequence that is believed to carry the genetic message. The coding problem can thus be stated more explicitly as the problem of how the sequence of the four bases in the nucleic acid determines the sequence of the 20 amino acids in the protein.

The problem has two major aspects, one general and one specific. Specifically one would like to know just what sequence of bases codes for each amino acid. Remarkable progress toward this goal was reported early this year by Marshall W. Nirenberg and J. Heinrich Matthaei of the National Institutes of Health and by Severo Ochoa and his colleagues at the New York University School of Medicine. [Editor's note: Brief accounts of this work appeared in "Science and the Citizen" for February and March 1962. This article is a companion to one by Nirenberg, which deals with the biochemical aspects of the genetic code].

The more general aspect of the coding problem, which will be my subject, has to do with the length of the genetic coding units, the way they are arranged in the DNA molecule and the way in which the message is read out. The experiments I shall report were performed at the Medical Research Council Laboratory of Molecular Biology in Cambridge, England. My colleagues were Mrs. Leslie Barnett, Sydney Brenner, Richard J. Watts-Tobin and, more recently, Robert Shulman.

The organism used in our work is the bacteriophage T4, a virus that infects the colon bacillus and subverts the biochemical machinery of the bacillus to make multiple copies of itself. The infective process starts when T4 injects its genetic core, consisting of a long strand of DNA, into the bacillus. In less than 20 minutes the virus DNA causes the manufacture of 100 or so copies of the complete virus particle, consisting of a DNA core and a shell containing at least six distinct protein components. In the process the bacillus is killed and the virus particles spill out. The great value of the T4 virus for genetic experiments is that many generations and billions of individuals can be produced in a short time. Colonies containing mutant individuals can be detected by the appearance of the small circular "plaques" they form on culture plates. Moreover, by the use of suitable cultures it is possible to select a single individual of interest from a population of a billion.

Using the same general technique, Seymour Benzer of Purdue University was able to explore the fine structure of the A and B genes (or cistrons, as he prefers to call them) found at the "*r*II" locus of the DNA molecule of T4 [see the article "The Fine Structure of the Gene," by Seymour Benzer, *beginning on page 108*]. He showed that the A and B genes, which are next to each other on the virus chromosome, each consist of some hundreds of distinct sites arranged in linear order. This is exactly what one would expect if each gene is a segment, say 500 or 1,000 bases long, of the very long DNA molecule that forms the virus chromosome [see illustration on opposite page]. The entire DNA molecule in T4 contains about 200,000 base pairs.

The Usefulness of Mutations

From the work of Benzer and others we know that certain mutations in the A and B region made one or both genes inactive, whereas other mutations were only partially inactivating. It had also been observed that certain mutations were able to suppress the effect of harmful mutations, thereby restoring the function of one or both genes. We suspected that the various—and often puzzling—consequences of different kinds of mutation might provide a key to the nature of the genetic code.

We therefore set out to re-examine the effects of crossing T4 viruses bearing mutations at various sites. By growing two different viruses together in a common culture one can obtain "recombinants" that have some of the properties

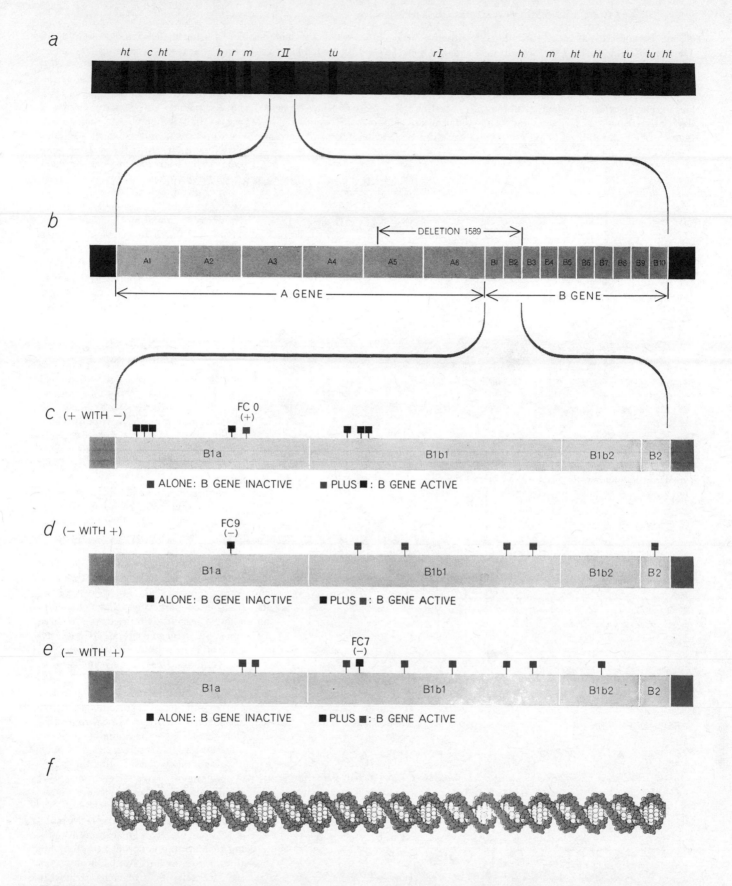

a

ht   c ht   h r m   rII   tu   rI   h m ht ht tu tu ht

b

|←— DELETION 1589 —→|

| A1 | A2 | A3 | A4 | A5 | A6 | B1 | B2 | B3 | B4 | B5 | B6 | B7 | B8 | B9 | B10 |

←————————————— A GENE —————————————→ ←——————— B GENE ———————→

c (+ WITH −)

FC 0
(+)

| B1a | B1b1 | B1b2 | B2 |

■ ALONE: B GENE INACTIVE   ■ PLUS ■: B GENE ACTIVE

d (− WITH +)

FC9
(−)

| B1a | B1b1 | B1b2 | B2 |

■ ALONE: B GENE INACTIVE   ■ PLUS ■: B GENE ACTIVE

e (− WITH +)

FC7
(−)

| B1a | B1b1 | B1b2 | B2 |

■ ALONE: B GENE INACTIVE   ■ PLUS ■: B GENE ACTIVE

f

rII REGION OF THE T4 VIRUS represents only a few per cent of the DNA (deoxyribonucleic acid) molecule that carries full instructions for creating the virus. The region consists of two genes, here called A and B. The A gene has been mapped into six major segments, the B gene into 10 (*b*). The experiments reported in this article involve mutations in the first and second segments of the B genes. The B gene is inactivated by any mutation that adds a molecular subunit called a base (*colored square*) or removes one (*black square*). But activity is restored by simultaneous addition and removal of a base, as shown in *c*, *d* and *e*. An explanation for this recovery of activity is illustrated on page 146. The molecular representation of DNA (*f*) is estimated to be approximately in scale with the length of the B1 and B2 segments of the B gene. The two segments contain about 100 base pairs.

of one parent and some of the other. Thus one defect, such as the alteration of a base at a particular point, can be combined with a defect at another point to produce a phage with both defects [*see upper illustration below*]. Alternatively, if a phage has several defects, they can be separated by being crossed with the "wild" type, which by definition has none. In short, by genetic methods one can either combine or separate different mutations, provided that they do not overlap.

Most of the defects we shall be considering are evidently the result of adding or deleting one base or a small group of bases in the DNA molecule and not merely the result of altering one of the bases [*see lower illustration on this page*]. Such additions and deletions can be produced in a random manner with the compounds called acridines, by a process that is not clearly understood. We think they are very small additions or deletions, because the altered gene seems to have lost its function completely; mutations produced by reagents capable of changing one base into another are often partly functional. Moreover, the acridine mutations cannot be reversed by such reagents (and vice versa). But our strongest reason for believing they are additions or deletions is that they can be combined in a way that suggests they have this character.

To understand this we shall have to go back to the genetic code. The simplest sort of code would be one in which a small group of bases stands for one particular acid. This group can scarcely be a pair, since this would yield only 4×4, or 16, possibilities, and at least 20 are needed. More likely the shortest code group is a triplet, which would provide $4 \times 4 \times 4$, or 64, possibilities. A small group of bases that codes one amino acid has recently been named a codon.

The first definite coding scheme to be proposed was put forward eight years ago by the physicist George Gamow, now at the University of Colorado. In this code adjacent codons overlap as illustrated on the following page. One consequence of such a code is that only certain amino acids can follow others. Another consequence is that a change in a single base leads to a change in three adjacent amino acids. Evidence gathered since Gamow advanced his ideas makes an overlapping code appear unlikely. In the first place there seems to be no restriction of amino acid sequence in any of the proteins so far examined. It has also been shown that typical mutations change only a single amino acid in the polypeptide chain of a protein. Although it is theoretically possible that the genetic code may be partly overlapping, it is more likely that adjacent codons do not overlap at all.

Since the backbone of the DNA molecule is completely regular, there is nothing to mark the code off into groups of three bases, or into groups of any other size. To solve this difficulty various ingenious solutions have been proposed. It was thought, for example, that the code might be designed in such a way that if the wrong set of triplets were chosen, the message would always be complete nonsense and no protein would

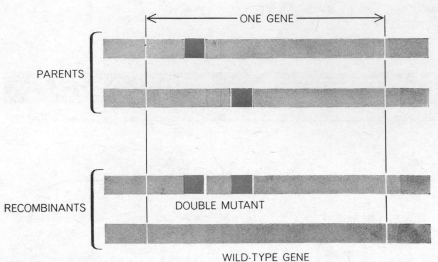

GENETIC RECOMBINATION provides the means for studying mutations. Colored squares represent mutations in the chromosome (DNA molecule) of the T4 virus. Through genetic recombination, the progeny can inherit the defects of both parents or of neither.

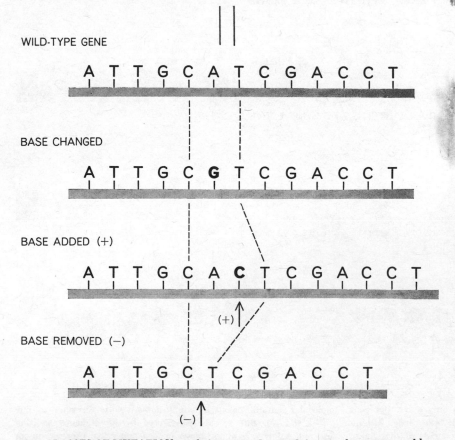

TWO CLASSES OF MUTATION result from introducing defects in the sequence of bases (A, T, G, C) that are attached to the backbone of the DNA molecule. In one class a base is simply changed from one into another, as A into G. In the second class a base is added or removed. Four bases are adenine (A), thymine (T), guanine (G) and cytosine (C).

be produced. But it now looks as if the most obvious solution is the correct one. That is, the message begins at a fixed starting point, probably one end of the gene, and is simply read three bases at a time. Notice that if the reading started at the wrong point, the message would fall into the wrong sets of three and would then be hopelessly incorrect. In fact, it is easy to see that while there is only one correct reading for a triplet code, there are two incorrect ones.

If this idea were right, it would immediately explain why the addition or the deletion of a base in most parts of the gene would make the gene completely nonfunctional, since the reading of the genetic message from that point onward would be totally wrong. Now, although our single mutations were always without function, we found that if we put certain pairs of them together, the gene would work. (In point of fact we picked up many of our functioning double mutations by starting with a nonfunctioning mutation and selecting for the rare second mutation that restored gene activity, but this does not affect our argument.) This enabled us to classify all our mutations as being either plus or minus. We found that by using the following rules we could always predict the behavior of any pair we put together in the same gene. First, if plus is combined with plus, the combination is nonfunctional. Second, if minus is combined with minus, the result is nonfunctional. Third, if plus is combined with minus, the combination is nonfunctional if the pair is too widely separated and functional if the pair is close together.

The interesting case is the last one. We could produce a gene that functioned, at least to some extent, if we combined a plus mutation with a minus mutation, provided that they were not too far apart.

To make it easier to follow, let us assume that the mutations we called plus really had an extra base at some point and that those we called minus had lost a base. (Proving this to be the case is rather difficult.) One can see that, starting from one end, the message would be read correctly until the extra base was reached; then the reading would get out of phase and the message would be wrong until the missing base was reached, after which the message would come back into phase again. Thus the genetic message would not be wrong over a long stretch but only over the short distance between the plus and the minus. By the same sort of argument one can see that for a triplet code the combination plus with plus or minus with

minus should never work [see illustration on following page].

We were fortunate to do most of our work with mutations at the left-hand end of the B gene of the rII region. It appears that the function of this part of the gene may not be too important, so that it may not matter if part of the genetic message in the region is incorrect. Even so, if the plus and minus are too far apart, the combination will not work.

Nonsense Triplets

To understand this we must go back once again to the code. There are 64 possible triplets but only 20 amino acids to be coded. Conceivably two or more triplets may stand for each amino acid. On the other hand, it is reasonable to expect that at least one or two triplets may not represent an amino acid at all but have some other meaning, such as "Begin here" or "End here." Although such hypothetical triplets may have a meaning of some sort, they have been named nonsense triplets. We surmised that sometimes the misreading produced in the region lying between a plus and a minus mutation might by chance give rise to a nonsense triplet, in which case the gene might not work.

We investigated a number of plus-with-minus combinations in which the distance between plus and minus was relatively short and found that certain combinations were indeed inactive when we might have expected them to function. Presumably an intervening nonsense triplet was to blame. We also found cases in which a plus followed by a minus worked but a minus followed by a plus did not, even though the two mutations appeared to be at the same sites, although in reverse sequence. As I have indicated, there are two wrong ways to read a message; one arises if the plus is to the left of the minus, the other if the plus is to the right of the minus. In cases where plus with minus gave rise to an active gene but minus with plus did not, even when the mutations evidently occupied the same pairs of sites, we concluded that the intervening misreading produced a nonsense triplet in one case but not in the other. In confirmation of this hypothesis we have been able to modify such nonsense triplets by mutagens that turn one base into another, and we have thereby restored the gene's activity. At the same time we have been able to locate the position of the nonsense triplet.

Recently we have undertaken one

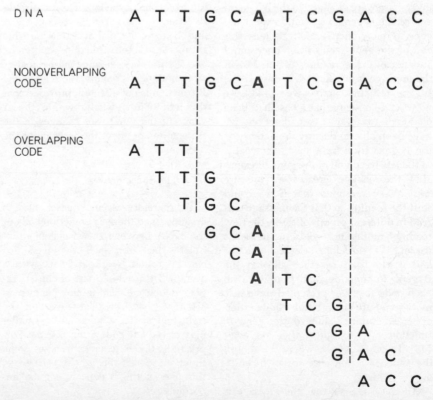

PROPOSED CODING SCHEMES show how the sequence of bases in DNA can be read. In a nonoverlapping code, which is favored by the author, code groups are read in simple sequence. In one type of overlapping code each base appears in three successive groups.

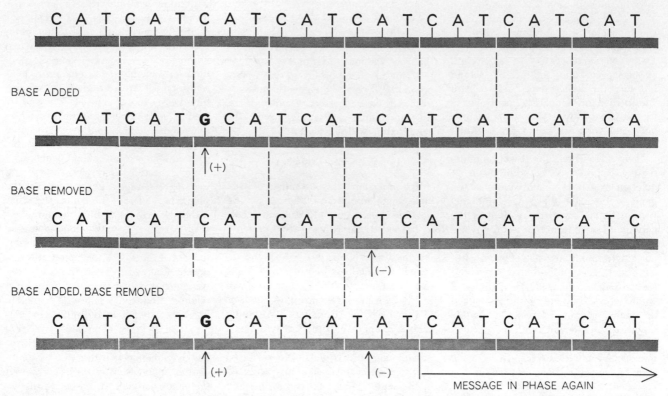

WILD-TYPE GENE

C A T C A T C A T C A T C A T C A T C A T C A T C A T

BASE ADDED

C A T C A T **G** C A T C A T C A T C A T C A T C A

↑ (+)

BASE REMOVED

C A T C A T C A T C A T C T C A T C A T C A T C

↑ (−)

BASE ADDED, BASE REMOVED

C A T C A T **G** C A T C A T A T C A T C A T C A T

↑ (+)   ↑ (−)   MESSAGE IN PHASE AGAIN →

EFFECT OF MUTATIONS that add or remove a base is to shift the reading of the genetic message, assuming that the reading begins at the left-hand end of the gene. The hypothetical message in the wild-type gene is CAT, CAT... Adding a base shifts the reading to TCA, TCA... Removing a base makes it ATC, ATC... Addition and removal of a base puts the message in phase again.

other rather amusing experiment. If a single base were changed in the left-hand end of the B gene, we would expect the gene to remain active, both because this end of the gene seems to be unessential and because the reading of the rest of the message is not shifted. In fact, if the B gene remained active, we would have no way of knowing that a base had been changed. In a few cases, however, we have been able to destroy the activity of the B gene by a base change traceable to the left-hand end of the gene. Presumably the change creates a nonsense triplet. We reasoned that if we could shift the reading so that the message was read in different groups of three, the new reading might not yield a nonsense triplet. We therefore selected a minus and a plus that together allowed the B gene to function, and that were on each side of the presumed nonsense mutation. Sure enough, this combination of three mutants allowed the gene to function [see top illustration on page 148]. In other words, we could abolish the effect of a nonsense triplet by shifting its reading.

All this suggests that the message is read from a fixed point, probably from one end. Here the question arises of how one gene ends and another begins,

since in our picture there is nothing on the backbone of the long DNA molecule to separate them. Yet the two genes A and B are quite distinct. It is possible to measure their function separately, and Benzer has shown that no matter what mutation is put into the A gene, the B function is not affected, provided that the mutation is wholly within the A gene. In the same way changes in the B gene do not affect the function of the A gene.

The Space between the Genes

It therefore seems reasonable to imagine that there is something about the DNA between the two genes that isolates them from each other. This idea can be tested by experiments with a mutant T4 in which part of the rII region is deleted. The mutant, known as T4 1589, has lost a large part of the right end of the A gene and a smaller part of the left end of the B gene. Surprisingly the B gene still shows some function; in fact this is why we believe this part of the B gene is not too important.

Although we describe this mutation as a deletion, since genetic mapping shows that a large piece of the genetic

information in the region is missing, it does not mean that physically there is a gap. It seems more likely that DNA is all one piece but that a stretch of it has been left out. It is only by comparing it with the complete version—the wild type —that one can see a piece of the message is missing.

We have argued that there must be a small region between the genes that separates them. Consequently one would predict that if this segment of the DNA were missing, the two genes would necessarily be joined. It turns out that it is quite easy to test this prediction, since by genetic methods one can construct double mutants. We therefore combined one of our acridine mutations, which in this case was near the beginning of the A gene, with the deletion 1589. Without the deletion present the acridine mutation had no effect on the B function, which showed that the genes were indeed separate. But when 1589 was there as well, the B function was completely destroyed [see top illustration on next page]. When the genes were joined, a change far away in the A gene knocked out the B gene completely. This strongly suggests that the reading proceeds from one end.

We tried other mutations in the A

gene combined with 1589. All the acridine mutations we tried knocked out the B function, whether they were plus or minus, but a pair of them (plus with minus) still allowed the B gene to work. On the other hand, in the case of the other type of mutation (which we believe is due to the change of a base and not to one being added or subtracted) about half of the mutations allowed the B gene to work and the other half did not. We surmise that the latter are nonsense mutations, and in fact Benzer has recently been using this test as a definition of nonsense.

Of course, we do not know exactly what is happening in biochemical terms. What we suspect is that the two genes, instead of producing two separate pieces of messenger RNA, produce a single piece, and that this in turn produces a protein with a long polypeptide chain, one end of which has the amino acid sequence of part of the presumed A protein and the other end of which has most of the B protein sequence—enough to give some B function to the combined molecule although the A function has been lost. The concept is illustrated schematically at the bottom of this page. Eventually it should be possible to check the prediction experimentally.

How the Message Is Read

So far all the evidence has fitted very well into the general idea that the message is read off in groups of three, starting at one end. We should have got the same results, however, if the message had been read off in groups of four, or indeed in groups of any larger size. To test this we put not just two of our acridine mutations into one gene but three of them. In particular we put in three with the same sign, such as plus with plus with plus, and we put them fairly close together. Taken either singly or in pairs, these mutations will destroy the function of the B gene. But when all three are placed in the same gene, the B function reappears. This is clearly a remarkable result: two blacks will not make a white but three will. Moreover, we have obtained the same result with several different combinations of this type and with several of the type minus with minus with minus.

The explanation, in terms of the ideas described here, is obvious. One plus will put the reading out of phase. A second plus will give the other wrong reading. But if the code is a triplet code, a third plus will bring the message back into phase again, and from then on to the end it will be read correctly. Only between

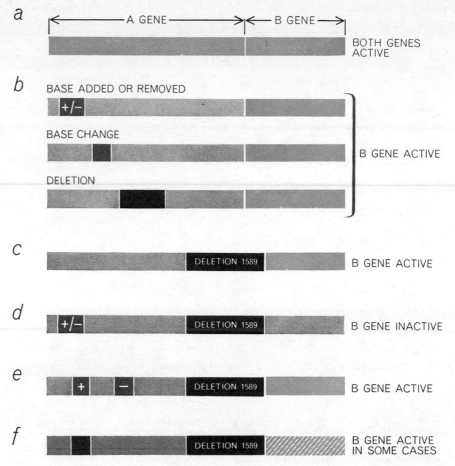

DELETION JOINING TWO GENES makes the B gene vulnerable to mutations in the A gene. The messages in two wild-type genes (a) are read independently, beginning at the left end of each gene. Regardless of the kind of mutation in A, the B gene remains active (b). The deletion known as 1589 inactivates the A gene but leaves the B gene active (c). But now alterations in the A gene will often inactivate the B gene, showing that the two genes have been joined in some way and are read as if they were a single gene (d, e, f).

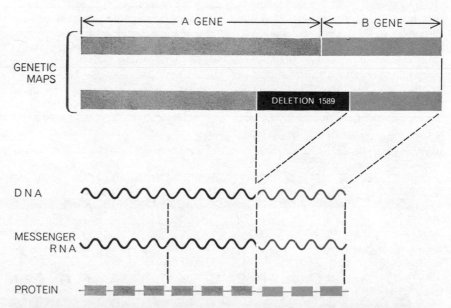

PROBABLE EFFECT OF DELETION 1589 is to produce a mixed protein with little or no A-gene activity but substantial B activity. Although the conventional genetic map shows the deletion as a gap, the DNA molecule itself is presumably continuous but shortened. In virus replication the genetic message in DNA is transcribed into a molecule of ribonucleic acid, called messenger RNA. This molecule carries the message to cellular particles known as ribosomes, where protein is synthesized, following instructions coded in the DNA.

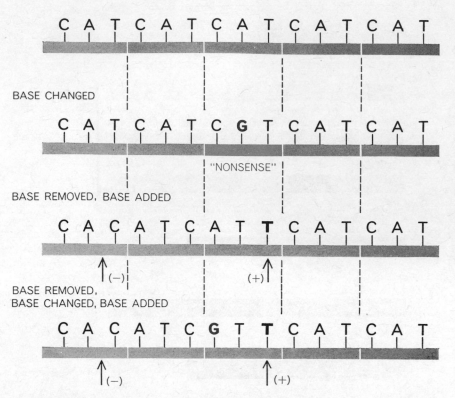

WILD-TYPE GENE

C A T C A T C A T C A T C A T

BASE CHANGED

C A T C A T C G T C A T C A T

"NONSENSE"

BASE REMOVED, BASE ADDED

C A C A T C A T T C A T C A T

(−)   (+)

BASE REMOVED,
BASE CHANGED, BASE ADDED

C A C A T C G T T C A T C A T

(−)   (+)

NONSENSE MUTATION is one creating a code group that evidently does not represent any of the 20 amino acids found in proteins. Thus it makes the gene inactive. In this hypothetical case a nonsense triplet, CGT, results when an A in the wild-type gene is changed to G. The nonsense triplet can be eliminated if the reading is shifted to put the G in a different triplet. This is done by recombining the inactive gene with one containing a minus-with-plus combination. In spite of three mutations, the resulting gene is active.

the pluses will the message be wrong [see illustration below].

Notice that it does not matter if plus is really one extra base and minus is one fewer; the conclusions would be the same if they were the other way around. In fact, even if some of the plus mutations were indeed a single extra base, others might be two fewer bases; in other words, a plus might really be minus minus. Similarly, some of the minus mutations might actually be plus plus. Even so they would still fit into our scheme.

Although the most likely explanation is that the message is read three bases at a time, this is not completely certain. The reading could be in multiples of three. Suppose, for example, that the message is actually read six bases at a time. In that case the only change needed in our interpretation of the facts is to assume that all our mutants have been changed by an even number of

bases. We have some weak experimental evidence that this is unlikely. For instance, we can combine the mutant 1589 (which joins the genes) with medium-sized deletions in the A cistron. Now, if deletions were random in length, we should expect about a third of them to allow the B function to be expressed if the message is indeed read three bases at a time, since those deletions that had lost an exact multiple of three bases should allow the B gene to function. By the same reasoning only a sixth of them should work (when combined with 1589) if the reading proceeds six at a time. Actually we find that the B gene is active in a little more than a third. Taking all the evidence together, however, we find that although three is the most likely coding unit, we cannot completely rule out multiples of three.

There is one other general conclusion we can draw about the genetic code. If we make a rough guess as to the actual size of the B gene (by comparing it with another gene whose size is known approximately), we can estimate how many bases can lie between a plus with minus combination and still allow the B gene to function. Knowing also the frequency with which nonsense triplets are created in the misread region between the plus and minus, we can get some idea whether there are many such triplets or only a few. Our calculation suggests that nonsense triplets are not too common. It seems, in other words, that most of the 64 possible triplets, or codons, are not nonsense, and therefore they stand for amino acids. This implies that probably more than one codon can stand for one amino acid. In the jargon of the trade, a code in which this is true is "degenerate."

In summary, then, we have arrived at three general conclusions about the genetic code:

1. The message is read in nonover-

WILD-TYPE GENE

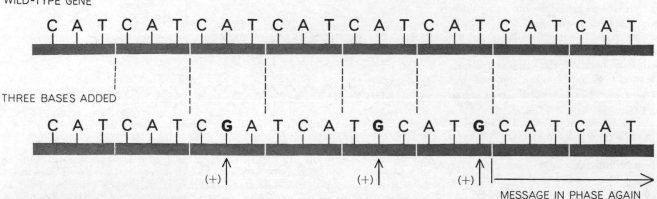

C A T C A T C A T C A T C A T C A T C A T C A T

THREE BASES ADDED

C A T C A T C G A T C A T G C A T G C A T C A T

(+)   (+)   (+)

MESSAGE IN PHASE AGAIN

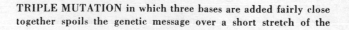

TRIPLE MUTATION in which three bases are added fairly close together spoils the genetic message over a short stretch of the gene but leaves the rest of the message unaffected. The same result can be achieved by the deletion of three neighboring bases.

lapping groups from a fixed point, probably from one end. The starting point determines that the message is read correctly into groups.

2. The message is read in groups of a fixed size that is probably three, although multiples of three are not completely ruled out.

3. There is very little nonsense in the code. Most triplets appear to allow the gene to function and therefore probably represent an amino acid. Thus in general more than one triplet will stand for each amino acid.

It is difficult to see how to get around our first conclusion, provided that the B gene really does code a polypeptide chain, as we have assumed. The second conclusion is also difficult to avoid. The third conclusion, however, is much more indirect and could be wrong.

Finally, we must ask what further evidence would really clinch the theory we have presented here. We are continuing to collect genetic data, but I doubt that this will make the story much more convincing. What we need is to obtain a protein, for example one produced by a double mutation of the form plus with minus, and then examine its amino acid sequence. According to conventional theory, because the gene is altered in only two places the amino acid sequences also should differ only in the two corresponding places. According to our theory it should be altered not only at these two places but also at all places in between. In other words, a whole string of amino acids should be changed. There is one protein, the lysozyme of the T4 phage, that is favorable for such an approach, and we hope that before long workers in the U.S. who have been studying phage lysozyme will confirm our theory in this way.

The same experiment should also be useful for checking the particular code schemes worked out by Nirenberg and Matthaei and by Ochoa and his colleagues. The phage lysozyme made by the wild-type gene should differ over only a short stretch from that made by the plus-with-minus mutant. Over this stretch the amino acid sequence of the two lysozyme variants should correspond to the same sequence of bases on the DNA but should be read in different groups of three.

If this part of the amino acid sequence of both the wild-type and the altered lysozyme could be established, one could check whether or not the codons assigned to the various amino acids did indeed predict similar sequences for that part of the DNA between the base added and the base removed.

The Genetic Code: III

by F. H. C. Crick
October 1966

The central theme of molecular biology is confirmed by detailed knowlege of how the four-letter language embodied in molecules of nucleic acid controls the 20-letter language of the proteins

The hypothesis that the genes of the living cell contain all the information needed for the cell to reproduce itself is now more than 50 years old. Implicit in the hypothesis is the idea that the genes bear in coded form the detailed specifications for the thousands of kinds of protein molecules the cell requires for its moment-to-moment existence: for extracting energy from molecules assimilated as food and for repairing itself as well as for replication. It is only within the past 15 years, however, that insight has been gained into the chemical nature of the genetic material and how its molecular structure can embody coded instructions that can be "read" by the machinery in the cell responsible for synthesizing protein molecules. As the result of intensive work by many investigators the story

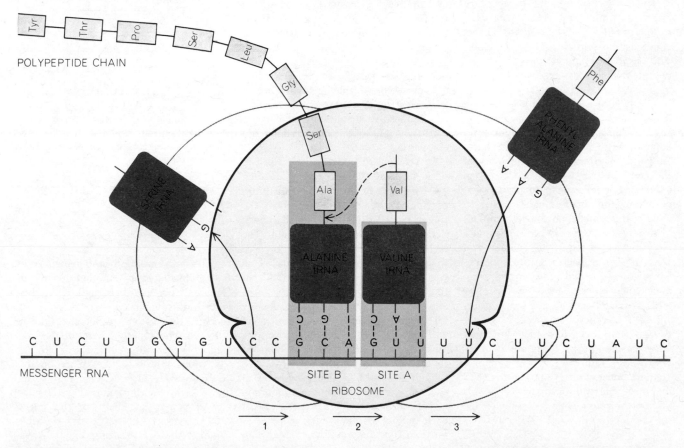

SYNTHESIS OF PROTEIN MOLECULES is accomplished by the intracellular particles called ribosomes. The coded instructions for making the protein molecule are carried to the ribosome by a form of ribonucleic acid (RNA) known as "messenger" RNA. The RNA code "letters" are four bases: uracil (U), cytosine (C), adenine (A) and guanine (G). A sequence of three bases, called a codon, is required to specify each of the 20 kinds of amino acid, identified here by their abbreviations. (A list of the 20 amino acids and their abbreviations appears on the next page.) When linked end to end, these amino acids form the polypeptide chains of which proteins are composed. Each type of amino acid is transported to the ribosome by a particular form of "transfer" RNA (tRNA), which carries an anticodon that can form a temporary bond with one of the codons in messenger RNA. Here the ribosome is shown moving along the chain of messenger RNA, "reading off" the codons in sequence. It appears that the ribosome has two binding sites for molecules of tRNA: one site (*A*) for positioning a newly arrived tRNA molecule and another (*B*) for holding the growing polypeptide chain.

of the genetic code is now essentially complete. One can trace the transmission of the coded message from its original site in the genetic material to the finished protein molecule.

The genetic material of the living cell is the chainlike molecule of deoxyribonucleic acid (DNA). The cells of many bacteria have only a single chain; the cells of mammals have dozens clustered together in chromosomes. The DNA molecules have a very long backbone made up of repeating groups of phosphate and a five-carbon sugar. To this backbone the side groups called bases are attached at regular intervals. There are four standard bases: adenine (A), guanine (G), thymine (T) and cytosine (C). They are the four "letters" used to spell out the genetic message. The exact sequence of bases along a length of the DNA molecule determines the structure of a particular protein molecule.

Proteins are synthesized from a standard set of 20 amino acids, uniform throughout nature, that are joined end to end to form the long polypeptide chains of protein molecules [see illustration at left]. Each protein has its own characteristic sequence of amino acids. The number of amino acids in a polypeptide chain ranges typically from 100 to 300 or more.

The genetic code is not the message itself but the "dictionary" used by the cell to translate from the four-letter language of nucleic acid to the 20-letter language of protein. The machinery of the cell can translate in one direction only: from nucleic acid to protein but not from protein to nucleic acid. In making this translation the cell employs a variety of accessory molecules and mechanisms. The message contained in DNA is first transcribed into the similar molecule called "messenger" ribonucleic acid—messenger RNA. (In many viruses—the tobacco mosaic virus, for example—the genetic material is simply RNA.) RNA too has four kinds of bases as side groups; three are identical with those found in DNA (adenine, guanine and cytosine) but the fourth is uracil (U) instead of thymine. In this first transcription of the genetic message the code letters A, G, T and C in DNA give rise respectively to U, C, A and G. In other words, wherever A appears in DNA, U appears in the RNA transcription; wherever G appears in DNA, C appears in the transcription, and so on. As it is usually presented the dictionary of the genetic code employs the letters found in RNA (U, C, A, G) rather than those found in DNA (A, G, T, C).

The genetic code could be broken easily if one could determine both the amino acid sequence of a protein and the base sequence of the piece of nucleic acid that codes it. A simple comparison of the two sequences would yield the code. Unfortunately the determination of the base sequence of a long nucleic acid molecule is, for a variety of reasons, still extremely difficult. More indirect approaches must be used.

Most of the genetic code first became known early in 1965. Since then additional evidence has proved that almost all of it is correct, although a few features remain uncertain. This article describes how the code was discovered and some of the work that supports it.

Scientific American has already presented a number of articles on the genetic code. In one of them ["The Genetic Code," *beginning on page 142*]. I explained that the experimental evidence (mainly indirect) suggested that the code was a triplet code: that the bases on the messenger RNA were read three at a time and that each group corresponded to a particular amino acid. Such a group is called a codon. Using four symbols in groups of three, one can form 64 distinct triplets. The evidence indicated that most of these stood for one amino acid or another, implying that an amino acid was usually represented by several codons. Adjacent amino acids were coded by adjacent codons, which did not overlap.

In a sequel to that article ["The Genetic Code: II," Offprint 153]. Marshall W. Nirenberg of the National Institutes of Health explained how the composition of many of the 64 triplets had been determined by actual experiment. The technique was to synthesize polypeptide chains in a cell-free system, which was made by breaking open cells of the colon bacillus (*Escherichia coli*) and extracting from them the machinery for protein synthesis. Then the system was provided with an energy supply, 20 amino acids and one or another of several types of synthetic RNA. Although the exact sequence of bases in each type was random, the proportion of bases was known. It was found that each type of synthetic messenger RNA directed the incorporation of certain amino acids only.

By means of this method, used in a quantitative way, the *composition* of many of the codons was obtained, but the *order* of bases in any triplet could not be determined. Codons rich in G were difficult to study, and in addition a few mistakes crept in. Of the 40 codon compositions listed by Nirenberg

AMINO ACID	ABBREVIATION
ALANINE	Ala
ARGININE	Arg
ASPARAGINE	AspN
ASPARTIC ACID	Asp
CYSTEINE	Cys
GLUTAMIC ACID	Glu
GLUTAMINE	GluN
GLYCINE	Gly
HISTIDINE	His
ISOLEUCINE	Ileu
LEUCINE	Leu
LYSINE	Lys
METHIONINE	Met
PHENYLALANINE	Phe
PROLINE	Pro
SERINE	Ser
THREONINE	Thr
TRYPTOPHAN	Tryp
TYROSINE	Tyr
VALINE	Val

TWENTY AMINO ACIDS constitute the standard set found in all proteins. A few other amino acids occur infrequently in proteins but it is suspected in each case that they originate as one of the standard set and become chemically modified after they have been incorporated into a polypeptide chain.

in his article we now know that 35 were correct.

The Triplet Code

The main outlines of the genetic code were elucidated by another technique invented by Nirenberg and Philip Leder. In this method no protein synthesis occurs. Instead one triplet at a time is used to bind together parts of the machinery of protein synthesis.

Protein synthesis takes place on the comparatively large intracellular structures known as ribosomes. These bodies travel along the chain of messenger RNA, reading off its triplets one after another and synthesizing the polypeptide chain of the protein, starting at the amino end (NH_2). The amino acids do not diffuse to the ribosomes by themselves. Each amino acid is joined chemically by a special enzyme to one of the codon-recognizing molecules known both as soluble RNA (sRNA) and transfer RNA (tRNA). (I prefer the latter designation.) Each tRNA molecule has its own triplet of bases, called an anticodon, that recognizes the relevant codon on the messenger RNA by pairing bases with it [*see illustration on page 151*].

Leder and Nirenberg studied which amino acid, joined to its tRNA molecules, was bound to the ribosomes in the presence of a particular triplet, that is, by a "message" with just three letters. They did so by the neat trick of passing the mixture over a nitrocellulose filter that retained the ribosomes. All the tRNA molecules passed through the filter except the ones specifically bound to the ribosomes by the triplet. Which they were could easily be decided by using mixtures of amino acids

in which one kind of amino acid had been made artificially radioactive, and determining the amount of radioactivity absorbed by the filter.

For example, the triplet GUU retained the tRNA for the amino acid valine, whereas the triplets UGU and UUG did not. (Here GUU actually stands for the trinucleoside diphosphate GpUpU.) Further experiments showed that UGU coded for cysteine and UUG for leucine.

Nirenberg and his colleagues synthesized all 64 triplets and tested them for their coding properties. Similar results have been obtained by H. Gobind Khorana and his co-workers at the University of Wisconsin. Various other

groups have checked a smaller number of codon assignments.

Close to 50 of the 64 triplets give a clearly unambiguous answer in the binding test. Of the remainder some evince only weak binding and some bind more than one kind of amino acid. Other results I shall describe later suggest that the multiple binding is often an artifact of the binding method. In short, the binding test gives the meaning of the majority of the triplets but it does not firmly establish all of them.

The genetic code obtained in this way, with a few additions secured by other methods, is shown in the table below. The 64 possible triplets are set out in a regular array, following a plan

SECOND LETTER

	U	C	A	G	
U	UUU ⎫ Phe UUC ⎬ UUA ⎫ Leu UUG ⎬	UCU ⎫ UCC ⎬ Ser UCA ⎪ UCG ⎭	UAU ⎫ Tyr UAC ⎬ UAA OCHRE UAG AMBER	UGU ⎫ Cys UGC ⎬ UGA ? UGG Tryp	U C A G
C	CUU ⎫ CUC ⎬ Leu CUA ⎪ CUG ⎭	CCU ⎫ CCC ⎬ Pro CCA ⎪ CCG ⎭	CAU ⎫ His CAC ⎬ CAA ⎫ GluN CAG ⎬	CGU ⎫ CGC ⎬ Arg CGA ⎪ CGG ⎭	U C A G
A	AUU ⎫ AUC ⎬ Ileu AUA ⎭ AUG Met	ACU ⎫ ACC ⎬ Thr ACA ⎪ ACG ⎭	AAU ⎫ AspN AAC ⎬ AAA ⎫ Lys AAG ⎬	AGU ⎫ Ser AGC ⎬ AGA ⎫ Arg AGG ⎬	U C A G
G	GUU ⎫ GUC ⎬ Val GUA ⎪ GUG ⎭	GCU ⎫ GCC ⎬ Ala GCA ⎪ GCG ⎭	GAU ⎫ Asp GAC ⎬ GAA ⎫ Glu GAG ⎬	GGU ⎫ GGC ⎬ Gly GGA ⎪ GGG ⎭	U C A G

FIRST LETTER (left margin) — THIRD LETTER (right margin)

GENETIC CODE, consisting of 64 triplet combinations and their corresponding amino acids, is shown in its most likely version. The importance of the first two letters in each triplet is readily apparent. Some of the allocations are still not completely certain, particularly for organisms other than the colon bacillus (*Escherichia coli*). "Amber" and "ochre" are terms that referred originally to certain mutant strains of bacteria. They designate two triplets, UAA and UAG, that may act as signals for terminating polypeptide chains.

that clarifies the relations between them.

Inspection of the table will show that the triplets coding for the same amino acid are often rather similar. For example, all four of the triplets starting with the doublet AC code for threonine. This pattern also holds for seven of the other amino acids. In every case the triplets XYU and XYC code for the same amino acid, and in many cases XYA and XYG are the same (methionine and tryptophan may be exceptions). Thus an amino acid is largely selected by the first two bases of the triplet. Given that a triplet codes for, say, valine, we know that the first two bases are GU, whatever the third may be. This pattern is true for all but three of the amino acids. Leucine can start with UU or CU, serine with UC or AG and arginine with CG or AG. In all other cases the amino acid is uniquely related to the first two bases of the triplet. Of course, the converse is often not true. Given that a triplet starts with, say, CA, it may code for either histidine or glutamine.

Synthetic Messenger RNA's

Probably the most direct way to confirm the genetic code is to synthesize a messenger RNA molecule with a strictly defined base sequence and then find the amino acid sequence of the polypeptide produced under its influence. The most extensive work of this nature has been done by Khorana and his colleagues. By a brilliant combination of ordinary chemical synthesis and synthesis catalyzed by enzymes, they have made long RNA molecules with various repeating sequences of bases. As an example, one RNA molecule they have synthesized has the sequence UGUG-UGUGUGUG.... When the biochemical machinery reads this as triplets the message is UGU–GUG–UGU–GUG.... Thus we expect that a polypeptide will be produced with an alternating sequence of two amino acids. In fact, it was found that the product is Cys–Val–Cys–Val.... This evidence alone would not tell us which triplet goes with which amino acid, but given the results of the binding test one has no hesitation in concluding that UGU codes for cysteine and GUG for valine.

In the same way Khorana has made chains with repeating sequences of the type XYZ... and also XXYZ.... The type XYZ... would be expected to give a "homopolypeptide" containing one amino acid corresponding to the triplet XYZ. Because the starting point is not clearly defined, however, the homo-polypeptides corresponding to YZX... and ZXY... will also be produced. Thus poly-AUC makes polyisoleucine, polyserine and polyhistidine. This confirms that AUC codes for isoleucine, UCA for serine and CAU for histidine. A repeating sequence of four bases will yield a single type of polypeptide with a repeating sequence of four amino acids. The general patterns to be expected in each case are set forth in the table on this page. The results to date have amply demonstrated by a direct biochemical method that the code is indeed a triplet code.

Khorana and his colleagues have so far confirmed about 25 triplets by this method, including several that were quite doubtful on the basis of the binding test. They plan to synthesize other sequences, so that eventually most of the triplets will be checked in this way.

The Use of Mutations

The two methods described so far are open to the objection that since they do not involve intact cells there may be some danger of false results. This objection can be met by two other methods of checking the code in which the act of protein synthesis takes place inside the cell. Both involve the effects of genetic mutations on the amino acid sequence of a protein.

It is now known that small mutations are normally of two types: "base substitution" mutants and "phase shift" mutants. In the first type one base is changed into another base but the total number of bases remains the same. In the second, one or a small number of bases are added to the message or subtracted from it.

There are now extensive data on base-substitution mutants, mainly from studies of three rather convenient proteins: human hemoglobin, the protein of tobacco mosaic virus and the A protein of the enzyme tryptophan synthetase obtained from the colon bacillus. At least 36 abnormal types of human hemoglobin have now been investigated by many different workers. More than 40 mutant forms of the protein of the tobacco mosaic virus have been examined by Hans Wittmann of the Max Planck Institute for Molecular Genetics in Tübingen and by Akita Tsugita and Heinz Fraenkel-Conrat of the University of California at Berkeley [see "The Genetic Code of a Virus," by Heinz Fraenkel-Conrat; Scientific American Offprint 193]. Charles Yanofsky and his group at Stanford University have characterized about 25 different mutations of the A protein of tryptophan synthetase.

RNA BASE SEQUENCE	READ AS	AMINO ACID SEQUENCE EXPECTED
(XY)n . . .	X Y X | Y X Y | X Y X | Y X Y	. . . αβαβ
(XYZ)n . . .	. . . X Y Z | X Y Z | X Y Z . . .	ααα
	. . . Y Z X | Y Z X | Y Z X . . .	βββ
	. . . Z X Y | Z X Y | Z X Y . . .	γγγ
(XXYZ)n . . .	X X Y Z | X X Y Z | X X Y Z	. . . αβγδαβγδ
(XYXZ)n . . .	X Y X Z | X Y X Z | X Y X Z	. . . αβγδαβγδ

VARIETY OF SYNTHETIC RNA's with repeating sequences of bases have been produced by H. Gobind Khorana and his colleagues at the University of Wisconsin. They contain two or three different bases (X, Y, Z) in groups of two, three or four. When introduced into cell-free systems containing the machinery for protein synthesis, the base sequences are read off as triplets (middle) and yield the amino acid sequences indicated at the right.

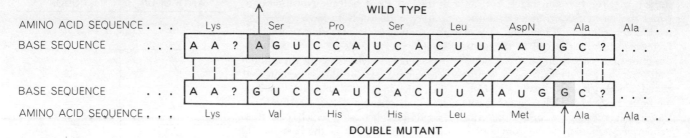

WILD TYPE

| AMINO ACID SEQUENCE... | Lys | Ser | Pro | Ser | Leu | AspN | Ala | Ala... |

BASE SEQUENCE ... A A ? | A G U | C C A | U C A | C U U | A A U | G C | ? ...

BASE SEQUENCE ... A A ? | G U C | C A U | C A C | U U A | U G | G C | ? ...

| AMINO ACID SEQUENCE... | Lys | Val | His | His | Leu | Met | Ala | Ala... |

DOUBLE MUTANT

"PHASE SHIFT" MUTATIONS help to establish the actual codons used by organisms in the synthesis of protein. The two partial amino acid sequences shown here were determined by George Streisinger and his colleagues at the University of Oregon. The sequences are from a protein, a type of lysozyme, produced by the bacterial virus T4. A pair of phase-shift mutations evidently removed one base, A, and inserted another, G, about 15 bases farther on. The base sequence was deduced theoretically from the genetic code.

The remarkable fact has emerged that in every case but one the genetic code shows that the change of an amino acid in a polypeptide chain could have been caused by the alteration of a single base in the relevant nucleic acid. For example, the first observed change of an amino acid by mutation (in the hemoglobin of a person suffering from sickle-cell anemia) was from glutamic acid to valine. From the genetic code dictionary on page 153 we see that this could have resulted from a mutation that changed either GAA to GUA or GAG to GUG. In either case the change involved a single base in the several hundred needed to code for one of the two kinds of chain in hemoglobin.

The one exception so far to the rule that all amino acid changes could be caused by single base changes has been found by Yanofsky. In this one case glutamic acid was replaced by methionine. It can be seen from the genetic code dictionary that this can be accomplished only by a change of *two* bases, since glutamic acid is encoded by either GAA or GAG and methionine is encoded only by AUG. This mutation has occurred only once, however, and of all the mutations studied by Yanofsky it is the only one not to back-mutate, or revert to "wild type." It is thus almost certainly the rare case of a double change. All the other cases fit the hypothesis that base-substitution mutations are normally caused by a single base change. Examination of the code shows that only about 40 percent of all the possible amino acid interchanges can be brought about by single base substitutions, and it is only these changes that are found in experiments. Therefore the study of actual mutations has provided strong confirmation of many features of the genetic code.

Because in general several codons stand for one amino acid it is not possible, knowing the amino acid sequence, to write down the exact RNA base sequence that encoded it. This is unfortunate. If we know which amino acid is changed into another by mutation, however, we can often, given the code, work out what that base change must have been. As an example, glutamic acid can be encoded by GAA or GAG and valine by GUU, GUC, GUA or GUG. If a mutation substitutes valine for glutamic acid, one can assume that only a single base change was involved. The only such change that could lead to the desired result would be a change from A to U in the middle position, and this would be true whether GAA became GUA or GAG became GUG.

It is thus possible in many cases (not in all) to compare the nature of the base change with the chemical mutagen used to produce the change. If RNA is treated with nitrous acid, C is changed to U and A is effectively changed to G. On the other hand, if double-strand DNA is treated under the right conditions with hydroxylamine, the mutagen acts only on C. As a result some C's are changed to T's (the DNA equivalent of U's), and thus G's, which are normally paired with C's in double-strand DNA, are replaced by A's.

If 2-aminopurine, a "base analogue" mutagen, is added when double-strand DNA is undergoing replication, it produces only "transitions." These are the same changes as those produced by hydroxylamine—plus the reverse changes. In almost all these different cases (the exceptions are unimportant) the changes observed are those expected from our knowledge of the genetic code.

Note the remarkable fact that, although the code was deduced mainly from studies of the colon bacillus, it appears to apply equally to human beings and tobacco plants. This, together with more fragmentary evidence, suggests that the genetic code is either the same or very similar in most organisms.

The second method of checking the code using intact cells depends on phase-shift mutations such as the addition of a single base to the message. Phase-shift mutations probably result from errors produced during genetic recombination or when the DNA molecule is being duplicated. Such errors have the effect of putting out of phase the reading of the message from that point on. This hypothesis leads to the prediction that the phase can be corrected if at some subsequent point a nucleotide is deleted. The pair of alterations would be expected not only to change two amino acids but also to alter all those encoded by bases lying between the two affected sites. The reason is that the intervening bases would be read out of phase and therefore grouped into triplets different from those contained in the normal message.

This expectation has recently been confirmed by George Streisinger and his colleagues at the University of Oregon. They have studied mutations in the protein lysozyme that were produced by the T4 virus, which infects the colon bacillus. One phase-shift mutation involved the amino acid sequence ...Lys—Ser—Pro—Ser—Leu—AspN—Ala—Ala—Lys.... They were then able to construct by genetic methods a double phase-shift mutant in which the corresponding sequence was ...Lys—Val—His—His—Leu—Met—Ala—Ala—Lys....

Given these two sequences, the reader should be able, using the genetic code dictionary on page 153, to decipher uniquely a short length of the nucleic acid message for both the original protein and the double mutant and thus deduce the changes produced by each of the phase-shift mutations. The correct result is presented in the illustration above. The result not only confirms several rather doubtful codons, such as UUA for leucine and AGU for serine, but also shows which codons are actually involved in a genetic message. Since the technique is difficult, however, it may not find wide application.

Streisinger's work also demonstrates what has so far been only tacitly assumed: that the two languages, both of which are written down in a certain direction according to convention, are in fact translated by the cell in the same direction and not in opposite directions. This fact had previously been established, with more direct chemical methods, by Severo Ochoa and his colleagues at the New York University School of Medicine. In the convention, which was adopted by chance, proteins are written with the amino (NH_2) end on the left. Nucleic acids are written with the end of the molecule containing a "5 prime" carbon atom at the left. (The "5 prime" refers to a particular carbon atom in the 5-carbon ring of ribose sugar or deoxyribose sugar.)

Finding the Anticodons

Still another method of checking the genetic code is to discover the three bases making up the anticodon in some particular variety of transfer RNA. The first tRNA to have its entire sequence worked out was alanine tRNA, a job done by Robert W. Holley and his collaborators at Cornell University [see "The Nucleotide Sequence of a Nucleic Acid," by Robert W. Holley; SCIENTIFIC AMERICAN Offprint 1033]. Alanine tRNA, obtained from yeast, contains 77 bases. A possible anticodon found near the middle of the molecule has the sequence IGC, where I stands for inosine, a base closely resembling guanine. Since then Hans Zachau and his colleagues at the University of Cologne have established the sequences of two closely related serine tRNA's from yeast, and James Madison and his group

at the U.S. Plant, Soil and Nutrition Laboratory at Ithaca, N.Y., have worked out the sequence of a tyrosine tRNA, also from yeast.

A detailed comparison of these three sequences makes it almost certain that the anticodons are alanine–IGC, serine–IGA and tyrosine–GΨA. (Ψ stands for pseudo-uridylic acid, which can form the same base pairs as the base uracil.) In addition there is preliminary evidence from other workers that an anticodon for valine is IAC and an anticodon for phenylalanine is GAA.

All these results would fit the rule that the codon and anticodon pair in an antiparallel manner, and that the pairing in the first two positions of the codon is of the standard type, that is, A pairs with U and G pairs with C. The pairing in the third position of the codon is more complicated. There is now good experimental evidence from both Nirenberg and Khorana and their co-workers that one tRNA can recognize several codons, provided that they differ only in the last place in the codon. Thus Holley's alanine tRNA appears to recognize GCU, GCC and GCA. If it recognizes GCG, it does so only very weakly.

The "Wobble" Hypothesis

I have suggested that this is because of a "wobble" in the pairing in the third place and have shown that a reasonable theoretical model will explain many of the observed results. The suggested rules for the pairing in the third position of the anticodon are presented in the table at the top of this page, but

ANTICODON	CODON
U	A G
C	G
A	U
G	U C
I	U C A

"WOBBLE" HYPOTHESIS has been proposed by the author to provide rules for the pairing of codon and anticodon at the *third* position of the codon. There is evidence, for example, that the anticodon base I, which stands for inosine, may pair with as many as three different bases: U, C and A. Inosine closely resembles the base guanine (G) and so would ordinarily be expected to pair with cytosine (C). Structural diagrams for standard base pairings and wobble base pairings are illustrated at the bottom of this page.

this theory is still speculative. The rules for the first two places of the codon seem reasonably secure, however, and can be used as partial confirmation of the genetic code. The likely codon-anticodon pairings for valine, serine, tyrosine, alanine and phenylalanine satisfy the standard base pairings in the first two places and the wobble hypothesis in the third place [see illustration on page 157].

Several points about the genetic code remain to be cleared up. For example, the triplet UGA has still to be allocated.

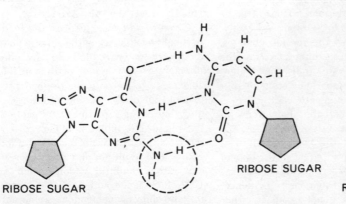
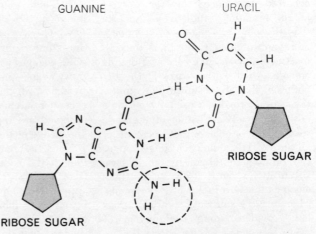

STANDARD AND WOBBLE BASE PAIRINGS both involve the formation of hydrogen bonds when certain bases are brought into close proximity. In the standard guanine-cytosine pairing (*left*) it is believed three hydrogen bonds are formed. The bases are shown as they exist in the RNA molecule, where they are attached to 5-carbon rings of ribose sugar. In the proposed wobble pairing (*right*) guanine is linked to uracil by only two hydrogen bonds. The base inosine (I) has a single hydrogen atom where guanine has an amino (NH_2) group (*broken circle*). In the author's wobble hypothesis inosine can pair with U as well as with C and A (*not shown*).

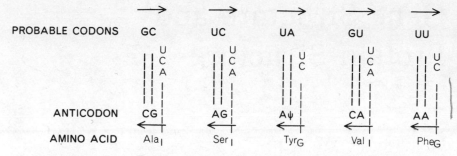

CODON-ANTICODON PAIRINGS take place in an antiparallel direction. Thus the anti-codons are shown here written backward, as opposed to the way they appear in the text. The five anticodons are those tentatively identified in the transfer RNA's for alanine, serine, tyrosine, valine and phenylalanine. Color indicates where wobble pairings may occur.

The punctuation marks—the signals for "begin chain" and "end chain"—are only partly understood. It seems likely that both the triplet UAA (called "ochre") and UAG (called "amber") can terminate the polypeptide chain, but which triplet is normally found at the end of a gene is still uncertain.

The picturesque terms for these two triplets originated when it was discovered in studies of the colon bacillus some years ago that mutations in other genes (mutations that in fact cause errors in chain termination) could "suppress" the action of certain mutant codons, now identified as either UAA or UAG. The terms "ochre" and "amber" are simply invented designations and have no reference to color.

A mechanism for chain initiation was discovered fairly recently. In the colon bacillus it seems certain that formyl-methionine, carried by a special tRNA, can initiate chains, although it is not clear if all chains have to start in this way, or what the mechanism is in mammals and other species. The formyl group (CHO) is not normally found on finished proteins, suggesting that it is probably removed by a special enzyme. It seems likely that sometimes the methionine is removed as well.

It is unfortunately possible that a few codons may be ambiguous, that is, may code for more than one amino acid. This is certainly not true of most codons. The present evidence for a small amount of ambiguity is suggestive but not conclusive. It will make the code more difficult to establish correctly if ambiguity can occur.

Problems for the Future

From what has been said it is clear that, although the entire genetic code is not known with complete certainty, it is highly likely that most of it is correct. Further work will surely clear up the doubtful codons, clarify the punctuation marks, delimit ambiguity and extend the code to many other species. Although the code lists the codons that *may* be used, we still have to determine if alternative codons are used equally. Some preliminary work suggests they may not be. There is also still much to be discovered about the machinery of protein synthesis. How many types of tRNA are there? What is the structure of the ribosome? How does it work, and why is it in two parts? In addition there are many questions concerning the control of the rate of protein synthesis that we are still a long way from answering.

When such questions have been answered, the major unsolved problem will be the structure of the genetic code. Is the present code merely the result of a series of evolutionary accidents, so that the allocations of triplets to amino acids is to some extent arbitrary? Or are there profound structural reasons why phenylalanine has to be coded by UUU and UUC and by no other triplets? Such questions will be difficult to decide, since the genetic code originated at least three billion years ago, and it may be impossible to reconstruct the sequence of events that took place at such a remote period. The origin of the code is very close to the origin of life. Unless we are lucky it is likely that much of the evidence we should like to have has long since disappeared.

Nevertheless, the genetic code is a major milestone on the long road of molecular biology. In showing in detail how the four-letter language of nucleic acid controls the 20-letter language of protein it confirms the central theme of molecular biology that genetic information can be stored as a one-dimensional message on nucleic acid and be expressed as the one-dimensional amino acid sequence of a protein. Many problems remain, but this knowledge is now secure.

16

Gene Structure and Protein Structure

by Charles Yanofsky
May 1967

A linear correspondence between these two chainlike molecules was postulated more than a dozen years ago. Here is how the correspondence was finally demonstrated

The present molecular theory of genetics, known irreverently as "the central dogma," is now 14 years old. Implicit in the theory from the outset was the notion that genetic information is coded in linear sequence in molecules of deoxyribonucleic acid (DNA) and that the sequence directly determines the linear sequence of amino acid units in molecules of protein. In other words, one expected the two molecules to be colinear. The problem was to prove that they were.

Over the same 14 years, as a consequence of an international effort, most of the predictions of the central dogma have been verified one by one. The results were recently summarized in these pages by F. H. C. Crick, who together with James D. Watson proposed the helical, two-strand structure for DNA on

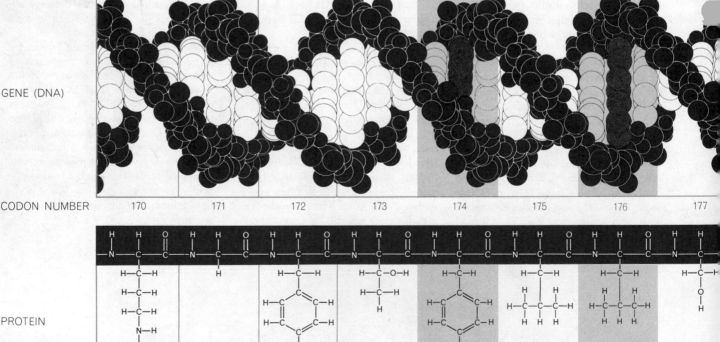

GENE (DNA)

CODON NUMBER   170   171   172   173   174   175   176   177

PROTEIN

AMINO ACID   ARG   GLY   TYR   THR   TYR→CYS   LEU   LEU→ARG   SER
              170   171   172   173   174   175   176   177

STRUCTURES OF GENE AND PROTEIN have been shown to bear a direct linear correspondence by the author and his colleagues at Stanford University. They demonstrated that a particular sequence of coding units (codons) in the genetic molecule deoxyribonucleic acid, or DNA (*top*), specifies a corresponding sequence of amino acid units in the structure of a protein molecule (*bottom*). In the DNA molecule depicted here the black spheres represent repeating units of deoxyribose sugar and phosphate, which form the helical backbone of the two-strand molecule. The white spheres connecting the two strands represent complementary pairs of the four kinds of base that provide the "letters" in which the genetic message is written. A sequence of three bases attached to one strand of DNA

which the central dogma is based [see "The Genetic Code: III," by F. H. C. Crick, *beginning on page 151*]. Here I shall describe in somewhat more detail how our studies at Stanford University demonstrated the colinearity of genetic structure (as embodied in DNA) and protein structure.

Let me begin with a brief review. The molecular subunits that provide the "letters" of the code alphabet in DNA are the four nitrogenous bases adenine (A), guanine (G), cytosine (C) and thymine (T). If the four letters were taken in pairs, they would provide only 16 different code words—too few to specify the 20 different amino acids commonly found in protein molecules. If they are taken in triplets, however, the four letters can provide 64 different code words, which would seem too many for the efficient specification of the 20 amino acids. Accordingly it was conceivable that the cell might employ fewer than the 64 possible triplets. We now know that na-

ture not only has selected the triplet code but also makes use of most (if not all) of the 64 triplets, which are called codons. Each amino acid but two (tryptophan and methionine) are specified by at least two different codons, and a few amino acids are specified by as many as six codons. It is becoming clear that the living cell exploits this redundancy in subtle ways. Of the 64 codons, 61 have been shown to specify one or another of the 20 amino acids. The remaining three can act as "chain terminators," which signal the end of a genetic message.

A genetic message is defined as the amount of information in one gene; it is the information needed to specify the complete amino acid sequence in one polypeptide chain. This relation, which underlies the central dogma, is sometimes expressed as the one-gene-one-enzyme hypothesis. It was first clearly enunciated by George W. Beadle and Edward L. Tatum, as a result of their studies with the red bread mold *Neurospora crassa* around 1940. In some cases

a single polypeptide chain constitutes a complete protein molecule, which often acts as an enzyme, or biological catalyst. Frequently, however, two or more polypeptide chains must join together in order to form an active protein. For example, tryptophan synthetase, the enzyme we used in our colinearity studies, consists of four polypeptide chains: two alpha chains and two beta chains.

How might one establish the colinearity of codons in DNA and amino acid units in a polypeptide chain? The most direct approach would be to separate the two strands of DNA obtained from some organism and determine the base sequence of that portion of a strand which is presumed to be colinear with the amino acid sequence of a particular protein. If the amino acid sequence of the protein were not already known, it too would have to be established. One could then write the two sequences in adjacent columns and see if the same codon (or its synonym) always appeared adjacent to a particular amino acid. If it

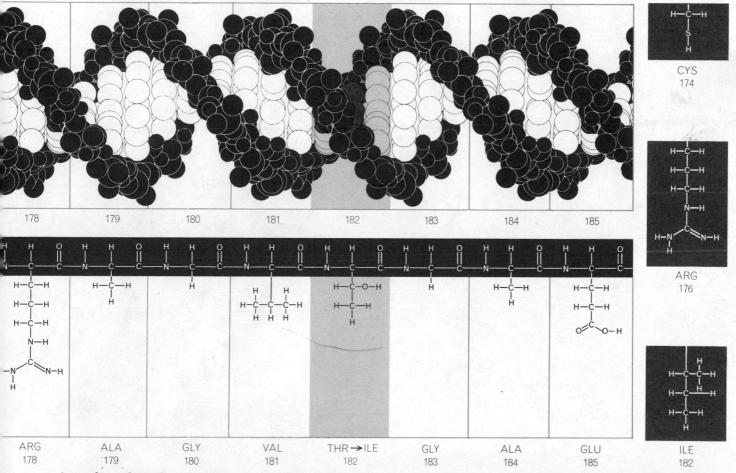

is a codon and specifies one amino acid. The amino acid sequence illustrated here is the region from position 170 through 185 in the *A* protein of the enzyme tryptophan synthetase produced by the bacterium *Escherichia coli*. It was found that mutations in the *A* gene of *E. coli* altered the amino acids at three places (174, 176 and 182) in this region of the *A* protein. (A key to the amino acid abbreviation can be found on page 161). The three amino acids that replace the three normal ones as a result of mutation are shown at the extreme right. Each replacement is produced by a mutation at one site (*dark color*) in the DNA of the *A* gene. In all, the author and his associates correlated mutations at eight sites in the *A* gene with alterations in the *A* protein.

did, a colinear relation would be established. Unfortunately this direct approach cannot be taken because so far it has not been possible to isolate and identify individual genes. Even if one could isolate a single gene that specified a polypeptide made up of 150 amino acids (and not many polypeptides are that small), one would have to determine the sequence of units in a DNA strand consisting of some 450 bases.

It was necessary, therefore, to consider a more feasible way of attacking the problem. An approach that immediately suggests itself to a geneticist is to construct a genetic map, which is a representation of the information contained in the gene, and see if the map can be related to protein structure. A genetic map is constructed solely on the basis of information obtained by crossing individual organisms that differ in two or more hereditary respects (a refinement of the technique originally

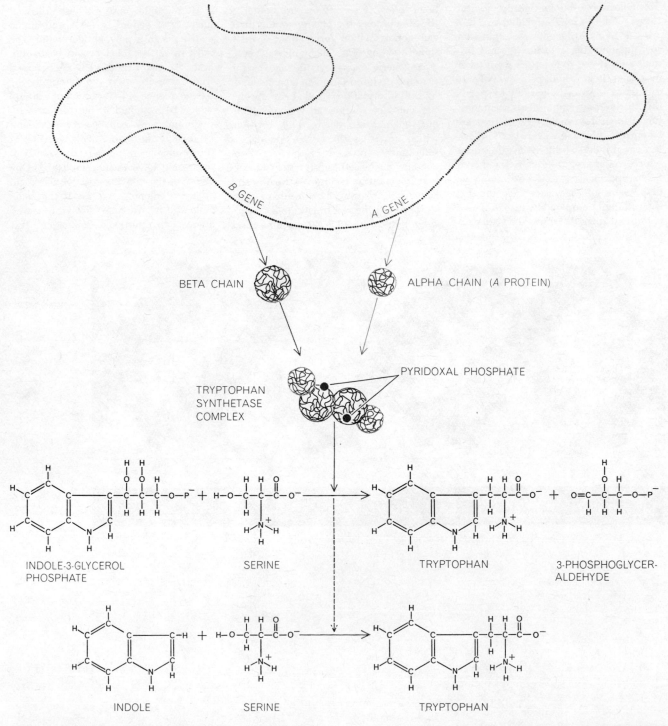

GENETIC CONTROL OF CELL'S CHEMISTRY is exemplified by the two genes in *E. coli* that carry the instructions for making the enzyme tryptophan synthetase. The enzyme is actually a complex of four polypeptide chains: two alpha chains and two beta chains. The alpha chain is the *A* protein in which changes produced by mutations in the *A* gene have provided the evidence for gene-pro-

tein colinearity. One class of *A*-protein mutants retains the ability to associate with beta chains but the complex is no longer able to catalyze the normal biochemical reaction: the conversion of indole-3-glycerol phosphate and serine to tryptophan and 3-phosphoglyceraldehyde. But the complex can still catalyze a simpler nonphysiological reaction: the conversion of indole and serine to tryptophan.

used by Gregor Mendel to demonstrate how characteristics are inherited).

By using bacteria and bacterial viruses in such studies one can catalogue the results of crosses involving millions of individual organisms and thereby deduce the actual distances separating the sites of mutational changes in a single gene. The distances are inferred from the frequency with which parent organisms, each with at least one mutation in the same gene, give rise to offspring in which neither mutation is present. As a result of the recombination of genetic material the offspring can inherit a gene that is assembled from the mutation-free portions of each parental gene. If the mutational markers lie far apart on the parental genes, recombination will frequently produce mutation-free progeny. If the markers are close together, mutation-free progeny will be rare [see bottom illustration on next page].

In his elegant studies with the "rII" region of the chromosome of the bacterial virus designated T4, Seymour Benzer, then at Purdue University, showed that the number of genetically distinguishable mutation sites on the map of the gene approaches the estimated number of base pairs in the DNA molecule corresponding to that gene. (Mutations involve pairs of bases because the bases in each of the two entwined strands of the DNA molecule are paired with and are complementary to the bases in the other strand. If a mutation alters one base in the DNA molecule, its partner is eventually changed too during DNA replication.) Benzer also showed that the only type of genetic map consistent with his data is a map on which the sites altered by mutation are arranged linearly. Subsequently A. D. Kaiser and David Hogness of Stanford University demonstrated with another bacterial virus that there is a linear correspondence between the sites on a genetic map and the altered regions of a DNA molecule isolated from the virus. Thus there is direct experimental evidence indicating that the genetic map is a valid representation of DNA structure and that the map can be employed as a substitute for information about base sequence.

This, then, provided the basis of our approach. We would pick a suitable organism and isolate a large number of mutant individuals with mutations in the same gene. From recombination studies we would make a fine-structure genetic map relating the sites of the mutations. In addition we would have to be able to isolate the protein specified by that gene and determine its amino acid sequence. Finally we would have to analyze the protein produced by each mutant (assuming a protein were still produced) in order to find the position of the amino acid change brought about in its amino acid sequence by the mutation. If gene structure and protein structure were colinear, the positions at which amino acid changes occur in the protein should be in the same order as the positions of the corresponding mutationally altered sites on the genetic map. Although this approach to the question of colinearity would require a great deal of work and much luck, it was logical and experimentally feasible. Several research groups besides our own set out to find a suitable system for a study of this kind.

The essential requirement of a suitable system was that a genetically

ALA	ALANINE	GLY	GLYCINE	PRO	PROLINE
ARG	ARGININE	HIS	HISTIDINE	SER	SERINE
ASN	ASPARAGINE	ILE	ISOLEUCINE	THR	THREONINE
ASP	ASPARTIC ACID	LEU	LEUCINE	TRP	TRYPTOPHAN
CYS	CYSTEINE	LYS	LYSINE	TYR	TYROSINE
GLN	GLUTAMINE	MET	METHIONINE	VAL	VALINE
GLU	GLUTAMIC ACID	PHE	PHENYLALANINE		

AMINO ACID ABBREVIATIONS identify the 20 amino acids commonly found in all proteins. Each amino acid is specified by a triplet codon in the DNA molecule (see below).

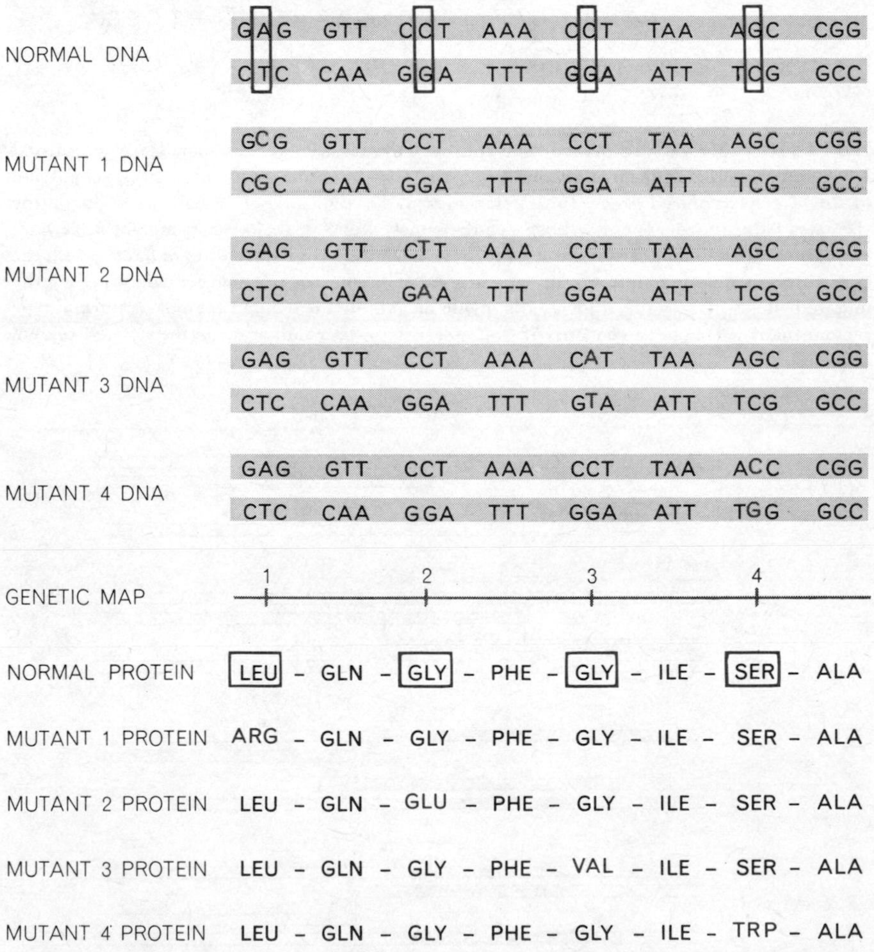

GENETIC MUTATIONS can result from the alteration of a single base in a DNA codon. The letters stand for the four bases: adenine (A), thymine (T), guanine (G) and cytosine (C). Since the DNA molecule consists of two complementary strands, a base change in one strand involves a complementary change in the second strand. In the four mutant DNA sequences shown here (top) a pair of bases (color) is different from that in the normal sequence. By genetic studies one can map the sequence and approximate spacing of the four mutations (middle). By chemical studies of the proteins produced by the normal and mutant DNA sequences (bottom) one can establish the corresponding amino acid changes.

a
NORMAL DNA
```
···GAAGTCG[T]GCAGCGTATAGCTGAGCCTGT···
···CTTCAGCA[A]CGTCGCATATCGACTCGGACA···
```

MUTANT *A* DNA
```
···GAAGTCG[C]GCAGCGTATAGCTGAGCCTGT···
···CTTCAGC[G]GCGTCGCATATCGACTCGGACA···
```

DELETION MUTANT 1 DNA
```
···GAAGTCG[T]GCAGCGCCTGT···
···CTTCAGCA[A]CGTCGCGGACA···
```

DELETION MUTANT 2 DNA
```
···GAAGTGT···
···CTTCACA···
```

b
MUTANT *A*
```
···GAAGTCG[C]GCAGCGTATAGCTGAGCCTGT···
···CTTCAGC[G]GCGTCGCATATCGACTCGGACA···
```

NORMAL RECOMBINANT
```
···GAAGTCG[T]GCAGCGTATAGCTGAGCCTGT···
···CTTCAGCA[A]CGTCGCATATCGACTCGGACA···
```

DELETION MUTANT 1
```
···GAAGTCG[T]GCAGCGCCTGT···
···CTTCAGCA[A]CGTCGCGGACA···
```

c
MUTANT *A*
```
···GAAGTCG[C]GCAGCGTATAGCTGAGCCTGT···
···CTTCAGC[G]GCGTCGCATATCGACTCGGACA···
```

NO NORMAL RECOMBINANTS
```
···GAAGTCG[C]GCAGCGTATAGCTGAGCCTGT···
···CTTCAGC[G]GCGTCGCATATCGACTCGGACA···
```

DELETION MUTANT 2
```
···GAAGTGT···
···CTTCACA···
```

"DELETION" MUTANTS provide one approach to making a genetic map. Here (*a*) normal DNA and mutant *A* differ by only one base pair (*C–G* has replaced *T–A*) in a certain portion of the *A* gene (*colored area*). In deletion mutant 1 a sequence of 10 base pairs, including six pairs from the *A* gene, has been spontaneously deleted. In deletion mutant 2, 22 base pairs, including 15 pairs from the *A* gene, have been deleted. By crossing mutant *A* with the two different deletion mutants in separate experiments (*b, c*), one can tell whether the mutated site (*C–G*) in the *A* gene falls inside or outside the deleted regions. A normal-type recombinant will appear (*b*) only if the altered base pair falls outside the deleted region.

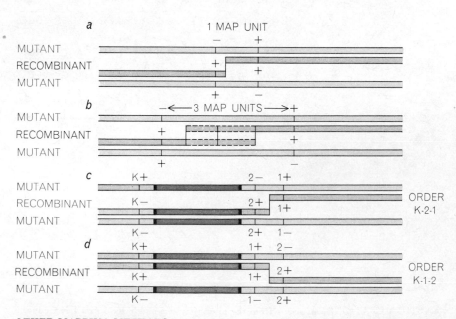

a   1 MAP UNIT

b   ←—— 3 MAP UNITS ——→

ORDER K-2-1

ORDER K-1-2

OTHER MAPPING METHODS involve determination of recombination frequency (*a, b*) and the distribution of outside markers (*c, d*). The site of a mutational alteration is indicated by "−," the corresponding unaltered site by "+." If the altered sites are widely spaced (*b*), normal recombinants will appear more often than if the altered sites are close together (*a*). In the second method the mutants are linked to another gene that is either normal (*K*+) or mutated (*K*−). Recombinant strains that contain 1+ and 2+ will carry the *K*− gene if the correct order is *K*-2-1. They will carry the *K*+ gene if the order is *K*-1-2.

mappable gene should specify a protein whose amino acid sequence could be determined. Since no such system was known we had to gamble on a choice of our own. Fortunately we were studying at the time how the bacterium *Escherichia coli* synthesizes the amino acid tryptophan. Irving Crawford and I observed that the enzyme that catalyzed the last step in tryptophan synthesis could be readily separated into two different protein species, or subunits, one of which could be clearly isolated from the thousands of other proteins synthesized by *E. coli*. This protein, called the tryptophan synthetase *A* protein, had a molecular weight indicating that it had slightly fewer than 300 amino acid units. Furthermore, we already knew how to force *E. coli* to produce comparatively large amounts of the protein—up to 2 percent of the total cell protein—and we also had a collection of mutants in which the activity of the tryptophan synthetase *A* protein was lacking. Finally, the bacterial strain we were using was one for which genetic procedures for preparing fine-structure maps had already been developed. Thus we could hope to map the *A* gene that presumably controlled the structure of the *A* protein.

To accomplish the mapping we needed a set of bacterial mutants with mutational alterations at many different sites on the *A* gene. If we could determine the amino acid change in the *A* protein of each of these mutants, and discover its position in the linear sequence of amino acids in the protein, we could test the concept of colinearity. Here again we were fortunate in the nature of the complex of subunits represented by tryptophan synthetase.

The normal complex consists of two *A*-protein subunits (the alpha chains) and one subunit consisting of two beta chains. Within the bacterial cell the complex acts as an enzyme to catalyze the reaction of indole-3-glycerol phosphate and serine to produce tryptophan and 3-phosphoglyceraldehyde [*see illustration, page 160*]. If the *A* protein undergoes certain kinds of mutations, it is still able to form a complex with the beta chains, but the complex loses the ability to catalyze the reaction. It retains the ability, however, to catalyze a simpler reaction when it is tested outside the cell: it will convert indole and serine to tryptophan. There are still other kinds of *A*-gene mutants that evidently lack the ability to form an *A* protein that can combine with beta chains; thus these strains are not able to catalyze even the simpler reaction. The first class of mutants—those that produce an *A* protein that is still able to combine with beta chains and exhibit catalytic activity when they are tested outside the cell—proved to be the most important for our study.

A fine-structure map of the *A* gene was constructed on the basis of genetic crosses performed by the process called transduction. This employs a particular bacterial virus known as transducing phage *P1kc*. When this virus multiplies in a bacterium, it occasionally incorporates a segment of the bacterial DNA within its own coat of protein. When the virus progeny infect other bacteria, genetic material of the donor bacteria is introduced into some of the recipient cells. A fraction of the recip-

ients survive the infection. In these survivors segments of the bacterium's own genetic material pair with like segments of the "foreign" genetic material and recombination between the two takes place. As a result the offspring of an infected bacterium can contain characteristics inherited from its remote parent as well as from its immediate one.

In order to establish the order of mutationally altered sites in the *A* gene we have relied partly on a set of mutant bacteria in which one end of a deleted segment of DNA lies within the *A* gene. In each of these "deletion" mutants a segment of the genetic material of the bacterium was deleted spontaneously.

Thus each deletion mutant in the set retains a different segment of the *A* gene. This set of mutants can now be crossed with any other mutant in which the *A* gene is altered at only a single site. Recombination can give rise to a normal gene only if the altered site does not fall within the region of the *A* gene that is missing in the deletion mutant [*see top illustration on opposite page*]. By crossing many *A*-protein mutants with the set of deletion mutants one can establish the linear order of many of the mutated sites in the *A* gene. The ordering is limited only by the number of deletion mutants at one's disposal.

A second method, which more closely

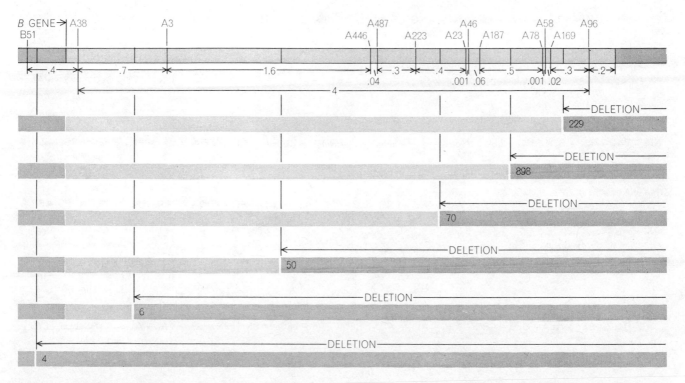

MAP OF *A* GENE shows the location of mutationally altered sites, drawn to scale, as determined by the three genetic-mapping methods illustrated on the opposite page. The total length of the *A* gene is slightly over four map units (probably 4.2). Below map are six deletion mutants that made it possible to assign each of the 12 *A*-gene mutants to one of six regions within the gene. The more sensitive mapping methods were employed to establish the order of mutations and the distance between mutation sites within each region.

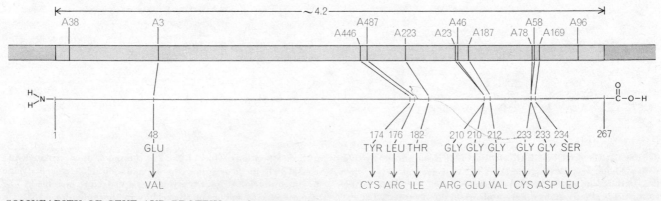

COLINEARITY OF GENE AND PROTEIN can be inferred by comparing the *A*-gene map (*top*) with the various amino acid changes in the *A* protein (*bottom*), both drawn to scale. The amino acid changes associated with 10 of the 12 mutations are also shown.

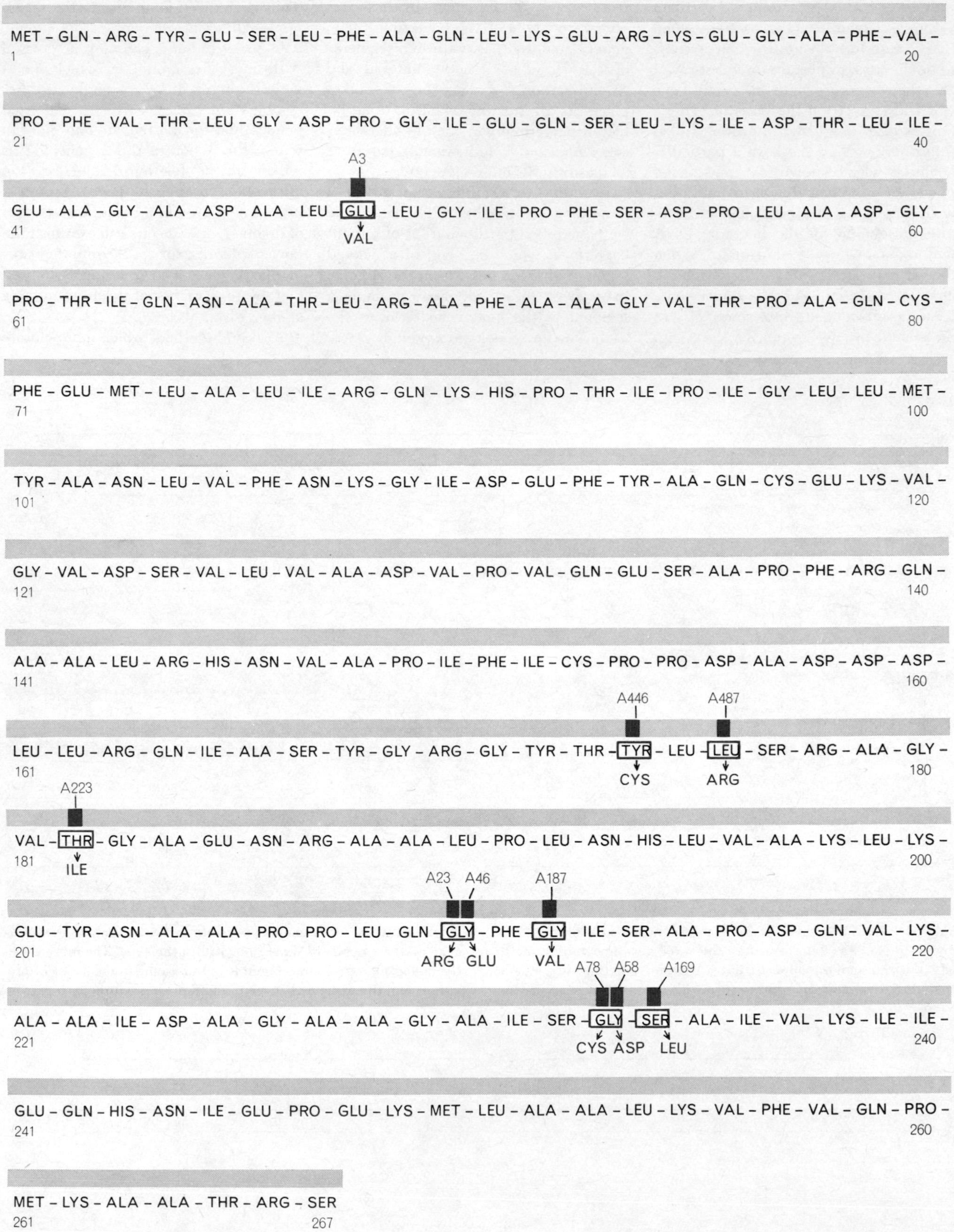

MET – GLN – ARG – TYR – GLU – SER – LEU – PHE – ALA – GLN – LEU – LYS – GLU – ARG – LYS – GLU – GLY – ALA – PHE – VAL –
1   20

PRO – PHE – VAL – THR – LEU – GLY – ASP – PRO – GLY – ILE – GLU – GLN – SER – LEU – LYS – ILE – ASP – THR – LEU – ILE –
21   40

A3

GLU – ALA – GLY – ALA – ASP – ALA – LEU – [GLU] – LEU – GLY – ILE – PRO – PHE – SER – ASP – PRO – LEU – ALA – ASP – GLY –
41   ↓   60
                                VAL

PRO – THR – ILE – GLN – ASN – ALA – THR – LEU – ARG – ALA – PHE – ALA – ALA – GLY – VAL – THR – PRO – ALA – GLN – CYS –
61   80

PHE – GLU – MET – LEU – ALA – LEU – ILE – ARG – GLN – LYS – HIS – PRO – THR – ILE – PRO – ILE – GLY – LEU – LEU – MET –
71   100

TYR – ALA – ASN – LEU – VAL – PHE – ASN – LYS – GLY – ILE – ASP – GLU – PHE – TYR – ALA – GLN – CYS – GLU – LYS – VAL –
101   120

GLY – VAL – ASP – SER – VAL – LEU – VAL – ALA – ASP – VAL – PRO – VAL – GLN – GLU – SER – ALA – PRO – PHE – ARG – GLN –
121   140

ALA – ALA – LEU – ARG – HIS – ASN – VAL – ALA – PRO – ILE – PHE – ILE – CYS – PRO – PRO – ASP – ALA – ASP – ASP – ASP –
141   160

                          A446   A487

LEU – LEU – ARG – GLN – ILE – ALA – SER – TYR – GLY – ARG – GLY – TYR – THR – [TYR] – LEU – [LEU] – SER – ARG – ALA – GLY –
161   ↓   ↓   180
                                  CYS   ARG

    A223

VAL – [THR] – GLY – ALA – GLU – ASN – ARG – ALA – ALA – LEU – PRO – LEU – ASN – HIS – LEU – VAL – ALA – LYS – LEU – LYS –
181 ↓   200
   ILE

                      A23 A46   A187

GLU – TYR – ASN – ALA – ALA – PRO – PRO – LEU – GLN – [GLY] – PHE – [GLY] – ILE – SER – ALA – PRO – ASP – GLN – VAL – LYS –
201   ↓   ↓   ↓   220
                          ARG GLU   VAL

                           A78 A58   A169

ALA – ALA – ILE – ASP – ALA – GLY – ALA – ALA – GLY – ALA – ILE – SER – [GLY] – [SER] – ALA – ILE – VAL – LYS – ILE – ILE –
221   ↓ ↓   ↓   240
                           CYS ASP LEU

GLU – GLN – HIS – ASN – ILE – GLU – PRO – GLU – LYS – MET – LEU – ALA – ALA – LEU – LYS – VAL – PHE – VAL – GLN – PRO –
241   260

MET – LYS – ALA – ALA – THR – ARG – SER
261   267

AMINO ACID SEQUENCE OF *A* PROTEIN is shown side by side with a ribbon representing the DNA of the *A* gene. It can be seen that 10 different mutations in the gene produced alterations in the amino acids at only eight different places in the *A* protein. The explanation is that at two of them, 210 and 233, there were a total of four alterations. Thus at No. 210 the mutation designated A23 changed glycine to arginine, whereas mutation A46 changed glycine to glutamic acid. At No. 233 glycine was changed to cysteine by one mutation (A78) and to aspartic acid by another mutation (A58). On the genetic map A23 and A46, like A78 and A58, are very close.

resembles traditional genetic procedures, relies on recombination frequencies to establish the order of the mutationally altered sites in the *A* gene with respect to one another. By this method one can assign relative distances—map distances—to the regions between altered sites. The method is often of little use, however, when the distances are very close.

In such cases we have used a third method that involves a mutationally altered gene, or genetic marker, close to the *A* gene. This marker produces a recognizable genetic trait unrelated to the *A* protein. What this does, in effect, is provide a reading direction so that one can tell whether two closely spaced mutants, say No. 58 and No. 78, lie in the order 58–78, reading from the left on the map, or vice versa [*see bottom illustration on page 162*].

With these procedures we were able to construct a genetic map relating the altered sites in a group of mutants responsible for altered *A* proteins that could themselves be isolated for study. Some of the sites were very close together, whereas others were far apart [*see upper illustration, page 163*]. The next step was to determine the nature of the amino acid changes in each of the mutationally altered proteins.

It was expected that each mutant of the *A* protein would have a localized change, probably involving only one amino acid. Before we could hope to identify such a specific change we would have to know the sequence of amino acids in the unmutated *A* protein. This was determined by John R. Guest, Gabriel R. Drapeau, Bruce C. Carlton and me, by means of a well-established procedure. The procedure involves breaking the protein molecule into many short fragments by digesting it with a suitable enzyme. Since any particular protein rarely has repeating sequences of amino acids, each digested fragment is likely to be unique. Moreover, the fragments are short enough—typically between two and two dozen amino acids in length—so that careful further treatments can release one amino acid at a time for analysis. In this way one can identify all the amino acids in all the fragments, but the sequential order of the fragments is still unknown. This can be established by digesting the complete protein molecule with a different enzyme that cleaves it into a uniquely different set of fragments. These are again analyzed in detail. With two fully analyzed sets of fragments in hand, it is not difficult to

SEGMENT OF PROTEIN	MUTANT										NORMAL
	H11	C140	B17	B272	H32	B278	C137	H36	A489	C208	
I	+	+	+	+	+	+	+	+	+	+	+
II	−	+	+	+	+	+	+	+	+	+	+
III	−	−	+	+	+	+	+	+	+	+	+
IV	−	−	−	+	+	+	+	+	+	+	+
V	−	−	−	−	+	+	+	+	+	+	+
VI	−	−	−	−	−	+	+	+	+	+	+
VII	−	−	−	−	−	−	+	+	+	+	+
VIII	−	−	−	−	−	−	−	+	+	+	+
IX	−	−	−	−	−	−	−	−	+	+	+
X	−	−	−	−	−	−	−	−	−	+	+
XI	−	−	−	−	−	−	−	−	−	−	+

GENETIC MAP   H11   C140   B17   B272   H32   B278   C137   H36   A489   C208

INDEPENDENT EVIDENCE FOR COLINEARITY of gene and protein structure has been obtained from studies of the protein that forms the head of the bacterial virus T4D. Sydney Brenner and his co-workers at the University of Cambridge have found that mutations in the gene for the head protein alter the length of head-protein fragments. In the table "+" indicates that a given segment of the head protein is produced by a particular mutant; "−" indicates that the segment is not produced. When the genetic map was plotted, it was found that the farther to the right a mutation appears, the longer the fragment of head protein.

find short sequences of amino acids that are grouped together in the fragment of one set but that are divided between two fragments in the other. This provides the clue for putting the two sets of fragments in order. In this way we ultimately determined the identity and location of each of the 267 amino acids in the unmutated *A* protein of tryptophan synthetase.

Simultaneously my colleagues and I were examining the mutants of the *A* protein to identify the specific sites of mutational changes. For this work we used a procedure first developed by Vernon M. Ingram, now at the Massachusetts Institute of Technology, in his studies of naturally occurring abnormal forms of human hemoglobin. This procedure also uses an enzyme (trypsin) to break the protein chain into peptides, or polypeptide fragments. If the peptides are placed on filter paper wetted with certain solvents, they will migrate across

the paper at different rates; if an electric potential is applied across the paper, the peptides will be dispersed even more, depending on whether they are negatively charged, positively charged or uncharged under controlled conditions of acidity. The former separation process is chromatography; the latter, electrophoresis. When they are employed in combination, they produce a unique "fingerprint" for each set of peptides obtained by digesting the *A* protein from a particular mutant bacterium. The positions of the peptides are located by spraying the filter paper with a solution of ninhydrin and heating it for a few minutes at about 70 degrees centigrade. Each peptide reacts to yield a characteristic shade of yellow, gray or blue.

When the fingerprints of mutationally altered *A* proteins were compared with the fingerprint of the unmutated protein, they were found to be remarkably similar. In each case, however, there was a difference. The mutant fingerprint usually lacked one peptide spot that appears in the nonmutant fingerprint and exhibited a spot that the nonmutant fingerprint lacks. The two peptides would presumably be related to each other with the exception of the change resulting from the mutational event. One can isolate each of the peptides and compare their amino acid composition. Guest, Drapeau, Carlton and I, together with D. R. Helinski and U. Henning, identified the amino acid substitutions in

each of a variety of altered *A* proteins.

The final step was to compare the locations of these changes in the *A* protein with the genetic map of the mutationally altered sites. There could be no doubt that the amino acid sequence of the *A* protein and the map of the *A* gene are in fact colinear [*see lower illustration on page 163*].

One can also see that the distances between mutational sites on the map of the *A* gene correspond quite closely to the distances separating the corresponding

amino acid changes in the *A* protein. In two instances two separate mutational changes, so close as to be almost at the same point on the genetic map, led to changes of the same amino acid in the unmutated protein. This is to be expected if a codon of three bases in DNA is required to specify a single amino acid in a protein. Evidently the most closely spaced mutational sites in our genetic map represent alterations in two bases within a single codon.

Thus our studies have shown that each

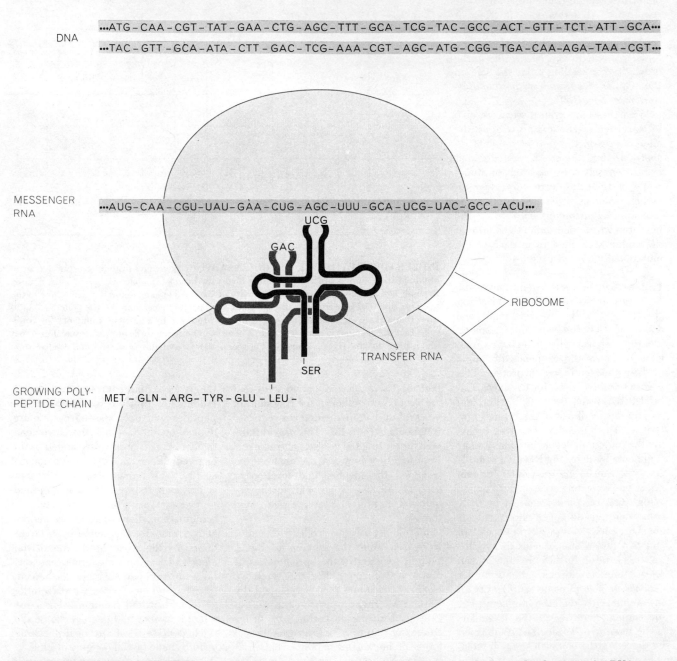

SCHEME OF PROTEIN SYNTHESIS, according to the current view, involves the following steps. Genetic information is transcribed from double-strand DNA into single-strand messenger ribonucleic acid (RNA), which becomes associated with a ribosome. Amino acids are delivered to the ribosome by molecules of transfer RNA, which embody codons complementary to the codons in mes-senger RNA. The next to the last molecule of transfer RNA to arrive (*color*) holds the growing polypeptide chain while the arriving molecule of transfer RNA (*black*) delivers the amino acid that is to be added to the chain next (serine in this example). The completed polypeptide chain, either alone or in association with other chains, is the protein whose specification was originally embodied in DNA.

```
 AGY                                                                      AGX AGY                                                                        AGY
 CGZ  GGZ  UAX  ACZ [UAX] XUZ [CUZ]      UCZ  CGZ  GCZ  GGZ  GUZ [ACW] GGZ  GCZ  GAY  AAX  CGZ  GCZ  GCZ  XUZ
-ARG - GLY - TYR - THR-[TYR]-LEU-[LEU]- SER - ARG - ALA - GLY - VAL -[THR]-GLY - ALA - GLU - ASN - ARG - ALA - ALA - LEU-
 170                                                                                                                                    190

      CCZ  XUZ  AAX  CAX  XUZ  GUZ  GCZ  AAY  XUZ  AAY  GAY  UAX  AAX  GCZ  GCZ  CCZ  CCZ  XUZ  CAY [GGA]
      PRO - LEU - ASN - HIS - LEU - VAL - ALA - LYS - LEU - LYS - GLU - TYR - ASN - ALA - ALA - PRO - PRO - LEU - GLN-[GLY]-
      191                                                                                                          210

                  AGX
 UUX [GGZ] AUW  UCZ  GCZ  CCZ  GAX  CAY  GUZ  AAY  GCZ  GCZ  AUW  GAX  GCZ  GGZ  GCZ  GCZ  GGZ  GCZ
 PHE-[GLY]- ILE - SER - ALA - PRO - ASP - GLN - VAL - LYS - ALA - ALA - ILE - ASP - ALA - GLY - ALA - ALA - GLY - ALA -
 211                                                                                                          230

              AGX
 AUW  UCZ [GGX][UCZ] GCZ  AUW  GUZ  AAY  AUW  AUW  GAY  CAY  CAX  AAX  AUW  GAY  CCZ  GAY  AAY  AUG
 ILE - SER-[GLY]-[SER]-ALA - ILE - VAL - LYS - ILE - ILE - GLU - GLN - HIS - ASN - ILE - GLU - PRO - GLU - LYS - MET -
 231                                                                                                          250
```

W = U, C or A   X = U or C   Y = A or G   Z = U, C, A or G

PROBABLE CODONS IN MESSENGER RNA that determines the sequence of amino acids in the *A* protein are shown for 81 of the protein's 267 amino acid units. The region includes seven of the eight mutationally altered positions (*colored boxes*) in the *A* protein. The codons were selected from those assigned to the amino acids by Marshall Nirenberg and his associates at the National Institutes of Health and by H. Gobind Khorana and his associates at the University of Wisconsin. Codons for the remaining 186 amino acids in the *A* protein can be supplied similarly. In most cases the last base in the codon cannot be specified because there are usually several synonymous codons for each amino acid. With a few exceptions the synonyms differ from each other only in the third position.

unique sequence of bases in DNA—a sequence constituting a gene—is ultimately translated into a corresponding unique linear sequence of amino acids—a sequence constituting a polypeptide chain. Such chains, either by themselves or in conjunction with other chains, fold into the three-dimensional structures we recognize as protein molecules. In the great majority of cases these proteins act as biological catalysts and are therefore classed as enzymes.

The colinear relation between a genetic map and the corresponding protein has also been convincingly demonstrated by Sydney Brenner and his co-workers at the University of Cambridge. The protein they studied was not an enzyme but a protein that forms the head of the bacterial virus T4. One class of mutants of this virus produces fragments of the head protein that are related to one another in a curious way: much of their amino acid sequence appears to be identical, but the fragments are of various lengths. Brenner and his group found that when the chemically similar regions in fragments produced by many mutants were matched, the fragments could be arranged in order of increasing length. When they made a genetic map of the mutants that produced these fragments, they found that the mutationally altered sites on the genetic map were in the same order as the termination points in the protein fragments. Thus the length of the fragment of the head protein produced by a mutant increased as the site of mutation was displaced farther from one end of the genetic map [*see illustration on page 165*].

The details of how the living cell translates information coded in gene structure into protein structure are now reasonably well known. The base sequence of one strand of DNA is transcribed into a single-strand molecule of messenger ribonucleic acid (RNA), in which each base is complementary to one in DNA. Each strand of messenger RNA corresponds to relatively few genes; hence there are a great many different messenger molecules in each cell. These messengers become associated with the small cellular bodies called ribosomes, which are the actual site of protein synthesis [*see illustration on page 166*]. In the ribosome the bases on the messenger RNA are read in groups of three and translated into the appropriate amino acid, which is attached to the growing polypeptide chain. The messenger also contains in code a precise starting point and stopping point for each polypeptide.

From the studies of Marshall Nirenberg and his colleagues at the National Institutes of Health and of H. Gobind Khorana and his group at the University of Wisconsin the RNA codons corresponding to each of the amino acids are known. By using their genetic code dictionary we can indicate approximately two-thirds of the bases in the messenger RNA that specifies the structure of the *A*-protein molecule. The remaining third cannot be filled in because synonyms in the code make it impossible, in most cases, to know which of two or more bases is the actual base in the third position of a given codon [*see illustration above*]. This ambiguity is removed, however, in two cases where the amino acid change directed by a mutation narrows down the assignment of probable codons. Thus at amino acid position 48 in the *A*-protein molecule, where a mutation changes the amino acid glutamic acid to valine, one can deduce from the many known changes at this position that of the two possible codons for glutamic acid, GAA and GAG, GAG is the correct one. In other words, GAG (specifying glutamic acid) is changed to GUG (specifying valine). The other position for which the codon assignment can be made definite in this way is No. 210. This position is affected by two different mutations: the amino acid glycine is replaced by arginine in one case and by glutamic acid in the other. Here one can infer from the observed amino acid changes that of the four possible codons for glycine, only one—GGA—can yield by a single base change either arginine (AGA) or glutamic acid (GAA).

Knowledge of the bases in the messenger RNA for the *A* protein can be translated, of course, into knowledge of the base pairs in the *A* gene, since each base pair in DNA corresponds to one of the bases in the RNA messenger. When the ambiguity in the third position of most of the codons is resolved, and when we can distinguish between two quite different sets of codons for arginine, leucine and serine, we shall be able to write down the complete base sequence of the *A* gene—the base sequence that specifies the sequence of the 267 amino acids in the *A* protein of the enzyme tryptophan synthetase.

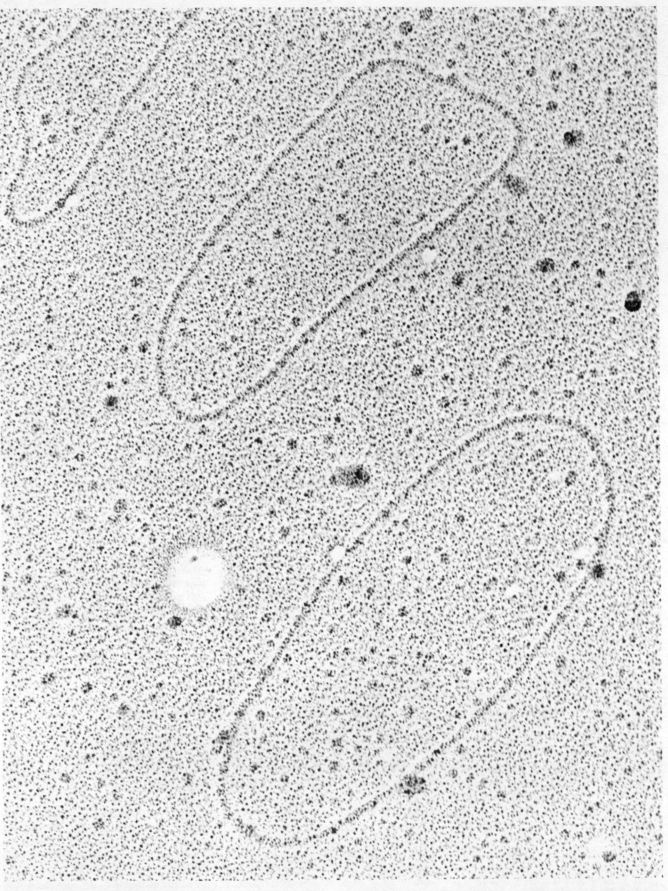

DNA OF THE BACTERIAL VIRUS φX174 is a circular molecule about 1.8 micrometers long. In this electron micrograph made by G. Nigel Godson of the Yale University School of Medicine dou- ble-strand molecules of the DNA are enlarged about 180,000 diameters. The DNA is present within a virus particle as a single strand. In an infected cell the single strand is copied to make two strands.

17

The Nucleotide Sequence of a Viral DNA

by John C. Fiddes
December 1977

Genetic information is encoded by the order in which nucleotides are arrayed to form a strand of DNA. Now that order has been established in full for the 5,375-nucleotide DNA of the bacterial virus φX174

Molecular biology has come a long way since DNA was identified in the 1940's as the genetic material. In the 1950's and 1960's the structure of the DNA molecule and its mode of replication were determined, revealing the chemical basis of heredity. In the past 20 years the mechanisms and processes by which the hereditary information encoded in the sequence of DNA nucleotides is translated into proteins, and thus into the substance of all living things, have become understood in increasing detail. In all this time molecular biologists have been at a serious disadvantage: they did not know the complete sequence of nucleotides constituting the genome, or total genetic message, of any organism.

Such a message is now available: the complete nucleotide sequence of the DNA of a small bacterial virus, φX174, has been established. It is a short message, with only 5,375 nucleotides compared with the millions in a bacterial chromosome or the billions in the chromosomes of a mammalian cell. Because the message is complete, however, it is now possible to relate the genetic information of an organism to its proteins and functions more directly than before. In this article I shall tell how the φX174 sequence was established and report some of the new perceptions of the remarkable coding ability of DNA that the analysis of the viral genome has already yielded.

A molecule of DNA is a long strand made up of four kinds of nucleotides: deoxyadenylate (A), deoxyguanylate (G), thymidylate (T) and deoxycytidylate (C). Each nucleotide in a DNA molecule has three components. Two of them, a five-carbon sugar ring (deoxyribose) and a phosphate group, alternate to form a continuous backbone. The other component is one of four nitrogenous bases: adenine, guanine, thymine

and cytosine. The bases are side chains extending from the backbone, and in the double-helix form in which DNA is present in most organisms the two strands of the helix are joined by hydrogen bonds linking the bases on each strand. (The DNA in some viruses, in-

cluding φX174, is single-strand until it is injected into a host cell.) The strands are complementary, because the size and bonding properties of the bases ordain that A be always opposite T and that G be opposite C. The complementarity achieved by base-pairing is the basis of

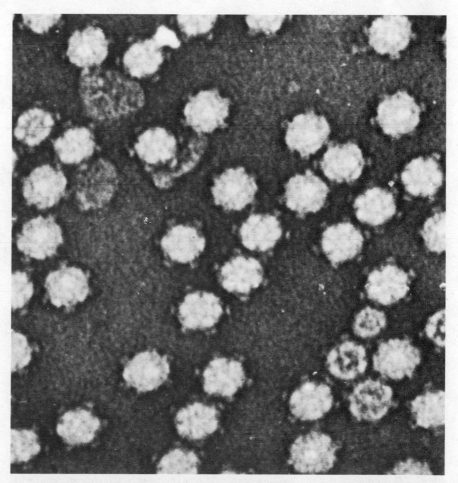

φX174 VIRUS PARTICLES are enlarged about 400,000 diameters in this electron micrograph made by John Finch of the British Medical Research Council's Laboratory of Molecular Biology in Cambridge. The virus's capsid, or protein coat, is an icosahedron, a polyhedron with 20 faces. Each of 12 vertexes bears a spike by which virus can attach itself to a bacterium.

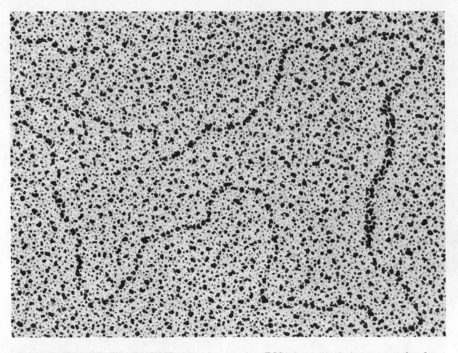

SINGLE-STRAND FRAGMENT cleaved from viral DNA by a restriction enzyme has been "annealed" to a single-strand viral-DNA molecule and is seen as a denser portion of the loop (*right*) in this micrograph made by Godson and W. Keegstra of the University of Utrecht. A "primer" fragment is annealed to viral DNA in this way to establish the nucleotide sequence.

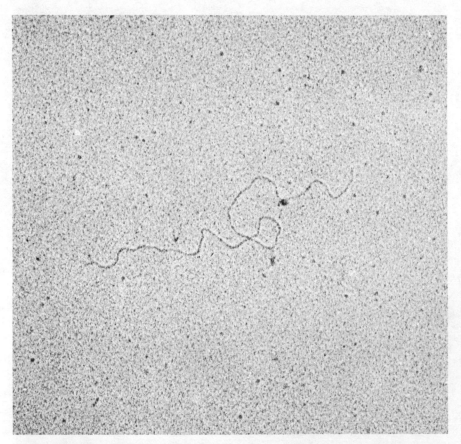

VIRAL PROTEIN encoded by gene *A* of the viral DNA plays a role in replicating the double-strand form within the bacterial cell. It begins by attaching itself to a particular site on the viral strand. The location of that site, and thus of the origin of DNA replication, was identified with respect to one end of the DNA by Shlomo Eisenberg, Jack D. Griffith and Arthur Kornberg of the Stanford University School of Medicine. The electron micrograph, made by Griffith, shows the gene-*A* protein (*black spot*) about a fifth of the way from the end of the strand. Although the strand normally forms a closed loop, here it has been opened by treatment with an enzyme.

the faithful replication of DNA and hence of the transmission of genetic information from one generation to the next. It is the particular sequence in which the four kinds of nucleotides are aligned along a strand of DNA that embodies the genetic message. Some stretches of the sequence specify, by means of the genetic code, the sequence in which amino acids are to be assembled to form the organism's proteins. Other stretches do not code for proteins but include signals that regulate the turning on and off of particular genes, or coding regions, and the rate at which those genes are to be expressed.

Until recently it was a formidable task to learn the nucleotide sequence of a DNA molecule because the techniques available for DNA-sequence analysis lagged far behind those that had been applied since the mid-1960's for analyzing RNA, the very similar nucleic acid that has several roles in the translation of the DNA message into proteins. Techniques for sequencing RNA rely on the ability of several enzymes to cleave the RNA at specific sites and thus to produce short pieces of RNA whose sequence can be established readily. It was difficult to apply a similar method to DNA. The smallest DNA molecules, those of certain viruses, are perhaps 70 times longer than the 75-nucleotide transfer-RNA molecules that were the subject of early RNA sequencing. And no enzymes are known that cleave DNA with the same degree of single-base specificity as the ribonucleases that are available to cleave RNA.

In the past few years, however, two new methods have been devised for sequencing DNA, both of them very different from the classical RNA-sequencing methods and both much faster. In 1975 Frederick Sanger and Alan R. Coulson of the British Medical Research Council's Laboratory of Molecular Biology in Cambridge published a report on their "plus-and-minus" method, and early this year Allan Maxam and Walter Gilbert of Harvard University published their "chemical" method. Both of the new techniques have since been exploited intensively. Here I shall be describing the plus-and-minus method, the principal one by which Sanger's group established the ϕX174 sequence. (The group consisted of Sanger, Gillian M. Air, Barclay G. Barrell, Nigel L. Brown, Coulson, myself, Clyde A. Hutchison III, Patrick M. Slocombe and Michael Smith.)

In broad outline the plus-and-minus method calls for synthesizing pieces of DNA that are complementary to various regions of the DNA being sequenced. Because of the way the synthesis is manipulated each piece ends either with or just before a particular, identifiable nucleotide (*A, G, T* or *C*). By performing the synthesis to produce a large number of pieces, each one ending with a successive nucleotide in the sequence, it is possible to establish the sequence of the entire DNA molecule.

The synthesis is brought about through the agency of a DNA polymerase, an enzyme that adds successive nucleotides to a DNA strand by linking the 3′ carbon atom on the sugar ring of the nucleotide at the end of the strand to the 5′ carbon atom on the sugar ring of the next nucleotide in the chain [*see illustration on next page*]. DNA polymerases cannot start copying a template DNA at just any point. They must be directed to a specific position by means of a "primer": a DNA molecule that has a sequence complementary to the sequence of a particular region of the template and that therefore "anneals" itself to the template by base-pairing. A free hydroxyl (OH) group at the 3′ end of the primer acts as a substrate for the DNA polymerase, which catalyzes the addition of the nucleotide called for by the base-pairing rules, thereby extending the chain in the 5′-to-3′ direction.

In the first step of the plus-and-minus procedure millions of molecules of the template DNA are incubated with DNA polymerase and with the four building blocks of DNA: the highly energetic deoxynucleoside triphosphates. (Two of the phosphates are dropped as the bond is formed.) At least one of the nucleoside triphosphates is labeled by incorporating into it a radioactive atom of phosphorus. The extension of the primer molecule is done under carefully controlled conditions at a low temperature calculated to retard the enzyme's rate of progress. Because under such conditions the strands do not all grow at exactly the same rate and because the mixture is sampled frequently, every possible chain length is represented among the products of the extension process.

That this is so can be established by electrophoresis, a process that can sort molecules according to their size. The pieces of double-strand DNA that include the extension products of the primed reaction are placed on a slab of polyacrylamide gel containing a high concentration of urea, which separates the newly synthesized extensions from their template DNA. Under the influence of an electric current the extension molecules migrate through the gel. The distance they travel varies with the size of the molecules, so that all the molecules of a given size end up in a small group. The groups are visualized by autoradiography: since some nucleotides added to form the extension molecules were labeled with radioactive phosphorus, when a piece of X-ray film is placed in contact with the gel, a pattern of

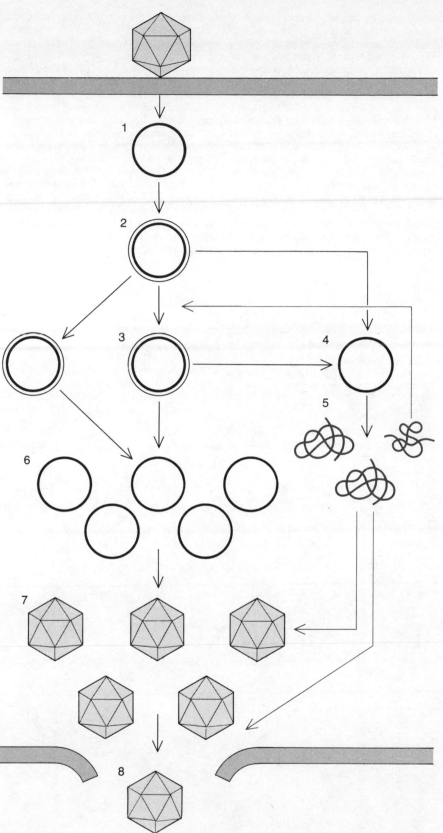

VIRAL DNA injected into a cell by an infecting virus (*1*) serves as the template for synthesis of a complementary strand (*2*). The double-strand DNA replicates to form about 20 copies (*3*). The complementary strand of the double-strand DNA serves as the template for the synthesis of messenger RNA (*4*) that is translated to make viral proteins (*5*); the proteins have enzymatic and structural roles in DNA replication, the formation of new virus particles and lysis (dissolution) of the cell wall (*colored arrows*). Some 200 copies of the single viral-DNA strand are produced (*6*), packaged in a protein capsid (*7*) and released (*8*) after lysis of the cell wall.

bands appears on the film corresponding to the positions in the gel of the groups of DNA molecules. Each successive band represents a molecule one nucleotide longer than the molecule represented by the preceding band. The purpose of this autoradiograph of the extension products is to create a standard, an array of bands representing every successive chain length. In essence the objective of the plus and the minus treatments is to create different populations of extension products: molecules that end not at every nucleotide position but specifically at or just before an *A*, a *G*, a *T* or a *C*.

First it is necessary to pass a sample of the products of the initial extension over a gel filtration column and thus to separate the relatively large extension chains

attached to their templates from the small individual nucleoside triphosphate building blocks that were not incorporated into new DNA. This is important because the results of the plus or the minus treatment will depend entirely on the absence of specific building blocks in the subsequent reactions. Then the reaction mixture is split into eight subsamples, one subsample for each of the four nucleotides in a plus system and one for each in a minus system.

Each subsample to be given the minus treatment is incubated with DNA polymerase and with only three of the four nucleotide building blocks. In the minus-*A* system, for example, the extension molecules, still bonded to their templates, are incubated with the en-

zyme and with the *G*, *T* and *C* nucleoside triphosphates but without the *A* nucleoside triphosphate. The synthesis resumes, but now the building blocks are added to each chain only up to the position just before the one where an *A* is called for by the base-pairing requirement; the synthesis stops at that point because no *A* building blocks are available. When these minus-*A* chains are subjected to the electrophoretic size-fractionation described for the original extension products, one does not observe a radioactive band at every position on the autoradiograph. Only a limited number of bands appear, each one representing the position in the sequence just preceding an *A* nucleotide. In the three other minus systems the three oth-

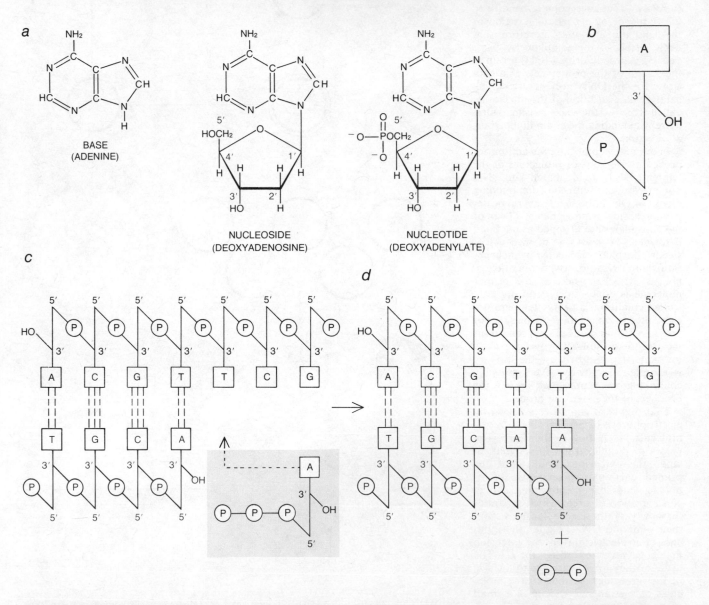

NUCLEOTIDE (*a*) consists of a base (adenine in this case), a sugar and a phosphate group. The base and the sugar constitute a nucleoside; the addition of a phosphate forms a nucleotide. A nucleotide can be represented as shown (*b*): the vertical line is the sugar, which carries the base (*A*) as a side chain and is linked by the phosphate (*P*) at its 5′ carbon to the 3′ carbon of the adjacent nucleotide's sugar. In a double-strand DNA molecule (*c*) hydrogen bonds (*broken lines*) between complementary bases link the two strands; *A* always pairs with *T* and *G* always pairs with *C*. A complementary strand (*color*) is synthesized along a template strand (*black*) by DNA polymerase; the enzyme links one phosphate of appropriate deoxynucleoside triphosphate (*colored panel*) to hydroxyl (*OH*) group at 3′ end of chain (*d*).

er subsamples of extension products are similarly extended in the absence of *G, T* and *C* nucleoside triphosphates; when the products of those reactions are fractionated, only the bands preceding each *G, T* and *C* are present on the autoradiograph.

In the four plus systems, also designed to yield chains ending at defined positions, extension-product subsamples are incubated with only one nucleoside triphosphate instead of with three and with a different DNA polymerase, one that is manufactured by the bacterial virus designated *T*4. This enzyme has a special property: in addition to synthesizing DNA chains in the 5'-to-3' direction under certain circumstances, in other circumstances it degrades the DNA, sequentially removing nucleotides from the 3' end of the chain. The *T*4 polymerase degrades the original extension molecules back toward their primer. In the plus-*A* system, however, the *A* nucleoside triphosphate is present, and so the degradation of each molecule ceases at the first *A* the polymerase encounters; at these positions the easy availability of *A*'s allows the enzyme's polymerizing activity to dominate its degradative activity, so that an *A* nucleotide that is removed is immediately replaced by a new *A* nucleotide. The plus-*A* reaction hence produces populations of molecules all of which end with an *A*. Similarly, the plus-*G*, plus-*T* and plus-*C* reactions produce molecules terminating in *G, T* and *C*. Again molecules of each length are grouped by electrophoresis, and the autoradiograph bands indicate each molecule's length and thus the position of each *A, G, T* and *C*.

The extension products of the four minus systems and the four plus systems are fractionated along with the extension products of the initial "zero" system in adjacent tracks on a single gel [*see illustrations on pages 175 and 176*]. The nucleotide sequence of the extensions can be read off the gel because most of the nucleotides are represented by a band at the appropriate position in one of the plus and one of the minus systems. (In the case of a "run" of a particular nucleotide, for example a sequence such as *AAAAA*, bands are not present at every position. In the plus system only the band representing the end of the run is present; in the minus system one observes only the product representing the start of the run. The precise length of the run can be determined easily, however, by reference to the number of bands in the corresponding area of the zero-system column.) Under particularly favorable circumstances the sequence can be read on a single gel out to 200 or more nucleotides from the 3' end of the primer. A sequence several hundred nucleotides long, which might have taken

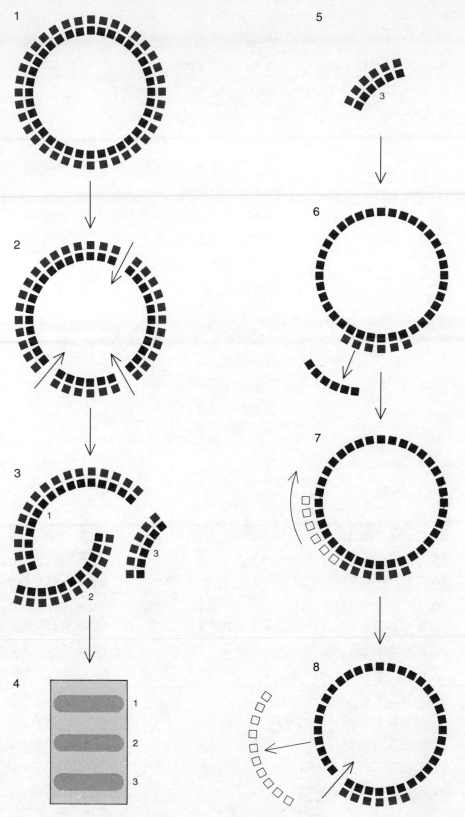

PLUS-AND-MINUS METHOD of determining a nucleotide sequence begins with double-strand DNA taken from infected cells (*1*). Cleavage by a restriction enzyme (*2*) produces fragments (*3*), which are separated according to size by gel electrophoresis (*4*). One of the fragments is selected (*5*) to serve as a primer for the sequencing procedure. Incubated with a single-strand viral-DNA template, the primer's complementary strand is annealed to the template by base-pairing (*6*). Now new DNA is synthesized: successive nucleotides (*open colored blocks*) are added (*7*) under various conditions to make extension molecules, which are separated from viral DNA by the restriction enzyme (*8*) before being analyzed for nucleotide sequence.

174

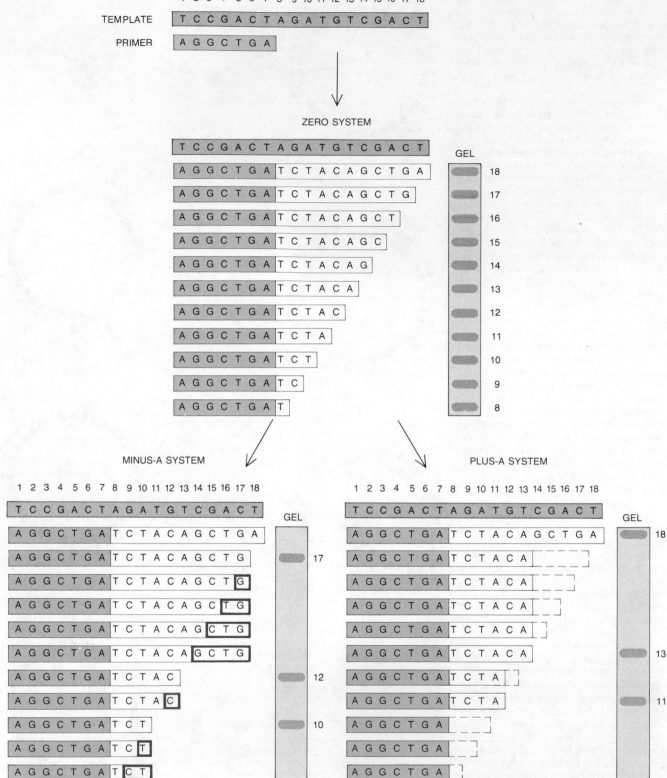

PRIMER IS EXTENDED under three different conditions after being annealed to the template (whose nucleotides are here arbitrarily numbered from 1 through 18). In the "zero" system the primer and template are incubated with DNA polymerase and all four nucleotide building blocks under conditions such that successive nucleotides are added, according to the base-pairing rules, to produce extension products (*open colored boxes*) of every possible chain length. A subsample of these extension molecules is separated according to size by electrophoresis on a polyacrylamide gel (*gray slab at right*). Other subsamples are subjected to "minus" or "plus" treatment. In the minus-*A* system, for example, the *A* building block is withheld from the mixture; the extension products are further extended (*heavy boxes*), but only up to the position just before an *A*. Now electrophoresis reveals only three bands, each band representing the position in the sequence one nucleotide before the next *A*. The same procedure is followed with each of the other three nucleotides withheld from the synthesis process to establish the positions one nucleotide before each appearance of a *G*, *T* or *C*. In the plus-*A* system only *A*'s are provided, and the polymerase is one that degrades DNA except in the presence of an excess of a nucleotide building block. Extension products are therefore degraded (*broken boxes*) until position of an *A* is reached. Electrophoresis again shows three bands, this time at position of *A*'s.

years to determine by the classical method of cleaving into small fragments, can now be established in a few days.

An important tool for plus-and-minus sequencing and for molecular biology in general was provided by the discovery several years ago of the enzymes called restriction endonucleases, which cleave large DNA molecules into discrete fragments. They cleave DNA not at the site of a particular single nucleotide, as the ribonucleases mentioned above do, but only at particular sites in specific sequences of four to six nucleotides, and so they generate much larger fragments than the ribonucleases do. Usually the sequence recognized by a restriction enzyme is symmetrical in double-strand DNA. For example, one endonuclease recognizes the double-strand sequence consisting of *GGCC* in one strand opposite *CCGG* in the other strand and cleaves the symmetrical tetranucleotides at their midpoint. The fragments produced by an endonuclease have an exposed hydroxyl group at the 3' end, so that they are suitable as primers for chain extension. More than 20 restriction endonucleases are known that cleave DNA at different nucleotide sequences. Treating different samples of a DNA with different endonucleases therefore produces fragments of various sizes and with known nucleotides at each end. Biochemical experiments can establish the position of each of these fragments with respect to one another on the original DNA, so that the choice of a particular fragment to be annealed as a primer determines the portion of the total DNA whose sequence will be elucidated.

The φX174 virus infects the bacterium *Escherichia coli.* It is a small virus and one that has been well studied in many laboratories, notably those of Robert L. Sinsheimer at the California Institute of Technology and of Irwin Tessman and Ethel S. Tessman at Purdue University. The infectious virus consists of a protective capsid, or protein shell, containing a single-strand DNA molecule that carries the genes coding for the virus's nine proteins. In electron micrographs the viral particle appears as a symmetrical icosahedron (a polygon with 20 faces) about 250 angstroms in diameter with a knobby spike at each of the 12 vertexes. The viral protein designated *F* is the major component of the capsid; the *G* and *H* proteins form the spikes.

An infecting virus particle attaches itself to a specific receptor site on the surface of the bacterium. The single strand of DNA is injected into the cell. It is copied (by host-cell enzymes) to produce a double-strand form, which then replicates (apparently with the help of viral protein *A* as well as host enzymes) to make about 20 copies, each one consisting of a viral strand like the one origi-

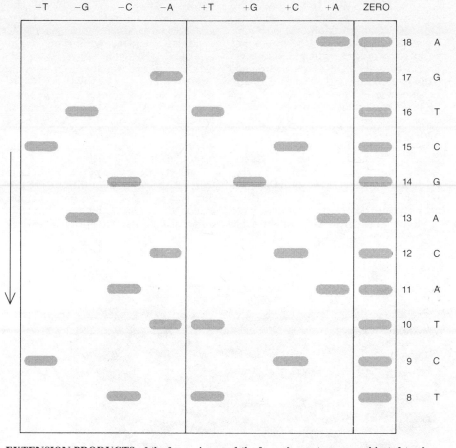

EXTENSION PRODUCTS of the four minus and the four plus systems are subjected to electrophoresis side by side along with the products of the zero system. This is a schematic representation of autoradiograph that would result from procedures illustrated on opposite page. Each group of molecules travels down gel a distance proportional to its length and shows up as a gray band on autoradiograph. Nucleotide sequence (*right*) is determined from band pattern.

nally injected and a complementary strand. The complementary strands serve as templates for transcription of the DNA into messenger RNA, which therefore has the same nucleotide sequence as the viral strand; translation of the messenger RNA manufactures the viral proteins. When enough proteins have been accumulated, the DNA replication is skewed so that only viral strands are formed—about 200 of them. Under the influence of viral proteins *B*, *C* and *D* these viral-DNA strands are packaged in capsids to produce mature virus particles, and then the cell wall is ruptured by the *E* protein, releasing the proliferated virus particles.

The φX174 virus was particularly well suited to DNA-sequence analysis because the relative order of the genes on its circular DNA molecule had been established by genetic-mapping experiments. Certain control sites had also been located, including sites where transcription into messenger RNA starts, the major site where transcription terminates and the approximate site where double-strand DNA replication begins. A physical map of the viral DNA was constructed by cleaving the strand into fragments with endonucleases of varying specificities, determining the order

of those fragments and correlating them with the genetic map, so that any gene or part of a gene could be isolated as required on a small piece of DNA; such pieces could be selected as primers that would direct DNA synthesis to specific parts of the viral strand.

By priming with a large number of restriction-enzyme fragments, and with either the viral or the complementary strand or both of them in turn as the template, we determined the nucleotide sequence of many pieces of the φX174 DNA molecule. Advantage was taken of earlier sequencing work done by different methods in a few regions of the DNA and of the fact that the amino acid sequence of some of the viral proteins was known and could, through the genetic code, yield some nucleotide sequences. By putting together all the bits and pieces we were able to establish the complete sequence of the 5,375 nucleotides [*see illustration on pages 178 and 179*].

As the complete sequence became available we studied it for new information on the overall organization of the viral genome. The major surprise came as we located the nucleotides that constitute the "start" and "stop" signals for each of the viral genes.

DNA has usually been thought to consist of coding and noncoding regions. The coding regions, the genes, are the stretches of nucleotide sequence that specify the order in which amino acids are to be assembled to form proteins. This is accomplished by means of the genetic code, in which successive groups of three nucleotides constitute the codons, or code "words," for particular amino acids. For example, the DNA codon *GGT* specifies the amino acid glycine, *TAC* specifies tyrosine, *CAT* specifies histidine. Since the number of three-nucleotide combinations of four nucleotides is 4^3, there are 64 codons. There are only 20 amino acids, so that the code is "degenerate": most of the amino acids are specified by several codons. The coding information carried by DNA is expressed through messenger RNA, which has a sequence complementary to that of the DNA from which it is copied. (The codons are usually given in terms of RNA, but in working with ϕX174, in which the messenger RNA has the same sequence as the viral strand, we speak of the codons in terms of DNA.) Messenger RNA is translated into proteins by a complicated procedure involving the cell organelles called ribosomes.

Each coding region of the DNA starts with the sequence *ATG*, which is the codon for methionine, the initial amino acid in all protein chains. Not all methionine is at the start of a protein, however, so that clearly not all *ATG* sequences can be signals for initiating a protein; for an *ATG* to start a protein it must be preceded by another signal that tells the ribosome to begin to translate. These ribosome-recognition signals are sequences that are complementary to, and presumably recognized by, a region on one of the small RNA components of ribosomes. Since the extent of this complementarity can vary, not all ribosome-recognition signals are the same, but they are similar enough to direct the investigator to the beginning of a coding region, which is to say to the *ATG* that follows a characteristic ribosome-recognition signal. The *ATG* also establishes the "reading frame" for the coding region. For example, in a sequence including *CGATGCACT* the codons could be *CGA, TGC, ACT* or ...*C, GAT, GCA, CT*... or ...*CG, ATG, CAC, T*.... If the *ATG* is a start signal, the correct reading frame for the coding region must be the last of the three possible frames. A coding region ends with one of three codons that do not specify a protein and therefore serve as termination codons: *TAA, TGA* or *TAG* in the correct reading frame.

It had been thought that the coding regions of DNA were clearly separated by noncoding regions that function as control signals. This assumption was substantiated when Walter Fiers and his colleagues at the University of Ghent recently established the nucleotide sequence of the small RNA bacteriophage *MS*2, whose genome is a single-strand RNA molecule about 3,300 nucleotides long that codes for just three proteins. The sequence analysis showed that the three genes are separated by two noncoding regions, respectively 26 and 36 nucleotides long.

The ϕX174 virus is very different. Some of the coding regions are indeed separated by untranslated regions. In three instances, however, two coding regions merge: the termination codon of one gene overlaps the initiation codon of the next gene, leaving no space for an untranslated region. This lapped joint is formed in one of two ways. Either a *TAA* termination codon overlaps the *ATG* start signal of the next gene or an *ATG* comes just before and overlaps the *TGA* termination codon of the preceding gene; in both cases part of the sequence thus has a dual function. As a result of this kind of overlap genes *A, C, D* and *J* form a continuous coding region.

Even more surprising, in two instances a small gene is completely overlapped by a larger one, that is, the same stretch of DNA carries the information for producing two proteins with an entirely different amino acid sequence. This remarkable achievement of genetic-coding economy is accounted for by

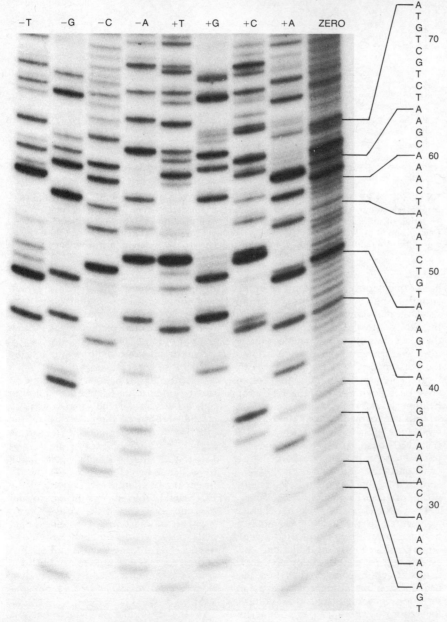

ACTUAL AUTORADIOGRAPH of a gel is interpreted here. A sequence extending from 21 to 73 nucleotides from the end of the primer can be determined by reading the band pattern. Bands representing successively larger products (*farther up the gel*) are closer together because the distance of migration is inversely proportional to the logarithm of the chain length.

an elegant trick: the overlapping genes are encoded in different reading frames.

The gene-within-a-gene phenomenon was discovered by looking for gene E. Genetic-mapping experiments had determined that in a certain portion of the φX174 genome the order of the genes was D, E, J. The complete amino acid sequence had been determined for both the D and J proteins. When the amino acid sequences were related to the newly available nucleotide sequence, it was found that the coding regions of the D and the J gene are contiguous. In fact, as was mentioned above, the termination codon of gene D overlaps the initiation codon of gene J. Apparently there was no room for E—which had been mapped between D and J.

In searching for gene E it was not possible, as it was in the case of genes D and J, to match amino acid and nucleotide sequences, because the virus produces such small amounts of the E protein that it has not been feasible to isolate the protein and determine its amino acid sequence. Instead a more indirect method was used: a search was carried out for a gene-E reading frame within the D, J sequence. The search was made with a φX174 mutant that does not make the E protein. The reason is not that these mutant viruses have lost the stretch of DNA that codes for the E protein but that a single-nucleotide mutation converts a codon for an amino acid into one of the three termination codons, and translation of the RNA transcribed from the mutated gene ends prematurely when ribosomes encounter the anomalous termination codon. There are bacterial strains, however, whose translating machinery contains a "suppressor" mutation that overcomes this premature termination codon by in effect reading the incorrect codon incorrectly, enabling the mutant virus to produce E protein and multiply.

The gene-E virus mutant was grown in one of these suppressor bacterial strains and the viral DNA was sequenced. By comparing the sequence with the normal viral sequence the nucleotide change that is responsible for the gene-E termination could be identified: in the mutant DNA a G is changed to an A. The change is in the region of the DNA that codes for the D protein. It was observed that in a particular reading frame the change would convert a codon for the amino acid tryptophan, TGG, into the termination codon TAG, thereby explaining the premature termination. The reading frame of this codon change was not the same as the clearly established gene-D reading frame, however, but one nucleotide to the right of it. This new reading frame accounts for the E gene. (It is interesting to note that in the D-gene reading frame the change from G to A converts the codon CTG

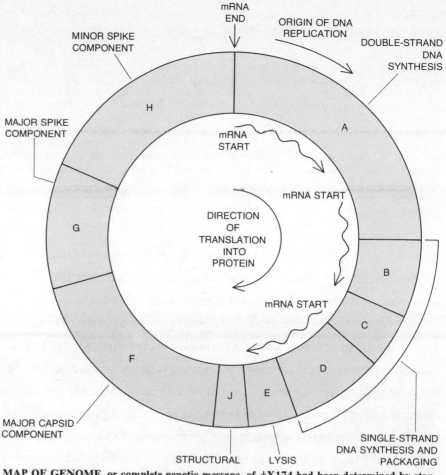

MAP OF GENOME, or complete genetic message, of φX174 had been determined by standard techniques before the sequence was established. Map showed relative length and position of each of the virus's nine genes (A–J). Function of each gene was known in a general way: they serve as enzymes for replication or packing of DNA or as structural components. Location was known of three start signals for messenger-RNA transcription and a termination signal, and of origin of DNA replication. Transcription, translation and replication go clockwise.

into CTA. Both codons happen to code for leucine, so that the E-gene mutation does not affect the D gene's protein product.)

It was not hard to locate the beginning of the E gene by searching the DNA sequence for a characteristic ribosome-recognition sequence followed by an ATG initiation codon in the correct reading frame. The end of the E gene had to be the first termination codon encountered in the same reading frame. The location of the start and stop signals made it clear that gene E is contained entirely within gene D. Given the nucleotide sequence and the E reading frame, the amino acid sequence of the E protein has been predicted even though the protein has yet to be isolated.

The possibility that genes might overlap in different reading frames had been discounted on theoretical grounds because it imposes a severe evolutionary constraint. If a stretch of DNA can be read in two different frames, a favorable mutation in one frame may well be unfavorable in the other frame; for example, it may introduce a premature ter-

mination codon. The two overlapping genes have to evolve in parallel, the only permissible nucleotide changes being changes that allow both reading frames still to code for a viable protein.

In the case of the D and E genes it is possible to speculate as to the order in which the two genes evolved. The reasoning begins with the finding that the viral DNA has a large proportion of T nucleotides, the extra T's appearing most frequently in the third position of the codons. This high incidence of T's in the third position is specifically true in the D-gene sequence. Since the E-gene reading frame is displaced one position to the right of the D frame, the E-gene codons must have a high level of T's in their second position. Such a codon tends to specify one of the hydrophobic, or water-repelling, amino acids. The E gene must therefore have an overall structure somewhat similar to that of a detergent molecule. Such a structure might account for the E protein's function: disrupting the host cell's outer membrane. (Bacterial cells are broken open in the laboratory by detergent solutions, and some secretory proteins in higher organisms that need to move through membranes also need to have a detergent structure.) We therefore speculate that the *E* gene arose after the *D* gene as the result of a mutational event that turned out to be highly adaptive: it allowed the virus to capitalize on its high content of *T*'s to break down the cells it infected.

Such an explanation of the origin of the *E* gene seems to imply that gene overlap might be the result of a rather special event exploiting a particular feature of the viral DNA. There is a second pair of overlapping genes in φX174,

however, and to them this kind of rationalization does not seem to apply.

The apparent inconsistency between the small size of the viral DNA and the fact that it codes for nine proteins had suggested that there might be more than one pair of overlapping genes, and specifically that gene *B* might be contained entirely within gene *A*. This was investigated in much the same way as the overlap of genes *D* and *E*. A ribosome-recognition sequence within gene *A* led to an *ATG* that seemed to be the initiating codon of gene *B*. This *ATG* was in a reading frame two nucleotides to the right (or one nucleotide to the left) of

gene *A*'s frame. By sequencing chain-termination mutants in both genes and finding the mutations that created termination codons, the two reading frames were confirmed and the fact that gene *B* is contained entirely within gene *A* was established.

The overlap of genes *A* and *B* is surprising. The proteins encoded by these genes appear to function as enzymes, whose activity depends on their precise structure, and one would expect that evolutionary constraints on changes in their amino acid sequence would be stronger than the constraints on

CCGTCAGGATTGACACCCTCCCAATTGTATGTTTTCATGCCTCCAAATCTTGGAGGCTTTTTTATGGTTCGTTCTTATTACCCTTCTGAA

TGTCACGCTGATTATTTTGACTTTGAGCGTATCGAGGCTCTTAAACCTGCTATTGAGGCTTGTGGCATTTCTACTCTTTCTCAATCCCCA

ATGCTTGGCTTCCATAAGCAGATGGATAACCGCATCAAGCTCTTGGAAGAGATTCTGTCTTTTCGTATGCAGGGCGTTGAGTTCGATAAT

GGTGATATGTATGTTGACGGCCATAAGGCTGCTTCTGACGTTCGTGATGAGTTTGTATCTGTTACTGAGAAGTTAATGGATGAATTGGCA

CAATGCTACAATGTGCTCCCCCAACTTGATATTAATAACACTATAGACCACCGCCCCGAAGGGGACGAAAAATGGTTTTTAGAGAACGAG

AAGACGGTTACGCAGTTTTGCCGCAAGCTGGCTGCTGAACGCCCTCTTAAGGATATTCGCGATGAGTATAATTACCCCAAAAAGAAAGGT

ATTAAGGATGAGTGTTCAAGATTGCTGGAGGCCTCCACTAAGATATCGCGTAGAGGCTTTGCTATTCAGCGTTTGATGAATGCAATGCGA

CAGGCTCATGCTGATGGTTGGTTTATCGTTTTTGACACTCTCACGTTGGCTGACGACCGATTAGAGGCGTTTTATGATAATCCCAATGCT

TTGCGTGACTATTTTCGTGATATTGGTCGTATGGTTCTTGCTGCCGAGGGTCGCAAGGCTAATGATTCACACGCCGACTGCTATCAGTAT

TTTTGTGTGCCTGAGTATGGTACAGCTAATGGCCGTCTTCATTTCCATGCGGTGCACTTTATGCGGACACTTCCTACAGGTAGCGTTGAC

CCTAATTTTGGTCGTCGGATACGCAATCGCCGCCAGTTAAATAGCTTGCAAAATACGTGGCCTTATGGTTACAGTATGCCCATCGCAGTT

CGCTACACGCAGGACGCTTTTTCACGTTCTGGTTGGTTGTGGCCTGTTGATGCTAAAGGTGAGCCGCTTAAAGCTACCAGTTATATGGCT

GTTGGTTTCTATGTGGCTAAATACGTTAACAAAAAGTCAGATATGGACCTTGCTGCTAAAGGTCTAGGAGCTAAAGAATGGAACAACTCA

CTAAAAACCAAGCTGTCGCTACTTCCCAAGAAGCTGTTCAGAATCAGAATGAGCCGCAACTTCGGGATGAAAATGCTCACAATGACAAAT

CTGTCCACGGAGTGCTTAATCCAACTTACCAAGCTGGGTTACGACGCGACGCCGTTCAACCAGATATTGAAGCAGAACGCAAAAAGAGAG

ATGAGATTGAGGCTGGGAAAAGTTACTGTAGCCGACGTTTTGGCGGCGCAACCTGTGACGACAAATCTGCTCAAATTTATGCGCGCTTCG

ATAAAAATGATTGGCGTATCCAACCTGCAGAGTTTTATCGCTTCCATGACGCAGAAGTTAACACTTTCGGATATTTCTGATGAGTCGAAA

AATTATCTTGATAAAGCAGGAATTACTACTGCTTGTTTACGAATTAAATCGAAGTGGACTGCTGGCGGAAAATGAGAAAATTCGACCTAT

CCTTGCGCAGCTCGAGAAGCTCTTACTTTGCGACCTTTCGCCATCAACTAACGATTCTGTCAAAAACTGACGCGTTGGATGAGGAGAAGT

GGCTTAATATGCTTGGCACGTTCGTCAAGGACTGGTTTAGATATGAGTCACATTTTGTTCATGGTAGAGATTCTCTTGTTGACATTTTAA

AAGAGCGTGGATTACTATCTGAGTCCGATGCTGTTCAACCACTAATAGGTAAGAAATCATGAGTCAAGTTACTGAACAATCCGTACGTTT

CCAGACCGCTTTGGCCTCTATTAAGCTCATTCAGGCTTCTGCCGTTTTGGATTTAACCGAAGATGATTTCGATTTTCTGACGAGTAACAA

AGTTTGGATTGCTACTGACCGCTCTCGTGCTCGTCGCTGCGTTGAGGCTTGCGTTTATGGTACGCTGGACTTTGTAGGATACCCTCGCTT

TCCTGCTCCTGTTGAGTTTATTGCTGCCGTCATTGCTTATTATGTTCATCCCGTCAACATTCAAACGGCCTGTCTCATCATGGAAGGCGC

TGAATTTACGGAAAACATTATTAATGGCGTCGAGCGTCCGGTTAAAGCCGCTGAATTGTTCGCGTTTACCTTGCGTGTACGCGCAGGAAA

CACTGACGTTCTTACTGACGCAGAAGAAAACGTGCGTCAAAAATTACGTGCGGAAGGAGTGATGTAATGTCTAAAGGTAAAAAACGTTCT

GGCGCTCGCCCTGGTCGTCCGCAGCCGTTGCGAGGTACTAAAGGCAAGCGTAAAGGCGCTCGTCTTTGGTATGTAGGTGGTCAACAATTT

TAATTGCAGGGGCTTCGGCCCCCTTACTTGAGGATAAATTATGTCTAATATTCAAACTGGCGCCGAGCGTATGCCGCATGACCTTTCCCAT

CTTGGCTTCCTTGCTGGTCAGATTGGTCGTCTTATTACCATTTCAACTACTCCGGTTATCGCTGGCGACTCCTTCGAGATGGACGCCGTT

GGCGCTCTCCGTCTTTCTCCATTGCGTCGTGGCCTTGCTATTGACTCTACTGTAGACATTTTTACTTTTTATGTCCCTCATCGTCACGTT

COMPLETE SEQUENCE of the 5,375 nucleotides of the φX174 viral DNA is listed on these two pages. Although the strand of DNA is a closed loop, here the sequence reads from left to right in each line, first down this page and then down the opposite page; the sequence

changes in a detergentlike molecule such as the *E* protein. It is possible, however, to form some idea of how these two genes evolved by studying the distribution of *T*'s in the third position of codons. In the region of gene *A* that precedes the overlap there is in the *A*-gene reading frame the usual high incidence of codons ending in *T*, but within the overlap region that high incidence becomes a feature of the *B* reading frame. Assuming that the high level of third-position *T*'s is a basic characteristic of φX174 DNA, this suggests that the *A* and *B* genes were once distinct, with *A* ending before the beginning of *B*, and

that a mutation arose in what had been the termination codon for the *A* gene, allowing it to read on into the *B* gene.

Whether overlapping will turn out to be a general phenomenon in the DNA's of various kinds of organisms is not yet clear. It can be argued that φX174 is a very special case, since the total amount of its DNA is severely limited by the physical size of the capsid in which the DNA is packaged. Presumably the only way such a physically constrained genome can evolve extra, advantageous functions is to develop overlapping genes. In the cells of higher organisms, including mammals, the problem seems

to be quite the reverse. Instead of there being not enough DNA for the required functions there appears to be a vast excess of DNA beyond what is required for coding and control functions. This does not necessarily mean, however, that overlapping genes will not be found in higher organisms.

The combination of the two pairs of overlapping genes and the three instances in which termination and initiation codons overlap means that from the start of the *A* gene all the way to the end of the *J* gene, a region encompassing more than 40 percent of the viral genome, there is no untranslated DNA.

TATGGTGAACAGTGGATTAAGTTCATGAAGGATGGTGTTAATGCCACTCCTCTCCCGACTGTTAACACTACTGGTTATATTGACCATGCC

GCTTTTCTTGGCACGATTAACCCTGATACCAATAAAATCCCTAAGCATTTGTTTCAGGGTTATTTGAATATCTATAACAACTATTTTAAA

GCGCCGTGGATGCCTGACCGTACCGAGGCTAACCCTAATGAGCTTAATCAAGATGATGCTCGTTATGGTTTCCGTTGCTGCCATCTCAAA

AACATTTGGACTGCTCCGCTTCCTCCTGAGACTGAGCTTTCTCGCCAAATGACGACTTCTACCACATCTATTGACATTATGGGTCTGCAA

GCTGCTTATGCTAATTTGCATACTGACCAAGAACGTGATTACTTCATGCAGCGTTACCATGATGTTATTTCTTCATTTGGAGGTAAAACC

TCATATGACGCTGACAACCGTCCTTTACTTGTCATGCGCTCTAATCTCTGGGCATCTGGCTATGATGTTGATGGAACTGACCAAACGTCG

TTAGGCCAGTTTTCTGGTCGTGTTCAACAGACCTATAAACATTCTGTGCCGCGTTTCTTTGTTCCTGAGCATGGCACTATGTTTACTCTT

GCGCTTGTTCGTTTTCCGCCTACTGCGACTAAAGAGATTCAGTACCTTAACGCTAAAGGTGCTTTGACTTATACCGATATTGCTGGCGAC

CCTGTTTTGTATGGCAACTTGCCGCCGCGTGAAATTTCTATGAAGGATGTTTTCCGTTCTGGTGATTCGTCTAAGAAGTTTAAGATTGCT

GAGGGTCAGTGGTATCGTTATGCGCCTTCGTATGTTTCTCCTGCTTATCACCTTCTTGAAGGCTTCCCATTCATTCAGGAACCGCCTTCT

GGTGATTTGCAAGAACGCGTACTTATTCGCAACCATGATTATGACCAGTGTTTCAGTCGTTCAGTTGTTGCAGTGGATAGTCTTACCTCA

                                                                   F END

TGTGACGTTTATCGCAATCTGCCGACCACTCGCGATTCAATCATGACTTCG[TGA]TAAAAGATTGAGTGTGAGGTTATAACCGAAGCGGTA

                                                                  G START

AAAATTTTAATTTTTGCCGCTGAGGGGGTTGACCAAGCGAAGCGCGGTAGGTTTTCTGCTT**AGGAG**TTTAATC[ATG]TTTCAGACTTTTATT

TCTCGCCACAATTCAAACTTTTTTTCTGATAAGCTGGTTCTCACTTCTGTTACTCCAGCTTCTTCGGCACCTGTTTTACAGACACCTAAA

GCTACATCGTCAACGTTATATTTTGATAGTTTGACGGTTAATGCTGGTAATGGTGGTTTTCTTCATTGCATTCAGATGGATACATCTGTC

AACGCCGCTAATCAGGTTGTTTCAGTTGGTGCTGATATTGCTTTTGATGCCGACCCTAAATTTTTTGCCTGTTTGGTTCGCTTTGAGTCT

TCTTCGGTTCCGACTACCCTCCCGACTGCCTATGATGTTTATCCTTTGGATGGTCGCCATGATGGTGGTTATTATACCGTCAAGGACTGT

GTGACTATTGACGTCCTTCCCCGTACGCCCGGCAATAACGTCTACGTTGGTTTCATGGTTTGGTCTAACTTTACCGCTACTAAATGCCGC

                                                   G END   H START

GGATTGGTTTCGCTGAATCAGGTTATTAAAGAGATTATTTGTCTCCAGCCACTTAAG[TGA]GGTGATTT[ATG]TTTGGTGCTATTGCTGGCG

GTATTGCTTCTGCTCTTGCTGGTGGCGCCATGTCTAAATTGTTTGGAGGCGGTCAAAAAGCCGCCTCCGGTGGCATTCAAGGTGATGTGC

TTGCTACCGATAACAATACTGTAGGCATGGGTGATGCTGGTATTAAATCTGCCATTCAAGGCTCTAATGTTCCTAACCCTGATGAGGCCG

CCCCTAGTTTTGTTTCTGGTGCTATGGCTAAAGCTGGTAAAGGACTTCTTGAAGGTACGTTGCAGGCTGGCACTTCTGCCGTTTCTGATA

AGTTGCTTGATTTGGTTGGACTTGGTGGCAAGTCTGCCGCTGATAAAGGAAAGGATACTCGTGATTATCTTGCTGCTGCATTTCCTGAGC

TTAATGCTTGGGAGCGTGCTGGTGCTGATGCTTCCTCTGCTGGTATGGTTGACGCCGGATTTGAGAATCAAAAAGAGCTTACTAAAATGC

AACTGGACAATCAGAAAGAGATTGCCGAGATGCAAAATGAGACTCAAAAAGAGATTGCTGGCATTCAGTCGGCGACTTCACGCCAGAATA

CGAAAGACCAGGTATATGCACAAAATGAGATGCTTGCTTATCAACAGAAGGAGTCTACTGCTCGCGTTGCGTCTATTATGGAAAACACCA

ATCTTTCCAAGCAACAGCAGGTTTCCGAGATTATGCGCCAAATGCTTACTCAAGCTCAAACGGCTGGTCAGTATTTTACCAATGACCAAA

TCAAAGAAATGACTCGCAAGGTTAGTGCTGAGGTTGACTTAGTTCATCAGCAAACGCAGAATCAGCGGTATGGCTCTTCTCATATTGGCG

CTACTGCAAAGGATATTTCTAATGTCGTCACTGATGCTGCTTCTGGTGTGGTTGATATTTTT CATGGTATTGATAAAGCTGTTGCCGATA

                                                                   H END

CTTGGAACAATTTCTGGAAAGACGGTAAAGCTGATGGTATTGGCTCTAATTTGTCTAGGAAA[TAA]

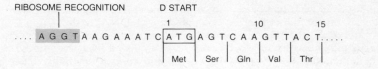

RIBOSOME RECOGNITION   D START

.... A G G T A A G A A A T C [A T G] A G T C A A G T T A C T .....

     Met | Ser | Gln | Val | Thr

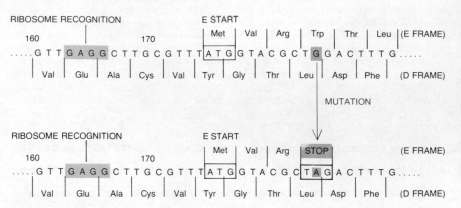

RIBOSOME RECOGNITION   E START

160   170   Met | Val | Arg | Trp | Thr | Leu | (E FRAME)

..... G T T GAGG C T T G C G T T T [A T G G] T A C G C T G G A C T T T G .....

     Val | Glu | Ala | Cys | Val | Tyr | Gly | Thr | Leu | Asp | Phe | (D FRAME)

MUTATION

RIBOSOME RECOGNITION   E START

160   170   Met | Val | Arg | STOP | (E FRAME)

..... G T T GAGG C T T G C G T T T [A T G G] T A C G C [T A G] A C T T T G .....

     Val | Glu | Ala | Cys | Val | Tyr | Gly | Thr | Leu | Asp | Phe | (D FRAME)

OVERLAP OF *D* AND *E* GENES is illustrated. Translation of the *D* gene is triggered by the *ATG* initiation codon following a ribosome-recognition signal (*top line*). Each three-nucleotide codon is labeled with the amino acid it specifies. In the midst of the *D*-gene sequence (*middle line*) there is another ribosome-recognition signal and then an *ATG* initiation codon in a new reading frame, which begins the *E* gene. The *E*-gene frame was identified by the discovery of the mutation of a *G* to an *A* in a mutant virus (*bottom line*), changing the codon for the amino acid tryptophan (*Trp*) to a premature termination codon, *TAG*, that stops *E*-gene translation. The reading frame for *D* gene is not affected, since both *CTG* and *CTA* specify leucine (*Leu*).

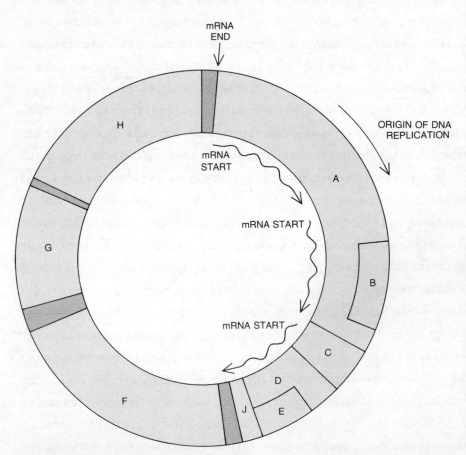

GENETIC MAP of the φX174 genome has now been refined on the basis of the nucleotide sequence, with the noncoding regions indicated (*gray*). From the start of the *A* gene to the end of gene *J* there is no untranslated DNA; gene *B* is within gene *A* and gene *E* is within gene *D*.

The situation is more typical, however, on the other side of the genetic map, where the structural genes are situated. These genes are separated by distinct untranslated regions: spaces respectively 39, 111, 11 and 66 nucleotides long between genes *J* and *F*, *F* and *G*, *G* and *H*, and *H* and *A*. Each of the three larger spaces contains a self-complementary sequence: a sequence such that the single strand of DNA can form a "hairpin" structure by looping and base-pairing to itself [*see top illustration on opposite page*]. Because such hairpin loops are not nearly as common in the coding regions of φX174 as they are in these noncoding spaces, we think they may have some regulatory function.

Particular control functions have in fact been ascribed to several sequences in these untranslated spaces. The region preceding the start of the *A* gene contains near its hairpin loop one of three sites on the φX174 DNA where the enzyme RNA polymerase recognizes the DNA and initiates the process of transcription into RNA. Several such sites, called promoters, have been identified in *E. coli* and its viruses and have been found to have a characteristic sequence, to which the gene-*A* promoter conforms. The *H-A* space also contains the major site in φX174 where the process of transcription terminates, the signal for which appears to be the sequence *TTTTTTA* preceded by a hairpin loop. (How can termination follow so quickly on initiation? It has been speculated on the basis of work done in other systems that it is hairpin plus *TTTTTTA* that halts transcription; an RNA just initiated at the peak of the hairpin would therefore not be terminated.) There is a weaker transcription-termination site between the *J* and *F* genes, where the hairpin structure alone may constitute the signal. As for the longest untranslated region, between the *F* and *G* genes, no particular function has yet been ascribed to it.

Not all the control signals in φX174 are in untranslated regions, however. Some nucleotide sequences serve as both a coding region and a recognition signal. For example, transcription promoters preceding genes *B* and *D* are respectively situated within the coding region for genes *A* and *C;* they have sequences characteristic of RNA-polymerase recognition sites even though their stretch of nucleotides codes for amino acids.

The most striking example of economy in genetic coding is to be found in the case of the ribosome-recognition sequence for the *J* gene. The sequence, *AAGGAG*, not only serves as a recognition site for ribosomes but also codes for the *D* protein in one reading frame and for the *E* protein in another frame [*see bottom illustration on opposite page*].

```
                                                    T
                                                  A   C
                                                  A–T
MESSENGER-RNA START                               A–T
                                                  C–G
                                                  C–G
                                                  T–A
                        PROMOTER                  C–G
                                                  C–G                        A START
H END
T A A C G T C A G G A T T G A C A C C C T C C C A A T T G T A T G T T T T C A T G–C T T T T T A T G

                                                           MESSENGER-RNA END
```

```
                  T C
                T     G
                C–G
                G–C
                G–C
                G–C
J END           G–C                  F START
T A A T T G C A–T T A C T T G A G G A T A A A T T A T G
```

```
                              T T
                            T     A
                            A–T
                            A–T
                            A–T
                            A–T
                            A–T               G
                            T G             A   C
                            G–C             A     G
                            G–C             G–C
                            C–G             C–G
                            G–C             G G
                            A–T             A–T
                            A G             A A
                            G A             C–G
F END                       C–G             C–G                              G START
T G A T A A A A G A T T G A G T G T G A G G T T A T A A C–G G G T T G A–T T T T C T G C T T A G G A G T T T A A T C A T G
```

THREE UNTRANSLATED REGIONS, between genes *H* and *A*, *J* and *F*, and *F* and *G*, contain self-complementary sequences that form "hairpin" loops by base-pairing. The region before the *A* gene has a promoter: a recognition site for the enzyme RNA polymerase, which mediates transcription of the DNA into messenger RNA; the transcription begins a few nucleotides later. The hairpin loops in *H–A* and *J–F* spaces are thought to be involved in the termination of transcription. The function of the *F–G* space has not been determined.

The ribosome-binding sites for genes *B, C, D* and *E* are also within coding regions, albeit regions that code for only one protein. In all these cases the DNA sequence must have had to evolve in such a way that none of its multiple functions was impaired.

The first fruits of the complete sequencing of φX174 have been these unexpected glimpses into the organization of the viral genome. The earlier model of DNA as having coding regions interspersed with largely regulatory spacer regions has had to be revised in view of the discovery of overlapping genes and the realization that a single stretch of DNA can have both a coding function and a control function. Moreover, such overlapping suggests that the evolutionary constraints on at least some nucleotide sequences must be much more stringent than has been believed.

Now that DNA can be sequenced readily and rapidly we can expect that in the next few years the precise composition of many DNA's will be established. In particular, new techniques for recombining the DNA of different species have already made it possible to determine the sequence of some regions of the DNA in systems other than viruses, and thus to begin to understand the complex mechanisms by which genes are expressed in higher organisms.

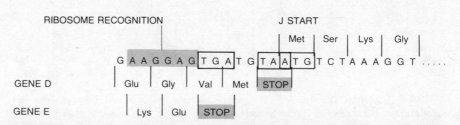
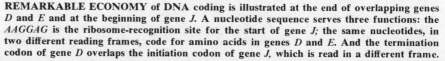

REMARKABLE ECONOMY of DNA coding is illustrated at the end of overlapping genes *D* and *E* and at the beginning of gene *J*. A nucleotide sequence serves three functions: the *AAGGAG* is the ribosome-recognition site for the start of gene *J*; the same nucleotides, in two different reading frames, code for amino acids in genes *D* and *E*. And the termination codon of gene *D* overlaps the initiation codon of gene *J*, which is read in a different frame.

18

A DNA Operator-Repressor System

by Tom Maniatis and Mark Ptashne
January 1976

An operator is a segment of DNA adjacent to a gene; a repressor is a protein that binds to the operator and controls the expression of the gene. How such a system works in a virus is explored in detail

A fundamental property of living cells is their ability to turn their genes on and off in response to extracellular signals. In the human body, for example, every cell (with the exception of a few cell types such as the red blood cell) has the same set of genes, yet in the course of embryonic development cells take on different shapes and functions as their genes are selectively switched on and off. How are the genes regulated? Is there a common mechanism underlying such regulation in different organisms?

Through the study of gene regulation in bacteria and viruses it has been learned in recent years that a fundamental mechanism of gene control depends on the interaction of protein molecules with specific regions on the long-chain molecule of DNA, the material that embodies the genetic instructions of all organisms from bacteria to man. As a result of this interaction genes are switched on or off. In the best-understood in-

stances genes are switched off by controlling molecules named repressors. The existence of repressors was first hypothesized in 1960 by François Jacob and Jacques Monod of the Pasteur Institute in Paris. Seven years later Walter Gilbert (in collaboration with Benno Müller-Hill) and one of us (Ptashne), working independently at Harvard University, succeeded in isolating repressors from bacteria [see "Genetic Repressors," by Mark Ptashne and Walter Gilbert; SCIENTIFIC AMERICAN Offprint 1179]. Later it was shown that repressors could bind tightly and specifically to sites on DNA called operators and that in so doing they could prevent genes adjacent to the operators from being transcribed and translated into proteins.

Since these early discoveries we and many others have pursued the molecular details of gene repression. This article is a brief progress report on some of the things that have been learned. We now know, for example, the sequence of

bases, or code units, in DNA that constitutes the operators to which a repressor binds. In the case we shall discuss here the operators have several nearly identical binding sites, each capable of being recognized by the same repressor molecule. We are only beginning to learn why several sites are provided when seemingly one would do the job.

Before we describe this recent work let us quickly review the molecular structure of the gene. In man, as in bacteria, a gene can be defined as a sequence of bases along a DNA molecule. (In certain viruses the gene consists of RNA rather than DNA.) The DNA molecule consists of two long chains of nucleotides wound in a double helix and linked to each other by hydrogen bonds. Each nucleotide consists of a deoxyribose sugar, a phosphate group and one of four nitrogenous bases: adenine, guanine, thymine or cytosine (abbreviated A, G, T and C). The sugar and phosphate groups form the backbone of each chain; the bases extend toward the central axis of the double helix and pair with the bases extending from the other chain. The sequences of bases along the chains are complementary: A always pairs with T and G always pairs with C. The information content of DNA is specified by the sequence of bases. A typical gene consists of roughly 1,000 base pairs.

The translation from gene to protein begins when the enzyme RNA polymerase copies the base sequence into a complementary sequence on the linear molecule of "messenger" RNA. The intracellular translating machines called ribosomes attach themselves to the messenger RNA and translate its base sequence into a sequence of amino acids, which are linked to form a protein molecule. Since there are only four different bases and 20 different amino acids, a sequence of three bases is needed to spec-

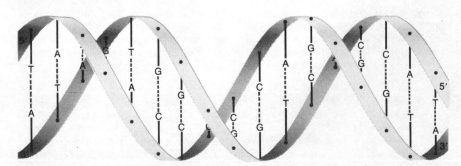

DOUBLE HELIX OF DNA encodes the genetic information of all cellular organisms and most viruses. (In some viruses the genetic material is RNA.) The genetic code is written in the particular sequences of nitrogenous bases that connect the two strands of the DNA molecule. The bases are of four kinds: adenine (A), thymine (T), guanine (G) and cytosine (C). A always pairs with T and G always pairs with C. The strands of the double helix consist of alternating subunits of phosphate and ribose, a five-carbon sugar. In one strand the phosphate links the No. 5 carbon in one sugar to the No. 3 carbon in the adjacent sugar, creating what is denoted a 5′–3′ linkage. In the opposing strand, proceeding in the same direction, the linkage is 3′–5′. Thus each strand of DNA molecule has a 5′ end and a 3′ end.

ify one amino acid. (For example, *ACA* specifies the amino acid threonine.) It is clear that the translation of a gene into a protein molecule might conceivably be repressed, or blocked, at any one of several stages along this complex pathway. It turns out in the case we have studied that repression takes place directly at the DNA molecule, so that the genetic information is not transcribed into messenger RNA unless the repressor is inactivated. Just how is this repression achieved?

Bacteriophage Lambda

The repressor we have studied is a protein molecule manufactured by bacteriophage lambda, a virus that infects the common colon bacterium *Escherichia coli*. The bacteriophage, or phage, particle consists of a single DNA molecule with about 47,000 base pairs, enough for some 50 genes, enclosed in a protein coat equipped with a tail through which the DNA is injected into the bacterial cell. Once the DNA is inside the cell it can follow either of two pathways. It can complete the course of infection by causing the machinery of the bacterial cell to translate the phage genes into proteins. Some of the proteins are enzymes that replicate several hundred

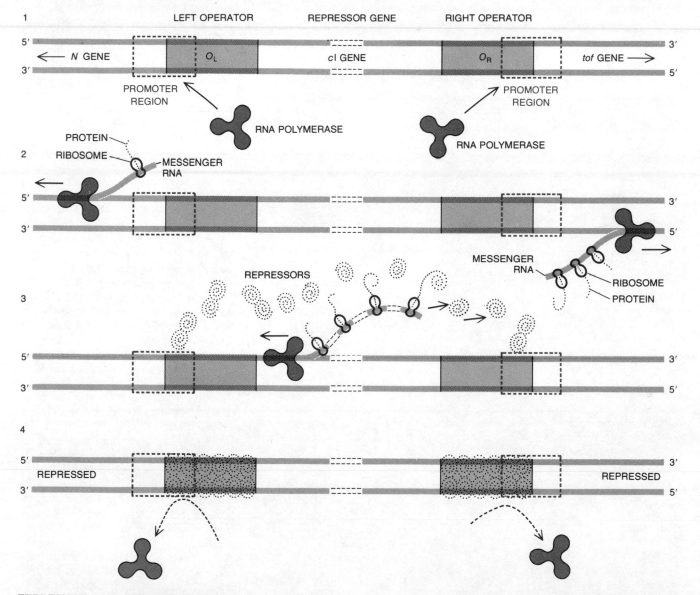

EXPRESSION AND REPRESSION OF GENES is conveniently studied in bacteriophage lambda, a virus that infects the bacterium *Escherichia coli*. When the 50 or so genes of the phage-lambda DNA are translated into protein inside a cell of *E. coli*, the phage multiplies and kills the cell in about 45 minutes. In some cases, however, repressor proteins specified by a particular gene in the lambda DNA, the *cI* gene, get the upper hand and block the transcription of genes on either side, forcing the lambda DNA into a state of dormancy. Normal transcription of lambda DNA into "messenger" RNA and its translation into protein molecules are depicted in *1* and *2*. The transcription is effectuated by the enzyme RNA polymerase, which attaches itself to a promoter region and assembles an RNA chain in the 5'-3' direction by copying a DNA strand of the opposite polarity. Thus the RNA-polymerase molecules travel in opposite directions, copying the different strands of the DNA as they transcribe into messenger RNA the complete instructions for replicating the phage, beginning with the *N* gene on the left and the *tof* gene on the right. Transcription of these two genes begins just outside the operator. The ribosomes fasten themselves to the emerging "tape" of messenger RNA and translate the encoded message into the protein molecule. (In this diagram the structures are not drawn to scale; the ribosomes in particular are much larger than they are shown.) Under other conditions (*3*) the *cI* repressor gene is transcribed by the same process and translated into repressor molecules. (Specific terminator signals in the DNA prevent RNA polymerase from continuing on into left operator.) Perhaps singly as monomers, but more likely as dimers depicted in *4* or even as tetramers, repressor molecules migrate to binding sites in two operator regions of lambda DNA, blocking access of RNA polymerase to promoter regions of *N* and *tof* genes.

copies of the phage-DNA molecule; other phage proteins package each DNA copy in a protein coat, thus creating multiple copies of the original phage particle. Typically within 45 minutes the bacterial cell, swollen with phage particles, bursts.

The other pathway is the more interesting one for our purposes. Occasionally after the phage genes enter the bacterial cell they are switched off and the phage DNA becomes integrated into the DNA of the host cell. There it remains dormant, replicating with the bacterial DNA at every cell division and giving rise to a population of E. coli cells each of which contains a chain of phage genes. The dormant phage genes are called a prophage.

It has been known for some years that the phage genes are turned off by a specific repressor molecule specified by one of the phage's own genes, the cI gene. The repressor actually binds to two separate operators on the phage-DNA molecule, thereby blocking the transcription of two different sets of genes. The turning off of these two sets of genes is sufficient to cause the 40-odd remaining genes, with the exception of cI, to stop functioning. The dormant phage genes can be switched on again by a suitable inducing agent such as a low dose of ultraviolet radiation, which causes the repressor to be inactivated.

The two operators, which are separated by some 2,000 base pairs (including the cI gene), are designated O_L and O_R, the subscripts denoting left and right. Repressor bound to O_L blocks the transcription of gene N and repressor bound to O_R blocks the transcription of gene tof. Gene N is transcribed to the left beginning near O_L and gene tof is transcribed to the right from the opposite DNA strand, beginning near O_R. (DNA chains have a polarity determined by the orientation of the sugar-phosphate linkage in their backbone. In each double helix the linkage is designated 5'–3' in one chain and 3'–5' in the other. The numbers 3 and 5 refer to the third and fifth carbon atoms in the five-carbon sugar molecule. RNA is assembled in the 5'–3' direction, copied from a DNA strand of the opposite polarity.)

One can speculate that it is clearly to the phages' advantage for some of them to go into the prophage, or dormant, stage and to be "revived" at a later time under conditions that may be more favorable for multiplication. The phage genes are reactivated by inactivation of the repressor. Although there are many conditions that result in repressor inactivation, the details of the process are not fully understood.

The lambda-phage repressor is a protein composed of subunits that have a molecular weight of about 27,000. When the concentration of the subunits in a suitable medium is increased, they form dimers and tetramers: two-subunit and four-subunit associations. When the concentration is reduced, they dissociate. Dimers or possibly tetramers must form before the repressor is able to interact strongly with DNA. We shall return to the possible significance of this fact.

The Isolation of Lambda Operators

Regions on the DNA molecule that are bound to specific proteins can be isolated by virtue of the fact that the enzyme DNAase digests any naked DNA it encounters but leaves intact any DNA that is covered by a protein. This property of DNAase was utilized by Allan Maxam and Gilbert to isolate an operator region from the DNA of E. coli and was

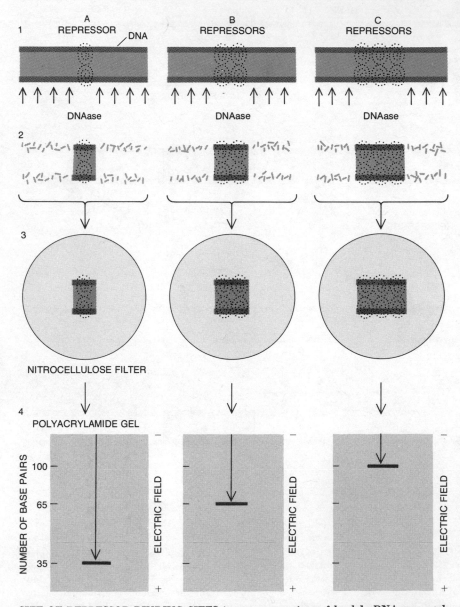

SIZE OF REPRESSOR BINDING SITES in operator regions of lambda DNA was established by mixing the DNA with repressor molecules. Wherever repressors bind to the DNA they protect it from digestion when the enzyme DNAase is added to the mixture (1, 2). As the ratio of repressor to DNA is increased (A, B, C) the segment of DNA protected increases in length, thus indicating that repressor binds to several adjacent sites. The pieces of protected DNA are collected on nitrocellulose filters (3), extracted and subjected to electrophoresis in a polyacrylamide gel (4). The fewer base pairs present in a fragment, the farther it will migrate when it is subjected to an electric field. Evidently this repressor can cover from 30 to 100 base pairs. If the DNA has been labeled with atoms of a radioactive isotope such as phosphorus 32, location of fragments can be visualized with autoradiogram.

later utilized by Vincenzo Pirrotta, then a postdoctoral fellow working with one of us (Ptashne), to isolate the operators on the DNA of the lambda phage to which the repressor becomes attached.

By growing phage particles in a nutrient medium of *E. coli* cells containing the radioactive isotope phosphorus 32 Pirrotta obtained molecules of phage DNA in which the radioactive atoms replaced many nonradioactive phosphorus atoms. The highly radioactive DNA was mixed with purified repressor and then with DNAase. The protected segments of operator DNA were recovered simply by passing the mixture through a nitrocellulose filter. The repressor, like many other proteins, binds tightly to the filter whereas free DNA and DNA-digestion products pass through it. The operator fragment bound to repressor is retained in the filter and can be washed out with a detergent solution. The operator fragment isolated in this way was found to be surprisingly large. A protein the size of the lambda-repressor dimer or tetramer should cover only 15 to 30 base pairs of DNA, but Pirrotta found that roughly 85 base pairs were protected from DNAase digestion by repressor.

We continued these studies of the lambda operators by trying to discover why such an unexpectedly large stretch of DNA is covered by repressor. One possibility we considered (the right one, it turned out) was that the lambda operators may have more than one repressor binding site. We reasoned that if the operators do have more than one binding site, the size of the fragment protected from DNAase digestion should depend on the ratio of repressor to operator in the digestion mixture.

We tested this possibility as follows. We repeated Pirrotta's procedure but varied the ratio of repressor to operator in the DNAase digestion mixture. We then determined the size of the operator fragments by subjecting them to electrophoresis in a polyacrylamide gel. The gel acts as a molecular sieve; under the influence of an electric field smaller DNA fragments migrate through the gel faster than larger ones. Since the operator fragments were labeled with radioactive phosphorus their position at the end of electrophoresis could be determined by placing the gel on photographic film and making an autoradiograph.

We discovered not only that the size of the operator fragment protected by repressor increases as the ratio of repressor is increased but also that the size increases in discrete steps. At low ratios of repressor to operator a single DNA

fragment about 30 base pairs long is recovered. At the highest ratios a single fragment about 100 base pairs long is obtained. At intermediate ratios the fragments are of several intermediate lengths. Moreover, this interesting result was obtained whether we used lambda DNA containing only the left operator or lambda DNA containing only the right operator. These experiments and related ones led us to conclude that each lambda operator does in fact have more than one binding site for repressor. Apparently repressor molecules can line up adjacent to each other on the operator, covering a minimum of 30 base pairs and a maximum of 100.

An analysis of the complexity of the DNA sequence in the various operator fragments identified in these experiments revealed two important additional facts about the structure of the two operators. First, the repressor does not bind randomly to any site within the 100-base-pair operator sequence. Rather, it binds initially at a site adjacent to the N gene (in the case of O_L) and at one or two sites adjacent to the *tof* gene (in the case of O_R). As the repressor-to-operator ratio is increased secondary sites adjacent to the first sites are filled. Second, the base sequences of the various repressor binding sites are similar but not identical. We obtained strong evidence supporting these two facts by certain experiments we need not review here and confirmed them by the work we shall now describe.

Host-Restriction Endonucleases

Experiments on the properties of DNA molecules have been revolutionized in recent years by the use of the enzymes known as host-restriction endonucleases. Most of these enzymes, which are widely distributed in the bacterial kingdom, have the remarkable property of recognizing certain base sequences in DNA molecules and cutting the DNA within those sequences. For example, an enzyme (abbreviated *Hin*) isolated from

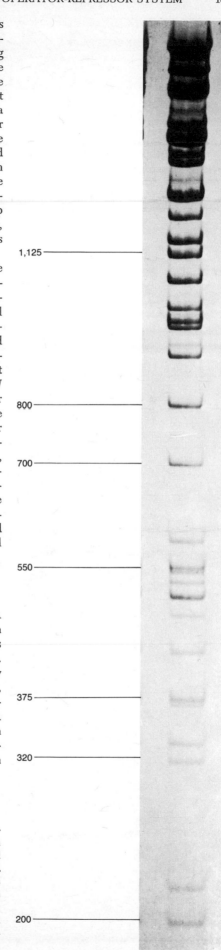

AUTORADIOGRAM shows the result of digesting lambda DNA with a host-restriction endonuclease, an enzyme (*Hin*) isolated from the bacterium *Haemophilus influenzae*. The enzyme cuts DNA when it encounters a particular sequence of six bases. It cuts lambda DNA into about 50 fragments whose sizes are established by electrophoresis. Numbers beside autoradiogram show number of base pairs in representative fragments.

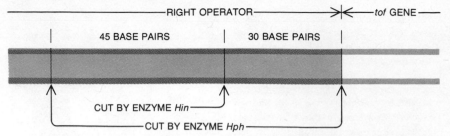

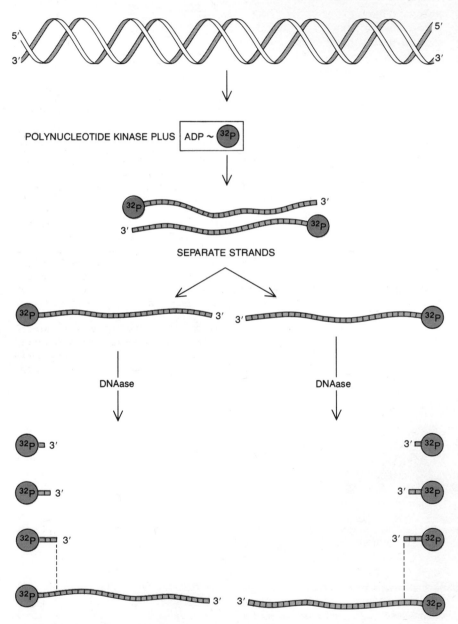

HOST-RESTRICTION ENDONUCLEASES were used to cut lambda-phage DNA in various ways until fragments were obtained in which the base sequence of the operator could be determined. An enzyme (*Hph*) that was isolated from the bacterium *Haemophilus parahaemolyticus* yields a DNA fragment consisting of 75 base pairs that include most of the right operator, which lies next to the *tof* gene. The fragment can be cut in two by enzyme *Hin*.

DETERMINATION OF BASE SEQUENCE in segments of operator DNA is established with the help of enzymes. First, polynucleotide kinase is used to attach an atom of phosphorus 32 (^{32}P) to the 5′ end of each strand in an operator segment including perhaps 30 bases. The strands are separated and mixed with exonuclease, a DNAase that degrades DNA starting from 3′ ends. By removing samples at intervals fragments from one base to 30 bases in length are obtained. Subsequent steps are described in illustration on opposite page.

the influenza bacterium *Haemophilus influenzae* cuts DNA in the middle of the sequence *GTTGAC* paired (on the other chain of the double helix) with *CAACTG*. Another enzyme (*Hpa*) isolated from the bacterial strain *Haemophilus parainfluenzae* cuts DNA in the middle of the sequence *CCGG* paired with *GGCC*. We need not be concerned here that such enzymes may help to protect bacterial species from foreign DNA's (hence the term restriction enzymes) or that some restriction endonucleases act in a more complex way than *Hin* and *Hpa*. The important point is that because the recognition sequences are short they tend to appear at many specific sites on DNA molecules. For example, the enzymes *Hin* and *Hpa* both cut lambda DNA at about 50 sites, generating about 50 specific DNA fragments, ranging in length from fewer than 100 base pairs to several thousand. (Remember that the total length of the lambda DNA is about 47,000 base pairs.)

We designed an experiment with lambda DNA and host-restriction enzymes that we hoped would yield two specific fragments: one containing the left operator and the other the right operator. We reasoned that since the enzymes *Hin* and *Hpa* cut lambda DNA into segments with an average length of 1,000 base pairs and since the operators are separated by some 2,000 base pairs, the enzymes should cut at least once between the operators and, of course, somewhere on either side of the operators. The result should be two fragments with one operator in each. We further reasoned that we could isolate the two fragments by mixing the digestion mixture with repressor and passing the mixture through a nitrocellulose filter. Presumably only those fragments bearing an operator would bind to the repressor molecules and would be trapped in the filter.

We performed this experiment using the enzyme *Hpa*. When we examined the trapped pieces of the lambda-DNA molecule, we found, as we had hoped, two specific fragments, one bearing the left operator and the other the right operator. When we repeated the experiment with the enzyme *Hin*, however, the results were strikingly different. Now we recovered not two fragments bound to repressor molecules but four. We interpreted this result to mean that *Hin*, unlike *Hpa*, cuts the lambda DNA within each operator, thereby splitting each operator into two fragments that independently bind repressor. Our conclusion implied that the largest fragment of operator, 100 base pairs long, isolated in the DNAase digestion experiment would be cut by the *Hin* enzyme. That prediction was soon verified.

The experiments with host-restriction enzymes enabled us to state conclusively not only that each lambda operator has multiple repressor binding sites but also that the sites can function independently. The latter conclusion was demonstrated by the sites' ability to bind repressor even when they are separated with the cutting of the lambda DNA by *Hin*.

The Base Sequence of Operators

The ability of the lambda repressor to recognize and to bind to a number of different sites within each of the two lambda-DNA operators provides an unusual opportunity for studying the molecular basis of specific interactions between protein and DNA. With the hope of being able to identify the sequence of bases within each operator that interact with repressor, we traveled to the Medical Research Council Laboratory of Molecular Biology in Cambridge, England, in the spring of 1973. There Frederick Sanger, George Brownlee, Bart Barrell and their co-workers had developed novel methods for determining the sequence of bases in RNA molecules. Just before our visit Sanger and his co-workers reported their first success at developing methods for doing the same thing with DNA molecules. In collaboration with Barrell and John Donelson, a visiting American postdoctoral fellow, we determined the base sequence of the binding site in the left operator of the lambda DNA that had the highest affinity for the repressor protein.

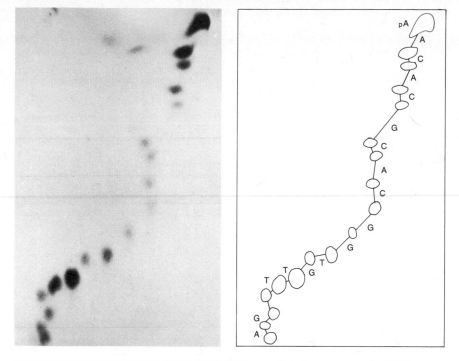

DIFFERENTIAL MOBILITY OF FRAGMENTS is exploited to separate the DNA pieces containing from one base to 30 or 40 bases produced by the method described in illustration on opposite page. The mixture of fragments is first subjected to a special type of electrophoresis (*not shown*), which separates them in one dimension according to base composition, with fragments rich in *T*'s traveling fastest and fragments rich in *C*'s traveling slowest. The separated fragments are then carefully transferred to one edge of a glass plate coated with an ionic resin to be separated in a second dimension by homochromatography. The edge of the plate is dipped in a solution containing RNA molecules of assorted lengths, which compete with the DNA fragments for ionic sites. The shorter DNA fragments are bumped along faster than the longer ones. The result of the two-dimensional separation is shown in the autoradiogram at the left. Reading upward, each spot represents a DNA fragment one base shorter than the fragment responsible for the spot below it. The angle of displacement between any two spots reveals which base is present in the lower fragment but absent in the upper one and hence also reveals the sequence in which bases were removed when the strands were digested one base at a time with DNAase. Thus the letters in the drawing at the right represent sequence of bases in a central region of right operator; if bases are read downward, they represent sequence along one strand in the 5'–3' direction. This method of determining base sequence was devised by Frederick Sanger and co-workers at Medical Research Council Laboratory of Molecular Biology in Cambridge, England.

On returning to Harvard some eight months later one of us (Maniatis) established the base sequence of most of the right operator with the help of Andrea Jeffrey, a research assistant, and Dennis Kleid, a postdoctoral fellow. The methods used to determine the base sequences can be described very briefly. First, we took advantage of the fact that Richard J. Roberts and his colleagues at the Cold Spring Harbor Laboratory had collected and characterized a substantial number of host-restriction endonucleases, samples of which they had isolated from various strains of bacteria. With these enzymes, which were generously provided by Roberts, we proceeded to dissect lambda DNA in and around the right operator. We found that *Hph*, an enzyme isolated from *Haemophilus parahaemolyticus*, neatly excises a 75-base-pair fragment that incorporates most of the right operator. As one would expect, the fragment is itself cleaved by the enzyme *Hin*. The cleavage specifically yields two DNA fragments, one of which is 45 base pairs in length and the other 30, each incorporating part of the right operator.

By combining some new methods of our own with methods developed by workers in Sanger's laboratory (including Edward Ziff and John Sedat, who were then visiting American postdoctoral fellows) we soon developed a fast and accurate technique for determining the base sequence in small lengths of DNA. Remember that DNA chains have a polarity depending on whether the backbone linkage is 5'–3' or 3'–5'. Our technique involves the use of the enzyme polynucleotide kinase to attach a radioactive phosphorus atom to the 5' end of each DNA chain. The chains are then separated from each other and subjected to electrophoresis under conditions such that the base composition of each chain determines its relative mobility in the electric field.

We now determine the sequence of bases along each chain with an ingenious method developed by Sanger's group. Each chain is mixed with exonuclease, a DNAase that degrades DNA one base at a time starting from the 3' end, the end that does not bear the radioactive phosphorus atom. Samples are removed from the digestion mixture at regular intervals and the enzyme is inactivated; hence we recover all the partial products of the degradation. Starting with a chain 30 bases long we end up with as many as 30 different DNA fragments having from

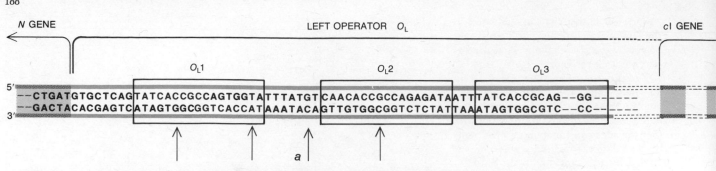

SEQUENCES OF BASES have now been determined for 60-odd base pairs in the left operator of the lambda-phage DNA and 70-odd base pairs in the right operator. The presumed repressor binding sites, each 17 base pairs long, are in colored rectangles. Base sequence of the O_L3 site has not been completely determined. Binding sites are similar in base sequence; moreover, each site has considerable rotational symmetry around its central base pair. This is most readily seen by reading the base sequence on opposite strands starting from opposite ends of a binding site and proceeding toward the middle. Within each site one set of these "half-site" sequences

one to 30 bases, all labeled at the 5′ end with radioactive phosphorus.

The fragments are now separated from one another by electrophoresis; they are placed along one edge of a sheet of cellulose acetate and subjected to the electric field. Under the influence of the field the fragments migrate at a speed influenced by their base composition. T's give rise to the fastest migration and C's to the slowest; the overall order is T, G, A, C. Thus at the end of the electrophoresis molecules rich in T's, regardless of their size, will have traveled farther than those rich in G's, and so on.

The distributed molecules are carefully transferred to one edge of a glass slide coated with a thin layer of ionic resin. There they are further separated along an axis at right angles to the electrophoresis axis by the procedure called homochromatography. The plate, clamped vertically with the fractionated DNA fragments lined up along its lower edge, is dipped in a solution containing RNA molecules of various lengths. As the RNA molecules compete with the DNA molecules for sites in the layer of ionic resin, the DNA molecules are displaced upward; the shorter the DNA molecule is, the farther it travels. The final position of each DNA fragment is revealed on photographic film by autoradiography.

Since the fragments are labeled with radioactive phosphorus, they appear as dark spots running up the film. Remarkably, from the angle of displacement of each fragment from its neighbor one can deduce the identity of the base removed by each step in the DNAase digestion [see illustration on preceding page]. By using both chains of the double-chain DNA fragment, each labeled at its 5′ end, we can determine the probable sequence of 30 to 40 base pairs quite quickly. Although the sequence assignments made in this way are not completely reliable, they can be verified and the ambiguities can be resolved by further manipulations. The important general point is that these methods enable us to determine the base sequence of any double DNA chain up to about 40 base pairs in length.

Our efforts in Cambridge and at Harvard have now yielded the base sequence of a large portion of the left and right operators of phage-lambda DNA [see illustration above]. The sequence of the right operator was also determined by Pirrotta at the University of Basel, who used a method different from ours. It is gratifying that the two approaches yield the same sequence.

What do we learn by determining the sequence of bases that constitutes the lambda operators? As was first noticed by Keith C. Backman, a graduate student working with us at Harvard, there are base sequences in both operators that are strikingly similar. Presumably they are the sites recognized by the repressor. Each of the sites is exactly 17 base pairs long; moreover, they are separated from one another by "spacers," strings of three to seven bases that contain only, or nearly only, the base pair AT.

Mutations in Binding Sites

One of the strongest indications that the closely related 17-base-pair sequences are indeed the sites recognized by repressor comes from a study of mutations that change a base pair in the operator region. Kleid, Zafri Humayun and Stuart M. Flashman, working with one of us (Ptashne), have now determined the sequence of 12 such mutants, many of which were selected and studied by Flashman. Ten of the mutants change the sequence of bases in the sites we have called O_L1, O_L2, O_R1 and O_R2; all of them decrease the affinity of that portion of the operator for repressor. Two mutations located in two different spacer regions, and a third probably located in another spacer, have no effect on repressor binding. Instead they drastically decrease the efficiency with which the enzyme RNA polymerase initiates transcription of the adjacent genes.

An interesting feature of the 17-base-pair repressor binding sites is that each has a partial internal symmetry. What is meant by this is as follows. One reads off the sequence of bases on opposite strands of the binding site starting at opposite ends and proceeding toward the middle. One finds that the two sequences of eight bases on either side of the central base pair are more similar than would be expected if the sequences were random. This partial twofold rotational symmetry, as it can be called, is most apparent in the site O_L1, where six of the eight positions are occupied by the identical base. If the sequence were random, one would expect only two of the eight bases to be the same. The other 17-base-pair binding sites are also more symmetrical than would be expected by chance, although they are less symmetrical than the O_L1 site. That site, perhaps significantly, is also the site with the highest affinity for repressor.

One way to compare the various repressor binding sequences with one another is to consider only the "half-site" sequences obtained by reading the opposing chains in the 5′–3′ direction, beginning at the ends of each 17-base sequence and proceeding toward the middle. If any site were perfectly symmetrical, the two half-site sequences compared in this way would be identical. In fact, certain symmetrically arranged positions in every half-site are identical [see top illustration on page 190]. For example, at position No. 2 we always find the base A; at positions No. 4 and No. 6 we always find the base C. Positions No. 5 and No. 8 are usually occupied by an A and a G respectively. We infer that the identity of the bases at these positions cannot be changed without strongly af-

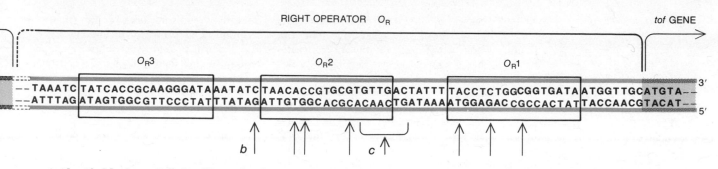

is identified by letters in color. Nine colored arrows identify base pairs within repressor binding sites that have been altered in mutant lambda DNA. In every case the mutation abolishes the affinity of the repressor for that binding site. Three black arrows (*a*, *b*, *c*) show location of mutations that do not affect repressor binding but do interfere markedly with the ability of RNA polymerase to recognize promoter regions for gene *N* and genes *cI* and *tof* respectively. (Exact location of the mutation indicated by the arrow *c* is uncertain.) The base sequence of the *cI* gene, which codes for the repressor, has not been determined; it is about 1,000 base pairs long.

fecting the affinity of the site for repressor. Evidently the requirements of the other positions are not so stringent, although the favored sequence is *TATCACCG*, with either *C* or *G* as the ninth base (that is, the middle base of the 17-base sequence). The 10 operator mutations that have been found, each of which abolishes affinity for repressor, are located at positions No. 2, No. 5, No. 6 and No. 8.

Is there any reason to expect that the base sequences recognized by repressor would be symmetrical? Some years ago Walter Gierer of the Max Planck Institute for Biophysical Chemistry in Göttingen suggested that symmetrical sequences in DNA might form looped structures and that the loops might be recognized by repressors. We now have good reason to believe repressors bind to DNA in linear form; thus we prefer to think that symmetry of the operator reflects a symmetry in the structure of the repressor. As we have mentioned, the lambda repressor binds to DNA as a dimer or tetramer made up of identical subunits. On the basis of general considerations of protein architecture we suspect the repressor is largely, if not entirely, symmetrical; therefore we would not be surprised to find symmetry in the operator to which it binds.

As was first noted by Hamilton O. Smith and Kent W. Wilcox of Johns Hopkins University, the sites recognized by some restriction endonucleases, although much shorter than those recognized by repressors, also have a twofold symmetry. Moreover, as Gilbert, Maxam and their colleagues have shown, the operator recognized by a different repressor, the *lac* repressor, is highly symmetrical. (The *lac* repressor turns off the genes in *E. coli* that specify enzymes for the utilization of the sugar lactose.)

Our analysis of the lambda operators has emphasized the internal symmetries in the repressor binding sites. It must be said, however, that there is no direct evidence that the repressor actually interacts with the DNA symmetrically. Moreover, we know of cases where proteins recognize DNA sequences without there being any apparent symmetries. In fact, there may be a systematic asymmetry in the various lambda-repressor binding sites. Each 17-base-pair site has on one side the sequence *TATCACCGC* or a sequence closely related to it, together with another half-site where the sequence is more variable. It is possible that repressor protein bound to its operator DNA is somewhat deformed and not perfectly symmetrical.

An appraisal of the role of symmetry in operator-repressor interaction may call for the chemical synthesis of various operator sequences and a study of their interactions with repressor. In any case we believe each repressor will recognize its own favored DNA sequence; we have no reason to believe the sequences of operators recognized by different repressors will be similar to the sequences of the lambda operators. In particular, the base sequence of the *lac* operator, the only other operator whose sequence is known, bears no obvious relation to the sequences of the lambda operators. Moreover, we cannot yet describe in any detail how the repressors recognize their operator targets in DNA. For example, the 17 base pairs of a binding site occupy nearly two full turns of the double-chain DNA helix. One can only guess at what the repressor protein "sees." Perhaps some simple rules govern the recognition of base sequences by proteins, but if they do, they remain to be discovered.

How the Repressor Works

Our experiments with lambda repressors and operators, taken together with the results from several other laboratories, have revealed how the repressor turns off its target genes. At the University of Basel, Alfred Walz and Pirrotta have shown that in the absence of repressor the enzyme RNA polymerase binds tightly to, and protects from DNAase digestion, a 45-base-pair sequence that includes most of one of the repressor binding sites. The same sequence includes about 20 base pairs of the beginning of the gene the RNA polymerase transcribes into messenger RNA. Many other cases have been found where RNA polymerase covers about 20 base pairs on either side of the beginning of a gene. (The covered region does not include all the bases required for RNA-polymerase recognition because, as was found by Russell A. Maurer, a graduate student working with us, and by others, mutations some seven bases to the left of that region severely impede the action of RNA polymerase. The entire DNA region required for RNA-polymerase binding is called the promoter, and its exact extent is not known.) RNA polymerase apparently recognizes some aspect of the base sequence near the beginning of many genes. When the enzyme is supplied with the appropriate substances on which to act, it copies one chain of the DNA sequence into messenger RNA.

The RNA-polymerase molecule is some 20 times heavier than the repressor molecule and is therefore several times larger in volume. From the DNA region covered by RNA polymerase one can see that the enzyme competes with the repressor for its binding site. Therefore when repressor is bound to the operator, RNA polymerase cannot bind to the promoter region. The same effect is seen with the *lac* repressor.

Why does each lambda operator have more than one repressor binding site? The *lac* operator has only one repressor binding site, so that multiple binding sites do not seem to be the general rule. One possibility is that multiple sites allow for graded control. For example, if

WILD-TYPE SITES									MUTANT BASES
O_L1	T A T C A C C G C								A OR T
	T A C C A C T G G								G
O_R1	T A T C A C C G C								T
	T A C C T C T G G								G, A
O_L2	C A A C A C C G C								T
	T A T C T C T G G								
O_R2	C A A C A C G C A								A
	T A A C A C C G T								G, A
O_R3	T A T C C C T T G								
	T A T C A C C G C								
O_L3	T A T C A C C G C								

$$T_9 \ A_{11} \ T_6 \ C_{11} \ A_8 \ C_{11} \ C_6 \ G_9 \ C_5$$
$$C_2 \ A_3 \ T_2 \qquad T_4 \ C_1 \ G_4$$
$$C_2 \qquad C_1 \qquad G_1 \ T_1 \ A_1$$
$$T_1$$

OPERATOR HALF-SITES are compared to reveal their similarity. The 11 sequences are those designated by colored letters in illustration on preceding two pages. Each sequence is written left to right in the 5'–3' direction. Letters in color at left indicate nine sites where 10 different mutations have been found. Mutant bases that replace normal bases and abolish repressor affinity are listed in the column at the right. Tabulation at lower left summarizes frequency with which each base appears at each position in the 11 half-sites. One can see that the second, fourth and sixth positions are invariant.

only one operator site is occupied by repressor, some gene transcription might occur, whereas if all sites were filled, transcription would be abolished. Flashman and Barbara Meyer, another graduate student, have recently shown that the maximum repression of transcription of the *tof* gene requires that repressor be bound to both O_R1 and O_R2. Thus mu-

tation of either O_R1 or O_R2 decreases the effect of repressor on the transcription of *tof,* and mutation of both sites has a stronger effect. Significantly, two mutations at one site do not have as strong an effect as two mutations at two sites. Similar experiments show that the maximum repression of gene *N* requires repressor bound to both O_L1 and O_L2. We imagine that occasional unnecessary expression of the *lac* genes is not harmful to the cell, whereas occasional expression of the lambda genes could be lethal. We may therefore speculate that the lambda system, involving multiple repressor binding sites at each operator, has evolved stricter controls than the *lac* system, involving only a single repressor binding site.

Recently Meyer has discovered a most remarkable function for O_R3: the third repressor binding site in the right operator. She found that RNA polymerase recognizes not one promoter region in the right operator but two regions. Moving to the right from one promoter region, RNA polymerase transcribes the *tof* gene. Moving to the left from the second promoter region, it transcribes the *cI* gene, the gene that codes for the repressor. Although the exact starting point for the transcription of the *cI* gene is not known, it is probably just to the left of O_R3. We have mentioned the fact that transcription of *tof* begins just to the right of O_R1. Therefore we have two genes, *cI* and *tof*, transcribed in opposite directions and separated by three repressor binding sites.

Meyer has found that repressor not only turns off the transcription of *tof* but also turns off transcription of *cI*. In other words, repressor regulates its own level in the cell. From analysis of the effects of mutations in the right operator we de-

duce that repressor turns off the transcription of *cI* primarily by binding to O_R3. Because repressor binds more weakly to O_R3 than to O_R1 and O_R2 it allows a higher level of *cI* transcription (and hence a higher level of repressor protein) than it does of *tof* transcription [*see illustration below*].

Even this description of the role of the reiterated repressor binding sites is probably incomplete. For example, Meyer has found that relatively large amounts of repressor turn off the transcription of both *tof* and *cI*, whereas smaller amounts suffice to turn off the transcription of *tof*. On the other hand, the smaller amounts of repressor actually enhance the transcription of *cI*, the repressor's own gene. This positive effect of repressor on the transcription of its own gene had been predicted by work on whole bacterial cells infected with bacteriophage lambda; the work was done by, among others, Louis Reichardt, who is now at the Harvard Medical School. We do not yet know what molecular mechanisms are involved.

We suspect that as investigations progress even more sophisticated roles in gene regulation will be assigned to the system we have been describing. We have described some features of the interaction of two proteins, lambda repressor and RNA polymerase, with the sequences in and around the lambda operators. We know, however, that at least two other proteins, the products of genes *N* and *tof*, are themselves regulatory proteins that almost certainly recognize sequences in the operators. Although our understanding is far from complete, these studies have begun to reveal how complex patterns of gene regulation can be described in terms of specific interactions between proteins and DNA.

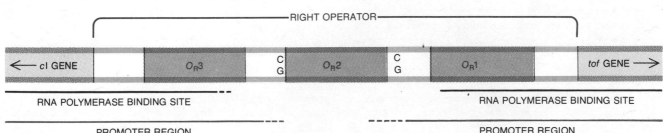

RNA-POLYMERASE BINDING SITES overlap at one end of the left lambda operator and both ends of the right operator. For simplicity only the right operator is shown. Repressors produced by the *cI* gene bind preferentially to O_R1 and O_R2, blocking expression of the *tof* gene. As long as no repressor is bound to O_R3, RNA-polymerase molecules have access to the binding site next to the *cI* gene and keep generating more repressors. (The exact location of the RNA-polymerase binding site next to the *cI* gene is not known.) Eventually, however, O_R3 is occupied by repressors and access to the *cI* gene is blocked. The ability of RNA polymerase to bind to lambda DNA is influenced by the integrity of bases lying somewhat outside the binding site. Thus mutations in the base pair *CG* to the left of O_R2 make it impossible for polymerase to transcribe the *cI* gene. Location of this gene within right operator was first noted by Gerald Smith and his colleagues at University of Geneva. A similar mutation in a *CG* pair between O_L1 and O_L2 (*not shown*) prevents transcription of *N*. A mutation believed to affect a *CG* pair to right of O_R2 blocks transcription of *tof*. Total sequence of bases required for polymerase recognition is termed a promoter; exactly how many bases are involved is not established.

GENETIC
TRANSACTIONS

GENETIC TRANSACTIONS

V

INTRODUCTION

If an organism is the means by which an ensemble of DNA molecules attempt to assure their continued propagation, then it is not surprising that some DNA molecules have acquired the capacity to propagate in cells as parasites.

The strategies that these parasite DNA molecules have evolved to serve their needs are various. Some involve a rather straightforward takeover of the cell whose substance they invade, such as is displayed by the virulent T4 bacteriophage described previously in Edgar and Epstein's article "The Genetics of a Bacterial Virus."

This takeover leads to the production of progeny virus with the subsequent lysis and death of the cell. In these instances, the cell-to-cell transmission of the parasitic DNA molecules of the virus requires the DNA molecules to encode a specific protein apparatus for the encapsulation of the DNA, absorption of the virus to specific host cell surface receptors, and delivery of the DNA molecule into the cytoplasm of a new host cell.

Other and more subtle molecular parasites (plasmids) arrange for their transfer from host to host by means of instigating a mating reaction between the donor plasmid-carrying cell and a recipient nonplasmid-carrying cell by a mechanism akin to that described by Jacob and Wollman in their article "Viruses and Genes"; this is the conjugal "chromosome" transfer that occurs in *Escherichia coli* and some related bacterial species.

The plasmids that have been discovered and described in any detail are those that confer, on the bacterial cells carrying them, resistance to one or more antibiotic drugs. Thus, the relationship that exists between the host cell and the plasmid is more accurately described as one of conditional symbiosis. The natural history and molecular biology of the drug-resistant plasmids are described in Royston C. Clowes's article "The Molecule of Infectious Drug Resistance."

The ultimate subtlety in molecular parasitism so far described is that enjoyed by certain temperate bacterial viruses and tumor viruses. In both of these virus types the parasitic molecule (or a DNA copy of it, if the virus is an RNA tumor virus) manages to intercalate itself into the continuity of the host cell DNA, where it can persist, replicating along with the host cell DNA.

Two examples of this variety of molecular parasitism are included in this collection. The first deals with the classic instance of this phenomenon—namely, that of the lambda bacteriophage, as is described in the article by Allan M. Campbell, "How Viruses Insert Their DNA into the Host Cell."

The second example describes the surprising strategy adopted by RNA tumor viruses that contain a polymerase capable of producing a DNA copy from the viral RNA template. The DNA so made is then intercalated into the host DNA, where it not only plays a critical role in the infection cycle of the virus but is believed to be the basis of the tumorogenicity of the virus, by some as yet undescribed mechanism. The RNA tumor virus story and the speculation that RNA-DNA transcription may be involved in some aspects of normal development and differentiation of higher organisms is presented in Howard M. Temin's article, "RNA-Directed DNA Synthesis."

DNA molecules are capable of interchanging their substance, and, thus, their information, by mechanisms of recombination. While you are no doubt familiar with the consequences of genetic recombination, and you may be familiar with the much rarer phenomenon of chromosome rearrangement leading to deletions, duplication, inversions, and translocation, you may not be aware of a phenomenon that seems to occupy a halfway position between regular recombination and structural rearrangement and is indicative of a surprising plasticity in the organization of a creature's genome. This plasticity is revealed by evidence that clusters of genes are not only capable of moving by conjugal interaction from cell to cell as described for the plasmids but are also capable of moving from place to place within a genome. Such "jumping genes" or DNA sequences were first described by the famous corn geneticist Barbara McClintock. The article by Stanley N. Cohen and James A. Shapiro, "Transposable Genetic Elements," summarizes the ubiquity and molecular basis of this phenomenon and concludes this section on genetic transactions.

19

The Molecule of Infectious Drug Resistance

by Royston C. Clowes
April 1973

R factors, which transmit resistance to antibiotics from one bacterial strain to another, are carried on extrachromosomal genetic elements called plasmids, some of which have now been isolated and measured

In 1971, only a few months after the antibiotic gentamycin had been introduced for the treatment of bacterial infections, strains of bacteria had developed resistance to the antibiotic. This development was not too surprising. Ever since sulfonamides were first administered to treat bacterial infection it had been known that repeated exposure of bacteria to an antibiotic results in the development of a resistant strain through the processes of mutation and selection. A bacterium that has spontaneously undergone a gene mutation that makes it no longer susceptible to the antibiotic's effect will live and multiply in the presence of the drug, and its similarly resistant progeny constitute the resistant strain. Mutations are rare, however; usually only one bacterial cell in a population of several hundred million will have mutated, so that the physician has a chance to treat the infection successfully by administering antibiotics before resistant cells develop in large numbers.

The resistance to gentamycin was different. The bacteria from the first patient in whom a gentamycin-resistant strain was detected were resistant not only to gentamycin but also to three other common kinds of antibacterial agent: chloramphenicol, ampicillin and sulfonamide. Then, from a second patient, investigators isolated four different bacterial strains, each of which carried resistance to the same four antibiotics. Moreover, when they grew any of these bacteria in laboratory glassware with a bacterial strain that was susceptible to the antibiotic's effects, the susceptible strain also became resistant to all four antibiotics in a matter of hours. From the investigators' knowledge of many similar cases reported during the past decade

they realized that they were dealing with a case of infectious antibiotic resistance. Each strain isolated from the patients carried an "infectious antibiotic resistance factor," or R factor: a complex consisting of a gene producing gentamycin resistance, genes determining resistance to the other three antibiotics and genes that facilitate the transfer of the R factor.

Intestinal bacteria that give rise to such diseases as enteritis or typhoid fever and carry R-factor-controlled resistance to all known antibiotics have now been isolated in most parts of the world. Because the genes of these R factors are transferred independently of the genes on the bacterial chromosome, and because they are "cured," or lost, if bacteria carrying them are grown in the presence of certain dyes or other agents, they are known to exist independently of the chromosome; they are carried on the extrachromosomal genetic elements called plasmids. A strain carrying an R factor is able to conjugate, or pair, with other bacterial cells. When that happens, a copy of the R plasmid is transmitted to the second cell, which thereby acquires all the drug resistances carried on the plasmid and also the potential of transmitting those resistances to still more cells.

During the 1960's genetic experiments in a number of laboratories established the mode of transference of R

factors and revealed something of their composition and structure [see "Infectious Drug Resistance," by Tsutomu Watanabe; SCIENTIFIC AMERICAN, December, 1967]. More recently the effort in my laboratory at the University of Texas at Dallas and in other laboratories has been to learn more about the physical nature of the plasmids that constitute the R factor: their size and structure, the ways in which they multiply and how they may evolve.

It seemed likely that R factors are composed of the genetic material DNA, and that had been confirmed by transferring R factors to *Proteus mirabilis,* a strain of bacteria that has DNA of unusually low density. When DNA extracted from a *Proteus* culture is suspended in a solution of the salt cesium chloride and spun at high speed in a centrifuge, the DNA molecules move to positions in the centrifuge tube where their density is equal to the density of the salt surrounding them. The contents of the tube are separated into a number of fractions, each from a different level in the tube. Fractions containing DNA are then identified either by measuring the extent to which they absorb ultraviolet radiation of a certain wavelength or by first incorporating radioactive subunits in the DNA of the live culture and then measuring the radioactivity of each fraction [*see illustrations on page 197*]. If the

R-FACTOR PLASMIDS, shadowed with platinum vapor, are enlarged 62,000 diameters in an electron micrograph made by Michiko Mitani. The plasmids are molecules of looped DNA. The extended one (*bottom*) is a "nicked duplex loop"; the distance around it can be measured to establish its molecular weight. The other one (*top*) is a "supercoiled" form.

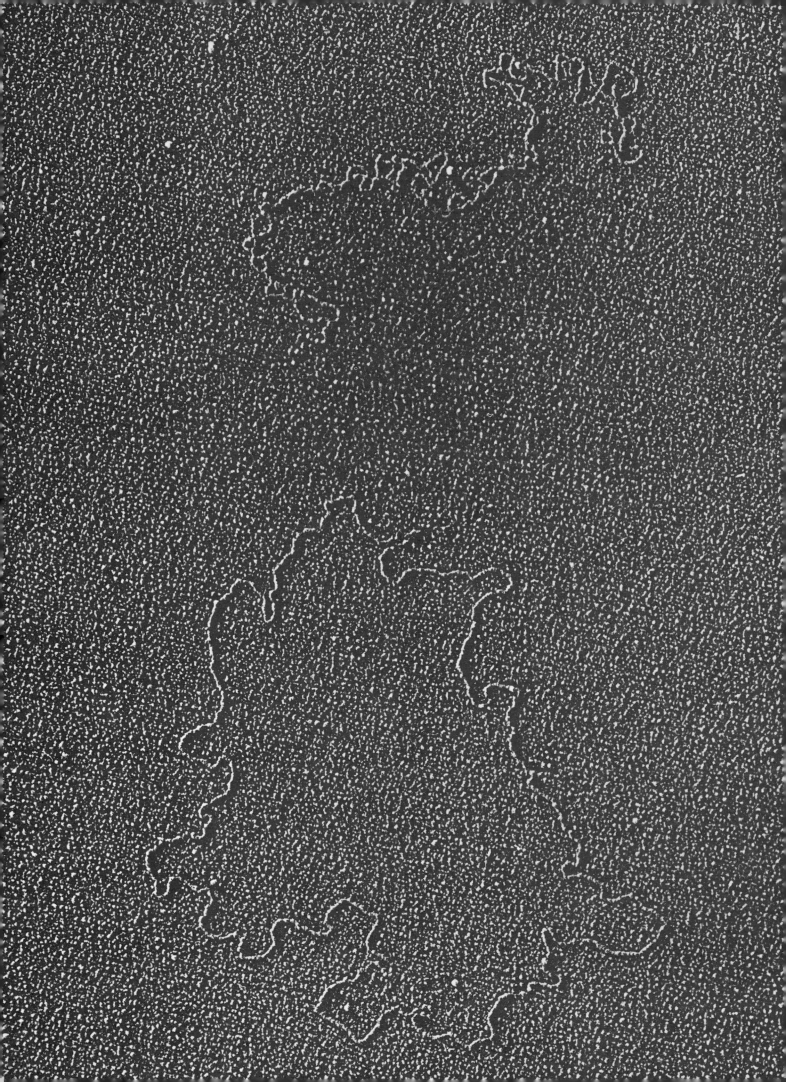

DNA is extracted from a *Proteus* strain to which an *R* factor has been transferred, one finds in addition to the main band of DNA characteristic of the *Proteus* chromosome a smaller "satellite" band at a greater density. If the strain of *Proteus* is then "cured" of its *R* factor, this denser satellite band is no longer seen, and so it is clear that the satellite band must be the DNA of the *R* plasmid.

The bacterial chromosome is a single giant DNA molecule in the form of a closed loop about a millimeter long (about 1,000 times as long as the cell). After DNA isolated from a culture of *Proteus* harboring an *R* factor had been centrifuged in cesium chloride, my colleagues Taizo Nisioka and Michiko Mitani and I took samples from the fractions containing the *R*-plasmid DNA and prepared them for examination in the electron microscope by what is called the microdrop technique. In this method the DNA samples are mixed with a solution of a basic protein such as cytochrome *c*. When small drops of the mixture are put on a material such as Teflon, the protein molecules spread over the surface of the drops as oil does on water, in a film one molecule thick. The acidic groups of the DNA become bound to the basic groups of the protein, so that the long DNA molecules, teased out gently without breaking, become extended at the surface of the film [*see top illustration on page 198*]. A sample is taken from this surface to an electron-microscope specimen grid and dried, and the DNA molecules can be viewed in the microscope.

We found that the DNA of the *R* factor consists of closed looped molecules, some of them extended and some of them tightly coiled [*see illustration on opposite page*]. The distance around all the extended loops was about the same. After storage over many months the proportion of tightly coiled molecules decreased, but the number of extended loops, which were the same size as the original ones, increased. It was clear, therefore, that the DNA of the *R* factor is made up of looped (or what are often called "circular") molecules of a specific size.

Knowing the length of other DNA molecules of known molecular weight, and thus the molecular weight per unit length of DNA, we could calculate the molecular weight of the *R*-plasmid molecules. By this method and others all *R* factors so far investigated have been shown to be looped DNA molecules, each with a characteristic size. Their molecular weights range from about one

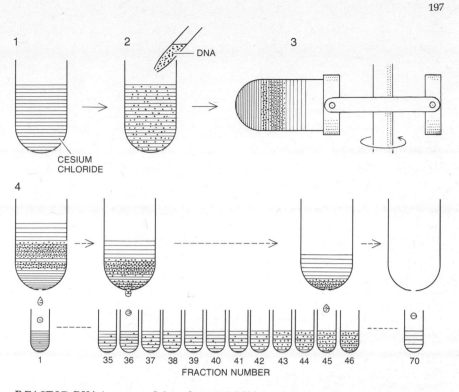

R-FACTOR DNA is separated from bacterial DNA by density-gradient centrifugation. A solution of cesium chloride (*1*) to which DNA from a culture of *Proteus mirabilis* carrying an *R* factor has been added (*2*) is spun in an ultracentrifuge for 48 hours (*3*). The DNA's move to points in the centrifuge tubes where their density is equivalent to that of the cesium chloride. The contents of a centrifuge tube are collected drop by drop in smaller tubes (*4*); each fraction contains material from a different level and hence of a different density.

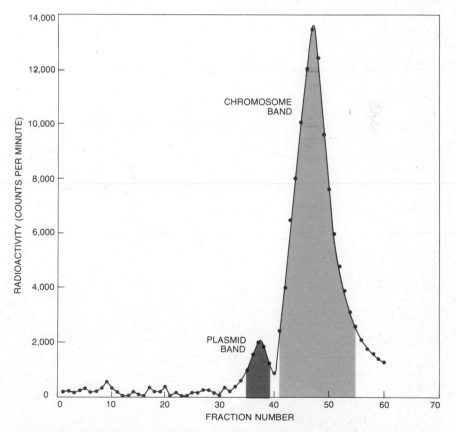

DNA-CONTAINING FRACTIONS can be identified if radioactive subunits were previously incorporated in the bacteria; the amount of DNA in each fraction is proportional to the radioactivity. In the case of *Proteus mirabilis* carrying an *R* factor there are two DNA peaks. The major, less dense band (*gray*) is the DNA of the *Proteus* chromosome. The satellite band (*color*), which comes from earlier fractions, represents the denser DNA of the *R* factor.

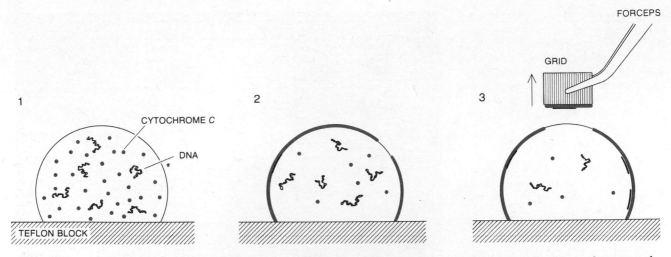

MICRODROP TECHNIQUE developed by Dimitrij Lang and Michiko Mitani at the University of Texas at Dallas mounts DNA molecules for electron microscopy without distorting them. A drop of water containing DNA (*black*) from a plasmid-rich centrifuge fraction, salt, formaldehyde and the protein cytochrome *c* (*color*) is placed on a water-repellent surface (*1*). The cytochrome *c* molecules form a monomolecular film at the surface of the drop and the positively charged protein attracts the negatively charged DNA molecules, which unfold against the film (*2*). A part of the film, with the DNA, is picked up on a microscope specimen grid (*3*).

megadalton to more than 100 megadaltons. (A dalton, the unit of molecular weight, is the weight of one hydrogen atom. One megadalton is a molecular weight of one million.) This spectrum of sizes is about the same as that found for the DNA of bacterial viruses and is at most only a small fraction of the size of the bacterial chromosome.

The fact that *R*-plasmid molecules are circular and tightly coiled was of great value in our subsequent work. When DNA molecules are extracted from cells, they are subjected to shearing forces that may break them. The small size and tight coiling of plasmid molecules make them resistant to these forces. More important, a circular molecule must by its nature have remained unbroken during extraction, and so its size and weight are true measures of what the native molecule was inside the cell. The special properties of various species of circular DNA have been investigated intensively by two groups headed by Robert L. Sinsheimer and Jerome Vinograd at the California Institute of Technology. The Cal Tech groups found circular DNA molecules of two kinds. In one the two strands of the DNA double helix are joined to form a duplex loop; these structures are "supercoiled," presumably so that the long DNA molecules can be fitted inside a cell. In the other kind only one of the DNA strands is joined; these molecules are not supercoiled. If a sin-

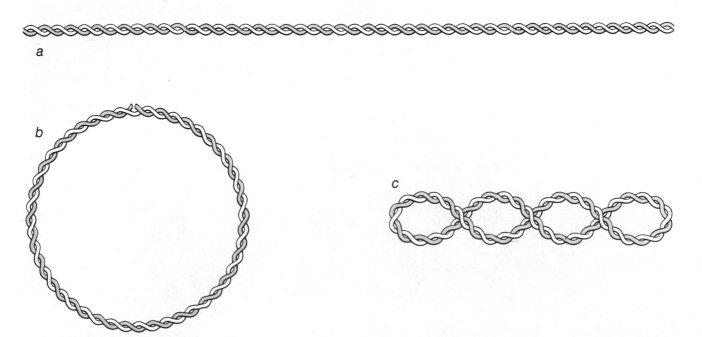

DNA MOLECULE is found in three forms. In each the DNA consists of two twisted strands that form a double helix, or duplex; the strands are connected, as by the rungs of a ladder, by paired subunits (nucleotides) whose particular sequence constitutes the genetic code. The basic form is the linear duplex (*a*). If the ends of the duplex are joined through one strand, the structure is a nicked duplex loop (*b*). If both strands are joined, the structure is an intact duplex loop, usually found in the supercoiled form (*c*).

gle break is made in one strand of a supercoiled molecule, the molecule loses its coils and is converted to the uncoiled structure, which is therefore called a "nicked duplex loop" [*see bottom illustration on page 198*]. The two kinds of circular DNA we had found in R factors were just such nicked duplex loops and supercoiled molecules.

There is an effective method of distinguishing supercoiled DNA from other DNA, which depends on the insertion of molecules of ethidium bromide, one of the dyes that can "cure" plasmids, between adjacent DNA subunits. This intercalation, as it is called, results in two changes: the length of the DNA molecule is increased, so that it becomes less dense, and the helix becomes slightly less twisted. Large numbers of ethidium bromide molecules can intercalate in a nicked duplex loop or a linear duplex, without the untwisting effect leading to coiling. In contrast, the number of dye molecules that can be inserted in a supercoiled loop is strictly limited because there is a limit to the further coiling that is produced as a result of untwisting. The limitation on intercalation limits the associated extension in length and thus leads to a smaller decrease in density for supercoiled loops than for nicked duplex loops or linear duplex DNA. Therefore if ethidium bromide is added to a mixture of these three types of DNA molecule, the supercoiled loops can be separated from the other forms as a denser band after centrifugation.

Since, as we have seen, a major portion of R-factor DNA exists in the cell as supercoiled loops, it can be isolated and separated by this procedure from the chromosomal DNA. That DNA, being about 100 times longer, is generally broken into linear fragments when it is extracted from the cell. The procedure provides a means of separating plasmid DNA even in host bacteria such as *Escherichia coli*, in which the chromosomal DNA does not differ in density from the R plasmid as it does so conveniently in *Proteus*. Since *E. coli* and related bacteria are the natural hosts of many R factors, the procedure enables one to study the plasmids as they are found in nature.

The same experiments can be used to measure the number of R-plasmid molecules in each cell. When bacteria are grown in the presence of a radioactive subunit of DNA such as thymine, the DNA in the cell takes up this radioactive label. If we assume that the uptake of radioactive thymine into plasmid DNA occurs to the same extent as it does into

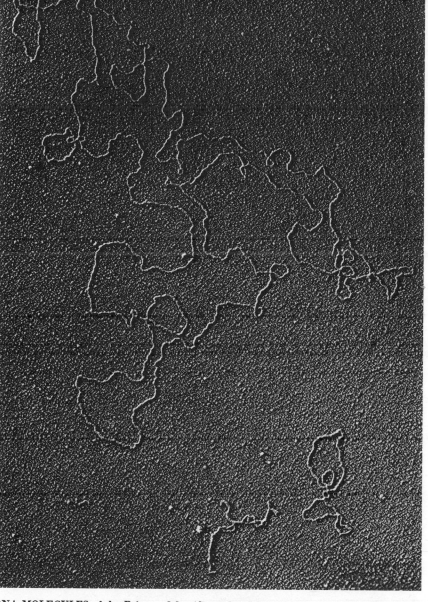

DNA MOLECULES of the R factor delta (S), a plasmid aggregate, are enlarged 40,000 diameters in an electron micrograph made by Christine Milliken. The large nicked duplex loop (*top*) is the delta plasmid, the transfer factor. The small molecules (*bottom*) are S plasmids, carrying streptomycin resistance. One is supercoiled, one is a nicked duplex loop.

chromosomal DNA, and if most of the plasmid DNA exists as supercoiled loops, then the relative amounts of DNA in the plasmid and in the chromosome can be measured by the relative amounts of label in the denser satellite peak and in the main peak. Since we know the molecular weights of a number of R plasmids, and other experiments have shown that the molecular weight of the *E. coli* chromosome is 2,500 megadaltons, we can estimate the relative number of R-plasmid molecules per chromosome from their relative DNA content in the cell. By this method and others many R-plasmid molecules have been found to exist in an approximate one-to-one relation with the chromosome. Although the DNA's of the R plasmid and of the chromosome are physically separated, then, there appears to be some regulatory mechanism in the bacterium such that the R plasmid replicates only once for every chromosomal replication. How this regulation is effected is not known and is the subject of intensive work in a number of laboratories.

Before 1955 infectious antibiotic resistance factors were rare. By 1965 it was found that up to 60 to 70 percent of all common intestinal bacteria from hospitals and other clinical sources carried *R* factors, usually determining resistance to three or more antibiotics. *R* factors can also be transferred to bacterial species responsible for cholera and plague, and similar elements are now common in staphylococci. The epidemic spread of these factors among pathogenic bacterial strains has made an understanding of how they have evolved—and how they evolve so quickly—an important problem in clinical bacteriology today.

One of the original hypotheses for *R*-factor evolution suggests that it depends on a process known as gene pickup. To explain this concept we must consider another bacterial plasmid, the *F* factor, which determines sexuality and conjugation in *E. coli*. Like the *R* factors, the *F* factor is infectious, can transmit itself sequentially by conjugation and is an intact duplex DNA loop (of about 62 megadaltons), and there is about one of these *F* plasmids for every chromosome. The infectious transfer of the *F* factor and of a number of *R* factors has recently been elegantly defined in molecular terms by Daniel Vapnek and W. Dean Rupp of the Yale University School of Medicine. They have shown that conjugation leads to a break in one of the two strands of the intact *F*-factor duplex loop and the transfer of that strand to the recipient cell. There it synthesizes its complementary strand and subsequently forms an intact duplex loop; meanwhile the

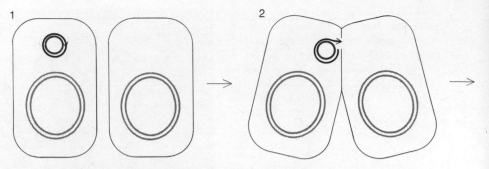

INFECTIOUS TRANSFER of a bacterial plasmid such as an *F* factor or one of the *R* plasmids is diagrammed. Two bacteria are shown about to conjugate, or pair (*1*); each of them has a circular chromosome (*gray*), and one of them, the donor cell, also has a

strand retained in the donor also forms its complementary strand and is converted into an intact duplex loop [*see top illustration on these two pages*].

From donor cells that harbor an *F* factor (called *F*+ donors) rare mutant cells can be selected in which the *F*-plasmid DNA has been inserted within the bacterial chromosome without causing a break in the chromosomal loop [*see bottom illustration on these two pages*]. Conjugation by one of these mutant strains results in a similar break within the DNA of the inserted *F* plasmid. Thereafter a strand of DNA consisting of a part of the *F* plasmid is transferred to the recipient cell, taking with it the attached chromosomal strand. This chromosomal transfer occurs very frequently, so that the mutant strains are called *Hfr* (for high-frequency recombination). Different *Hfr* cells are formed after the insertion of the *F* plasmid at different sites along the chromosome.

Still a third type of donor is derived

from *Hfr* donors by a reversal of the insertion process, leading to the release of the integrated *F* factor from the chromosome. This release usually occurs at a point slightly different from the point of insertion, in which case a segment of the bacterial chromosome is incorporated within the DNA of the released *F* plasmid and some of the plasmid DNA remains in the chromosome. These released *F* factors that have "picked up" chromosomal DNA are known as *F*-prime factors, and bacterial strains in which they are present are called *F*-prime donors. A similar mechanism of gene pickup was suggested as the basis for the evolution of *R* plasmids. The idea was that a "transfer factor" resembling *F* was integrated in the chromosome and then released, together with a segment of adjacent chromosomal DNA on which a gene controlling drug resistance was located.

Some recent findings make the idea of evolution of *R* plasmids through gene

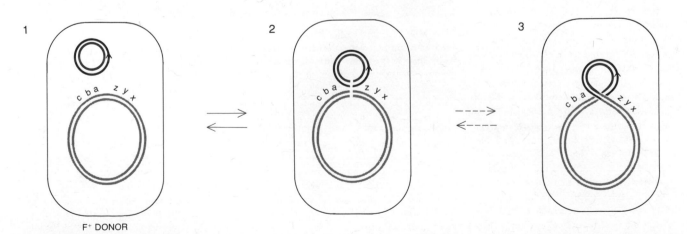

F+ DONOR

DONOR CELLS can be of three types, the relations among which are illustrated here with the *F* factor as a model. An *F*+ donor (*1*) usually transfers *F* as in the illustration at the top of the page. Sometimes, however, a break occurs in both strands of the *F* plasmid and of the chromosome (*2*) in a region where the nucleotide sequences of plasmid and chromosome are similar enough so that the two sets of ends join (*3*) and the plasmid is inserted into the chromosome (in this case between genes *x, y, z* and *a, b, c*) to form

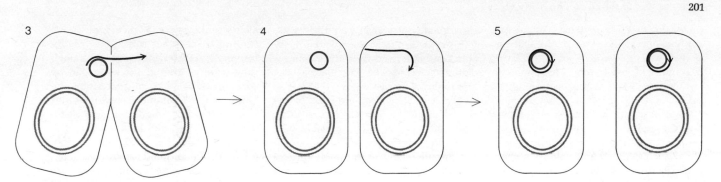

plasmid, an intact duplex loop (*color and black*). On pair formation (2) one strand of the plasmid DNA breaks at a particular point (*arrow*) and passes into the recipient cell (3, 4). After transfer each strand of DNA synthesizes its complementary strand, so that two intact duplex loops are formed. Each cell now has a copy of the original plasmid and is therefore a potential donor cell (5).

pickup seem less likely. The most telling argument comes from discoveries about the mode of action of antibiotic resistance mediated by *R* factors. Many antibiotic molecules exert their lethal effect by becoming attached to bacterial ribosomes: the cellular particles where proteins are synthesized. In this way they interfere with protein synthesis. In all bacteria where they have been analyzed resistance to these antibiotics that arises from mutations of chromosomal genes is effected by changes in the structure of the ribosomes such that the antibiotic can no longer interfere with protein synthesis [*see illustration on page 202*]. In contrast, *R*-factor resistance to these same antibiotics involves the direct modification of the antibiotic molecule in such a way that it can no longer interact with the ribosome, which remains unchanged. Since the mode of action of *R*-factor resistance appears to be so different from that of chromosomal resistance, *R*-factor genes (unless they arise from chromosomal genes as yet unidentified) must be extrachromosomal in origin.

Although the DNA of any one of a number of *R* factors isolated from either *E. coli* or *Proteus mirabilis* consists of

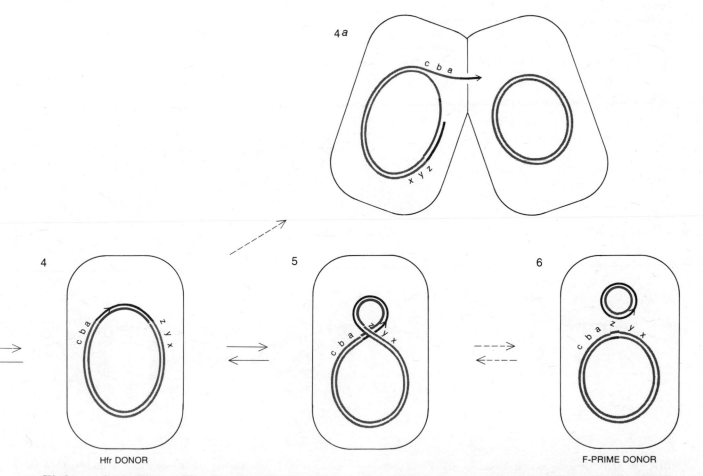

an *Hfr* donor cell (4). When an *Hfr* cell conjugates, one strand of the *F* plasmid breaks and the first part of the plasmid strand is transferred to the recipient along with the chromosome (4a). Occasionally the *F* plasmid is released from the chromosome of an *Hfr* donor. If this occurs by breakage and reunion at a point different from where the plasmid was incorporated, segments of chromosomal and plasmid DNA are interchanged. The plasmid, having picked up a chromosomal region, becomes an *F*-prime factor (5, 6).

molecules of the same size, certain exceptions are found. One of the first reported was in the case of the 222/R plasmid, which determines resistance to tetracycline, chloramphenicol, streptomycin and sulfonamide. DNA from a *Proteus* host harboring this plasmid produces a satellite band with three distinct peaks [*see illustration on opposite page*]. When we examined a sample from each peak in the electron microscope, we found molecules of 12 megadaltons in the peak of greatest density, molecules of 70 megadaltons in the middle peak and molecules of 58 megadaltons in the lowest-density peak. The sizes and densities of these molecules suggested that some of the 222/R-factor plasmids (which in *E. coli* were all 70 megadaltons in size) were divided into two smaller molecules of 12 and 58 megadaltons in *P. mirabilis*. Since the peak of greatest density is about the same size as the other two peaks (that is, it has the same amount of DNA) in spite of the fact that the 12-megadalton molecules in this peak are much smaller than the molecules in the other peaks, there must be many more of the 12-megadalton molecules than of the larger molecules. It is therefore likely that after the 70-megadalton molecule subdivides into two smaller ones the 12-megadalton molecule can replicate independently. Other

experiments show that a modified plasmid similar to the 58-megadalton molecule can be derived from the 70-megadalton molecule, and this plasmid can also replicate normally.

The natural replication of the bacterial chromosome (or any other DNA molecule) appears to be possible only beginning at a unique point of origin on the chromosome, and this point appears to be essential for replication. If an R factor such as 222/R had in fact evolved by sequential chromosomal pickup, it would be unlikely that enough of the chromosome would have been picked up to have much chance of including the unique point of origin. The R factor would therefore be expected to have only the one point of origin for DNA replication that was present in the original transfer-factor molecule. Since in the *Proteus* host the 222 factor segregates into two molecules, both of which may replicate independently, this again does not fit the idea of evolution by sequential chromosomal pickup. On the other hand, these facts can be reconciled with an alternative idea: that the genes on R factors arise from molecules that were originally extrachromosomal and capable of independent replication.

The R factors first studied by genetic and physical methods gave results similar to those of 222/R in *E. coli*, that is,

all the antibiotic-resistance genes were transmitted together with the genes determining transfer properties as though they were all part of the same genetic structure. The physical analysis of the DNA of these plasmids confirmed this unitary structure, disclosing only a single molecular species for each R factor, as shown in the case of 222/R in *E. coli*. Since all the antibiotic-resistance genes and the genes determining transfer are integrated in a single structure, these R factors are called plasmid cointegrates. A number of other factors have recently been shown to have different properties, in particular an R factor studied in *E. coli* by E. S. Anderson and Malcolm Lewis at the Central Enteric Reference Laboratory in London. That factor controls resistance to streptomycin, ampicillin and tetracycline but transfers only some of its properties to some recipient cells [*see illustration on page 204*]. After conjugation some of the recipient cells carry infectious resistance to only one of the three antibiotics; others carry either streptomycin resistance or ampicillin resistance that is no longer infectious. Still other cells have lost all antibiotic resistance but, when they are incubated with the noninfectious streptomycin- or ampicillin-resistant cells, can potentiate the transfer of this noninfectious resistance. Anderson and Lewis therefore suggested

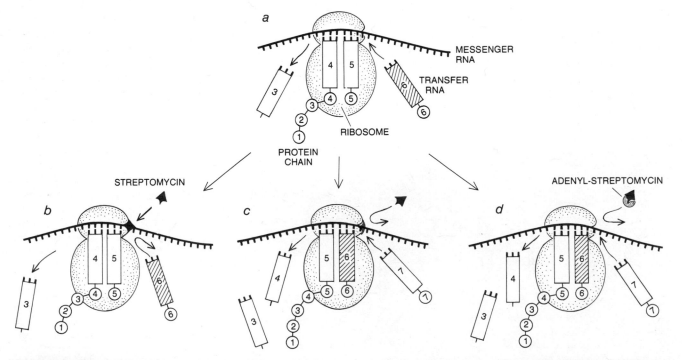

MODE OF ACTION is different in chromosomal and in plasmid-mediated antibiotic resistance. Streptomycin inhibits protein synthesis in bacteria. Ordinarily transfer RNA assembles amino acids on ribosomes according to instructions recorded in messenger RNA (*a*). Streptomycin attaches itself to a site on the ribosome and prevents protein synthesis, perhaps by blocking transfer-RNA attachment (*b*). Mutation of a chromosomal gene changes the ribosome's structure so that the streptomycin cannot bind (*c*), but R-factor genes make an enzyme that attaches an adenine molecule to the streptomycin, which no longer fits the unchanged ribosome (*d*).

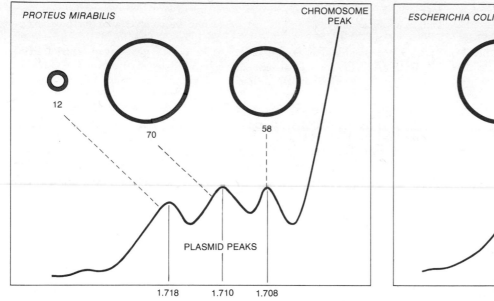

PLASMIDS representing the 222/R factor in *Proteus mirabilis* and *Escherichia coli* are compared after centrifugation and fractionation, with ethidium bromide present in the case of *E. coli*. In *Proteus* there are three peaks in the satellite band. In each peak a different plasmid, measuring respectively 12, 70 and 58 megadaltons, is found. In *Escherichia*, on the other hand, there is only one plasmid peak, containing 70-megadalton molecules. The genes for resistance to four antibiotics are on the 12-megadalton plasmid, those for the transfer factor on the 58-megadalton plasmid. The two plasmids are integrated in the 70-megadalton plasmid cointegrate.

that these antibiotic-sensitive cells contain a plasmid (called delta) that controls only transfer-factor properties.

In my laboratory I undertook a biophysical study of the DNA molecules isolated from these cultures in collaboration with Christine Milliken. This study has fully confirmed the conclusion, reached by Anderson and Lewis from their genetic studies, that the original strain harbors several independent plasmids. The delta plasmid controls transfer-factor properties, a second plasmid (called S) controls noninfectious streptomycin resistance and a third (called A) controls noninfectious ampicillin resistance. The delta plasmid was found to be joined with a tetracycline factor (T) to form a composite delta-T plasmid controlling infectious tetracycline resistance. When A or S is present in the same cell as a transfer factor (delta or delta-T), it too can be transferred to recipient cells.

Satellite DNA separated from *E. coli* host cells by centrifugation showed the following features: DNA from delta+ cells (cells with transfer-factor properties but no drug resistance) contained a single species of duplex looped DNA molecules of 60 megadaltons; DNA from S or A cells (respectively carrying noninfectious streptomycin or ampicillin resistance) was also a single species but was a tenth as large (six megadaltons); DNA from delta (S) or delta (A) cells (respectively carrying infectious streptomycin or ampicillin resistance) was found to contain two distinct species of molecules, 60 megadaltons and six megadaltons in size.

It appears, then, that an R factor can be composed of an aggregate of two or more plasmid molecules. That is, delta (S) has the properties of an R factor, since it can transfer infectious streptomycin resistance, but those properties are due to two independent molecules, one (delta) controlling transfer-factor properties and the other (S) controlling noninfectious streptomycin resistance. As a result of conjugation brought about by the presence of the delta factor, delta itself is transferred and the probability of simultaneous transfer of S is very high. The majority of recipient cells therefore inherit both delta and S molecules and show the characteristics of infectious streptomycin resistance. Occasionally some of the recipient cells acquire only one or the other of the two molecules, and then they acquire either transfer-factor properties or noninfectious streptomycin resistance. Similar properties are proposed for delta (A) cells.

In distinction to delta and delta-T, which like all other R plasmids were found in a one-to-one molecular ratio with the chromosome, there were more than 10 copies of either the A- or the S-plasmid for each chromosome. This new type of R factor, delta (S) or delta (A), is termed a "plasmid aggregate." Its evolution is difficult to explain by gene pickup and is clearly more consistent with extra-chromosomal evolution. The plasmid aggregate may be a more primitive evolutionary type that can evolve, through a joining of the separate molecules, into a plasmid cointegrate such as delta-T.

E. coli is a normal inhabitant of the human alimentary tract; it seems unlikely that during its evolutionary history in the preantibiotic era such an organism would ever have come into contact with an antibiotic such as streptomycin, which is produced by a very different genus of microorganism usually found only in soil. How then can the bacterium produce an enzyme that recognizes streptomycin and inactivates it? (The evolution of chromosomal, as opposed to R-factor, resistance is easier to explain, since chromosomal resistance stems from a change in the structure of a normal bacterial product, the ribosome, through mutation.)

William Shaw of the·University of Miami has pointed out that many R-factor resistances depend on the inactivation of antibiotics through such enzymic processes as acetylation, phosphorylation or adenylation, processes that normal metabolic compounds undergo in the normal bacterial cell. Shaw suggests that enzymes coded for by R-factor genes and operating to inactivate antibiotics may have evolved from enzymes that originally acted on normal bacterial molecules. He has found that resistance to chloramphenicol by a number of plasmids of different origin is due in all cases to its acetylation by enzymes that are similar in size and constitution and appear to differ only in the efficiency with which they can bind to chloramphenicol. He suggests that these enzymes may have evolved from one that originally acetylated a normal cell component. Mutations of the gene determining this enzyme might have modified the enzyme so that it bound to chloramphenicol with increasing efficiency, until now it binds to chloramphenicol rather than to the original molecule and can

inactivate it effectively enough to prevent its antibiotic activity.

The rapid proliferation of R factors and the identification of R-plasmid aggregates suggest that similar entities could have importance in the normal regulatory control of bacteria. Genes on the bacterial chromosome determine the activities that enable the bacterial cell to grow and multiply in many very different environments. Under most circumstances most of these genes are not active; they are "switched off" except in the particular environment in which their function is needed [see "The Control of Biochemical Reactions," by Jean-Pierre Changeux; SCIENTIFIC AMERICAN Offprint 1008]. Now, if a certain function determined by several sequentially acting enzymes is required very infrequently, the continual replication of the genes that determine it may impose

more of an evolutionary burden than would be compensated for by the rare need for the function. If, however, these determinant genes were carried on a plasmid present in only a few cells of a population, they would need to replicate only in those rare cells. Any time there arose a need for the function in question, these special genes could be acquired rapidly by all the cells in a progeny population if the plasmid also had transfer-factor properties or (even more economically) if other rare cells contained transfer factors that could interact to establish a plasmid aggregate.

The control of a number of activities in this way by their distribution to a number of different plasmids carried by different cells in the population would enormously increase the total gene pool and extend the overall metabolic potential of the species. This would appear to represent a logical extension of genetic

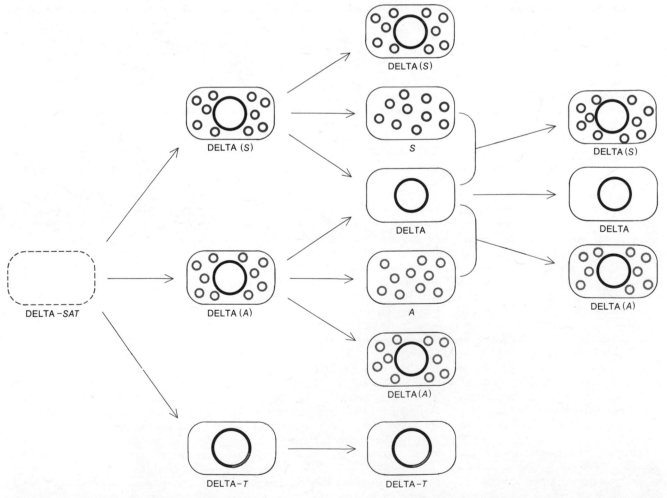

PLASMID AGGREGATE is an R factor composed of several independent plasmids. Here the original donor is resistant to streptomycin (S), ampicillin (A) and tetracycline (T) and has a transfer factor (delta). On conjugating it gives rise to three kinds of new donor, each carrying infectious resistance to one antibiotic. When these donor cells conjugate, the strains that result either

have the same characteristics as the donor or have noninfectious resistance (either S or A) or have only the transfer factor (delta); tetracycline resistance, however, is always infectious. These properties have now been explained by the presence of four distinct plasmids: delta, measuring 60 megadaltons; S and A, each of six megadaltons, and delta-T, a plasmid cointegrate of 67 megadaltons.

regulation: from the present concept of individual cellular control by the switching on and switching off of chromosomal genes to a species-wide control of gene-pool synthesis.

As a matter of fact, entities that are similar to R factors but that control enzymes involved in metabolism have recently been recognized. For example, genes for enzymes catalyzing the breakdown of complex organic molecules to enable them to be used as a source of carbon have recently been shown to be situated on plasmids in *Pseudomonas,* a genus of bacteria well known for its ability to grow on almost any kind of organic substance. Certain strains of *Pseudomonas* are enabled to grow on camphor by a number of enzymes determined by one plasmid, and a plasmid controlling a sequence of genes that allow another strain to grow on salicylic acid has been identified. Since R factors have recently been found in *Pseudomonas* strains (which are also being seen more often in postoperative infections), it would not be too surprising if much of the nutritional versatility of *Pseudomonas* turns out to depend on a series of plasmids, each controlling enzymes that make it possible for the organism to grow on unusual organic compounds.

Plasmids, which are being found in an increasing number of bacterial species, share many properties with bacterial viruses, particularly the "temperate" type of bacterial virus, which in the latent state establishes a stable symbiotic relation with its host bacterium. Bacterial plasmids can be thought of as symbionts without even the limited lethal effects of temperate viruses, whose spread depends on the eventual destruction of their host cells. Since it has been recognized that certain components of plant and animal cells, such as chloroplasts and mitochondria, also rely for their maintenance partly on extrachromosomal DNA, the possibility exists that plasmids not only may be common in many bacterial species and be concerned with activities other than antibiotic resistance but also may be closely related to other extrachromosomal elements in all cells.

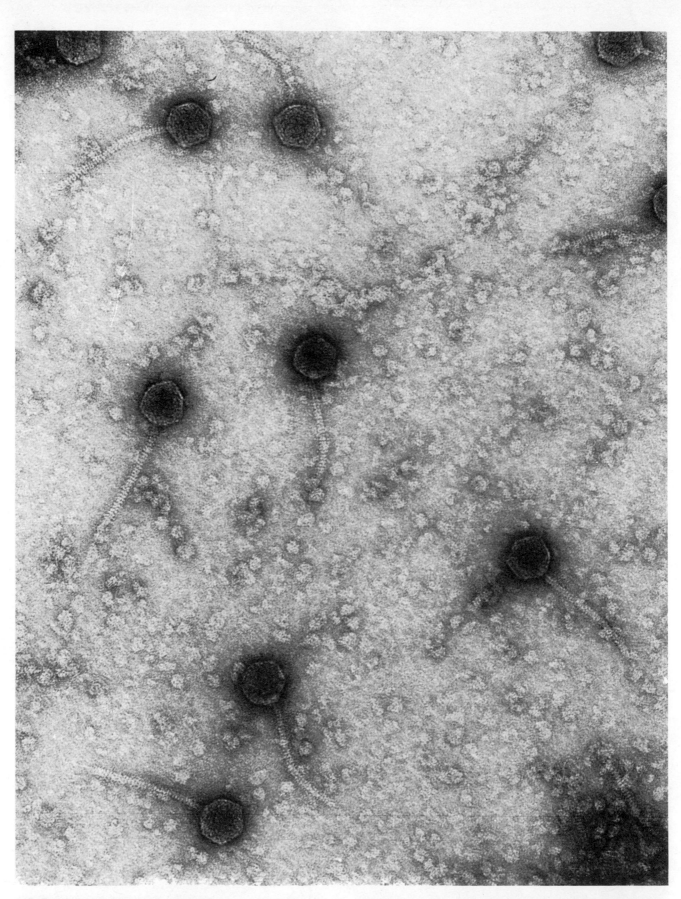

BACTERIAL VIRUSES, or phages, of the strain designated lambda are magnified 200,000 diameters in this electron micrograph made by William C. Earnshaw and Philip A. Youdarian of the Massachusetts Institute of Technology. Each virus particle has an icosahedral head containing a molecule of infectious viral DNA, and a flexible hollow tail with which the phage attaches itself to the outer mem- brane of its bacterial-cell host and injects its DNA into the interior of the cell. Negative staining with uranyl acetate clearly reveals the subunits that make up the virus's protein shell. The phage-lambda particles shown here were extracted from an infected strain of the colon bacillus *Escherichia coli*, together with a large number of bacterial ribosomes (*lightly stained particles scattered over background*).

20

How Viruses Insert Their DNA into the DNA of the Host Cell

by Allan M. Campbell
December 1976

Some viruses are able to coexist peaceably with their host cell for long periods, incorporating their genes into the host chromosome. The details of the insertion process are now fairly well understood

The imagery of medicine often seems to portray viruses as aggressive organisms bent on human destruction. In actuality their role in disease is merely a by-product of their parasitic existence. Unable to reproduce by themselves, viruses must invade living cells and redirect some of the cellular machinery to the production of new virus particles. In the process many of the host cells are destroyed, leaving the tissue of which they are a part damaged and giving rise to disease.

Although virus production and cell destruction are the most dramatic outcomes of virus infection, they are by no means the only outcomes possible. Like other parasites, many viruses find it advantageous to persist innocuously within their host cell for an indefinite period, actively multiplying only when the host weakens or stops growing. The existence of such latent viruses was first suspected in the 1920's, but it was not until the early 1950's that André Lwoff and his colleagues at the Pasteur Institute established unequivocally that a latent virus, or "provirus," can be transmitted from one cell generation to the next without external reinfection. Working with bacteriophages, the viruses that infect bacteria, they showed that the provirus exists in a "lysogenic" state. The term lysogenic refers to the fact that the provirus can come out of the dormant state and give rise to mature virus particles that lyse, or dissolve, the bacterial cells. Among the agents that can induce the provirus to resume its former mode of multiplication are ultraviolet radiation, X rays and carcinogenic chemical compounds. In the intervening 20 years the further study of bacteriophages has revealed much about the mechanism of lysogeny.

A typical bacteriophage consists of a single linear molecule of nucleic acid enclosed in a protein coat. Resembling a minute hypodermic syringe, the phage attaches itself to a bacterial cell and injects its strand of DNA into the interior.

Once inside the cell, the viral DNA may begin directing the manufacture of new virus particles or become a provirus by incorporating itself into the DNA of the bacterial cell's long, threadlike chromosome.

What then switches the provirus from the lysogenic state to the active production of virus particles? A mechanism was put forward by François Jacob and Jacques Monod of the Pasteur Institute in 1961 and later demonstrated by Mark Ptashne of Harvard University. In the provirus only a few genes are expressed, and the product of one of them is a "repressor" protein that combines with the viral DNA and prevents the expression of the other viral genes, particularly those responsible for the independent replication of the viral chromosome. Inducing agents such as X rays alter the metabolism of the bacterium in such a way that a substance is produced that inactivates the repressor. If all the repressor molecules in a given lysogenic bacterium are simultaneously inactivated, the viral genes are sequentially expressed. Proteins needed for the replication of the viral DNA are made first, followed by the head and tail proteins of the virus particle, which spontaneously assemble inside the bacterium. Unit segments of viral DNA are then packaged into the phage heads. Finally, some 60 minutes after the cell was exposed to the inducing agent, the cell bursts, releasing about 100 virus particles that are capable of infecting other cells.

When a single lysogenic bacterium multiplies to form a colony, every cell of that colony is potentially capable of manufacturing virus particles and can express that potentiality when its repressor is destroyed. The viral genes have therefore been added to the bacterial genes in such a way that both sets of genetic information are inherited by the cell's descendants. In order for this to happen the viral chromosome must replicate like any normal cellular component, and it must be partitioned at cell division so that each daughter cell receives at least one copy of it.

Such a result can be achieved in at least three possible ways. One is for the viral chromosome to insert itself into the host chromosome, after which it can be passively replicated and distributed at cell division as part of the host DNA. Another way is for the viral chromosome to establish itself independently; it replicates and is distributed like the host chromosome but is separate from it. A third way is for the viral chromosome to replicate separately from the host chromosome and in multiple copies. In that case the viral chromosome might not be distributed regularly when the bacterial cell divides, but if the number of copies is large enough, the chance of a daughter cell's receiving no copies whatever is small. If we survey all known viruses, we find that the first two mechanisms exist in nature, and the third is observed in certain mutant viruses produced in the laboratory. One property all three mechanisms have in common is that once the virus has become established as a hereditary component of the cell, there is nothing particularly novel in its mode of inheritance from then on.

Bacteriophage lambda, which was discovered by Esther M. Lederberg of Stanford University as a provirus carried by the K12 strain of the colon bacillus *Escherichia coli*, is the best-understood genetically of the bacteriophages and remains one of the favorite experimental organisms of molecular biologists. Soon after Lwoff's demonstration of heritable lysogeny several investigators were able to achieve genetic crosses between strains of *E. coli* that were lysogenic for phage lambda and strains that were nonlysogenic for it. They got the surprising result that the characteristic of being lysogenic or nonlysogenic was distributed among the progeny of the cross exactly as one would expect if the trait were being determined by a gene at a specific location on the bacterial chro-

208

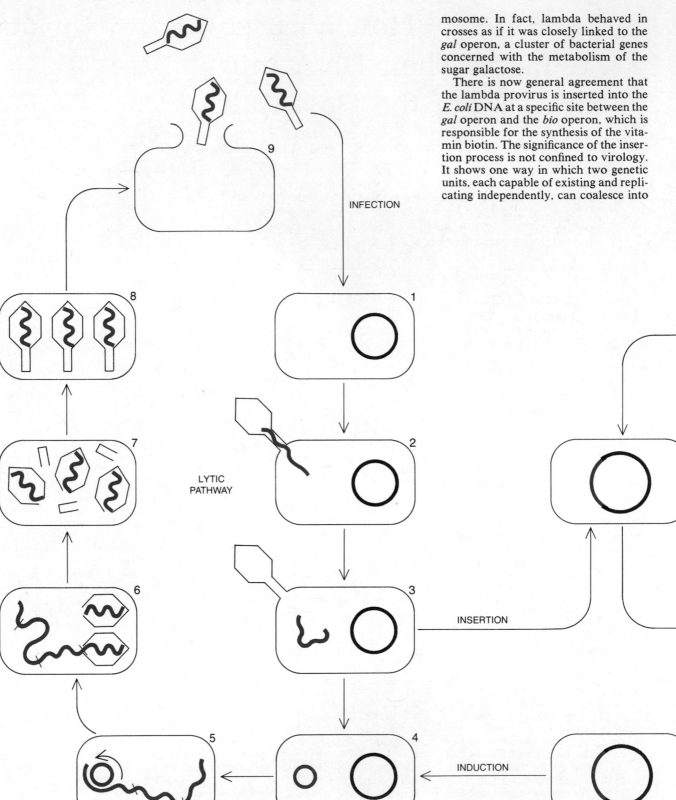

INFECTION

LYTIC
PATHWAY

INSERTION

INDUCTION

mosome. In fact, lambda behaved in crosses as if it was closely linked to the *gal* operon, a cluster of bacterial genes concerned with the metabolism of the sugar galactose.

There is now general agreement that the lambda provirus is inserted into the *E. coli* DNA at a specific site between the *gal* operon and the *bio* operon, which is responsible for the synthesis of the vitamin biotin. The significance of the insertion process is not confined to virology. It shows one way in which two genetic units, each capable of existing and replicating independently, can coalesce into

LIFE CYCLE OF PHAGE LAMBDA shows that lysis (dissolution) and death of the infected bacterial cell are not inevitable. After a molecule of viral DNA is injected into a healthy *E. coli* cell (2) the host may take either of two paths. In the productive, or lytic, pathway the viral DNA forms a circle (3 and 4) and replicates by the "rolling circle" process to give rise to a long "sausage string" of DNA containing multiple copies of the viral genes (5). Next the viral genes direct the synthesis and assembly of the head and tail proteins of the virus particles and the packaging of unit DNA segments into the heads **(6 and 7). Heads and tails then assemble spontaneously inside bacterium, giving rise to mature phage particles (8). Finally, some 60 minutes after infection, the host cell bursts, releasing about 100 progeny virus particles that can then infect other *E. coli* cells (9 and 1). Alternatively, depending on the conditions of infection, the phage-lambda DNA can live quietly within the host, establishing itself as a semipermanent part of the bacterial chromosome (10). It is then replicated and segregated at cell division (11 and 12) and passed on to succeeding cell generations for an indefinite period. The latent virus,**

a larger unit. Phage lambda is only one of a number of elements for which insertion is known or suspected. Most of these elements are viruses or the small pieces of bacterial DNA called plasmids, some of which can transfer resistance to antibiotics and other properties from one bacterial cell to another. In addition there are the "insertion sequences" of bacteria and the "transposable elements" of maize, which are known only for their ability to move occasionally from one chromosomal location to another and may well have no independent existence.

Although the widespread occurrence of genetic elements that can insert themselves into preexisting chromosomes is incontestable, there is a considerable divergence of opinion about their importance in the normal life of the host or in evolution. One extreme viewpoint is that these elements are basically foreign to the cells that harbor them; they are invaders like viruses that have somehow got into cells and chromosomes and do not really belong there. The alternative view is that much of the DNA of present-day chromosomes is derived from elements that were originally foreign but that through a series of small evolutionary steps gradually became naturalized citizens of the intracellular community. A somewhat intermediate position is that the mobility of these elements plays some essential role in the normal development of multicellular organisms from the fertilized egg into the adult.

Studies of the mechanism by which new genetic elements are added to existing chromosomes cannot provide any direct information about the origin or function of such elements. The studies are, however, relevant in one respect. If it turned out that the DNA of the added elements was hooked onto the rest of the chromosome in some unusual way, by connections that were not normal features of chromosome structure, then a clear distinction between foreign DNA and indigenous DNA would be implied. It therefore seemed particularly important to firmly establish whether or not the foreign DNA of some prototypical examples such as the lambda provirus is directly inserted into the linear sequence of genes on the host cell's chromosome.

The concept of genes' being arranged in a linear sequence, and the experimental basis of that concept, have a long history. In 1913 A. H. Sturtevant of Columbia University crossed fruit flies differing from each other by several genetic traits and analyzed the frequencies of progeny exhibiting new combinations of the parental traits. He concluded that the determinants for these traits (all of which happened to be on the same chromosome) distributed themselves as though they were on some one-dimensional structure, with new combinations arising from the redistribution of connected segments of that structure. It became possible to extend this kind of formal linkage analysis in the 1950's, when Seymour Benzer of Purdue University, working with small segments of phage chromosome, proved that the linkage maps of mutational sites within genes were also one-dimensional. These purely genetic studies were followed by cytological and biochemical work showing that the one-dimensionality of linkage maps was associated with the linear arrangement of the genes along the chromosome and ultimately with the linear

sequence of nucleotides along the double helix of DNA.

The question of whether all the linear DNA segments in a chromosome are joined end to end in one continuous double helix is still not completely settled, but the evidence is increasingly strong that they are. The uncertainty surrounding the question through the mid-1960's was such, however, that many bacterial geneticists found nothing particularly bizarre in the notion that the provirus might lie alongside the host chromosome rather than being inserted into it.

One way to investigate the question is to cross two lines of lysogenic bacteria, in which both the host genes and the viral genes are marked with mutations. In such bacterial crosses pieces of DNA from the donor cells are introduced into the recipient cells, where they pair with corresponding segments of the recipient cells' DNA and, in some fraction of those cells, replace them. Progeny in which a specific gene has been replaced can be recognized and selected for because they express a genetic trait characteristic of the donor, such as the ability to metabolize galactose. The recognition of the replacement of specific genes in the provirus has been made easier in recent years by the isolation of conditionally lethal mutant viruses, which fail to multiply in some condition under the control of the experimenter but which grow normally when that condition is changed. For example, temperature-sensitive mutants survive and multiply at one temperature (25 degrees Celsius) but not at another (42 degrees C.).

Such experiments throw some light on the provirus's mode of attachment. If, for example, the provirus is not linearly inserted into the host chromosome but instead projects from it sideways, one would expect that an exchange of genetic material occurring along the viral segment would not redistribute the genes lying on the main axis of the host chromosome. The fact is that redistribution of the host genes does occur, suggesting that insertion is indeed linear.

A second and more informative method of determining the topology of the lysogenic chromosome is deletion mapping, in which mutants marked by chromosomal deletions are crossed with strains that have other genetic markers. A deletion mutation involves the permanent loss from the chromosome of a string of neighboring nucleotides, numbering from one to many thousands. Since each deletion eliminates a continuous segment of chromosome, the order of the genes along the chromosome can be deduced by piecing together the information provided by the characteristics of the various deletion mutants observed. Deletion mapping provides a strict "betweenness" criterion for locating genes with respect to one another. The condition that every observed dele-

LYSOGENIC
PATHWAY

or "provirus," still retains the potential to grow lytically, however. Bacteria harboring a provirus are called lysogenic, meaning that they carry a property that can lead to viral multiplication and the fatal lysis of the host cell. Exposure of lysogenic bacterium to ultraviolet radiation, X rays or chemicals such as nitrogen mustard and organic peroxides can induce the provirus to return to lytic pathway.

INSERTED ELEMENT	APPROXIMATE LENGTH IN NUCLEOTIDES	INSERTED INTO	SPECIAL PROPERTIES
BACTERIOPHAGE LAMBDA	50,000	E. COLI CHROMOSOME	SPECIFIC SITES ON VIRUS AND HOST CHROMOSOMES
BACTERIOPHAGE MU-1	37,000	E. COLI CHROMOSOME	SPECIFIC SITE ON VIRUS, ANY SITE ON HOST CHROMOSOME
BACTERIAL SEX FACTOR F	100,000	E. COLI CHROMOSOME	MANY SITES ON HOST CHROMOSOME
DRUG-RESISTANCE PLASMID	20,000	BACTERIAL FACTOR RESEMBLING F	TRANSMITS RESISTANCE TO ANTIBIOTICS BETWEEN BACTERIAL STRAINS
TUMOR VIRUS SV-40	5,000	HUMAN CHROMOSOME	DERIVED ORIGINALLY FROM MONKEY CELLS
TRANSPOSABLE ELEMENTS	(UNKNOWN)	MAIZE CHROMOSOMES	NO DETECTABLE EXTRA-CHROMOSOMAL PHASE
INSERTION SEQUENCE IS2	1,400	E. COLI CHROMOSOME	

ADDED GENETIC ELEMENTS, small pieces of DNA that can exist as part of the main chromosome or independently, have been observed in bacterial, maize and human cells. Some are viruses; others are not. When they are inserted, they introduce into the cell instructions governing additional biochemical reactions that may be superimposed on the cell's metabolism.

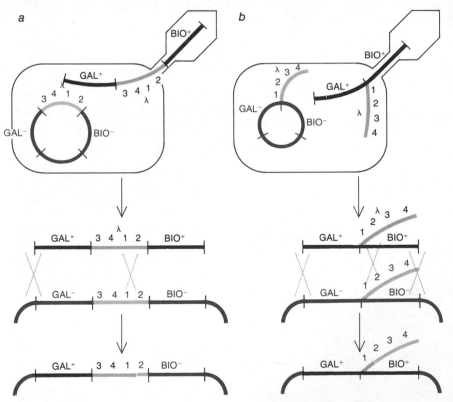

TYPES OF GENETIC EXCHANGE, or recombination, observed between two lysogenic E. coli chromosomes containing different genetic markers shed light on the mode of attachment of the phage-lambda genes. In the experiment depicted here a fragment of DNA from a bacterial cell capable of metabolizing galactose and synthesizing biotin (gal⁺bio⁺) and containing a lambda provirus with mutations in genes 1, 2, 3 and 4 is introduced by means of an infective phage coat into a recipient bacterium that is unable to utilize galactose or synthesize biotin because of genetic mutations (gal⁻bio⁻) and that harbors a nonmutant provirus. Type of recombination that might occur if the viral genes were linearly inserted into each chromosome is shown in a. The mechanism of insertion proposed by the author requires the permutation of the order of genes along the viral chromosome, so that a genetic exchange between viral genes 1 and 2 would serve to recombine the flanking bacterial markers gal and bio. The resulting recombinant bacterium would be capable of utilizing galactose but incapable of synthesizing biotin, a prediction that agrees with observed results. For comparison, b indicates the expectation for one kind of nonlinear topology, in which the provirus joins to the chromosome as a branch.

tion should be representable as a linear segment is a highly restrictive one; a segment represented by two deletion markers must necessarily include all points that lie between them. For example, if the provirus is inserted between two identified host genes, then every deletion removing both of those genes must remove the provirus as well. This turns out to be the case in experiments, and it is again consistent with the model that the provirus is continuous with the host DNA.

Until the beginning of the 1970's genetic analysis of the type I have been describing provided the only precise information on the relation between the provirus and the chromosome; no direct physical information was available on the relevant nucleotide sequences within the DNA of the provirus. In principle the simplest approach would be to use direct methods of determining the sequence of nucleotides along the DNA chains of the virus and the lysogenic chromosome. Although such sequencing methods are improving rapidly, the identification of all 50,000 nucleotides in the lambda provirus would be a time-consuming and costly task. For many purposes adequate information can be obtained by exploiting the fact that single DNA chains with complementary nucleotide sequences can find each other and form double helixes in the test tube. Electron micrographs of DNA molecules formed this way, notably in the laboratory of Norman R. Davidson at the California Institute of Technology and that of Waclaw T. Szybalski of the University of Wisconsin, have demonstrated that the structures inferred from genetic results have a physical reality.

For these experiments double-strand DNA molecules extracted from virus particles are dissociated into single chains by heating. If the solution is then cooled slowly, double helixes of the complementary chains will re-form. When single-strand DNA's from two viruses with some nucleotide sequences in common (such as phage lambda and a deletion mutant of it) are mixed before the cooling step, new helixes can form not only between complementary chains from one virus but also between complementary chains from both viruses; the latter kind of chain is known as a heteroduplex. The nucleotides in the complementary segments of these hybrid chains pair up and form double helixes but the noncomplementary sequences do not, leaving single-strand loops that can be seen in electron micrographs.

By some ingenious manipulations Davidson and his colleagues Phillip A. Sharp and Ming-Ta Hsu were able to examine heteroduplexes between viral

DNA and the DNA of an inserted provirus. Although the most straightforward approach would be to make heteroduplexes from one DNA chain of phage lambda and the complementary chain of a chromosome from a lysogenic bacterium, that experiment is not yet feasible because of the difficulties in handling a DNA molecule the size of the bacterial chromosome. The same end was achieved by letting the phage-lambda DNA insert itself not into the entire bacterial chromosome but into a smaller DNA molecule: a derivative of the bacterial sex factor that had picked up from the bacterial chromosome the specific DNA segment into which lambda inserts. Strands of the sex factor with the lambda provirus inserted into them could be readily isolated intact and used to form heteroduplexes with DNA extracted from virus particles.

The combined results of genetic and physical studies make us quite confident

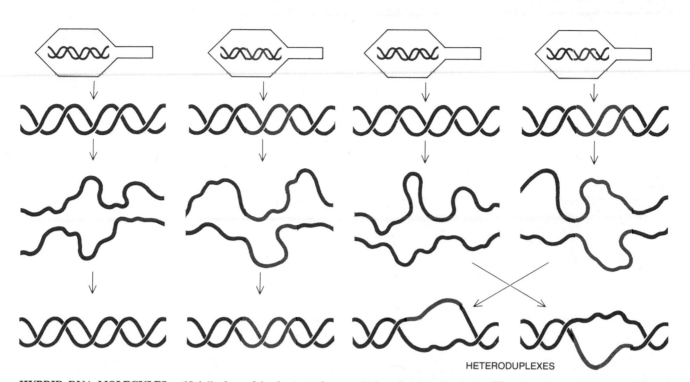

HETERODUPLEXES

HYBRID DNA MOLECULES artificially formed in the test tube can be used to map viral genes physically. When the DNA double helix is heated, it unwinds, giving rise to two single-strand chains. If single chains having complementary and noncomplementary regions are mixed together at this stage and slowly cooled, some "duplexes" will form between the chains. When these heteroduplexes are viewed in the electron microscope, the two DNA chains will be double helixes where they have the same sequence of nucleotides and unpaired where they differ in sequence. With this method one can precisely map position of a given marker mutation along the DNA molecule.

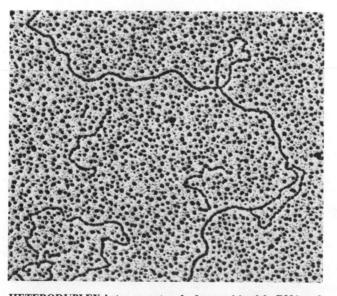

DELETION LOOP

DOUBLE-STRAND REGIONS

SUBSTITUTION BUBBLE

HETERODUPLEX between a strand of normal lambda DNA and a second strand incorporating two mutations is clearly visible in this electron micrograph made by Elizabeth A. Raleigh of M.I.T. (Only a small segment of the long viral DNA molecule is shown.) The loop of single-strand DNA at top right results from a deletion mutation that removed an entire segment of DNA from one strand. The "collapsed bubble" of single-strand DNA results from the substitution of several nucleotides in one strand by different ones, making that region of the two strands noncomplementary and hence unable to pair. The remaining portions of molecule shown here are double-helical.

that we now know the structure of the lysogenic chromosome. The steps by which this structure is formed and dissociated into its component parts are the subject of current research. During the life cycle of phage lambda, DNA must be cut and rejoined at the ends of the viral chromosome and at the ends of the provirus. The lambda DNA injected into the bacterial cell is in linear form, but before it is inserted into the bacterial chromosome its ends are joined so that it makes a circle. During insertion the circle is opened at a different point. As a result, although the provirus and virus chromosomes are both linear structures, the order of the genes along the two is not identical.

How the ends of the viral DNA are joined to form a circle is known, thanks largely to the work of A. D. Kaiser and

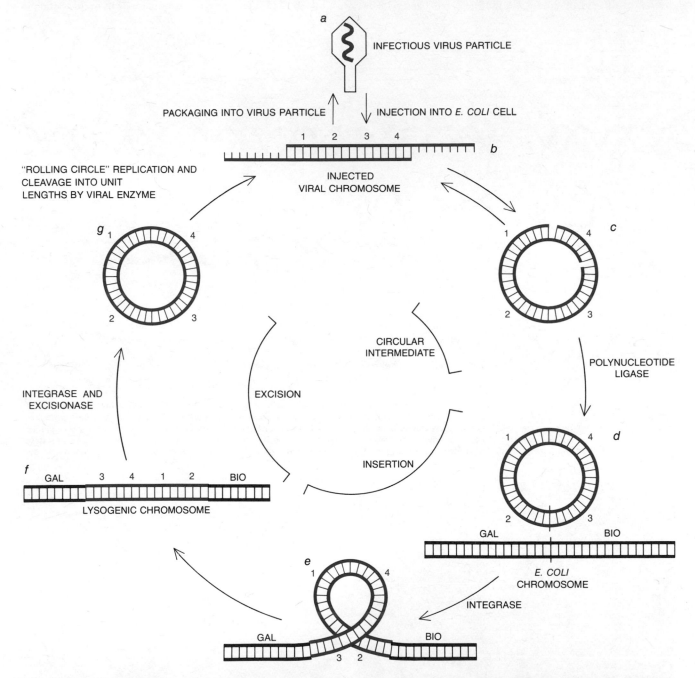

INSERTION AND EXCISION of the phage-lambda genes into the *E. coli* chromosome require the action of both bacterial and viral enzymes. As the DNA of phage lambda is packaged in the viral particle (*a*) it is linear and double-helical, except for complementary unpaired segments 12 nucleotides long at the ends of the two nucleotide chains (*b*). In solution this linear form comes to equilibrium with a circular form that has staggered "nicks" 12 nucleotides apart in the two complementary chains (*c*). When viral DNA is injected into the bacterial cell in the course of infection, the two nicks in the open circle are sealed by the bacterial enzyme polynucleotide ligase, so that both chains of the circle are now closed throughout their length (*d*). This circular intermediate then interacts with a particular segment of the *E. coli* chromosome (between the *gal* and *bio* genes). Viral and bacterial chromosomes break and rejoin at unique sites on each partner, so that viral DNA is spliced into the host DNA, a reaction catalyzed by the viral enzyme integrase (*e*). (Note that the gene order in the provirus is *3, 4, 1, 2*, a cyclic permutation of the viral gene order *1, 2, 3, 4*.) The *E. coli* chromosome is now lysogenic for phage lambda (*f*). After several cell generations radiation or chemically active compounds may induce the provirus to enter the lytic state. When this happens, the lambda repressor, which has so far blocked the expression of most of the viral genes, is inactivated, allowing the synthesis of the viral enzyme excisionase. Together with integrase, excisionase catalyzes the excision of the provirus from the host chromosome, converting it back into the circular form with the original gene order (*g*). The circle of viral DNA replicates, producing multiple copies that are then cleaved by a specific viral enzyme to give rise to the linear form with "sticky" ends (*b*). Each linear DNA segment·is then packaged in a virus coat (*a*). When the host cell ruptures, the liberated phages infect healthy cells and the lysogenic cycle begins anew.

his collaborators at Stanford. Lambda DNA is a double helix throughout most of its length, but one end of each polynucleotide chain extends for 12 nucleotides beyond the double helix. These two single-strand chains are complementary to each other and are called "sticky ends." In solution the linear DNA molecules can come to equilibrium with circular molecules formed by the pairing of the two ends. When lambda infects an *E. coli* cell, the open circle formed by the viral DNA is closed by the action of polynucleotide ligase, a bacterial enzyme that seals breaks in one chain of a double helix. This step requires no viral enzymes, and it is not specific to the nucleotide sequences involved. On the other hand, the insertion of viral DNA into the bacterial chromosome requires the recognition and cutting of highly specific nucleotide sequences in both the lambda and the *E. coli* DNA.

Little is known of the biochemistry of insertion, although its genetic control has been intensively explored. At the time I proposed the circular-molecule-intermediate model for the insertion of phage-lambda DNA in 1962, the only known mechanism for breaking or rejoining two DNA molecules at corresponding points was homologous recombination, which requires that the two molecules have similar or identical base sequences in the recombining region. The chemical steps by which homologous recombination takes place are still largely conjectural, but Alvin J. Clark of the University of California at Berkeley and others have isolated bacterial mutants that are unable to carry out this process. Under conditions where homologous recombination is blocked by such mutations, however, phage lambda can still insert its DNA with the normal frequency. Hence the insertion of viral DNA seems to be accomplished not by the same bacterial enzymes that are responsible for homologous recombination but by viral enzymes that cut and join DNA molecules at highly specific sites.

Direct evidence for the existence of such viral enzymes has been provided by the genetic studies of James F. Zissler of the University of Minnesota Medical School and the biochemical investigations of Howard A. Nash of the National Institute of Mental Health, which have shown that the enzyme product of a specific viral gene (dubbed integrase) is required for the insertion of viral DNA; mutant viruses lacking this enzyme are unable to enter the lysogenic state. Similar studies of the reverse process—the excision of viral DNA from the bacterial chromosome—by Gabriel Guarneros and Harrison Echols of the University of California at Berkeley and Susan Gottesman of the National Cancer Institute—have shown that excision requires in addition to integrase the

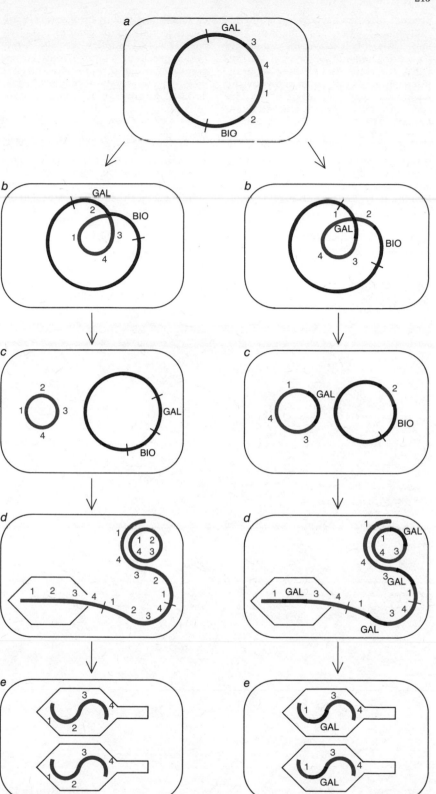

NOVEL VIRUS VARIANTS arise when certain proviruses turn into infectious viruses, killing the bacterium that harbors them and carrying away with them pieces of the DNA of the dead host. The "transduced" bacterial genes linked to the viral chromosome can subsequently replicate at the unrestricted viral pace. Transduction is believed to result from rare errors in the excision of the provirus from the host chromosome. Normal excision (*left panel*) takes place in the vast majority of cells. Viral DNA separates from the bacterial DNA (*b*) to give rise to a circular viral chromosome and a nonlysogenic bacterial chromosome (*c*). The viral chromosome replicates in several stages, ultimately as a rolling circle that generates a long sausage string of DNA in which the entire viral sequence is repeated many times (*d*). Unit DNA lengths are then packaged into infectious virus particles (*e*). In one cell out of 100,000 (*right panel*) abnormal excision generates a circular molecule including some host DNA. Infectious particles are thus formed in which a segment of bacterial DNA has replaced a segment of viral DNA.

product of a second viral gene (called excisionase). The virus thus introduces into the host cell enzymatic machinery for cutting and joining the viral and host DNA at specific sites to bring about the insertion and excision of the provirus. As long as the transcription of the gene coding for excisionase is blocked by the lambda repressor the provirus will remain inserted. When repression is released, the excisionase gene will be expressed and there will be a reciprocal exchange within the lysogenic chromosome, re-creating a circular molecule of lambda DNA and a nonlysogenic bacterial chromosome.

Excision is generally precise: more than 99 percent of the virus parti-

cles manufactured by lysogenic cells are identical with the original infecting virus. This fact implies that the DNA breaks at exactly the same point when it comes out of the chromosome as it does when it goes in. About one excision in 100,000, however, is abnormal. Instead of breaking away cleanly the host DNA and the viral DNA break and rejoin to create a circular DNA molecule incorporating some viral DNA and some host DNA. If the size and physical characteristics of the molecule allow it to be recognized by the viral proteins as being suitable for replication and packaging as a virus, it can then give rise to infectious particles in which a segment of host DNA has replaced part of the viral DNA.

M. Laurance Morse of the University of Colorado Medical Center first discovered the existence of these "transducing" virus variants when he found that some of the phage-lambda particles liberated from lysogenic bacteria contained the *gal* genes of the host. When lambda and lambda-*gal* DNA's were hybridized in the test tube, the heteroduplexes showed that nucleotides at the two ends of the molecules were complementary to one another but that in the middle of the duplex there was an unpaired region where the picked-up segment of bacterial DNA (including the *gal* operon) was not complementary to the viral DNA.

The theory that transducing phages are produced by errors in the excision of

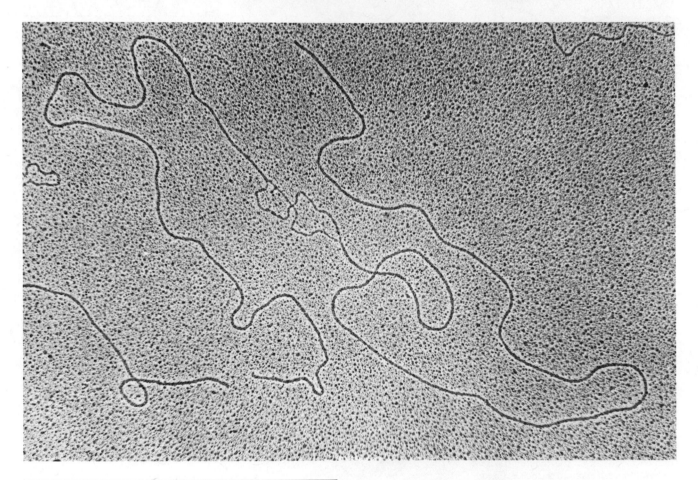

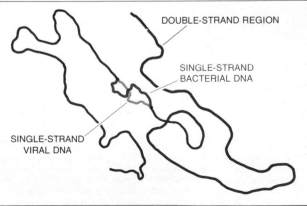

HETERODUPLEX made in the test tube between DNA from a transducing phage designated $\phi80psu_3$ and DNA from the parental phage $\phi80$ reveals the location of the inserted bacterial genes in the middle of the viral chromosome. The black lines on the map represent the double-helical regions where the nucleotide sequence on the DNA strands of both partners is complementary. The colored segment is the piece of *E. coli* DNA 3,000 nucleotides long carried by the transducing phage; the gray segment is the piece of viral DNA 2,000 nucleotides long that is present in the normal phage $\phi80$ but replaced by bacterial DNA in $\phi80psu_3$. Viral and bacterial DNA sequences on opposing strands are not complementary and cannot pair, forming a substitution bubble. Total length of duplex molecule is about 43,000 nucleotides. Electron micrograph was made by Madeline C. Wu and Norman R. Davidson of California Institute of Technology.

the provirus from the host chromosome was supported by the observation that under normal circumstances lambda can incorporate only genes, such as *gal*, that are within a few thousand nucleotides of its insertion site. This distance is a small fraction (less than 1 percent) of the total length of the host chromosome. Recently K. Shimada and his co-workers at the National Institutes of Health have studied rare bacterial lines in which lambda DNA has inserted itself into a chromosomal site other than the normal one. From such abnormal strains virus variants carrying *gal* are not obtained, but variants carrying genes close to the new attachment site are.

Why has the virus evolved such a complex and specific mechanism for getting its DNA into and out of chromosomes? The obvious answer is that the ability to do so at appropriate times plays an important role in the virus's survival. Little is known about the selective forces operating on viruses in nature, but one can imagine that it is to the virus's advantage for its DNA to be inserted soon after infection, for the DNA to remain stably inserted while the lysogenic bacterium is growing and for the DNA to be excised while the bacterial genes are repressed. Since insertion and excision have different enzymatic requirements, the virus can control both the direction and the extent of these activities by regulating integrase and excisionase.

The integrase reaction seen in phage lambda and similar viruses is the first case known where two DNA molecules are cut and rejoined at specific sites as part of the normal life cycle of an organism. Enzymatic cleavage and rejoining of DNA molecules in the test tube has become a common pastime of biochemists, but the bacterial restriction enzymes used for this purpose ordinarily function in DNA degradation rather than in genetic recombination. We do not yet know the actual nucleotide sequences recognized and acted on by integrase. The in vivo results require that the viral and bacterial sequences differ from each other, since the genetic requirements for insertion and excision are not the same.

The study of how the phage-lambda DNA inserts itself has provided some useful dividends, among them knowledge of the specific process of the breaking and joining of DNA molecules, which is becoming amenable to biochemical study and opens up new possibilities for the controlled translocation of DNA segments in other organisms. It has also given us the transducing virus variants, which have become workhorses of molecular biologists because they enable one to replicate specific segments of host DNA apart from the rest of the chromosome. In addition understanding

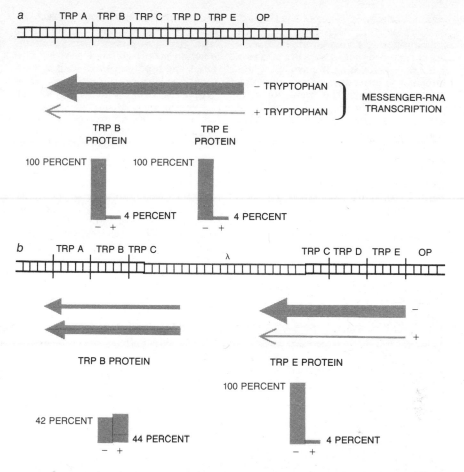

ABNORMAL INSERTION of the phage-lambda chromosome into the *trp* operon (a group of bacterial genes coding for five enzymes in the biosynthetic pathway of the amino acid tryptophan) disrupts the operon's genetic control mechanism. In the normal operon (*a*) transcription of the *trp* genes into messenger RNA is regulated by the tryptophan concentration. When tryptophan levels in the cell are high, transcription is repressed, and when tryptophan levels are low, transcription and synthesis of the *trp* enzymes is high. If phage lambda inserts its DNA into the middle of the *trpC* gene (*b*), the operon as a transcriptional unit will be disrupted. Genes "upstream" from the viral DNA (*trpD* and *trpE*) will continue to be transcribed normally, but the transcript cannot pass through the viral DNA to the genes downstream from the provirus (*trpA* and *trpB*). These genes are expressed to some extent, however, because of transcription arising within the provirus that cannot be repressed by tryptophan. (Protein levels shown are for a mutant of lambda in which the rate of this transcription is abnormally high.)

the mode of viral DNA insertion has helped to define the ways in which inserted elements and chromosomes can interact functionally. The simplest examples come from instances of abnormal insertion within known genes, such as the insertion of lambda DNA into the *trpC* gene.

The work of Charles Yanofsky of Stanford and others has shown that the *trp* operon consists of five genes, each of which codes for a different enzyme catalyzing a specific step in the biosynthesis of the amino acid tryptophan; the genes are designated *trpA, trpB* and so on. This entire stretch of DNA is transcribed into messenger RNA as a unit starting from *trpE* and continuing to *trpA*. The messenger RNA then attaches itself to the ribosomes, the subcellular particles where the enzymes are synthesized. Near the beginning of the transcribed stretch is the specific nucleotide sequence known as an operator. In the

presence of high concentrations of tryptophan a repressor protein binds to this sequence and prevents the transcription of the entire DNA segment. All five proteins are hence synthesized together when tryptophan is needed by the cell, but their synthesis is shut off by large amounts of the end product, a feedback control mechanism common to many operons.

How does the insertion of phage-lambda DNA change things? First, since the lambda DNA goes into the middle of the *trpC* gene, the complete protein product of that gene can no longer be formed. A bacterium that carries the abnormally inserted provirus is thus unable to make tryptophan, since one of the enzymes in the biosynthetic pathway is not synthesized. It is possible to recover descendants of this lysogenic bacterium that have lost the provirus. In these bacteria the two halves of the *trpC* gene are rejoined, and active enzyme is again synthesized. None of the DNA of the *trpC* gene has been damaged or permanently lost; it simply cannot code for its normal product when it is cut in two.

Besides disrupting the *trpC* gene, the provirus interrupts the *trp* operon as a functional unit. The transcription of RNA ordinarily proceeds along DNA segments such as the *trp* cluster from a fixed starting point to some stop signal. The precise nature of transcriptional stop signals is not known, but somewhere within the lambda DNA there must be one or more of them. *TrpE* and *trpD* proteins are synthesized normally in these lysogenic bacteria, but the transcription from the *trpE* end never reaches the *trpA* or *trpB* genes.

The provirus can hence constitute a barrier to RNA transcription, although the junction points between provirus DNA and bacterial DNA are not themselves barriers. Whereas bacterial transcription that can be repressed by the product tryptophan does not reach the *trpA* or *trpB* genes, transcription arising from within the provirus does cross the junction between viral DNA and host DNA and produces a low level of tryptophan-independent expression of these genes. The viral transcript that extends across the junction includes only one known gene, the integrase gene of the provirus, which is expressed at a low rate even when the other viral genes are repressed. Thus insertion can not only break up units of transcription but also create new ones.

The transcription of RNA across boundaries in the abnormal lysogenic bacterium illustrates some of the consequences of the viral insertion of DNA both for the regulation of cell function and for evolution. If new regulatory units can be created by insertion, we can be sure that natural selection will then act on them to maximize their selective value to the cell. How extensively DNA of viral origin may have become incorporated into the regulatory systems of existing chromosomes is not known.

There is an old dichotomy between those virologists who view the virus basically as a foreign invader and those who view it more as a cellular component that escapes normal regulatory controls. The argument frequently concerns matters of definition rather than of substance, but it tends to recur at different levels of sophistication as knowledge increases. Lysogenic bacteria have long constituted a prime example for the cellular-component school. At present one can say this much: The conception of the provirus as a normal cellular constituent is at least not a superficial one. The provirus not only behaves like an integral part of the host chromosome; it really is an integral part of the host chromosome. In its manner of attachment there is nothing to distinguish the DNA of the virus from the DNA of the host.

RNA-Directed DNA Synthesis

by Howard M. Temin
January 1972

*The discovery that in certain cancer-causing animal
viruses genetic information flows "in reverse"—from
RNA to DNA—has important implications for studies
of cancer in humans*

A major goal of present-day biology is to learn how information is coded in molecular structures and how it is transmitted from molecule to molecule in biological systems. Discovery of the rules governing this transmission is an integral part of understanding how embryonic cells differentiate into the hundreds of distinct types of cell observed in plants and animals and how normal healthy cells become cancerous.

It has now been known for nearly 20 years that the genetic information in all living cells is encoded in molecules of deoxyribonucleic acid (DNA) consisting of two long strands of DNA wound in a double helix. The genetic information for each organism is written in a four-letter alphabet, the "letters" being the four different chemical units called bases. In the normal cell short passages of the genetic message (individual genes) are transcribed from DNA into the closely related single-strand molecule ribonucleic acid (RNA). A length of RNA representing a gene is then translated into a particular protein, a molecule constructed with a 20-letter alphabet, the 20 amino acids. When a cell divides, the information contained in each of the two strands of DNA is replicated, thereby equipping the daughter cell with the full genetic blueprint of the parent.

Francis Crick, one of the codiscoverers of the helical structure of DNA, originally proposed that information can be transferred from nucleic acid to nucleic acid and from nucleic acid to protein, but that "once information has passed into protein it cannot get out again," that is, information cannot be transferred from protein to protein or from protein to nucleic acid. These concepts were simplified into what came to be known as the "central dogma" of molecular bi-

ology, which held that information is sequentially transferred from DNA to RNA to protein [see illustration on page 219]. Although Crick's original formulation contained no proscription against a "reverse" flow of information from RNA to DNA, organisms seemed to have no need for such a flow, and many molecular biologists came to believe that if it were discovered, it would violate the central dogma.

I shall describe experiments that originally hinted at a flow of information from RNA to DNA and that since have provided strong evidence that the "reverse" flow of information not only takes place but also accounts for the puzzling behavior of a sizable group of animal viruses whose genetic information is encoded in RNA rather than in DNA. Many of these viruses also produce cancer in animals. Although they have not yet been linked to cancer in man, their ability to transmit information from RNA to DNA inside the living cell makes it attractive to unify two hypotheses of the cause of human cancer that had previously seemed separate: the genetic hypothesis and the viral hypothesis.

There are two broad classes of viruses: viruses whose genome, or complete set of genes, consists of DNA and viruses whose genome consists of RNA. In the cells that they infect the DNA viruses replicate their DNA into new DNA and transmit information from DNA to RNA and thence into protein. Most RNA viruses, such as the viruses that cause poliomyelitis, the common cold and influenza, replicate RNA directly into new copies of RNA and translate information from RNA into protein; no DNA is directly involved in their replication.

In the past few years it has become apparent that a group of viruses, variously called the RNA tumor viruses, the

leukoviruses or the rousviruses (after their discoverer, Peyton Rous), replicate by another mode of information transfer. The rousviruses use information transfer from RNA to DNA in addition to the modes of information transfer (DNA to DNA, DNA to RNA and RNA

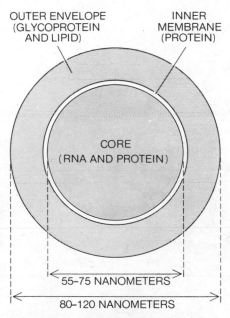

VIRIONS, or individual particles, of an "RNA-DNA virus," an animal-tumor virus that transfers genetic information from RNA (ribonucleic acid) to DNA (deoxyribonucleic acid) in addition to the normal modes of information transfer used by cells and other viruses, are enlarged about 700,000 diameters in the electron micrograph on following page. The particular RNA-DNA virions shown in thin section in the micrograph cause leukemia in mice; they are similar in structure and function to the Rous sarcoma virions discussed in this article. The electron micrograph was made by N. Sarkar of the Institute for Medical Research in Camden, N.J. A diagram of structure of a virion of this type is given above.

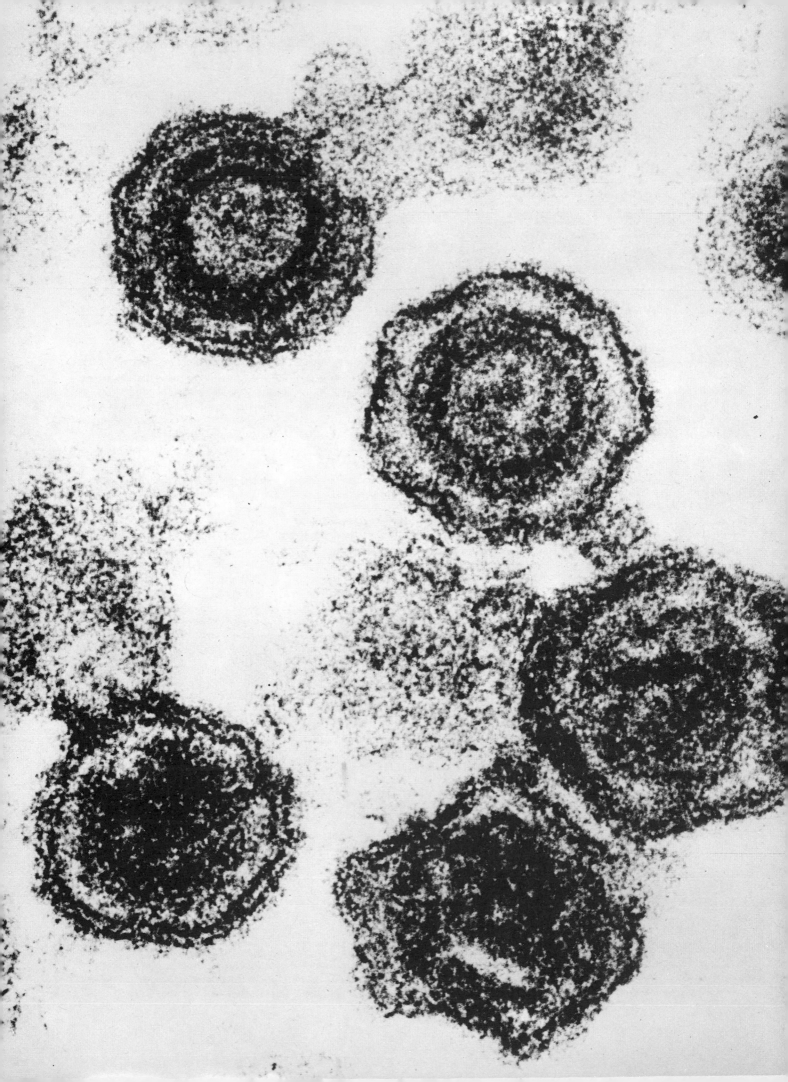

to protein) that are found in cells and in DNA viruses. The rousviruses do not transfer information from RNA to RNA, as other RNA viruses do. The existence of the RNA-to-DNA mode of information transfer in the replication of rousviruses has led some to suggest that there should be three major classes of viruses: DNA viruses, RNA viruses and RNA-DNA viruses [see top illustration on page 224].

The prototype RNA tumor virus, the Rous chicken sarcoma virus, was discovered by Rous 61 years ago at the Rockefeller Institute for Medical Research. An RNA tumor virus had actually been found earlier by V. Ellerman and O. Bang of Copenhagen, but their virus was little studied because it caused leukemia in chickens and was harder to work with than Rous's virus. Rous was studying a transplantable tumor of the barred Plymouth Rock hen. Originally he observed that he could transfer the tumor by the transfer of cells. In 1911 he found that the tumor could also be transferred by means of fluid from which the cells had been filtered. Demonstration that a disease can be transmitted by a cell-free filtrate is commonly accepted as evidence that it is caused by a virus. Descendants of the virus originally discovered by Rous are still being worked on in laboratories all over the world. At the time, however, Rous's discovery was met with disbelief, and after 10 years Rous himself stopped working with the tumor. It was not until nearly 30 years later, when Ludwik Gross of the Veterans Administration Hospital in the Bronx discovered that RNA tumor viruses cause leukemia in mice, that the study of rousviruses became popular.

It is now known that viruses in the same group as the virus originally discovered by Rous, or closely related to it, can cause tumors not only in chickens and mice but also in rats, hamsters, monkeys and many other species of animals. Moreover, viruses of the same group have been isolated from nonmammalian species, including snakes. As yet no bona fide human rousvirus has been discovered. It also appears that some members of this group, for example some of the "associated viruses," do not produce cancer.

In the 1950's, with the beginning of the application of cell-culture methods to animal virology, a tissue-culture assay for the Rous sarcoma virus was developed, first by Robert A. Manaker and Vincent Groupé at Rutgers University and subsequently by Harry Rubin and me at the California Institute of Technology. The assay involves adding suspensions of the virus to sparse cultures of cells taken from the body wall of chicken embryos. The Rous sarcoma virus infects some cells and transforms them into tumor cells. The transformed tumor cells differ in morphology and in growth properties from normal cells and therefore create a focus of altered cells. Assays of the same type have been developed for infections that the Rous sarcoma virus causes in cells taken from turkeys, ducks, quail and rats. Similar assays have also been developed for other transforming rousviruses.

The number of foci of transformed cells is proportional to the number of infectious units of the virus added to the cell culture and provides a rapid and reproducible assay for the Rous sarcoma virus. The use of this assay led to the discovery that the Rous sarcoma virus differs from the other viruses that had been studied up to that time in the way it interacts with the cell. The replication of most viruses is incompatible with cell division; in other words, the virus causes the infected cells to die. Chicken cells infected with the Rous sarcoma virus not only survive but also continue to divide and produce new virus particles [see middle illustration on page 224]. When the Rous sarcoma virus infects rat cells, there is a slightly different interaction of the cell and the virus. The rat cells are transformed into cancer cells, which divide, but the transformed cells do not produce the Rous sarcoma virus even though the genome (DNA) of the virus can be shown to be present. Production of the Rous sarcoma virus can be induced if the transformed rat cells are fused with normal chicken cells.

In the early 1960's the antibiotic actinomycin D was found to be very useful in unraveling the flow of genetic information in cells infected with RNA viruses. The antibiotic inhibits the synthesis of RNA made on a DNA template but not the synthesis of RNA made on an RNA template. The antibiotic therefore stops all RNA synthesis in cells infected by RNA viruses except for RNA specifically related to the viral genome. With this new tool it became easy to determine which RNA's were specific for the viruses.

When I added actinomycin D to cultures of cells producing Rous sarcoma virus, however, I found that the antibiotic inhibited the production of all RNA. One would have expected the replication of RNA on the template of an RNA viral genome to continue without hindrance [see bottom illustration on page 224]. This result was the first direct evidence that the molecular biology of the replication of Rous sarcoma virus was different from that of other RNA viruses. Since that observation was made the inhibition of the replication of rousviruses by actinomycin D has been recognized as one of their defining characteristics. The actinomycin D experiments suggested to me that the Rous sarcoma virus might replicate through a DNA intermediate. This hypothesis is called the DNA provirus hypothesis.

Further experiments, carried out by me and by John P. Bader at the National Cancer Institute, demonstrated that if one inhibits the synthesis of DNA in cells immediately after they have been inoculated with Rous sarcoma virus, one can protect the cells from infection. Here the inhibitors were amethopterin, fluorodeoxyuridine and cytosine arabinoside. These experiments appeared to support the idea that infection requires the synthesis of new viral DNA pro-

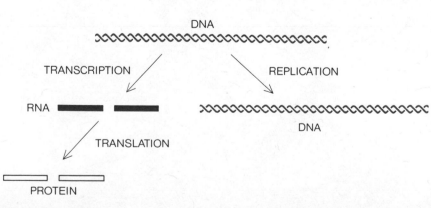

"CENTRAL DOGMA" of molecular biology, originally formulated by Francis Crick, states that within an organism genetic information can be transferred from DNA to DNA or from DNA to RNA to protein, but that it cannot be transferred from protein to protein or from protein to either DNA or RNA. Although a "reverse" flow of genetic information from RNA to DNA was not proscribed in Crick's original formulation, many molecular biologists came to believe that if such a flow were ever discovered, it would violate the central dogma.

duced on an RNA template. This interpretation was not unequivocal, however, because successful production of Rous sarcoma virus requires that the cells divide normally after infection. Therefore the inhibition of DNA synthesis after infection could inhibit production of Rous sarcoma virus not only by blocking possible new viral DNA synthesis but also by preventing normal cell division.

To get around this problem I introduced the idea of infecting cultures of stationary, or nondividing, cells with Rous sarcoma virus. Cells in culture usually require specific factors in blood serum to support their multiplication. If the serum is removed from the medium of the cell cultures, the cells stop dividing. If they are then exposed to Rous sarcoma virus, they become infected but there is no virus production or morphological transformation until serum is added back and the cells divide once again. When such stationary cells are exposed to inhibitors of DNA synthesis, the cells are not killed because they are not making DNA. When the stationary cells are exposed simultaneously to Rous sarcoma virus and to inhibitors of DNA synthesis, the cells are not killed but neither are they infected [see illustration at right].

If one now removes the inhibitor of DNA synthesis and adds serum, enabling the cells to divide once more, one finds that the cells remain free of infection. They do not become transformed and they do not produce virus. These experiments supported the hypothesis that after cells are infected by the Rous sarcoma virus new viral DNA is synthesized at a time different from the cell's normal synthesis of DNA. The new viral DNA is evidently synthesized on a template of viral RNA.

A further extension of this approach to understanding the replication of Rous sarcoma virus was carried out by one of my students, David E. Boettiger, and independently by Piero Balduzzi and Herbert R. Morgan at the University of Rochester School of Medicine and Dentistry. It had been found by others that if 5-bromodeoxyuridine, an analogue of the DNA constituent thymidine, is incorporated into DNA, the DNA becomes sensitized so that it can be inactivated by light. Under the same conditions normal thymidine-containing DNA is not affected by light. Boettiger therefore exposed stationary cells to Rous sarcoma virus in the presence of bromodeoxyuridine and then exposed the cells to light. Although the cells were not killed, the treatment prevented their

being infected by the virus. When serum was again added to enable the cells to divide, they did not become transformed and did not produce virus [see illustration on page 222].

In a related experiment Boettiger showed that the rate of inactivation of the infection by Rous sarcoma virus was dependent on the number of viruses infecting a cell. As he raised the number of viruses infecting each cell, he found that the infection became increasingly resistant to inactivation by light. We interpreted these experiments as showing that each infecting virus makes a new specific DNA, and that the more viruses that infect a cell, the more molecules of new viral DNA that are produced. The experiment seemed to effectively rule out the alternative hypothesis, which was that the infecting virus provokes a new synthesis of some preexisting cellular DNA.

Unfortunately no one has yet been able to unequivocally demonstrate the existence of newly synthesized viral DNA in cells infected with the Rous sarcoma virus. The available techniques are evidently too crude to detect the tiny

amounts of new viral DNA expected to be present. Certain results have been reported, however, with transformed cells. One approach has been to bring DNA from infected cells together with labeled viral RNA to see if single strands of the two molecules would coalesce into a double-strand hybrid molecule. Such hybrids are readily created when the base sequences in the DNA are complementary to the base sequences in the RNA, indicating that both carry the same genetic message and hence that each could arise from the transcription of the other.

The hybridization experiments reported thus far have aroused a great deal of controversy. Although some experiments, notably those of Marcel A. Baluda and Debi P. Nayak of the University of California at Los Angeles, have seemed to demonstrate the presence in infected cells of DNA complementary to viral RNA, the results have not been universally accepted. The finding of an intermediate viral DNA is an essential link in the chain of evidence that is still needed to establish firmly the DNA provirus hypothesis.

Meanwhile strong support for the hy-

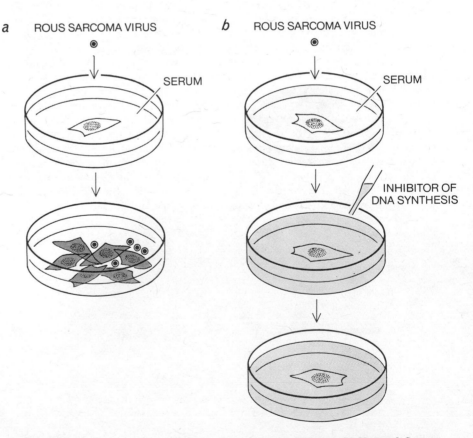

EXPERIMENTS carried out by the author and by John P. Bader at the National Cancer Institute supported the hypothesis that the infection of cells with Rous sarcoma virus requires the synthesis of new viral DNA produced on an RNA template. When the virus is added to cultures of normally dividing cells (a), the cells are transformed into cancer cells, which divide and produce new Rous sarcoma virus. By adding a substance that inhibits the synthesis of DNA in the cells immediately after they have been inoculated with Rous sar-

pothesis has come from experiments of a different kind. In 1969 Satoshi Mizutani, who had written his doctoral thesis on bacterial viruses, came to my laboratory for postdoctoral training. We decided to ask the question: What is the origin of the enzyme (a protein) responsible for forming proviral DNA using the viral RNA as template? When Mizutani exposed stationary cells to Rous sarcoma virus in the presence of inhibitors of protein synthesis, he found that the cells still became infected. We interpreted this experiment to mean that the enzyme that synthesizes DNA from the viral RNA template is already in existence before the infection.

Somewhat earlier other workers had fractionated virions—the actual virus particles as distinct from the forms assumed by the virus inside cells—and had found RNA polymerases, enzymes that catalyze the synthesis of RNA from its building blocks: four different ribonucleoside triphosphates. In 1967 Joseph Kates and B. R. McAuslan of Princeton University and William Munyon, E. Paoletti and J. T. Grace, Jr., of the Roswell Park Institute had found RNA polymer-

ases in a poxvirus, a large DNA virus. Other workers had found another RNA polymerase in a reovirus, a double-strand RNA virus. Therefore we decided to look in the virions of Rous sarcoma virus for a DNA polymerase capable of using the viral RNA as a template. After several months of preliminary experiments we succeeded in showing the existence of a DNA polymerase in purified virions of Rous sarcoma virus.

Before discussing this result I should digress briefly to describe the structure of the Rous sarcoma virus [see illustration, page 218]. The virion of the Rous sarcoma virus has a diameter of about 100 nanometers, which makes it larger than the particles of the viruses that cause poliomyelitis and smaller than the particles of the viruses that cause smallpox. The virion of the Rous sarcoma virus consists of a lipid-containing envelope (derived by budding from the cell membrane), an inner membrane and a nucleoid, or core, that contains the viral RNA and certain proteins.

In order to demonstrate that the Rous sarcoma virus contains a polymerase capable of producing DNA on an RNA

template, we first treated the virion with a detergent to disrupt its lipid-containing envelope. We then added to the disrupted virus the four deoxyribonucleoside triphosphates that are the building blocks of DNA. One of the deoxyribonucleoside triphosphates was radioactively labeled.

When the mixture was incubated at 40 degrees Celsius, it incorporated the radioactive label into an acid-insoluble substance that met the usual tests for DNA. The substance was stable in the presence of alkali and the enzyme ribonuclease, treatments that are known to destroy RNA, whereas it was attacked and fragmented by an enzyme that destroys DNA. When we repeated the experiment with disrupted virions pretreated with ribonuclease, an enzyme that destroys RNA, little or no DNA was produced, indicating that intact viral RNA was needed as the template for the synthesis of DNA [see top illustration on page 223].

After we had announced these results at the Tenth International Cancer Congress in Houston in May, 1970, we learned that David Baltimore of the

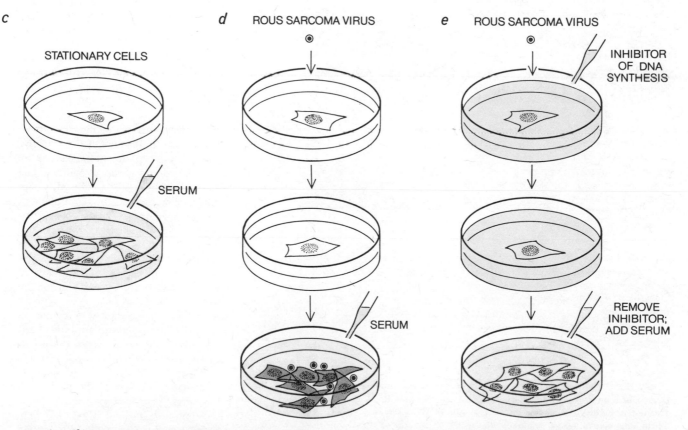

coma virus (b), one can protect the cells from infection. In subsequent experiments by the author cultures of stationary, or nondividing, cells were used; when blood serum is added to such cultures (c), they divide normally. If such stationary cells are first exposed to Rous sarcoma virus (d), however, they become infected but there is no virus production or morphological transformation until serum is added back and the cells divide once again. When the stationary cells are exposed simultaneously to Rous sarcoma virus and to an inhibitor of DNA synthesis (e), the cells are not killed but neither are they infected; when the inhibitor of DNA synthesis is removed and serum is added, cells divide normally, are not infected, do not become transformed and do not produce virus.

Massachusetts Institute of Technology had independently made similar observations with the virion of a mouse leukemia virus. The two papers describing these findings were published together in the June 27, 1970, issue of *Nature,* the British scientific weekly. The two publications stimulated an enormous amount of work whose peak is not yet in sight.

In our early papers we called the new viral enzyme RNA-dependent DNA polymerase because the template was RNA and the product was DNA. Subsequently we and others found that the enzyme could also use DNA as a template for DNA synthesis. We therefore decided to change the word "dependent" to "directed," so that we now refer to the enzyme as RNA-directed DNA polymerase. The revised name makes no statement about the origin of the enzyme or its relation to other DNA polymerases. Independently *Nature* began referring to the enzyme as "reverse

transcriptase," a name that I do not like because of its ambiguity but that has gained wide currency.

All the later studies confirm the original finding that the virions of RNA tumor viruses contain a DNA polymerase system that is activated by treating the virion with a detergent and that is sensitive to ribonuclease. Moreover, the virion enzyme functions only as a DNA polymerase; it will not act as an RNA polymerase. As I have mentioned, however, other unrelated RNA viruses do contain an RNA polymerase.

If the DNA produced by the RNA-directed DNA polymerase is isolated free of protein, the size of its molecule can be estimated by spinning it at high speed in a sucrose gradient in an ultracentrifuge. The molecule is surprisingly small: less than a tenth as long as one would expect a copy of the complete viral RNA to be. The reason for the small size is still elusive. If the isolated DNA product is centrifuged in a cesium sulfate density gradient, which separates RNA from

DNA on the basis of their different densities, one finds that the product has the density of DNA [*see bottom illustration on opposite page*]. Further characterization, for example by treatment with enzymes that specifically attack either single- or double-strand DNA, shows that the product of the DNA polymerase system is a double strand. From such studies one can conclude that the DNA polymerase system of the virion makes short pieces of double-strand DNA.

Many workers have demonstrated that the DNA product of the RNA-directed DNA polymerase system has a base sequence complementary to the viral RNA [*see top illustration on page 225*]. This conclusion is drawn from annealing, or molecular hybridization, experiments. Labeled DNA from the virion polymerase reaction is treated so that the strands of the DNA dissociate. The single-strand DNA is added to unlabeled viral RNA, and the mixture is incubated so that complementary strands can form a hybrid combination. The mixture is then centrifuged in a cesium sulfate density gradient. About half of the product DNA forms a band at a density characteristic of RNA or of hybrid RNA-DNA molecules rather than at a density characteristic of DNA. The test is quite specific and indicates that the DNA polymerase of the virion copies the sequence of the bases of the viral RNA into DNA. This experiment, however, still does not demonstrate that such a copying process takes place in cells infected by Rous sarcoma virus.

The viral DNA polymerase was shown to be present in the core of the virion by the following experiment carried out by George Todaro's group at the National Cancer Institute and by John M. Coffin in my laboratory at the University of Wisconsin. Rousvirus virions were treated with a detergent to disrupt the envelope. Then the disrupted virus was centrifuged in a sucrose density gradient. Most of the viral RNA, about 20 percent of the protein and most of the RNA-directed DNA polymerase activity were found to sediment together in "cores," a term given to structures that are denser than whole virions [*see bottom illustration on page 225*]. Further studies showed that with more extensive disruption of the virion the viral DNA polymerase can be freed from the viral RNA and then purified. The purified enzyme is capable of directing the synthesis of DNA on a variety of templates: synthetic and natural DNA, RNA and RNA-DNA hybrids.

The general conclusion from studies in a number of laboratories is that the

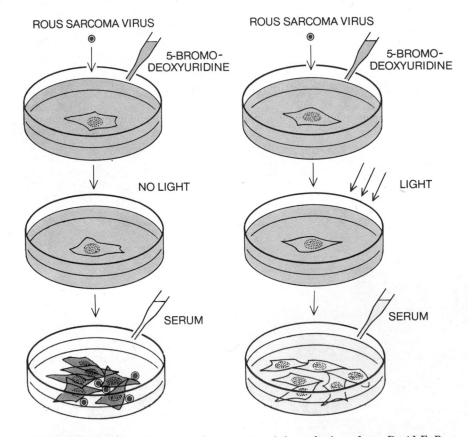

FURTHER EXPERIMENTS, carried out by one of the author's students, David E. Boettiger, and independently by Piero Balduzzi and Herbert R. Morgan at the University of Rochester, involved exposing stationary cells to Rous sarcoma virus in the presence of 5-bromodeoxyuridine, an analogue of the DNA constituent thymidine that, when incorporated into DNA, sensitizes the DNA to inactivation by light. As a control some of the treated cells were first not exposed to light (*left*); after serum was added to these cells to enable them to divide they were transformed into cancer cells and began to produce virus. When another culture of treated cells was exposed to light (*right*), the cells were not killed, but the treatment prevented their infection by the virus. When serum was again added to enable these cells to divide, they did not become transformed and did not produce virus.

rousvirus DNA polymerase closely resembles the other DNA polymerases described above that are present in more familiar biological systems and that catalyze the synthesis of DNA on a DNA template. In other words, it is not a unique property of the rousvirus DNA polymerase to be able to use RNA as a template for DNA synthesis. (This was first proposed several years ago by Sylvia Lee Huang and Liebe F. Cavalieri of the Sloan-Kettering Institute.) What is unique so far is the apparent biological role of RNA-directed DNA synthesis in the replication of rousviruses.

Further work in my laboratory has shown that preparations of purified virions of the Rous sarcoma virus contain other enzymes related to DNA replication. The most unusual of them is an enzyme that is named polynucleotide ligase, which repairs breaks in DNA molecules. It is an attractive hypothesis that the function of the ligase is to join the viral DNA to the chromosomal DNA of the host cell, thus integrating the viral genome with the cell genome. After this integration the genetic information of the virus would be replicated with that of the host and passed from the parent cell to the daughter cell. The Rous sarcoma virus virion also contains many other enzymes whose role is completely unknown. We do not know whether they participate in the life cycle of the virus or whether they are merely accidental contaminants picked up in the formation of the virion.

After the first discovery of a DNA polymerase in the virions of RNA tumor viruses, a great many other RNA viruses were examined to see if they contain a similar DNA polymerase system. First it was found that all the viruses previously classified in the RNA tumor virus group contain such an enzyme system. This group of RNA viruses includes both the rousviruses that cause tumors and those that do not cause tumors. Even more interesting, it was found that two types of virus that had not been classified in the same group with RNA tumor viruses also contain a DNA polymerase system. One of these viruses is Visna virus, which causes a slowly developing neurological disease in sheep. After the demonstration of a DNA polymerase in virions of the Visna virus, Kenneth Kaname Takemoto and L. B. Stone at the National Institutes of Health showed that the same virus could cause cancerous transformation of mouse cells in culture. Therefore Visna virus can now be considered a transforming rousvirus. The other type of virus that

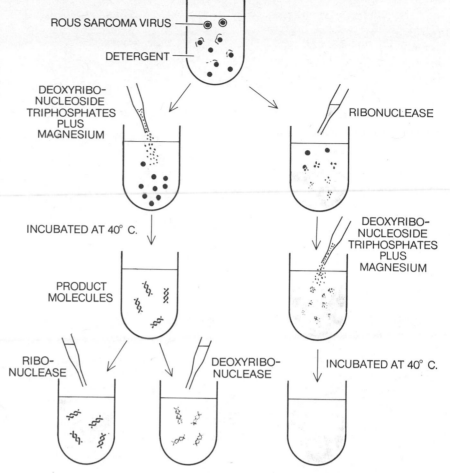

EXISTENCE OF A POLYMERASE capable of producing DNA on an RNA template in RNA tumor viruses was demonstrated by the author and his colleague Satoshi Mizutani (and also independently by David Baltimore of the Massachusetts Institute of Technology). In the experiment conducted by Mizutani and the author purified virions of Rous sarcoma virus were first treated with a detergent to disrupt their lipid-containing envelope. Four deoxyribonucleoside triphosphates, the "building blocks" of DNA, were then added to the disrupted virions. When the mixture was incubated, it incorporated the radioactive label associated with one of the building blocks into an acid-insoluble substance that was stable in the presence of ribonuclease (an enzyme known to destroy RNA), whereas it was fragmented by deoxyribonuclease (an enzyme that destroys DNA). When the experiment was repeated with disrupted virions pretreated with ribonuclease, little or no DNA was produced, indicating that intact viral RNA was needed as template for synthesis of DNA.

has been found to have a DNA polymerase system is the "foamy," or syncytium-forming, viruses. These viruses, isolated from monkeys and cats, have not been connected with any particular disease but are common contaminants of cell cultures. They have not yet been shown to cause tumors or cancerous transformation.

The DNA polymerase present in RNA tumor viruses may not only explain how these viruses produce stable cancerous transformations in the cells they infect but also account for some viral latency,

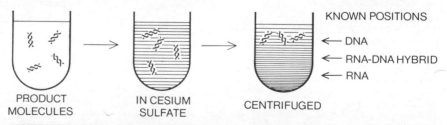

CENTRIFUGATION of the isolated DNA product of the RNA-directed DNA polymerase system in a cesium sulfate density gradient (which separates RNA from DNA on the basis of their different densities) resulted in the finding that the product has the density of DNA. In combination with other findings this result led to the conclusion that the DNA polymerase system of the Rous sarcoma virus virion makes short pieces of double-strand DNA.

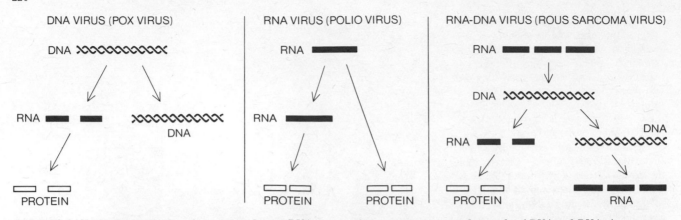

DNA VIRUS (POX VIRUS) RNA VIRUS (POLIO VIRUS) RNA-DNA VIRUS (ROUS SARCOMA VIRUS)

VIRUSES CAN BE GROUPED into three major classes: DNA viruses (*left*), whose genome, or complete set of genes, consists of DNA; RNA viruses (*middle*), whose genome consists of RNA, and RNA-DNA viruses (*right*), the most recently discovered group, whose genome consists alternately of RNA and DNA. A prototype virus in each major class is indicated in parentheses next to the class name. The diagrams illustrate the mode of information transfer that characterizes the replication of viruses in each class.

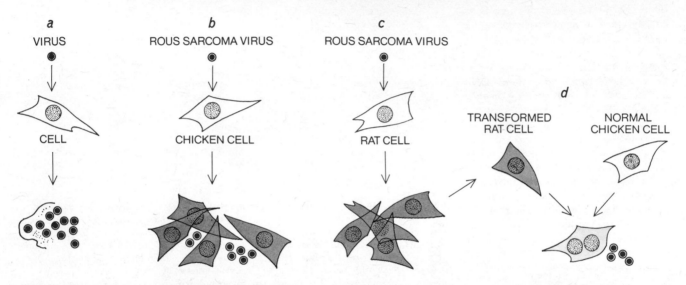

VIRUS-CELL INTERACTION usually leads to the death of the infected cell (*a*), since the replication of most viruses is incompatible with cell division. The Rous sarcoma virus, however, interacts with cells in a different way. Chicken cells infected with the Rous sarcoma virus (*b*) not only survive but also are transformed into cancer cells, which continue to divide and produce new virions. Rat cells infected with the Rous sarcoma virus (*c*) are transformed into cancer cells, which divide but do not produce new virions. By fusing the transformed rat cells with normal chicken cells the production of Rous sarcoma virions can be induced (*d*).

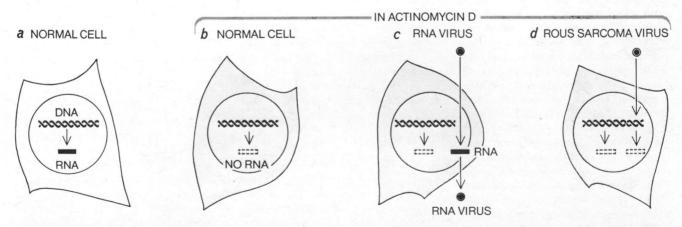

SYNTHESIS OF RNA on a DNA template in normal cells (*a*) is inhibited by the addition of the antibiotic actinomycin D (*b*). Since the antibiotic does not affect the synthesis of RNA made on an RNA template, however, it does not stop RNA synthesis specifically related to the viral genome in cells infected by most RNA viruses (*c*). The finding that actinomycin D inhibited the production of *all* RNA in cells producing Rous sarcoma virus (*d*) was the first direct evidence that the molecular biology of the replication of Rous sarcoma virus was different from that of other RNA viruses. The actinomycin D experiments led the author to propose the DNA provirus hypothesis, which holds that rousviruses such as the Rous sarcoma virus replicate through a DNA intermediate.

the phenomenon in which a virus disappears after infecting an organism only to reappear months or years later. Once an RNA virus has transferred its genetic information to DNA, it would be able to remain latent in a cell and be replicated by the cellular enzyme systems that replicate and repair the cell DNA. After some later activation the virus could appear again as infectious virus particles [*see top illustration on following page*].

About a year ago considerable public excitement was generated by the reported discovery of "RNA-dependent DNA polymerases" in human tumor cells. The general conclusion I would draw now from most of this work has been stated above: All DNA polymerases are capable, under the appropriate conditions, of transcribing information from RNA into DNA. At present we lack generally accepted criteria for determining whether or not such syntheses have any biological role or any relation to rousviruses.

In my laboratory we have taken a slightly different approach to the question of RNA-directed DNA synthesis in cells. We have used detergent activation and ribonuclease sensitivity as criteria in a broad search for DNA polymerase systems in a variety of animal cells. That is, we have looked in cells for a DNA polymerase system similar to viral "cores." Coffin has found such a DNA polymerase system in normal, uninfected rat embryo cells. So far we do not know the full significance of this discovery, but it suggests that ribonuclease-sensitive DNA polymerase systems are present in cells other than tumor cells or virus-infected cells.

For many years I have favored the idea that RNA-directed DNA synthesis may be important in normal cellular processes, particularly those involved in the embryonic differentiation of cells. This idea has been expanded in the form of the protovirus hypothesis [*see bottom illustration on following page*]. The general idea is that in normal cells there are regions of DNA that serve as templates for the synthesis of RNA, and that this RNA serves in turn as a template for the synthesis of DNA that subsequently becomes integrated with the cellular DNA. By this means certain regions of DNA can be amplified. With additional processes that introduce changes in the DNA, the DNA of different cells can be made different. This difference might serve as a means of distinguishing different cells.

What, then, are the general implications of this work for the prevention or treatment of human cancer? We can

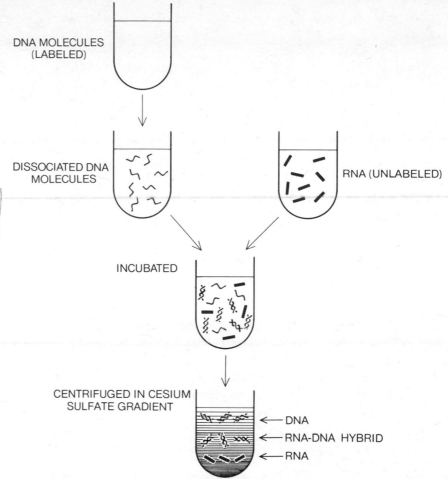

MOLECULAR-HYBRIDIZATION EXPERIMENTS demonstrated that the DNA product of the RNA-directed DNA polymerase system within the virion copies the sequence of bases of the viral RNA into DNA. Labeled DNA from the virion polymerase reaction was first treated so that strands of the DNA dissociated. The single-strand DNA was then added to unlabeled viral RNA, and the mixture was incubated at high temperature so that complementary strands could form a hybrid combination. When the resulting "annealed" mixture was centrifuged in a cesium sulfate density gradient, about half of the product DNA was observed to form a band at a density characteristic of hybrid RNA-DNA molecules.

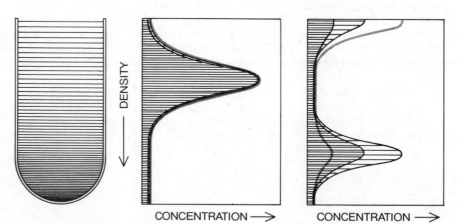

PRESENCE OF VIRAL DNA POLYMERASE in the cores of the Rous sarcoma virus virions was demonstrated by John M. Coffin in the author's laboratory. The curves at center show the density distribution of various radioactively labeled constituents of the whole virions as determined by centrifugation in a sucrose density gradient (*left*). The curves at right show the density distribution of the same constituents determined by centrifugation after the virions were treated with a detergent to disrupt their envelopes. Most of the viral RNA (*black curves*), about 20 percent of the protein (*gray curves*) and most of the RNA-directed DNA polymerase activity (*colored curves*) of the disrupted virions were found to sediment together at a higher density than the corresponding constituents of the whole virions, indicating that these constituents are concentrated in the cores of the virions.

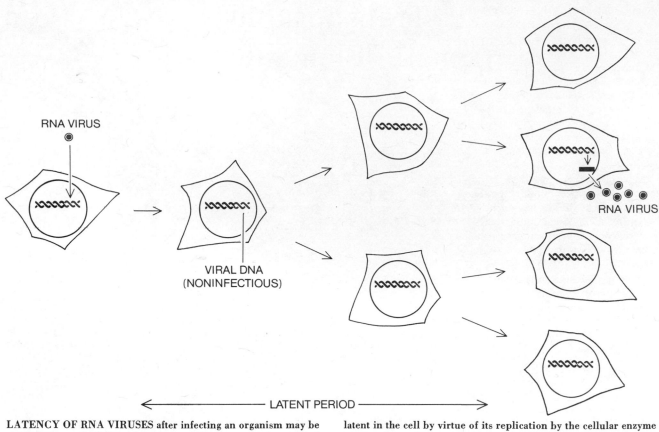

RNA VIRUS

VIRAL DNA
(NONINFECTIOUS)

RNA VIRUS

←———————————— LATENT PERIOD ————————————→

LATENCY OF RNA VIRUSES after infecting an organism may be attributable to the DNA polymerase system present in the cores of such viruses. After transferring its genetic information to DNA in the cell nucleus (*left*), the RNA virus would disappear, remaining latent in the cell by virtue of its replication by the cellular enzyme systems that replicate and repair the cell's DNA (*left center, right center*). Months or years later some form of activation could then cause the infectious RNA virions to appear again (*upper right*).

conclude only that some biological systems utilize a previously undescribed mode of information transfer: from RNA to DNA. It is an interesting coincidence that this new mode of information transfer was first discovered in tumor-causing viruses. We cannot say, however, that RNA-directed DNA synthesis is an exclusive property of such viruses. What the discovery of RNA-directed DNA synthesis does mean is that we now have some simple biochemical tests to determine whether or not newly discovered human viruses are members of the same group as the RNA viruses that produce tumors and cancerous transformations in animal cells, and to look for information related to these viruses in human cancers. We cannot now say that inhibitors of RNA-directed DNA synthesis would have any effect on human cancer. In rousvirus-induced tumors in animals the synthesis of new viral DNA appears to be important only at the initial stage of cancerous transformation, not thereafter.

Probably the most important implication of this discovery for the understanding of cancer in man has been the removal of the dichotomy between viral and genetic theories of the origin of cancer. At a time when genes were thought to consist of DNA alterable only by mutation, and when most of the known cancer-causing animal viruses were of the RNA type, it was hard to imagine common features of genetic and viral theories. Now that we have uncovered evidence that cancer-causing RNA viruses can produce a DNA transcript of the viral RNA, one can readily formulate hypotheses in which elements related to viral RNA are attached to the genome of the cell and transmitted genetically to become activated at some future time and cause "spontaneous" cancer. Experiments designed to test this idea are now in progress in a number of laboratories around the world.

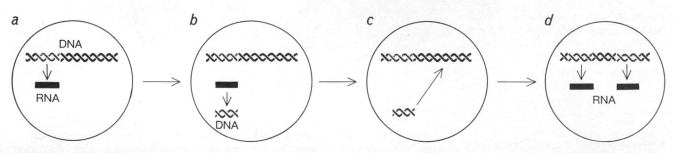

a DNA *b* DNA *c* *d* RNA

RNA

PROTOVIRUS HYPOTHESIS, put forward by the author, embodies the idea that RNA-directed DNA synthesis may be important in normal cellular processes. According to this view, there are regions of DNA in normal cells that serve as templates for the synthesis of RNA (*a*). This RNA serves in turn as a template for the synthesis of DNA (*b*), which later becomes integrated with the cellular DNA (*c*). The amplification of certain regions of DNA resulting from the repetition of the process (*d*) may, in conjunction with additional processes that introduce changes in the DNA, play an important role in the embryonic differentiation of cells.

Transposable Genetic Elements

by Stanley N. Cohen and James A. Shapiro
February 1980

*They bypass the rules of ordinary genetic
recombination and join together segments of DNA
that are unrelated, transferring groups of genes among
plasmids, viruses and chromosomes in living cells*

Natural selection, as Darwin recognized more than a century ago, favors individuals and populations that acquire traits conducive to survival and reproduction. The generation of biological variation, which gives rise to new and potentially advantageous combinations of genetic traits, is therefore a central requirement for the successful evolution of species in diverse and changing environments.

Hereditary information is encoded in the sequence of the building blocks, called nucleotides, that constitute a molecule of DNA, the genetic material. The basic step in the creation of genetic variation is the mutation, or alteration, of the DNA within a gene of a single individual. Mutations involve changes in nucleotide sequence, usually the replacement of one nucleotide by another. This can lead to a change in the chain of amino acids constituting the protein encoded by the gene, and the resulting change in the properties of the protein can influence the organism's biological characteristics. Spontaneous mutations are too rare, however, for genetic variation to depend on new mutations that arise in each generation. Instead variation is generated primarily by the reshuffling of large pools of mutations that have been accumulated within a population in the course of many generations.

In higher organisms this reshuffling is done in the process of sexual reproduction. The genes are arrayed on two sets of chromosomes, one set inherited from the female parent and the other set from the male parent, so that there are two copies of each gene. Sometimes the nucleotides of a genetic sequence differ slightly as a result of earlier mutation, producing alleles, or variant forms of a gene. In the formation of gametes (egg or sperm cells) the breakage of structurally similar pairs of chromosomes can result in the reciprocal exchange of alleles between the two members of a pair of chromosomes. Such genetic recombination requires that the segments of DNA undergoing exchange be homolo-

gous, that is, the sequence of nucleotides on one segment of DNA must be very similar to the sequence on the other segment, differing only at the sites where mutations have occurred.

The ability of segments of DNA on different chromosomes to recombine makes it likely that in complex plants or animals the particular collection of genes contained in each egg or sperm cell is different. An individual produces many eggs or sperms, which can potentially interact with sperms or eggs from many other individuals, so that there is a vast opportunity for the generation of genetic diversity within the population. In the absence of intentional and extended inbreeding the possibility that any two plants or animals will have an identical genetic composition is vanishingly small.

Genetic variation is also important in the evolution of lower organisms such as bacteria, and here too it arises from mutations. Bacteria have only one chromosome, however, so that different alleles of a gene are not normally present within a single cell. The reshuffling of bacterial genes therefore ordinarily requires the introduction into a bacterium of DNA carrying an allele that originated in a different cell. One mechanism accomplishing this interbacterial transfer of genes in nature is transduction: certain viruses that can infect bacterial cells pick up fragments of the bacterial DNA and carry the DNA to other cells in the course of a later infection. In another process, known as transformation, DNA released by cell death or other natural processes simply enters a new cell from the environment by penetrating the cell wall and membrane. A third mechanism, conjugation, involves certain of the self-replicating circular segments of DNA called plasmids, which can be transferred between bacterial cells that are in direct physical contact with each other.

Whether the genetic information is introduced into a bacterial cell by transduction, transformation or conjugation,

it must be incorporated into the new host's hereditary apparatus if it is to be propagated as part of that apparatus when the cell divides. As in the case of higher organisms, this incorporation is ordinarily accomplished by the exchange of homologous DNA; the entering gene must have an allelic counterpart in the recipient DNA. Because homologous recombination requires overall similarity of the two DNA segments being exchanged, it can take place only between structurally and ancestrally related segments. And so, in bacteria as well as in higher organisms, the generation of genetic variability by this mechanism is limited to what can be attained by exchanges between different alleles of the same genes or between different genes that have stretches of similar nucleotide sequences. This requirement imposes severe constraints on the rate of evolution that can be attained through homologous recombination.

Until recently mutation and homologous recombination nevertheless appeared to be the only important mechanisms for generating biological diversity. They seemed to be able to account for the degree of diversity observed in most species, and the implicit constraints of homologous recombination—which prevent the exchange of genetic information between unrelated organisms lacking extensive DNA-sequence similarity—appeared to be consistent with both a modest rate of biological evolution and the persistence of distinct species that retain their basic identity generation after generation.

Within the past decade or so, however, it has become increasingly apparent that there are various "illegitimate" recombinational processes, which can join together DNA segments having little or no nucleotide-sequence homology, and that such processes play a significant role in the organization of genetic information and the regulation of its expression. Such recombination is often effected by transposable genetic elements: structurally and genetically discrete segments of DNA that have the ability to move around among the chromosomes and the extrachromosomal DNA molecules of bacteria and higher organisms. Although transposable elements have been studied largely in bacterial cells, they were originally discovered in plants and are now known to exist in animals as well. Because illegitimate recombination can join together DNA segments that have little, if any, ancestral relationship, it can affect evolution in quantum jumps as well as in small steps.

In the late 1940's Barbara McClintock of the Carnegie Institution of Washington's Department of Genetics at Cold Spring Harbor, N.Y., first reported a genetic phenomenon in the common corn plant, *Zea mays*, that would later be found to have parallels in other biological systems. While studying the inheritance of color and the distribution of pigmentation in plants that had undergone repeated cycles of chromosome breakage she found that the activity of particular genes was being turned on or off at abnormal times. Because some of these genes were associated with the development of pigments in kernels as well as in the plant itself, certain kernels were mottled, showing patches of pigmentation against an otherwise colorless background. The patterns of this variegation were reproduced in successive generations and could be analyzed like other heritable traits. After painstaking study of many generations of corn plants McClintock concluded that the variegation she observed was the result of the action of distinct genetic units, which she called controlling elements, that could apparently move from site to site on different maize chromosomes; as they did so they sometimes served as novel biological switches, turning the expression of genes on or off.

McClintock's genetic analysis showed that some patterns of variegation affected three or more genes simultaneously, suggesting that the structure of one of the plant's chromosomes had been rearranged at the site of a controlling element. Direct microscopic examination of maize chromosomes containing controlling elements confirmed that these genetic elements did in fact serve as specific sites for the breakage and resealing of DNA, thereby giving rise to either minute or gross changes in chromosome structure.

Almost 20 years after McClintock reported her earliest studies on controlling elements in the corn plant Michael Malamy, who is now at the Tufts University School of Medicine, Elke Jordan, Heinz Saedler and Peter Starlinger of the University of Cologne and one of us (Shapiro), who was then at the University of Cambridge, found a new class of mutations in genes of a laboratory strain of the common intestinal bacterium *Escherichia coli*. They were unusual in that their effects were detectable beyond the borders of the mutated genes them-

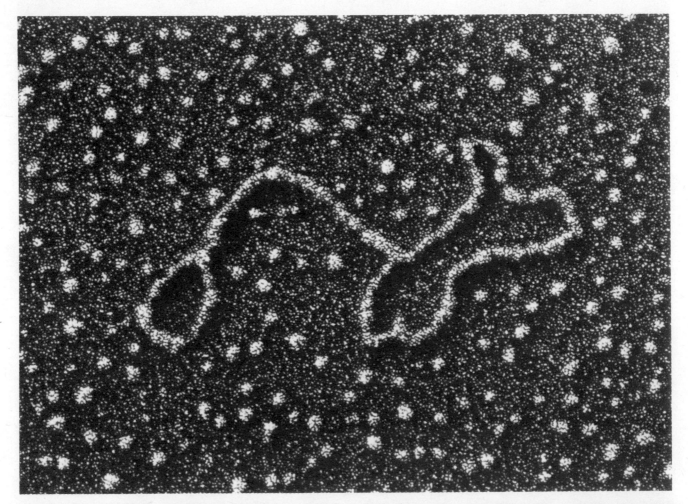

DNA OF TRANSPOSABLE GENETIC ELEMENT (a transposon) forms a characteristic stem-and-loop structure, which is seen here in an electron micrograph made by one of the authors (Cohen). The structure results from the "inverted repeat" nature of the nucleotide sequences at the two ends of the transposon DNA (*see upper illustration on page 230*). The double-strand DNA of the plasmid pSC105, into which the transposon had been inserted, was denatured and complementary nucleotide sequences on each strand were allowed to "re-anneal." The joining of the complementary nucleotides constituting the transposon's inverted-repeat termini formed the double-strand stem. The smaller loop was formed by the segment of single-strand transposon DNA between the inverted repeats, a segment that includes a gene conferring resistance to the antibiotic kanamycin. The larger loop represents the single-strand DNA of a miniplasmid derivative of the host plasmid. DNA was spread with formamide and shadowed with platinum-palladium. Enlargement is 230,000 diameters.

selves; this property could not be explained by any known mutational mechanism.

When the DNA segments carrying these mutations were inserted into particles of a bacterial virus and the density of the virus was compared with that of viruses carrying normal genes, it became clear that the mutated DNA was longer than the normal DNA: the mutations had been caused by the insertion of sizable DNA fragments into the mutated gene. It further developed that a limited number of other kinds of distinguishable DNA segments, which were up to 2,000 nucleotides in length, could also insert themselves within many different genes, interrupting the continuity of the gene and turning off its activity. These elements were named insertion sequences, or IS elements. The observation that a small number of specific DNA segments could be inserted at a large number of different sites in the bacterial chromosome suggested that some type of nonhomologous recombination was taking place; it seemed to be unlikely that an IS element could be homologous with the nucleotide sequences at so many different insertion sites.

At about the same time that IS elements were discovered other microbiologists and geneticists made observations hinting that certain genes known to be responsible for resistance to antibiotics by bacteria were capable of transfer from one molecule of DNA to another. Results obtained by Susumu Mitsuhashi and his colleagues at the University of Tokyo in the mid-1960's suggested that a gene encoding a protein that inactivates the antibiotic chloramphenicol could move from its normal site on a plasmid-DNA molecule to the chromosome of a bacterium or to the DNA of a virus.

Similar instances of the apparent transfer of antibiotic-resistance genes between different DNA molecules in the same cell were reported from the U.S. and Britain. The first direct evidence that such transfer is by a process analogous to the insertion of IS elements was published in 1974. R. W. Hedges and A. E. Jacob of the Hammersmith Hospital in London found that the transfer from one plasmid to another of a gene conferring resistance to antibiotics such as penicillin and ampicillin was always accompanied by an increase in the size of the recipient plasmid; the recipient could donate the resistance trait to still other plasmids, which thereupon showed a similar increase in size.

Hedges and Jacob postulated that the gene for ampicillin resistance was carried by a DNA element that could be "transposed," or could move from one molecule to another, and they called such an element a transposon. Their discovery of a transposable element that carries an antibiotic-resistance gene was

DIFFERENCES IN PIGMENTATION in kernels of the corn plant Zea mays (see page 228) reflect the action of a two-element control system discovered by Barbara McClintock of the Cold Spring Harbor Laboratory. Both elements are transposable. One element is at the locus of a gene whose action it modulates to yield the faintly and homogeneously pigmented kernels. The other element acts on the first one to produce the variegated pattern that is seen in many of the kernels.

an important advance. In earlier studies the movement of IS elements had been tracked only indirectly by genetic techniques: by observing the effects of insertions on various genetic properties of the host organism. It now became possible to track a transposable element's intermolecular travels directly by observing the inheritance of the antibiotic-resistance trait.

While the Hedges and Jacob experiments were being carried out Dennis J. Kopecko and one of us (Cohen), at the Stanford University School of Medicine, were studying the acquisition of a gene for resistance to ampicillin by still other plasmids. It emerged, as Hedges and Jacob had found, that the ampicillin-resistance trait present on one plasmid could be acquired by another plasmid. Surprisingly, however, it also developed that such transfer could take place in mutated bacteria lacking a particular protein, the product of a gene

designated recA, known to be necessary for homologous recombination. Examination of the plasmid DNA with the electron microscope revealed that a 4,800-nucleotide segment carrying the ampicillin-resistance trait was being transferred as a characteristic and discrete structural unit. Moreover, the segment could become inserted at many different sites on the recipient plasmid DNA.

Electron microscopy also showed that the two ends of the transposable DNA segment had a unique feature: they consisted of nucleotide sequences that were complementary to each other but in the reverse order. This finding calls for some explanation. The four nitrogenous bases that characterize DNA nucleotides are linked in complementary pairs by hydrogen bonds to form the double helix of DNA: adenine (A) is linked to thymine (T) and guanine (G) to cytosine (C). The nucleotide sequence $AGCTT$, for example, is complementary

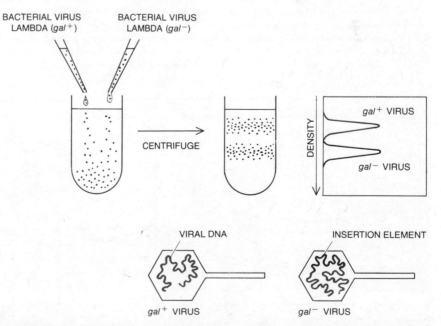

MUTATION BY INSERTION was demonstrated by one of the authors (Shapiro) with phage-lambda particles carrying the bacterial gene for galactose utilization ($gal+$) and particles carrying the mutant gene $gal-$. The viruses were centrifuged in a cesium chloride solution. The $gal-$ particles were found to be the denser. Because the virus particles all have the same volume and their outer shells all have the same mass, increased density of $gal-$ particles showed they must contain a larger DNA molecule: $gal-$ mutation was caused by insertion of DNA.

to the sequence *TCGAA*. The nucleotide sequence at one end of the transposable DNA segment was complementary in reverse order to a sequence on the same strand at the other end of the element [*see upper illustration on this page*]. These "inverted repeats" were revealed when the two strands of the double-strand plasmid DNA carrying the transposon were separated in the laboratory and each of the strands was allowed to "reanneal" with itself: a characteristic stem-and-loop structure was formed by the complementary inverted repeats.

The result of the transposition process is that a segment of DNA originally present on one molecule is transferred to a different molecule that has no genetic homology with the transposable element or with the donor DNA. The fact that the process does not require a bacterial gene product known to be necessary for homologous recombination indicates that transposition is accomplished by a mechanism different from the usual recombinational processes.

Subsequent experiments done in numerous laboratories have shown that DNA segments carrying genes encoding a wide variety of antibiotic-resistance traits can be transferred between DNA molecules as discrete units. Moreover, genes encoding other traits, such as resistance to toxic mercury compounds, synthesis of bacterial toxins and the capacity to ferment sugars or metabolize hydrocarbons, have been shown to be capable of transposition. All the transposons studied so far have ends consisting of inverted-repeat sequences, which range in length from only a few nucleotides to as many as 1,400. The ends of at least two transposons actually consist of two copies of the insertion sequence IS1 (which itself has been found to have terminal inverted-repeat sequences). Recent evidence has suggested that the insertion of any gene between two trans-

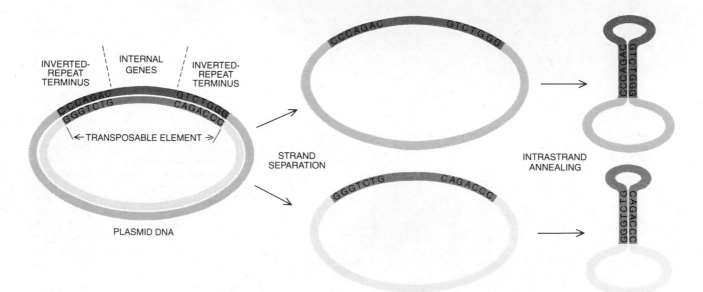

STEM-AND-LOOP STRUCTURES demonstrate the inverted-repeat nucleotide sequences of the ends of transposable elements. The four bases adenine (*A*), guanine (*G*), thymine (*T*) and cytosine (*C*) of DNA's four nucleotide building blocks are linked to form a helix (shown here schematically as a double strand); *A* always pairs with *T* and *G* pairs with *C*. The termini of a transposable element have sequences (seven nucleotides long here) that are bidirectionally and rotationally symmetrical. When the two strands of a plasmid containing an element are separated and each strand is allowed to self-anneal, the complementary nucleotides at the termini pair with each other, forming a double-strand stem (*right and in electron micrograph on page 228*). The remainder of the DNA is seen as single-strand loops.

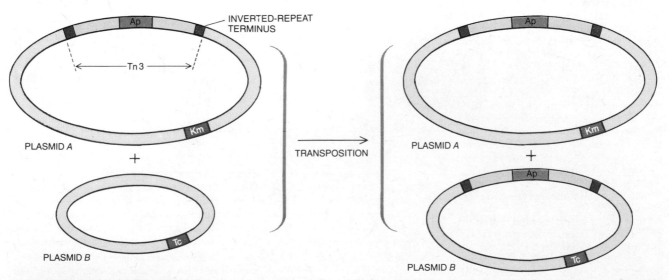

TRANSPOSITION of the transposon Tn3, which carries a gene conferring resistance to the antibiotic ampicillin (*Ap*), is diagrammed. It is shown as originally being part of plasmid *A*, which also includes a gene for resistance to kanamycin (*Km*). A plasmid *B*, which confers resistance to tetracycline (*Tc*), acquires a copy of the transposon. The new plasmid *B* confers resistance to ampicillin and tetracycline.

posable elements makes possible the transfer of the gene to a structurally unrelated DNA molecule by nonhomologous recombination.

Since transposable elements are transferred as discrete and characteristic genetic units, there must be some highly specific enzymatic mechanism capable of recognizing their inverted-repeat ends and cleaving DNA precisely at these locations. The first evidence that genes carried by the transposable elements themselves can encode such enzymes came from a series of experiments carried out by Frederick L. Heffron, Craig Rubens and Stanley Falkow at the University of Washington and continued by Heffron and his colleagues at the University of California at San Francisco. When these investigators introduced mutations that interrupted the continuity of genes at various locations within the ampicillin-resistance transposon designated Tn3, they found alterations in the ability of the element to function as a transposon. Mutation in the inverted-repeat ends or in a particular region of the DNA segment between the ends prevented transposition. On the other hand, mutations within another region of Tn3 actually increased the frequency of movement of Tn3 between the different plasmids, suggesting that this region might contain a gene modulating the ability of Tn3 to undergo transposition.

Recently published work by Joany Chou, Peggy G. Lemaux and Malcolm J. Casadaban in the laboratory of one of us (Cohen) and by Ronald Gill in Falkow's laboratory has shown that the Tn3 transposon does in fact encode both a "transposase"—an enzyme required for transposition—and a repressor substance that regulates both the transcription into RNA of the transposase gene and the repressor's own synthesis. Analogous experiments at the University of Chicago, the University of Wisconsin and Harvard University have shown that other transposable elements also encode proteins needed for their own transposition.

Even though transposons can insert themselves at multiple sites within a recipient DNA molecule, their insertion is not random. It has been recognized for several years that certain regions of DNA are "hot spots" prone to multiple insertions of transposons. Experiments recently reported by David Tu and one of us (Cohen) have shown that Tn3 is inserted preferentially in the vicinity of nucleotide sequences similar to sequences within its inverted-repeat ends, even in a bacterial cell that does not make the recA protein required for ordinary homologous recombination. It therefore appears that recognition of homologous DNA sequences may play some role in determining the frequency and site-specificity of transposon-asso-

ciated recombination, even though the actual recombinational mechanism differs from the one commonly associated with the exchange of homologous segments of DNA.

The discovery of the process of transposition explains a puzzling phenomenon in bacterial evolution that has serious implications for public health: the rapid spread of antibiotic resistance among bacteria. Under the selective pressure of extensive administration of antibiotics in human and veterinary medicine and their use as a supplement in animal feeds, bacteria carrying resistance genes have a great natural advantage. For some time it has been known that resistance to several different antibiotics can be transmitted simultaneously to a new bacterial cell by a plasmid, but until transposition was discovered it was not known how a number of genes conferring resistance to different antibiotics were accumulated on a single plasmid-DNA molecule. The explanation seems to be that the resistance-determi-

nant segments of drug-resistance plasmids have evolved as collections of transposons, each carrying a gene that confers resistance to one antibiotic or to several of them.

Work carried out at Stanford and by Phillip A. Sharp and others in the laboratory of Norman R. Davidson at the California Institute of Technology has made it clear that certain bacterial plasmids are constructed in a modular fashion. Plasmids isolated in different parts of the world show extensive sequence homology in certain of their DNA segments, whereas in other segments there is no structural similarity at all. In some instances plasmids can dissociate reversibly at specific sites. Transposable IS elements are found both at these sites and at sites where the plasmid interacts with chromosomal DNA to promote the transfer of chromosomes between different bacterial cells.

Identical transposons are commonly found in bacterial species that exchange genes with one another in nature. In addition antibiotic-resistance transposons

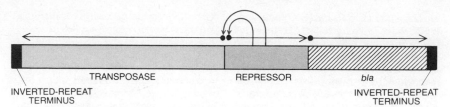

FUNCTIONAL COMPONENTS of the transposon Tn3 are diagrammed (not to scale). Genetic analysis shows there are at least four kinds of regions: the inverted-repeat termini; a gene for the enzyme beta-lactamase (*bla*), which confers resistance to ampicillin and related antibiotics; a gene encoding an enzyme required for transposition (a transposase), and a gene for a repressor protein that controls the transcription of the genes for transposase and for the repressor itself. The arrows indicate the direction in which DNA of various regions is transcribed.

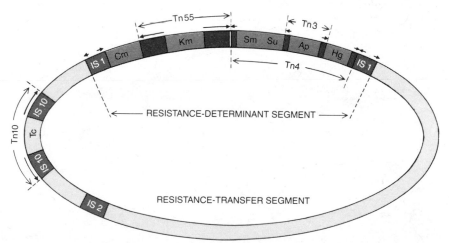

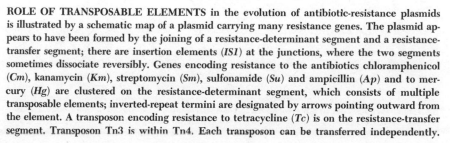

ROLE OF TRANSPOSABLE ELEMENTS in the evolution of antibiotic-resistance plasmids is illustrated by a schematic map of a plasmid carrying many resistance genes. The plasmid appears to have been formed by the joining of a resistance-determinant segment and a resistance-transfer segment; there are insertion elements (*IS1*) at the junctions, where the two segments sometimes dissociate reversibly. Genes encoding resistance to the antibiotics chloramphenicol (*Cm*), kanamycin (*Km*), streptomycin (*Sm*), sulfonamide (*Su*) and ampicillin (*Ap*) and to mercury (*Hg*) are clustered on the resistance-determinant segment, which consists of multiple transposable elements; inverted-repeat termini are designated by arrows pointing outward from the element. A transposon encoding resistance to tetracycline (*Tc*) is on the resistance-transfer segment. Transposon Tn3 is within Tn4. Each transposon can be transferred independently.

appear to be able to move among very different bacterial species that have not previously been known to exchange genes. For example, DNA sequences identical with part of Tn3 have recently been found to be responsible for penicillin resistance in two bacterial species unrelated to those commonly harboring Tn3 and in which such resistance had not previously been observed. Transposable elements seem, in other words, to accomplish in nature gene manipulations akin to the laboratory manipulations that have been called genetic engineering.

The effects of transposable genetic elements extend beyond their ability to join together unrelated DNA segments and move genes around among such segments. These elements can also promote both the rearrangement of genetic information on chromosomes and the deletion of genetic material. An awareness of these effects has emerged most clearly from studies of a peculiar phage, or bacterial virus, discovered in 1963 by Austin L. Taylor of the University of Colorado. Like other "temperate" bacterial viruses Taylor's phage could insert its DNA into a bacterial chromosome, creating a latent "prophage" that coexists with the bacterial cell and is transmitted to the bacterial progeny when the cell divides. Unlike other temperate phages, however, this one could become inserted at multiple sites within the chromosome, thereby causing many different kinds of mutation in the host bacterium. Because of this property Taylor called his phage Mu, for "mutator."

Further studies have shown that Mu is actually a transposable element that can also exist as an infectious virus. In the virus particle the Mu DNA is sandwiched between two short segments of bacterial DNA it has picked up from a bacterial chromosome. When the Mu virus infects a new cell, it sheds the old bacterial DNA and is transposed to a site in the new host chromosome. Ahmad I. Bukhari and his colleagues at the Cold Spring Harbor Laboratory have shown that Mu's ability to replicate is closely associated with its ability to be transposed; the virus has apparently evolved in such a way that its life span is dependent on transposition events.

While the structure of Mu's DNA and the details of the phage's life cycle were being unraveled Michel Faelen and Arianne Toussaint of the University of Brussels were doing genetic experiments aimed at understanding how the Mu DNA interacts with other DNA in a bacterial cell. The results of experiments carried out over a period of almost 10 years have demonstrated that Mu can catalyze a remarkable series of chromosome rearrangements. These include the fusion of two separate and independently replicating DNA molecules ("replicons"), the transposition of segments of the bacterial chromosome to plasmids, the deletion of DNA and the inversion of segments of the chromosome. Significantly, all these rearrangements seem to involve the nucleotide sequences at the ends of Mu DNA and to require the expression of a Mu gene that had been found earlier to be necessary both for transposition and for virus replication. Experiments done by Hans-Jorg Reif and Saedler at the University of Freiburg and by other groups have shown that many other transposable elements can, like Mu, promote the deletion of DNA; Nancy E. Kleckner and David Botstein and their associates at Harvard and at the Massachusetts Institute of Technology have shown that such elements can also bring about the inversion of DNA sequences. Indeed, there is evidence that some transposable elements may participate in specific re-

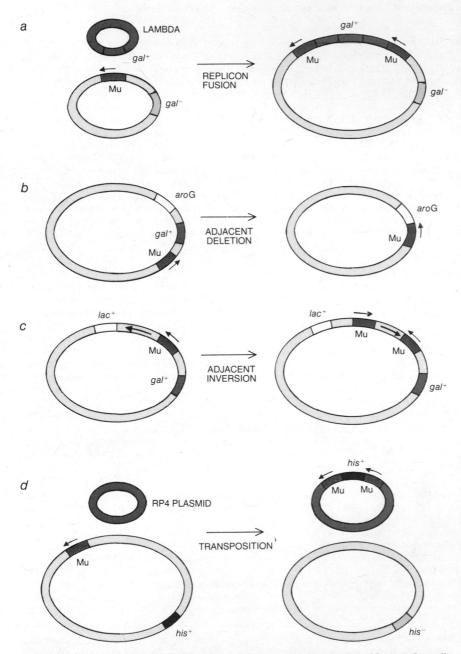

CHROMOSOME REARRANGEMENTS mediated by the bacterial virus Mu include replicon fusion, adjacent deletion, adjacent inversion and transposition of chromosome segments to a plasmid. Mu is shown in color; a small arrow gives its orientation. In a cell lysogenic for Mu (having Mu DNA integrated in its chromosome) and containing DNA of a lambda *gal*⁺ virus (*a*) the viral DNA becomes integrated into the chromosome between two copies of Mu. In a lysogenic cell in which Mu is near integrated *gal*⁺ genes (*b*) the *gal*⁺ genes are deleted. In a lysogenic male bacterium (*c*) with Mu near the origin of chromosome transfer (*large arrow*) the origin becomes inverted between two oppositely oriented copies of Mu. In a lysogenic bacterium carrying a plasmid (*d*) a bacterial *his*⁺ gene is transposed to plasmid between copies of Mu.

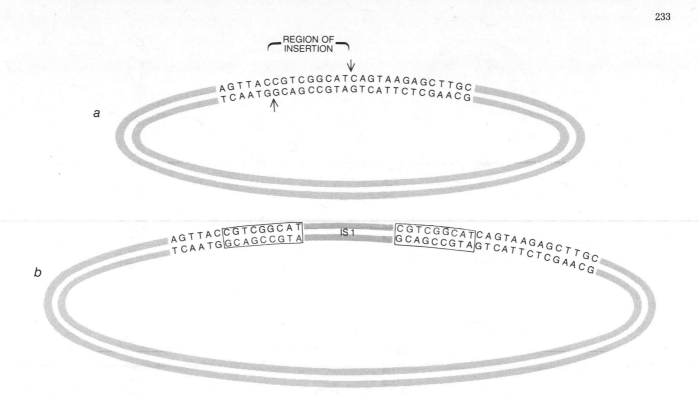

DUPLICATION of five, nine or 11 pairs of nucleotides in the recipient DNA is associated with the insertion of a transposable element; the two copies bracket the inserted element. Here the duplication that attends the insertion of IS1 is illustrated in a way that indicates how the duplication may come about. IS1 insertion causes a nine-nucleotide duplication. If the two strands of the recipient DNA are cleaved (*colored arrows*) at staggered sites that are nine nucleotides apart (*a*), then the subsequent filling in of single strands on each side of the newly inserted element (*b*) with the right complementary nucleotides (*color*) could account for the duplicated sequences (*colored boxes*).

arrangements of DNA more frequently than they do in transposition events.

New methods for determining DNA nucleotide sequences rapidly and simply have provided an important tool for elucidating the structure of transposable elements as well as the biochemical mechanisms involved in transposition and in chromosome rearrangements. The sequence of a transposable element (IS1) has been determined in its entirety by Hisako Ohtsubo and Eiichi Ohtsubo of the State University of New York at Stony Brook. An important insight into the mechanism of transposition has resulted from DNA-sequence observations initially made by Nigel Grindley of Yale University and the University of Pittsburgh and by Michele Calos and Lorraine Johnsrud of Harvard, working with Jeffrey Miller of the University of Geneva. Both groups examined the DNA sequences at the sites of several independently occurring insertions of the IS1 element. They found that the insertion of IS1 results in the duplication of a sequence of nine nucleotide pairs in the recipient DNA. The duplicated sequences bracket the insertion element and are immediately adjacent to its inverted-repeat ends. Since the sequence of the recipient DNA was different at each of the various insertion sites studied, different nucleotides were duplicated for each insertion.

Subsequent reports from many laboratories have shown that similar duplications of a short DNA sequence result from the insertion of other transposable elements. Some elements generate nine-nucleotide duplications and others generate duplications five or 11 nucleotides long. As Calos and her colleagues and Grindley have pointed out, these observations suggest that a step in the insertion process involves staggered cleavage (at positions five, nine or 11 nucleotides apart) of opposite DNA strands at the target site for transposition. The filling in of the single-strand segments following such cleavage would require the synthesis of short single-strand stretches of complementary DNA and would result in the nucleotide-sequence duplication. Faelen and Toussaint had also concluded that DNA synthesis is required in the generation of chromosome rearrangements by Mu: they had noted that the rearranged bacterial chromosome often included two copies of the prophage, the inserted form of Mu.

On the basis of these observations one of us (Shapiro) has proposed a model to explain transposition, chromosome rearrangements and the replication of transposable elements such as phage Mu as variations of a single biochemical pathway. The pathway is such that transposable elements can serve two functions in the structural reorganization of cellular DNA: they specifically duplicate themselves while remaining inserted in the bacterial chromosome, and they bring together unrelated chromosomal-DNA segments to form a variety of structural rearrangements, including fusions, deletions, inversions and transpositions.

If this model is at all close to reality, then the nonhomologous recombination events associated with transposable elements are rather different from other types of illegitimate recombination, such as the integration of the phage-lambda DNA into the bacterial chromosome, that do not involve DNA synthesis. It seems likely that bacterial cells will turn out to have several different systems for carrying out nonhomologous recombination, just as they have multiple pathways for homologous recombination.

The potential for multiple mechanisms of illegitimate recombination is important to bear in mind when comparing phenomena that appear to be similar in bacteria and higher cells. Transposition phenomena that are analogous genetically may not be similar biochemically. There is some genetic evidence indicating that the movement of controlling elements in maize from one chromosomal site to another may be brought about by a mechanism different from that of transposition in bacteria.

Genetic rearrangements can have biological importance on two time scales: on an evolutionary scale, where the effects of the rearrangement are seen

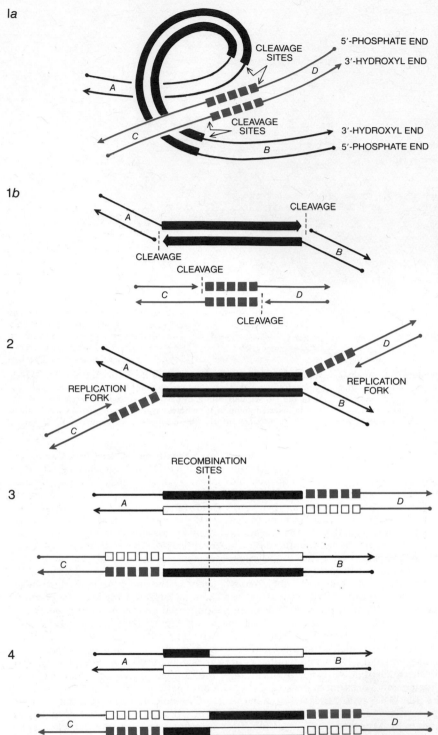

1a

CLEAVAGE
SITES

5'-PHOSPHATE END
3'-HYDROXYL END

A

D

C

CLEAVAGE
SITES

3'-HYDROXYL END

B

5'-PHOSPHATE END

1b

CLEAVAGE

A

CLEAVAGE

B

CLEAVAGE

C

D

CLEAVAGE

2

D

A

REPLICATION
FORK

REPLICATION
FORK

C

B

RECOMBINATION
SITES

3

A

D

C

B

4

A

B

C

D

POSSIBLE MOLECULAR PATHWAY is suggested to explain transposition and chromosome rearrangements. The donor DNA, including the transposon (*thick bars*), is in black, the recipient DNA in color. Arrowheads indicate the 3'-hydroxyl ends of DNA chains, dots the 5'-phosphate ends; the letters *A, B, C* and *D* identify segments of the two DNA molecules. The pathway has four steps, beginning with single-strand cleavage (*1a*) at each end of the transposable element and at each end of the "target" nucleotide sequence (*colored squares*) that will be duplicated. The cleavages expose (*1b*) the chemical groups involved in the next step: the joining of DNA strands from donor and recipient molecules in such a way that the double-strand transposable element has a DNA-replication fork at each end (*2*). DNA synthesis (*3*) replicates the transposon (*open bars*) and the target sequence (*open squares*), accounting for the observed duplication. This step forms two new complete double-strand molecules; each copy of the transposable element joins a segment of the donor molecule and a segment of the recipient molecule. (The copies of the element serve as linkers for the recombination of two unrelated DNA molecules.) In the final step (*4*) reciprocal recombination between copies of the transposable element inserts the element at a new genetic site and regenerates the donor molecule. The mechanism of this recombination is not known; it does not require proteins needed for homologous recombination, and at least in Tn3 it is mediated by sequences within element.

after many generations, and on a developmental time scale, where the effects are apparent within a single generation. It is known that transposable genetic elements can serve as biological switches, turning genes on or off as a consequence of their insertion at specific locations. In some instances the insertion of an IS element in one orientation turns off nearby genes, whereas an unexpressed gene can be turned on when the element is inserted in the opposite orientation.

An analogous regulation of gene expression through chromosome rearrangement is "phase variation," which is seen in certain disease-producing bacteria that can invade the gastrointestinal tract. The phase, or immunological specificity, of a hairlike flagellum on these bacteria can change suddenly within a single bacterial generation. Melvin Simon of the University of California at San Diego and his colleagues have recently shown that the choice between the expression of one *Salmonella* flagellum gene and the expression of its counterpart, which specifies a different phase, is controlled by the inversion of a particular segment of the bacterial chromosome. The inversion takes place in the absence of proteins needed for homologous recombination, and so it appears to depend on recombination enzymes that recognize the ends of the invertible segment. Whether the switching mechanism responsible for phase variation operates by a molecular process similar to transposition remains to be determined, but the process clearly falls within the category of recombination events that were considered "illegitimate" a few years ago.

Although molecular studies on transposable elements have so far been carried out primarily in bacteria, there has been extensive genetic evidence for the existence of similar elements in higher organisms for years. The pioneering work of Barbara McClintock not only established the existence of transposable genetic elements in the corn plant but also showed by genetic analysis that the movement of a controlling element from one site to another in the maize chromosome depends on the action of genes on certain of the elements themselves, genes presumably analogous to those encoding the transposases of Tn elements and of phage Mu. McClintock also showed that some controlling elements (called regulators) regulate the expression of distant genes carrying insertions of other controlling elements (called receptors). Groups of genes are expressed synchronously at specific times during plant development, and McClintock suggested that the transposition of receptor elements could provide a mechanism for the rapid evolution of control mechanisms in situations in which several genes must be switched on or off at the same time, as they are in the course of development.

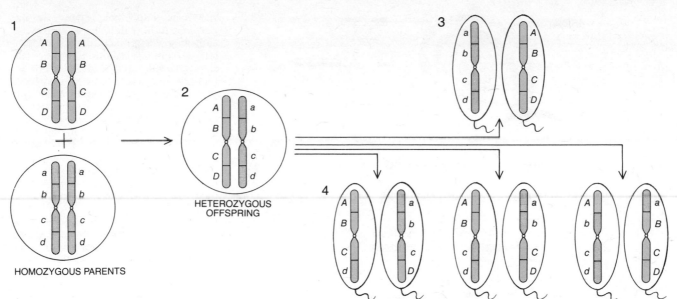

HOMOLOGOUS RECOMBINATION is accomplished in higher organisms by the "crossing over" of structurally similar chromosome segments during sexual reproduction. Here the process is shown for a hypothetical animal each of whose somatic (body) cells has a single chromosome pair carrying four genes, each of which may be present in either of two variant forms (alleles). Homozygous parents having the same set of alleles on both paired chromosomes (*1*) give rise to heterozygous offspring (*2*), which in turn can produce gametes (sperms or eggs) containing copies of the original chromosomes (*3*). As a result of crossing over and reciprocal homologous recombination, alleles can be reshuffled in various ways (*4*), producing gametes containing chromosomes that are different from either of original chromosomes.

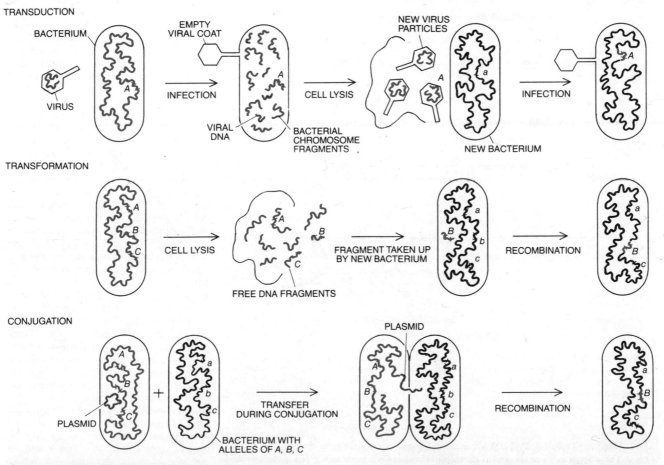

RECOMBINATION IN BACTERIA requires the introduction into a bacterial cell of an allele obtained from another cell. In transduction an infecting phage, or bacterial virus, picks up a bacterial-DNA segment carrying allele *A* and incorporates it instead of viral DNA into the virus particle. When such a particle infects another cell, the bacterial-DNA segment recombines with a homologous segment, thereby exchanging allele *A* for allele *a*. In transformation a DNA segment bearing allele *B* is taken up from the environment by a cell whose chromosome carries allele *b*; the alleles are exchanged by homologous recombination. In conjugation a plasmid inhabiting one bacterial cell can transfer the bacterium's chromosome, during cell-to-cell contact, to another cell whose chromosome carries alleles of genes on the transferred chromosome; again allele *B* is exchanged for allele *b* by recombination between homologous DNA segments.

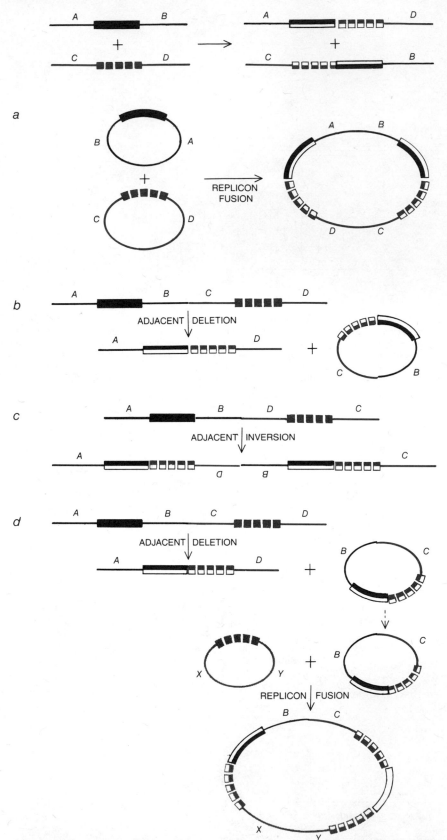

FIRST THREE STEPS OF PATHWAY are summarized schematically at the top of this illustration. These steps achieve reciprocal recombination between unrelated DNA molecules and explain all rearrangements shown in the illustration on page 232, as follows. If the donor and recipient molecules are circular, the three steps result in replicon fusion (*a*). If the donor and recipient regions are part of a single molecule, the steps generate an adjacent deletion (*b*) or an adjacent inversion (*c*), depending on the positions of regions **A, B, C** and **D**. Two successive events (deletion and then replicon fusion) can result in the transposition to a plasmid of a DNA segment adjacent to the transposable element, along with two copies of the element (*d*).

As often happens in science, the significance of McClintock's work was not entirely understood or appreciated until later studies carried out with the much simpler bacterial systems provided actual physical evidence for the existence of insertion sequences and transposons as discrete DNA segments and also established that transposition is brought about by a mechanism different from previously understood recombinational processes. Numerous other examples of transposable elements have now been recognized in higher organisms, such as the fruit fly *Drosophila* and the yeast *Saccharomyces cerevisiae*. The possible role of these elements in the generation of chromosome rearrangements is being actively investigated. Recent work on the control of immunoglobulin synthesis in mice by Susumu Tonegawa and his associates at the Basel Institute for Immunology has shown that the ability of mammalian cells to produce specific antibody molecules in response to injected foreign proteins also involves chromosome rearrangements. There is little doubt that additional instances will soon be found in which illegitimate recombination events play a major role in the expression of genes during cellular differentiation.

Even in bacteria much remains to be learned about the basic molecular mechanisms that accomplish the transposition of genetic elements and the associated rearrangement of DNA molecules. The various biochemical steps in the transposition pathway need to be more fully defined. What is the mechanism for recognition of the inverted-repeat ends of transposable elements? What proteins other than those encoded by the transposon play a role in transposition? What are the additional genetic aspects of the regulation of transposition? In a broader sense, what is the role of illegitimate recombination in the organization and expression of genes, not only in bacteria but also in higher organisms? Although the mechanisms that have been studied in bacteria provide a working model for the mechanisms of similar events in higher organisms, the parallels are probably incomplete.

It is already clear that the joining of structurally and ancestrally dissimilar DNA segments by transposable elements is of great importance for the production of genetic diversity and the evolution of biological systems. The discovery of such a fundamentally different recombinational process at a time when many molecular biologists believed virtually all the important aspects of bacterial genetics were understood in principle—with only the details of particular instances remaining to be learned—leads one to wonder whether still other fundamentally new and significant basic biological processes remain to be discovered.

EVOLUTION

EVOLUTION

INTRODUCTION

Paradoxically, belief in order is the article of faith implicitly professed by those who elect to call themselves scientists, who are consequently presumed to be free of the need for faith. Nowhere is this faith more clearly demonstrated than by evolutionists who seek to account for the manifestation of life in its great variety in terms of the adaptive meaningfulness of structure and strategy.

The adaptation of organisms to their environments and the intricacy of their strategies for survival and reproduction have always been a source of wonder and delight to man. So perfect do these adaptations seem that they are seen by many not only as the supreme example of God's handiwork but as the most persuasive evidence for His existence. Contrariwise, a man's moral imperfections, which are seen by the nondeterminists and optimists as sociological maladaptations, are not taken as evidence for the nonexistence of an omnipotent God but as a manifestation of man's often perverse and powerful free will.

Einstein, when confronted with stochastic implications of the quantum theory, turned away from these with the utterance, "I shall never believe that God plays dice with the world." While evolutionists have come to recognize the scope for chance in their accountings, no accommodation for God is made in modern evolutionary theory.

The uncharitable among you may believe that the growing and almost boundless literature on evolution and evolutionary theory is more the outcome of an unquenchable but unproductive dedication to description by devotees of natural history—who will not stop until their description is as complete and complex as that of nature itself—than it is the reflection of real and significant problems in the field. This position is ably challenged by Ernst Mayr in his article "Evolution," in which he sets out what he believes to be the cogent but as yet unsolved problems for evolutionists to tackle. His article originally served as an introduction to the September 1978 issue of *Scientific American*, entirely devoted to the topic of evolution.

Investigations on the relation between structure and function in protein have revealed that there are many alternate and virtually equivalent ways for amino acid sequences to be the solution of a particular functional requirement in an organism. As Mayr points out in his article, these discoveries have done much to eradicate the last vestiges of "essentialist" thinking from evolutionary theory—thinking which was conducive to the idea of created and immutable species and later to the notion of the "wild type" gene or allele. Just as it was realized that species represent transient gene assemblages which evolve, split, and, on occasion, fuse, so it is now realized that the assemblages of nucleotides that specify the genes are themselves transient and evolving. Thus, the concept of the "wild type" gene is now seen as an essentialist myth.

Because the phenomenon of adaptation is so basic to evolution and such a ground for contention between the materially and spiritually inclined, I have elected to include Richard Lewontin's article "Adaptation" in this series. Appropriately, Lewontin's article formulated the concluding thoughts in the *Scientific American* anthology on evolution, for, in his typically iconoclastic fashion, he demolishes the notion that adaptation is an obvious process or condition, a notion that advocates of Darwin's evolutionary theory sometimes hold. Like the Cheshire cat in *Alice in Wonderland,* the concept of adaptation tends to vanish on close examination.

The two final articles in this section follow Lewontin's article appropriately: they are about apparently nonadaptive phenomena manifested in the genetic material itself.

To this point in the series, the reader may have been lulled into a false sense of complacency about progress in molecular biology, and about how it has served to explain the mechanisms underlying the manifestation of classical genetics. However, one should be aware of disturbing recent findings in molecular genetics that seem to indicate the existence of unnecessary, baroque ornaments on the genetic system.

The first well-characterized example of what appears to be an unnecessary ornament is the occurrence of repeated sequences of DNA, variable both in degree and proportion, especially among the eucaryotes. There seems to be no pattern in the occurrence of this seemingly gratuitous excess of DNA, and its existence presents a formidable puzzle for evolutionists who seek to account for it. The discovery of this DNA and the progress in its characterization within and among species are presented in Roy J. Britten and David E. Kohne's article, "Repeated Segments of DNA."

The article by Motoo Kimura, "The Neutral Theory of Molecular Evolution," presents the case that much of the evolution in terms of changes in nucleotides within a gene coding for a functional polypeptide is neutral. Thus one must distinguish between adaptive evolution based on the forces of natural selection and those changes in nucleotide sequence within genes (and consequently, in some instances, changes in the amino acid sequence of their product that merely reflect the inescapable passage of time in the presence of mutation pressure) that provides the opportunity for trivial change.

In Kimura's model, mutation pressure averaged through time and opportunities for allele fixation through genetic drift are essentially constant for all lineages over the vast time span considered. The opportunity for trivial change is presented by the redundancy inherent in the genetic code and by the functional equivalence of subsets of the twenty amino acids used by nature to say her piece.

There are more recently described phenomena whose adaptive significance is obscure. These include the transposition of DNA sequences around the genome, as described by Stanley N. Cohen and James A. Shapiro in their article "Transposable Genetic Elements," which was included in the section on genetic transactions. In addition, there is the occurrence of intercalated noncoding sequences within the nucleotide sequence specifying a polypeptide product. As mentioned before, these sequences or "introns" are edited out of the messenger RNA sequence prior to its translation by an enzyme-catalysed RNA-sequence-specific splicing process.

Some evidence suggests that "gene jumping" may have some (ongoing rather than relict) utility to the organisms or cells in which this is found. For example, the assembly of the gene that is to specify a particular antibody may capitalize on the jumping gene mechanism to juxtapose the fragment coding for the constant region of the antibody against that of the fragment coding for its specificity.

The existence of these phenomena poses a formidable series of problems for the evolutionist and the molecularist alike—the evolutionist hopes to account for their existence and the molecularist hopes to describe their nature fully so that the evolutionist's accounting will seem more likely.

Some of the questions that might be asked are the following:

- Are new genes assembled from coding fragments of preexisting genes?

- Are introns—the noncoding space segments that separate coding segments within the gene—relics of that assembly process, or do they have contemporary utility for the organism?

- If new genes are assembled from old ones, do the coding segments correspond to the functional domains into which a polypeptide chain can be subdivided?

- If there is such an assembly process, where and how does it take place?

- Does the noncoding, repetitive DNA in higher organisms provide the place in which gene assembly can occur by means of a "jumping gene fragment" mechanism, thus avoiding the disruption of necessary genetic information that would occur if such a place were not available?

- Or is it possible, as Leslie Orgel and Francis H. C. Crick have suggested, that "intron" DNA sequences are no more than parasitic DNA sequences equivalent to "selfish" genes that are not functional or relevant parts of host genome?

These and many related questions will animate the increasingly convergent fields of reductionist molecular biology and the synthesist preoccupations of evolutionary biology.

Evolution

by Ernst Mayr
September 1978

Introducing a volume devoted to the history of life on the earth as it is understood in the light of the modern "synthetic" theory of evolution through natural selection, the organizing principle of biology today

The most consequential change in man's view of the world, of living nature and of himself came with the introduction, over a period of some 100 years beginning only in the 18th century, of the idea of change itself, of change over long periods of time: in a word, of evolution. Man's world view today is dominated by the knowledge that the universe, the stars, the earth and all living things have evolved through a long history that was not foreordained or programmed, a history of continual, gradual change shaped by more or less directional natural processes consistent with the laws of physics. Cosmic evolution and biological evolution have that much in common.

Yet biological evolution is fundamentally different from cosmic evolution in many ways. For one thing, it is more complicated than cosmic evolution, and the living systems that are its products are far more complex than any nonliving system; other differences will emerge in the course of this article. This *Scientific American* article deals with the origin, history and interrelations of living systems as they are understood in the light of the currently accepted general theory of life: the theory of evolution through natural selection, which was propounded more than 100 years ago by Charles Darwin, has since been modified and explicated by the science of genetics and stands today as the organizing principle of biology.

The creation myths of primitive peoples and of most religions had in common an essentially static concept of a world that, once it had been created, had not changed—and that indeed had not been in existence for very long. Bishop Ussher's 17th-century calculation that the world had been created in 4004 B.C. was noteworthy only for its misplaced precision in an age when the reach of history was still foreshortened by the limited arm's length of written records and tradition. It remained for the naturalists and philosophers of the 18th-century Enlightenment and the geologists and biologists of the 19th century to begin to extend the time dimension. In 1749 the French naturalist the Comte de Buffon first undertook to calculate the age of the earth. He reckoned it was at least 70,000 years (and suggested an age of as much as 500,000 years in his unpublished notes). Immanuel Kant was even more daring in his *Cosmogony* of 1755, in which he wrote in terms of millions or even hundreds of millions of years. Clearly both Buffon and Kant conceived of a physical universe that had evolved.

"Evolution" implies change with continuity, usually with a directional component. Biological evolution is best defined as change in the diversity and adaptation of populations of organisms. The first consistent theory of evolution was proposed in 1809 by the French naturalist and philosopher Jean Baptiste de Lamarck, who concentrated on the process of change over time: on what appeared to him to be a progression in nature from the smallest visible organisms to the most complex and most nearly perfect plants and animals and thence to man.

To explain the particular course of evolution Lamarck invoked four principles: the existence in organisms of a built-in drive toward perfection; the capacity of organisms to become adapted to "circumstances," that is, to the environment; the frequent occurrence of spontaneous generation, and the inheritance of acquired characters, or traits. The belief in the heritability of acquired characters, the error for which Lamarck is mainly remembered, was not new with him. It was a universal belief in his time, firmly grounded in folklore (one expression of which was the biblical story of Jacob and the division of the striped and speckled livestock). The belief persisted. Darwin, for example, assumed that the use or disuse of a structure by one generation would be reflected in the next generation, and so did many evolutionists until late in the century, when the German biologist August Weismann demonstrated the impossibility, or at least the improbability, of the inheritance of acquired characters. Lamarck's assumptions of a drive toward perfection and of frequent spontaneous generation were also not confirmed, but he was right in recognizing that much of evolution is what we now call adaptive. He understood, moreover, that one could explain the great diversity of living organisms only by postulating a great age for the earth, and that evolution was a gradual process.

Lamarck's main interest was evolution in the time dimension—in vertical

CHARLES DARWIN was 31 years old and had already published his journal of the round-the-world voyage of H.M.S. *Beagle* when he sat in 1840 for the watercolor portrait by George Richmond reproduced on the opposite page. By this time, judging from his notebooks, Darwin had already worked out the major features of his theory of evolution through natural selection. Recently married, he was living in London, writing a monograph on coral reefs and turning from time to time to the notes on species that were to lead in 1859 to *On the Origin of Species*.

evolution, so to speak. Darwin, in contrast, was initially intrigued by the problem of the origin of diversity, and more specifically by the origin of species through diversification in a geographical dimension—in horizontal evolution. His interest in diversification and speciation was aroused, as is well known, during his five-year voyage around the world, beginning in 1831, as naturalist on H.M.S. *Beagle*. In the Galápagos Islands, for example, he learned that each island had its own form of tortoise, of mocking bird and of finch; the various forms were closely related and yet distinctly different. Pondering his observations after his return to England, he came to the conclusion that each island population was an incipient species, and thus to the concept of the "transmutation," or evolution, of species. In 1838 he conceived of the mechanism that could account for evolution: natural selection. After more years of observation and experiment, informed by wide reading in geology, zoology and other fields, a preliminary statement of Darwin's theory of evolution through natural se-

lection was announced in 1858 in a report to the Linnean Society of London. Alfred Russel Wallace, a young English naturalist doing fieldwork in the East Indies, had come independently to the concept of natural selection and had set down his ideas in a manuscript he mailed to Darwin; his paper was read at the meeting along with Darwin's.

Darwin's full theory, buttressed with innumerable personal observations and carefully argued, was published on November 24, 1859, in *On the Origin of Species*. His broad explanatory scheme comprised a number of component subtheories, or postulates, of which I shall single out what I take to be the four principal ones. Two of them were consistent with Lamarck's thinking. The first was the postulate that the world is not static but is evolving. Species change continually, new ones originate and others become extinct. Biotas, as reflected in the fossil record, change over time, and the older they are the more they are seen to have differed from living organisms. Wherever one looks in living na-

ture one encounters phenomena that make no sense except in terms of evolution. Darwin's second Lamarckian concept was the postulate that the process of evolution is gradual and continuous; it does not consist of discontinuous saltations, or sudden changes.

Darwin's two other main postulates were essentially new concepts. One was the postulate of common descent. For Lamarck each organism or group of organisms represented an independent evolutionary line, having had a beginning in spontaneous generation and having constantly striven toward perfection. Darwin postulated instead that similar organisms were related, descended from a common ancestor. All mammals, he proposed, were derived from one ancestral species; all insects had a common ancestor, and so did all the organisms of any other group. He implied, in fact, that all living organisms might be traced back to a single origin of life.

Darwin's inclusion of man in the common descent of mammals was considered by many to be an unforgivable insult to the human race, and it aroused a storm of protest. The idea of common descent had such enormous explanatory power, however, that it was almost immediately adopted by most biologists. It explained both the Linnaean hierarchy of taxonomic categories and the finding by comparative anatomists that all organisms could be assigned to a limited number of morphological types.

Darwin's fourth subtheory was that of natural selection, and it was the key to his broad scheme. Evolutionary change, said Darwin, is not the result of any mysterious Lamarckian drive, nor is it a simple matter of chance; it is the result of selection. Selection is a two-step process. The first step is the production of variation. In every generation, according to Darwin, an enormous amount of variation is generated. Darwin did not know the source of this variation, which could not be understood until after the rise of the science of genetics. All he had was his empirical knowledge of a seemingly inexhaustible reservoir of large and small differences within species.

The second step is selection through survival in the struggle for existence. In most species of animals and plants a set of parents produces thousands if not millions of offspring. Darwin's reading of Thomas Malthus told him that very few of the offspring could survive. Which ones would have the best chance of surviving? They would be those individuals that have the most appropriate combination of characters for coping with the environment, including climate, competitors and enemies; they would have the greatest chance of surviving, of reproducing and of leaving survivors, and their characters would

IN ABOUT 1854, the year in which he published a large monograph on barnacles that had occupied him for some eight years, Darwin sat for this photograph. He continued meanwhile with what he called his "species work": reading, corresponding, collecting, experimenting and making notes on the subject of his major work but delaying the writing until 1856. The realization two years later that Alfred Russel Wallace had independently developed the concept of natural selection led Darwin to prepare the "abstract" we know as *On the Origin of Species*.

therefore be available for the next cycle of selection.

The concept of an evolving world rather than a static one was almost universally accepted by serious scientists even before Darwin's death in 1882, and those who accepted evolution also accepted the concept of common descent (although there were those who insisted on exempting man from the common lineage). The situation was very different, however, for Darwin's two other postulates, both of which were bitterly resisted by many learned and able men for the next 50 to 80 years.

One of the postulates was the concept of gradualism. Even T. H. Huxley, who was known as "Darwin's bulldog" for his vigorous championing of most aspects of the new theory, could not accept the gradual origin of higher types and new species; he proposed a saltational origin instead. Saltationism was also popular with such biologists as Hugo De Vries, one of the rediscoverers of Gregor Mendel's laws of inheritance. He proposed a theory in 1901 according to which new species originate by mutation. As late as 1940 the geneticist Richard B. G. Goldschmidt was defending "systemic mutations" as the source of new higher types.

Three developments eventually resulted in the abandonment of such saltational theories. One development was the gradual adoption of a new attitude toward the physical world and its variation. Since the time of Plato the dominant view had been what the philosopher Karl Popper has called "essentialism": the world consisted of a limited number of unvarying essences (Plato's *eide*), of which the visible world's variable manifestations are merely incomplete and imprecise reflections. In such a view genuine change could arise only through the origin of a new essence either by creation or through a spontaneous saltation (mutation). Classes of physical objects do consist of identical entities, and physical constants are unvarying under identical conditions, and so (in the 19th century) there was no conflict between mathematics or the physical sciences and the philosophy of essentialism.

Biology required a different philosophy. Living organisms are characterized by uniqueness; every population of organisms consists of uniquely distinct individuals. In "population thinking" the mean values are the abstractions; only the variant individual has reality. The importance of the population lies in its being a pool of variations (a gene pool, in the language of genetics). Population thinking makes gradual evolution possible, and it now dominates every aspect of evolutionary theory.

The second development that led to

IN ABOUT 1880 DARWIN was photographed at Down House in Kent, where he had lived and worked since 1842. When he died in 1882 at 73, he was buried in Westminster Abbey.

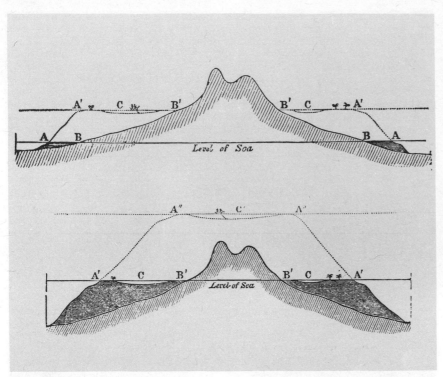

BIRTH OF AN ATOLL through subsidence of the ocean floor was illustrated by these woodcuts in Darwin's journal of the voyage of the *Beagle*. In the first stage (*top*) a fringing reef of coral (*A–B, B–A*) is built up at sea level around an island in the Pacific Ocean. As the island subsides, the coral polyps, which can survive only in shallow water, keep building the reef upward, forming a fringing reef (*A′–B′, B′–A′*) that encloses a lagoon (*C*). The island continues to subside (*bottom*) until it is below sea level; the barrier reef, growing, becomes an atoll (*A″–A″*).

DARWIN'S FINCHES, which he observed in the Galápagos Islands and some of which were shown in this woodcut from his published journal, provided him with a major insight. Seeing the wide range of beak sizes and shapes in "one small, intimately related group of birds," he wrote, "one might really fancy that... one species had been taken and modified for different ends."

the rejection of saltation was the discovery of the immense variability of natural populations and the realization that a high variability of discontinuous genetic factors, provided there are enough of them and provided the gaps between them are sufficiently small, can manifest itself in continuous variation of the organism. The third development was the demonstration by naturalists that processes of gradual evolution are entirely capable of explaining the origin of discontinuities such as new species and new types and of evolutionary novelties such as the wings of birds and the lungs of vertebrates.

The other Darwinian concept that was long resisted by most biologists and philosophers was natural selection. At first many rejected it because it was not deterministic, and hence predictive, in the style of 19th-century science. How could a proposed "natural law" such as natural selection be entirely a matter of chance? Others attacked its "crass materialism." In the 19th century to attribute the harmony of the living world to the arbitrary workings of natural selection was to undermine the natural theologian's "argument from design," which held that the existence of a Creator could be inferred from the beautiful design of his works. Those who rejected natural selection on religious or philosophical grounds or simply because it seemed too random a process to explain evolution continued for many years to put forward alternative schemes with such names as orthogenesis, nomogenesis, aristogenesis or the "omega principle" of Teilhard de Chardin, each scheme relying on some built-in tendency or drive toward perfection or progress. All these theories were finalistic: they postulated some form of cosmic teleology, of purpose or program.

The proponents of teleological theories, for all their efforts, have been unable to find any mechanisms (except supernatural ones) that can account for their postulated finalism. The possibility that any such mechanism can exist has now been virtually ruled out by the findings of molecular biology. As the late Jacques Monod argued with particular force, the genetic material is constant; it can change only through mutation. Finalistic theories have also been refuted by the paleontological evidence, as George Gaylord Simpson has shown most clearly. When the evolutionary trend of any character—a trend toward larger body size or longer teeth, for example—is examined carefully, the trend is found not to be consistent but to change direction repeatedly and even to reverse itself occasionally. The frequency of extinction in every geological period is another powerful argument against any finalistic trend toward perfection.

As for the objection to the presumed random aspect of natural selection, it is not hard to deal with. The process is not at all a matter of pure chance. Although variations arise through random processes, those variations are sorted by the second step in the process: selection by survival, which is very much an anti-chance factor. And if it is nonetheless true that some evolution is the result of chance, it is now known that physical processes in general have a far larger probabilistic component than was recognized 100 years ago.

Even so, can natural selection explain the long evolutionary progression up to the "highest" plants and animals, including man, from the origin of life between three and four billion years ago [see "Chemical Evolution and the Origin of Life," by Richard E. Dickerson; SCIENTIFIC AMERICAN Offprint 1401]? How can natural selection account not only for differential survival and adaptive changes within a species but also for the rise of new and differently adapted species? Again it was Darwin who suggested the right answer. An organism competes not only with other individuals of the same species but also with individuals of other species. A new adaptation or general physiological improvement will make an individual and its descendants stronger interspecific competitors and so contribute to diversification and specialization. Such specialization may often be a dead-end street, as it is in the case of adaptation to life in caves or hot springs. Many specializations, however, and particularly those that were acquired early in evolutionary history, opened up entirely new levels of adaptive radiation. These ranged from the invention of membranes and an organized cell nucleus [see "The Evolution of the Earliest Cells," by J. William Schopf; SCIENTIFIC AMERICAN Offprint 1402] and the aggregation of cells to form multicellular organisms [see "The Evolution of Multicellular Plants and Animals," by James W. Valentine; SCIENTIFIC AMERICAN Offprint 1403] to the advent of highly developed central nervous systems and the invention of long-continued parental care.

Evolution, as Simpson has emphasized, is recklessly opportunistic: it favors any variation that provides a competitive advantage over other members of an organism's own population or over individuals of different species. For billions of years this process has automatically fueled what we call evolutionary progress. No program controlled or directed this progression; it was the result of the spur-of-the-moment decisions of natural selection.

Darwin's uncertainty concerning the source of the genetic variability that supplies raw material for natural selection left a major hole in his argument. That hole was plugged by the science of genetics. Mendel discovered in 1865 that the factors transmitting hereditary information are discrete units transmitted by each parent to the offspring, preserved uncontaminated and reassorted in each generation. Darwin never knew of Mendel's findings, which were largely ignored until they were rediscovered in 1900.

We now know that DNA in the cell nucleus is organized in numerous self-replicating genes (Mendel's hereditary units), which can mutate to form different alleles, or alternative forms. There are structural genes that encode the information for making a specific protein and there are regulatory genes that turn the structural genes on and off. A mutated structural gene can code for a variant

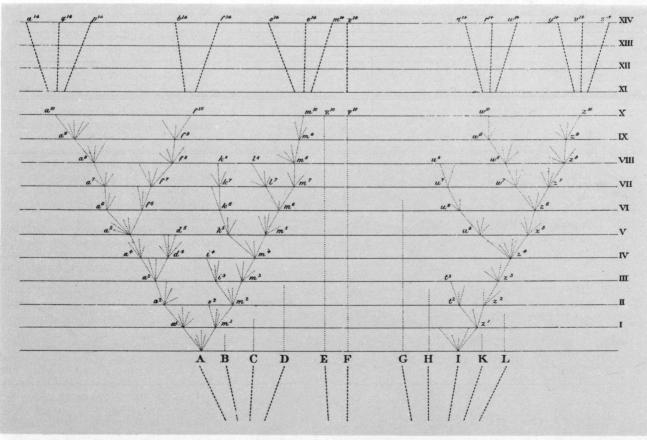

FORMATION OF NEW SPECIES through the divergence of characters and natural selection was illustrated in *On the Origin of Species*. The capital letters (*bottom*) represent species of the same genus. Horizontal lines marked by Roman numerals (*right*) represent, say, a 1,000-generation gap. Branching, diverging dotted lines represent varying offspring, the "profitable" ones of which are "preserved or naturally selected." Some species (*B, C and so on*) die out; some (*E, F*) remain essentially unchanged. Some (*A, I*) diverge widely, giving rise after many generations to new varieties (a^1, m^1, z^1) that diverge in turn, giving rise to increasingly divergent varieties that eventually become distinct new species (a^{14}, q^{14}, p^{14} and so on). After longer intervals these may become new genera or even higher categories.

protein, leading to a variant character. The genes are arrayed on chromosomes and may recombine with one another during meiosis, the cellular process that precedes the formation of germ cells in sexually reproducing species. The diversity of genotypes (full sets of genes) that can be produced during meiosis is almost unimaginably great, and much of that diversity is preserved in populations in spite of natural selection [see "The Mechanisms of Evolution," by Francisco J. Ayala; SCIENTIFIC AMERICAN Offprint 1407].

Strangely, the early Mendelians did not accept the theory of natural selection. They were essentialists and saltationists, and they looked on mutation as the probable driving force in evolution. That began to change with the development of population genetics in the 1920's. Eventually, during the 1930's and 1940's, a synthesis was achieved, expressed in and largely brought about by books written by Theodosius Dobzhansky, Julian Huxley, Bernhard Rensch, Simpson, G. Ledyard Stebbins and me. The new "synthetic theory" of evolution amplified Darwin's theory in the light of the chromosome theory of heredity, population genetics, the biological concept of a species and many other concepts of biology and paleontology. The new synthesis is characterized by the complete rejection of the inheritance of acquired characters, an emphasis on the gradualness of evolution, the realization that evolutionary phenomena are population phenomena and a reaffirmation of the overwhelming importance of natural selection.

The understanding of the evolutionary process achieved by the synthetic theory has had a profound effect on all biology. It led to the realization that every biological problem poses an evolutionary question, that it is legitimate to ask with respect to any biological structure, function or process: Why is it there? What was its selective advantage when it was acquired? Such questions have had an enormous impact on every area of biology, notably molecular biology, behavioral studies and ecology [see "The Evolution of Ecological Systems," by Robert M. May; SCIENTIFIC AMERICAN Offprint 1404].

Philosophers and physical scientists as well as lay people continue to have trouble understanding the modern theory of organic evolution through natural selection. At the risk of repeating some points I have already made in a historical context, let me outline the special features of the current theory, in particular drawing attention to what distinguishes organic evolution from cosmic evolution and other processes dealt with by physical scientists.

Evolution through natural selection is (I repeat!) a two-step process. The first step is the production (through recombination, mutation and chance events) of genetic variability; the second is the ordering of that variability by selection. Most of the variation produced by the first step is random in that it is not caused by, and is unrelated to, the current needs of the organism or the nature of its environment.

Natural selection can operate successfully because of the inexhaustible supply of variation made available to it owing to the high degree of individuality of biological systems. No two cells within an organism are precisely identical; each individual is unique, each species is unique and each ecosystem is unique. Many nonbiologists find the extent of organic variability incomprehensible. It is totally incompatible with traditional essentialist thinking and calls for a very different conceptual framework: population thinking. (The individuality of biological systems and the fact that there are multiple solutions for almost any environmental problem combine to make organic evolution nonrepeatable. Deterministically inclined astronomers are convinced by statistical reasoning that what has happened on the earth must also have happened on planets of stars other than the sun. Biologists, impressed by the inherent improbability of every single step that led to the evolution of man, consider what Simpson called "the prevalence of humanoids" exceedingly improbable.)

Uniquely different individuals are organized into interbreeding populations and into species. All the members are "parts" of the species, since they are derived from and contribute to a single gene pool. The population or species as a whole is itself the "individual" that undergoes evolution; it is not a class with members.

Every biological individual has a peculiarly dualistic nature. It consists of a genotype (its full complement of genes, not all of which may be expressed) and a phenotype (the organism that results from the translation of genes in the genotype). The genotype is part of the gene pool of the population; the phenotype competes with other phenotypes for reproductive success. This success (which defines the "fitness" of the individual) is not determined intrinsically but is the result of multiple interactions with enemies, competitors, pathogens and other selection pressures. The constellation of such pressures changes with the seasons, through the years and geographically.

JEAN BAPTISTE DE LAMARCK, the French naturalist and philosopher who was the first consistent evolutionist, understood that the earth is very old, that evolution is gradual and that organisms adapt. Lamarck also believed, however, in the inheritance of acquired characters.

The second step of natural selection, selection itself, is an extrinsic ordering principle. In a population of thousands or millions of unique individuals some will have sets of genes that are better suited to the currently prevailing assortment of ecological pressures. Such individuals will have a statistically greater probability of surviving and of leaving survivors than other members of the population. It is this second step in natural selection that determines evolutionary direction, increasing the frequency of genes and constellations of genes that are adaptive at a given time and place, increasing fitness, promoting specialization and giving rise to adaptive radiation and to what may be loosely described as evolutionary progress [see "Adaptation," by Richard C. Lewontin, *page 252*].

Selectionist evolution, in other words, is neither a chance phenomenon nor a deterministic phenomenon but a two-step tandem process combining the advantages of both. As the pioneering population geneticist Sewall Wright wrote: "The Darwinian process of continued interplay of a random and a selective process is not intermediate between pure chance and pure determinism, but in its consequences qualitatively utterly different from either."

No Darwinian I know questions the fact that the processes of organic evolution are consistent with the laws of the physical sciences, but it makes no sense to say that biological evolution has been "reduced" to physical laws. Biological evolution is the result of specific processes that impinge on specific systems, the explanation of which is meaningful only at the level of complexity of those processes and those systems. And the classical theory of evolution has not been reduced to a "molecular theory of evolution," an assertion based on such reductionist definitions of evolution as "a change in gene frequencies in natural populations." This reductionist definition omits the crucial aspects of evolution: changes in diversity and adaptation. (Once I gave a lump of sugar to a raccoon in a zoo. He ran with it to his water basin and washed it vigorously until there was nothing left of it. No complex system should be taken apart to the extent that nothing of significance is left.)

After the new synthesis of the 1930's and 1940's was achieved a few non-evolutionists asked whether it did not mark the end of research in evolution, whether all the questions had not been answered. The answer to both questions is decidedly no, as is made clear by the exponential increase in the number of publications in evolutionary biology. Let me mention some problems that currently interest workers in the field.

One major subject of inquiry is the role of chance. As far back as 1871 it was proposed that perhaps only some evolutionary change is due to selection, with much or even most change being

ALFRED RUSSEL WALLACE, as a young naturalist working in the East Indies, independently developed a theory of natural selection; his paper on the subject was read along with Darwin's in 1858. Later he differed with Darwin about the mechanisms of human evolution: Wallace believed that natural selection alone could not account for man's higher capacities.

T. H. HUXLEY, distinguished for brilliant work in many areas of biology, took on himself the role of Darwin's "general agent" and "bulldog," explicating and praising *On the Origin of Species* in a book review published in *The Times* of London and in many articles and lectures.

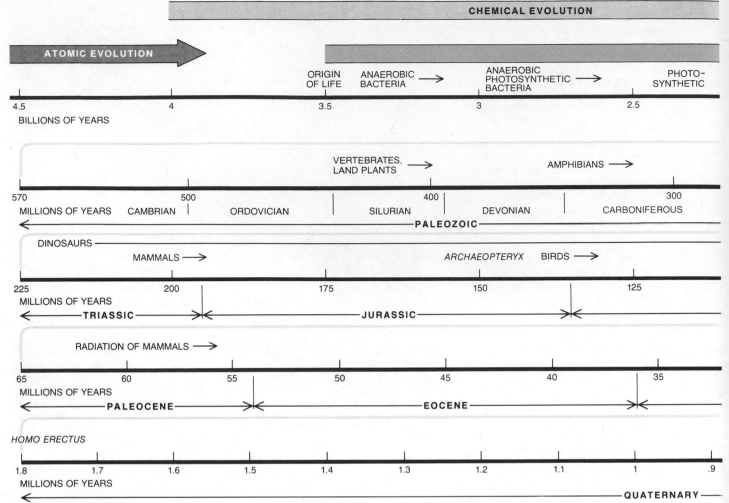

CHEMICAL EVOLUTION

ATOMIC EVOLUTION

ORIGIN OF LIFE   ANAEROBIC BACTERIA →   ANAEROBIC PHOTOSYNTHETIC BACTERIA →   PHOTO-SYNTHETIC

4.5   4   3.5   3   2.5
BILLIONS OF YEARS

VERTEBRATES, LAND PLANTS →   AMPHIBIANS →

570   500   400   300
MILLIONS OF YEARS   CAMBRIAN | ORDOVICIAN | SILURIAN | DEVONIAN | CARBONIFEROUS
← ———————————————————————————— PALEOZOIC ——————————

DINOSAURS —
MAMMALS →   *ARCHAEOPTERYX*   BIRDS →

225   200   175   150   125
MILLIONS OF YEARS
← ——— TRIASSIC ——— ×———————————— JURASSIC ————————×——

RADIATION OF MAMMALS →

65   60   55   50   45   40   35
MILLIONS OF YEARS
← ——— PALEOCENE ——— ×———————————— EOCENE ————————————×——

HOMO ERECTUS

1.8   1.7   1.6   1.5   1.4   1.3   1.2   1.1   1   .9
MILLIONS OF YEARS
← ——— QUATERNARY ———

GEOLOGIC TIME is charted on these two pages. The top line shows the full sweep from the origin of the earth some 4.6 billion years ago to the present day. The relatively short span of Phanerozoic time (*col-or*), the time during which the fossils of living things have been abundant in the geological record, is enlarged in the second line from the top, and successively shorter periods (*color*) are enlarged in the next

due to accidental variation, or to what are now called "neutral" mutations; the suggestion has been repeated many times since then. The problem acquired a new dimension when the technique of electrophoresis made it possible to detect small differences in the composition of a particular enzyme in a large random sample of individuals, thereby revealing the enormous extent of allelic variability. What part of that variability is evolutionary "noise" and what part is due to selection? How can one partition the variability into neutral and into relatively significant alleles?

The discovery of molecular biology that there are regulatory genes as well as structural ones poses new evolutionary questions. Is the rate of evolution of the two kinds of genes the same? Are they equally susceptible to natural selection? Is one kind of gene more important than the other in speciation or in the origin of higher taxa? (For example, the structural genes of the chimpanzee and of man appear to be remarkably similar. Is it perhaps the regulatory genes that make for most of the difference between us

and them?) Are there still other kinds of genes?

Darwin's favorite problem, that of the multiplication of species, has again become a focus of research. In certain groups of organisms, such as birds, new species seem to originate exclusively by geographical speciation: through the genetic restructuring of populations isolated from the remainder of a species' range, as on an island. In plants and in a few groups of animals, however, a different form of speciation can be effected through polyploidy, the doubling of the set of chromosomes, because polyploid individuals are immediately isolated reproductively from their parents. Another mode of speciation is "sympatric" speciation in parasites or in insects that are adapted to life on a specific host plant. Occasionally a new host species is colonized accidentally, and the descendants of the immigrant, perhaps aided by having favorable genes, come to constitute a well-established colony. In such a case there will be strong selection of genes that favor reproduction with other individuals living on the new host spe-

cies, so that conditions may favor the development of a new race adapted to the new host, and eventually of a new host-specific species. The frequency of sympatric speciation is still a matter of controversy. The respective role of genes and chromosomes in speciation is yet another controversial area.

In few areas of biology has the introduction of evolutionary thinking been as productive as it has in behavioral biology. The classical ethologists showed that such behavior patterns as the signaling displays of courtship can be as indicative of taxonomic relations as structural characters are. Classifications based on behavior have been worked out that agree remarkably well with systems based on structure, and the behavioral data have often provided decisive clues where the morphological evidence was ambiguous. More important has been the demonstration that behavior often—perhaps invariably—serves as a pacemaker in evolution. A change in behavior, such as the selection of a new habitat or food source, sets up new selective pressures and may lead to important

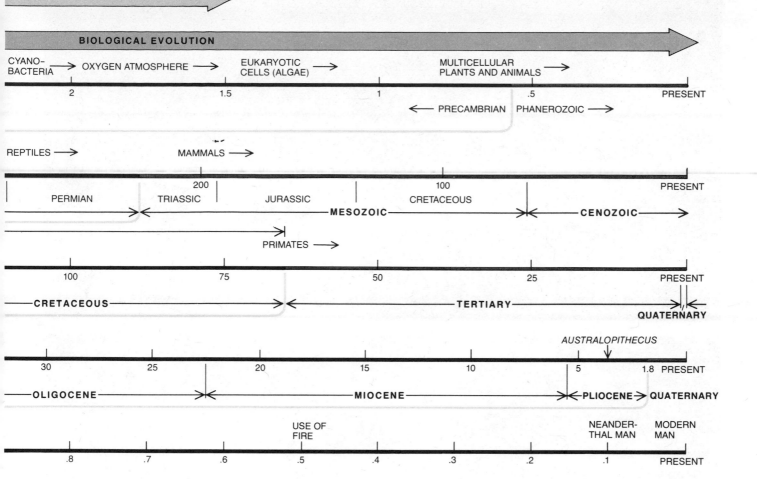

three lines. Three stages of evolution are shown at the top, with biological evolution beginning some 3.5 billion years ago with the appearance of the first living cells known. The three eras of Phanerozoic time (Paleozoic, Mesozoic and Cenozoic) are divided in turn into 11 periods; the Tertiary period is divided into five epochs, and the Quaternary period comprises the Pleistocene epoch and recent time.

adaptive shifts. There is little doubt that some of the most important events in the history of life, such as the conquest of land or of the air, were initiated by shifts in behavior. The selection pressures that potentiate such evolutionary progress are now receiving special attention [see "The Evolution of Behavior"; SCIENTIFIC AMERICAN Offprint 1405].

The perception that the world is not static but forever changing and that our own species is the product of evolution has inevitably had a fundamental impact on human understanding. We now know that the evolutionary line to which we belong arose from apelike ancestors over the course of millions of years, with the crucial steps having taken place during the past million years or so [see "The Evolution of Man," by Sherwood L. Washburn; SCIENTIFIC AMERICAN Offprint 1406]. We know that natural selection must have been responsible for this advance. What do past events enable one to predict with regard to the future of mankind? Since there is no finalistic element in organic evolution and no inheritance of ac-

quired characters, selection is obviously the only mechanism potentially capable of influencing human biological evolution.

That conclusion poses a dilemma. Eugenics, or deliberate selection, would be in conflict with cherished human values. Even if there were no moral objections, the necessary information on which to base such selection is simply not yet available. We know next to nothing about the genetic component of nonphysical human traits. There are innumerable and very different kinds of "good," "useful" or adapted human beings. Even if we could select a set of momentarily ideal characteristics, the changes generated in society by technological advances come so rapidly that no one could predict what particular blend of talents would lead in the future to the most harmonious human society. "Mankind is still evolving," Dobzhansky said, but we cannot know where it is headed biologically.

There is another kind of evolution, however: cultural evolution. It is a uniquely human process by which man

to some extent shapes and adapts to his environment. (Whereas birds, bats and insects became fliers by evolving genetically for millions of years, Dobzhansky pointed out, "man has become the most powerful flier of all, by constructing flying machines, not by reconstructing his genotype.") Cultural evolution is a much more rapid process than biological evolution. One of its aspects is the fundamental (and oddly Lamarckian) ability of human beings to evolve culturally through the transmission from generation to generation of learned information, including moral—and immoral—values. Surely in this area great advances can still be made, considering the modest level of moral values in mankind today. Even though we have no way of influencing our own biological evolution, we can surely influence our cultural and moral evolution. To do so in directions that are adaptive for all mankind would be a realistic evolutionary objective, but the fact remains that there are limits to cultural and moral evolution in a genetically unmanaged human species.

Adaptation

by Richard C. Lewontin
September 1978

The manifest fit between organisms and their environment is a major outcome of evolution. Yet natural selection does not lead inevitably to adaptation; indeed, it is sometimes hard to define an adaptation

The theory about the history of life that is now generally accepted, the Darwinian theory of evolution by natural selection, is meant to explain two different aspects of the appearance of the living world: diversity and fitness. There are on the order of two million species now living, and since at least 99.9 percent of the species that have ever lived are now extinct, the most conservative guess would be that two billion species have made their appearance on the earth since the beginning of the Cambrian period 600 million years ago. Where did they all come from? By the time Darwin published *On the Origin of Species* in 1859 it was widely (if not universally) held that species had evolved from one another, but no plausible mechanism for such evolution had been proposed. Darwin's solution to the problem was that small heritable variations among individuals within a species become the basis of large differences between species. Different forms survive and reproduce at different rates depending on their environment, and such differential reproduction results in the slow change of a population over a period of time and the eventual replacement of one common form by another. Different populations of the same species then diverge from one another if they occupy different habitats, and eventually they may become distinct species.

Life forms are more than simply multiple and diverse, however. Organisms fit remarkably well into the external world in which they live. They have morphologies, physiologies and behav-iors that appear to have been carefully and artfully designed to enable each organism to appropriate the world around it for its own life.

It was the marvelous fit of organisms to the environment, much more than the great diversity of forms, that was the chief evidence of a Supreme Designer. Darwin realized that if a naturalistic theory of evolution was to be successful, it would have to explain the apparent perfection of organisms and not simply their variation. At the very beginning of the *Origin of Species* he wrote: "In considering the Origin of Species, it is quite conceivable that a naturalist... might come to the conclusion that each species... had descended, like varieties, from other species. Nevertheless, such a conclusion, even if well founded, would be unsatisfactory, until it could be shown how the innumerable species inhabiting this world have been modified, so as to acquire that perfection of structure and coadaptation which most justly excites our admiration." Moreover, Darwin knew that "organs of extreme perfection and complication" were a critical test case for his theory, and he took them up in a section of the chapter on "Difficulties of the Theory." He wrote: "To suppose that the eye, with all its inimitable contrivances for adjusting the focus to different distances, for admitting different amounts of light, and for the correction of spherical and chromatic aberration, could have been formed by natural selection, seems, I freely confess, absurd in the highest degree."

These "organs of extreme perfection" were only the most extreme case of a more general phenomenon: adaptation. Darwin's theory of evolution by natural selection was meant to solve both the problem of the origin of diversity and the problem of the origin of adaptation at one stroke. Perfect organs were a difficulty of the theory not in that natural selection could not account for them but rather in that they were its most rigorous test, since on the face of it they seemed the best intuitive demonstration that a divine artificer was at work.

The modern view of adaptation is that the external world sets certain "problems" that organisms need to "solve," and that evolution by means of natural selection is the mechanism for creating these solutions. Adaptation is the process of evolutionary change by which the organism provides a better and better "solution" to the "problem," and the end result is the state of being adapted. In the course of the evolution of birds from reptiles there was a successive alteration of the bones, the muscles and the skin of the forelimb to give rise to a wing; an increase in the size of the breastbone to provide an anchor for the wing muscles; a general restructuring of bones to make them very light but strong, and the development of feathers to provide both aerodynamic elements and lightweight insulation. This wholesale reconstruction of a reptile to make a bird is considered a process of major adaptation by which birds solved the problem of flight. Yet there is no end to adaptation. Having adapted to flight, some birds reversed the process: the penguins adapted to marine life by changing their wings into flippers and their feathers into a waterproof covering, thus solving the problem of aquatic existence.

The concept of adaptation implies a preexisting world that poses a problem to which an adaptation is the solution. A key is adapted to a lock by cutting and filing it; an electrical appliance is adapted to a different voltage by a transform-

ADAPTATION is exemplified by "industrial melanism" in the peppered moth (*Biston betularia*). Air pollution kills the lichens that would normally colonize the bark of tree trunks. On the dark, lichenless bark of an oak tree near Liverpool in England the melanic (*black*) form is better adapted: it is better camouflaged against predation by birds than the light, peppered wild type (*top photograph on opposite page*), which it largely replaced through natural selection in industrial areas of England in the late 19th century. Now air quality is improving. On a nearby beech tree colonized by algae and the lichen *Lecanora conizaeoides*, which is itself particularly well adapted to low levels of pollution, the two forms of the moth are equally conspicuous (*middle*). On the lichened bark of an oak tree in rural Wales the wild type is almost invisible (*bottom*), and in such areas it predominates. The photographs were made by J. A. Bishop of the University of Liverpool and Laurence M. Cook of the University of Manchester.

REPTILES

BIRDS

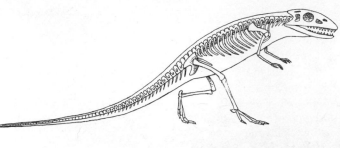

BONE

FORE LIMB

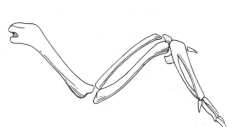

STERNUM

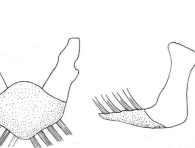

BOTTOM VIEW

SIDE VIEW

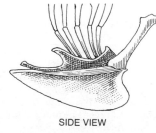

BOTTOM VIEW

SIDE VIEW

SKIN
COVERING

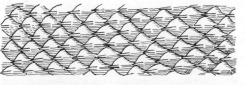

EVOLUTION OF BIRDS from reptiles can be considered a process of adaptation by which birds "solved" the "problem" of flight. At the top of the illustration the skeleton of a modern pigeon (*right*) is compared with that of an early reptile: a thecodont, a Triassic ancestor of dinosaurs and birds. Various reptile features were modified to become structures specialized for flight. Heavy, dense bone was restruc- tured to become lighter but strong; the forelimb was lengthened (and its muscles and skin covering were changed) to become a wing; the reptilian sternum, or breastbone, was enlarged and deepened to anchor the wing muscles (even in *Archaeopteryx,* the Jurassic transi- tion form between reptiles and birds whose sternum is pictured here, the sternum was small and shallow); scales developed into feathers.

er. Although the physical world certainly predated the biological one, there are certain grave difficulties for evolutionary theory in defining that world for the process of adaptation. It is the difficulty of defining the "ecological niche." The ecological niche is a multidimensional description of the total environment and way of life of an organism. Its description includes physical factors, such as temperature and moisture; biological factors, such as the nature and quantity of food sources and of predators, and factors of the behavior of the organism itself, such as its social organization, its pattern of movement and its daily and seasonal activity cycles.

The first difficulty is that if evolution is described as the process of adaptation of organisms to niches, then the niches must exist before the species that are to fit them. That is, there must be empty niches waiting to be filled by the evolution of new species. In the absence of organisms in actual relation to the environment, however, there is an infinity of ways the world can be broken up into arbitrary niches. It is trivially easy to describe "niches" that are unoccupied. For example, no organism makes a living by laying eggs, crawling along the surface of the ground, eating grass and living for several years. That is, there are no grass-eating snakes, even though snakes live in the grass. Nor are there any warm-blooded, egg-laying animals that eat the mature leaves of trees, even though birds inhabit trees. Given any description of an ecological niche occupied by an actual organism, one can create an infinity of descriptions of unoccupied niches simply by adding another arbitrary specification. Unless there is some preferred or natural way to subdivide the world into niches the concept loses all predictive and explanatory value.

A second difficulty with the specification of empty niches to which organisms adapt is that it leaves out of account the role of the organism itself in creating the niche. Organisms do not experience environments passively; they create and define the environment in which they live. Trees remake the soil in which they grow by dropping leaves and putting down roots. Grazing animals change the species composition of herbs on which they feed by cropping, by dropping manure and by physically disturbing the ground. There is a constant interplay of the organism and the environment, so that although natural selection may be adapting the organism to a particular set of environmental circumstances, the evolution of the organism itself changes those circumstances. Finally, organisms themselves determine which external factors will be part of their niche by their own activities. By building a nest the phoebe makes the availability of dried grass an important part of its

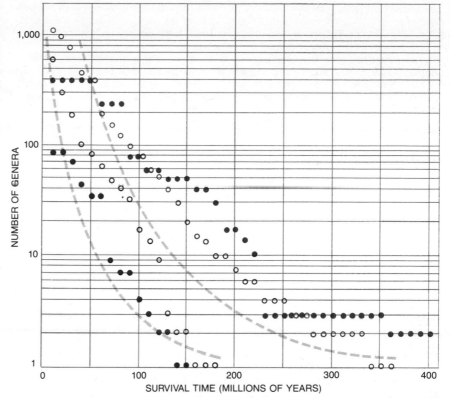

EXTINCTION RATES in many evolutionary lines suggest that natural selection does not necessarily improve adaptation. The data, from Leigh Van Valen of the University of Chicago, show the duration of survival of a number of living (*solid dots*) and extinct (*open circles*) genera of Echinoidea (*black*) and Pelecypoda (*color*), two classes of marine invertebrates. If natural selection truly fitted organisms to environments, the points should fall along concave curves (*broken-line curves*) indicating a lower probability of extinction for long-lived genera. Actually, points fall along rather straight lines, indicating constant rate of extinction for each group.

niche, at the same time making the nest itself a component of the niche.

If ecological niches can be specified only by the organisms that occupy them, evolution cannot be described as a process of adaptation because all organisms are already adapted. Then what is happening in evolution? One solution to this paradox is the Red Queen hypothesis, named by Leigh Van Valen of the University of Chicago for the character in *Through the Looking Glass* who had to keep running just to stay in the same place. Van Valen's theory is that the environment is constantly decaying with respect to existing organisms, so that natural selection operates essentially to enable the organisms to maintain their state of adaptation rather than to improve it. Evidence for the Red Queen hypothesis comes from an examination of extinction rates in a large number of evolutionary lines. If natural selection were actually improving the fit of organisms to their environments, then we might expect the probability that a species will become extinct in the next time period to be less for species that have already been in existence for a long time, since the long-lived species are presumably the ones that have been im-

proved by natural selection. The data show, however, that the probability of extinction of a species appears to be a constant, characteristic of the group to which it belongs but independent of whether the species has been in existence for a long time or a short one. In other words, natural selection over the long run does not seem to improve a species' chance of survival but simply enables it to "track," or keep up with, the constantly changing environment.

The Red Queen hypothesis also accounts for extinction (and for the occasional dramatic increases in the abundance and range of species). For a species to remain in existence in the face of a constantly changing environment it must have sufficient heritable variation of the right kind to change adaptively. For example, as a region becomes drier because of progressive changes in rainfall patterns, plants may respond by evolving a deeper root system or a thicker cuticle on the leaves, but only if their gene pool contains genetic variation for root length or cuticle thickness, and successfully only if there is enough genetic variation so that the species can change as fast as the environment. If the genetic variation is inadequate, the species will become extinct. The genetic resources of a species are finite, and eventually the environment will change so rapidly that the species is sure to become extinct.

The theory of environmental tracking seems at first to solve the problem of adaptation and the ecological niche. Whereas in a barren world there is no clear way to divide the environment into preexisting niches, in a world already occupied by many organisms the terms of the problem change. Niches are already defined by organisms. Small changes in the environment mean small changes in the conditions of life of those organisms, so that the new niches to which they must evolve are in a sense very close to the old ones in the multidimensional niche space. Moreover, the organisms that will occupy these slightly changed niches must themselves come from the previously existing niches, so that the kinds of species that can evolve are stringently limited to ones that are extremely similar to their immediate ancestors. This in turn guarantees that the changes induced in the environment by the changed organism will also be small and continuous in niche space. The picture of adaptation that emerges is the very slow movement of the niche through niche space, accompanied by a slowly changing species, always slightly behind, slightly ill-adapted, eventually becoming extinct as it fails to keep up with the changing environment because it runs out of genetic variation on which natural selection can operate. In this view species form when two populations of the same species track environments that diverge from each other over a period of time.

The problem with the theory of environmental tracking is that it does not predict or explain what is most dramatic in evolution: the immense diversification of organisms that has accompanied, for example, the occupation of the land from the water or of the air from the land. Why did warm-blooded animals arise at a time when cold-blooded animals were still plentiful and come to coexist with them? The appearance of entirely new life forms, of ways of making a living, is equivalent to the occupation of a previously barren world and brings us back to the preexistent empty niche waiting to be filled. Clearly there have been in the past ways of making a living that were unexploited and were then "discovered" or "created" by existing organisms. There is no way to explain and predict such evolutionary adaptations unless a priori niches can be described on the basis of some physical principles before organisms come to occupy them.

That is not easy to do, as is indicated by an experiment in just such a priori predictions that has been carried out by probes to Mars and Venus designed to detect life. The instruments are designed to detect life by detecting growth in nutrient solutions, and the solutions are prepared in accordance with knowledge of terrestrial microorganisms, so that the probes will detect only organisms whose ecological niches are like those on the earth. If Martian and Venusian life partition the environment in totally unexpected ways, they will remain unrecorded. What the designers of those instruments never dreamed of was that the reverse might happen: that the nature of the physical environment on Mars might be such that when it was provided with a terrestrial ecological niche, inorganic reactions might have a lifelike appearance. Yet that may be exactly what happened. When the Martian soil was dropped into the nutrient broth on the lander, there was a rapid production of carbon dioxide and then—nothing. Either an extraordinary kind of life began to grow much more rapidly than any terrestrial microorganism and then was poisoned by its own activity in a strange environment, or else the Martian soil is such that its contact with nutrient broths results in totally unexpected catalytic processes. In either case the Mars life-detection experiment has foundered on the problem of defining ecological niches without organisms.

Much of evolutionary biology is the working out of an adaptationist program. Evolutionary biologists assume that each aspect of an organism's morphology, physiology and behavior has been molded by natural selection as a solution to a problem posed by the

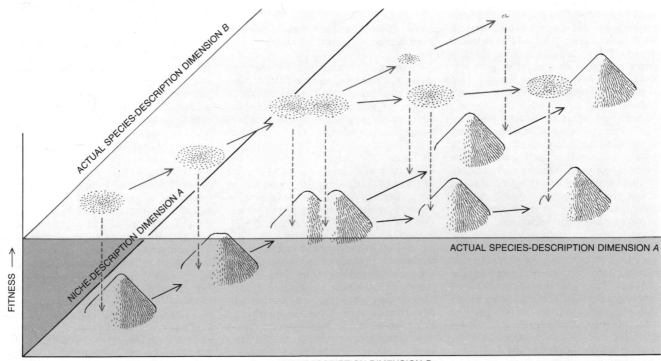

SPECIES TRACK ENVIRONMENT through niche space, according to one view of adaptation. The niche, visualized as an "adaptive peak," keeps changing (moving to the right); a slowly changing species population (*colored dots*) just manages to keep up with the niche, always a bit short of the peak. As the environment changes, the single peak becomes two distinct peaks, and two populations diverge to form distinct species. One species cannot keep up with its rapidly changing environment, becomes less fit (lags farther behind changing peak) and extinct. Here niche space and actual-species space have only two dimensions; both of them are actually multidimensional.

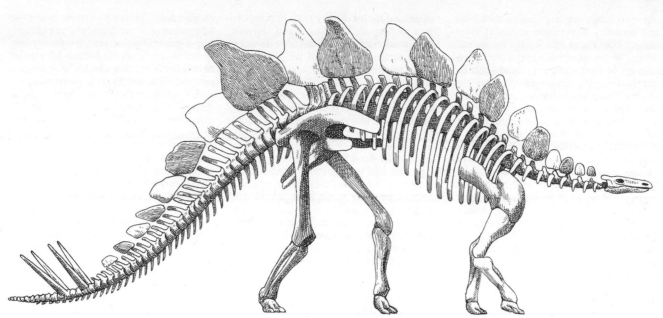

STEGOSAURUS, a large herbivorous dinosaur of the Jurassic period, had an array of bony plates along its back. Were they solutions to the problem of defense, courtship recognition or heat regulation? An engineering analysis reveals features characteristic of heat regulators: porous structure (suggesting a rich blood supply), particularly large plates over the massive part of the body, staggered arrangement along the midline, a constriction near the base and so on. This skeleton in the American Museum of Natural History is 18 feet long.

environment. The role of the evolutionary biologist is then to construct a plausible argument about how each part functions as an adaptive device. For example, functional anatomists study the structure of animal limbs and analyze their motions by time-lapse photography, comparing the action and the structure of the locomotor apparatus in different animals. Their interest is not, however, merely descriptive. Their work is informed by the adaptationist program, and their aim is to explain particular anatomical features by showing that they are well suited to the function they perform. Evolutionary ethologists and sociobiologists carry the adaptationist program into the realm of animal behavior, providing an adaptive explanation for differences among species in courting pattern, group size, aggressiveness, feeding behavior and so on. In each case they assume, like the functional anatomist, that the behavior is adaptive and that the goal of their analysis is to reveal the particular adaptation.

The dissection of an organism into parts, each of which is regarded as a specific adaptation, requires two sets of a priori decisions. First one must decide on the appropriate way to divide the organism and then one must describe what problem each part solves. This amounts to creating descriptions of the organism and of the environment and then relating the descriptions by functional statements; one can either start with the problems and try to infer which aspect of the organism is the solution or start with the organism and then ascribe adaptive functions to each part.

For example, for individuals of the same species to recognize each other at mating time is a problem, since mistakes about species mean time, energy and gametes wasted in courtship and mating without the production of viable offspring; species traits such as distinctive color markings, special courtship behavior, unique mating calls, odors and restricted time and place of activity can be considered specific adaptations for the proper recognition of potential mates. On the other hand, the large, leaf-shaped bony plates along the back of the dinosaur *Stegosaurus* constitute a specific characteristic for which an adaptive function needs to be inferred. They have been variously explained as solutions to the problem of defense (by making the animal appear to be larger or by interfering directly with the predator's attack), the problem of recognition in courtship and the problem of temperature regulation (by serving as cooling fins).

The same problems that arose in deciding on a proper description of the ecological niche without the organism arise when one tries to describe the organism itself. Is the leg a unit in evolution, so that the adaptive function of the leg can be inferred? If so, what about a part of the leg, say the foot, or a single toe, or one bone of a toe? The evolution of the human chin is an instructive example. Human morphological evolution can be generally described as a "neotenic" progression. That is, human infants and adults resemble the fetal and young forms of apes more than they resemble adult apes; it is as if human beings are born at an earlier stage of physical development than apes and do not

mature as far along the apes' development path. For example, the relative proportion of skull size to body size is about the same in newborn apes and human beings, whereas adult apes have much larger bodies in relation to their heads than we do; in effect their bodies "go further."

The exception to the rule of human neoteny is the chin, which grows relatively larger in human beings, whereas both infant and adult apes are chinless. Attempts to explain the human chin as a specific adaptation selected to grow larger failed to be convincing. Finally it was realized that in an evolutionary sense the chin does not exist! There are two growth fields in the lower jaw: the dentary field, which is the bony structure of the jaw, and the alveolar field, in which the teeth are set. Both the dentary and the alveolar fields do show neoteny. They have both become smaller in the human evolutionary line. The alveolar field has shrunk somewhat faster than the dentary one, however, with the result that a "chin" appears as a pure consequence of the relative regression rates of the two growth fields. With the recognition that the chin is a mental construct rather than a unit in evolution the problem of its adaptive explanation disappears. (Of course, we may go on to ask why the dentary and alveolar growth fields have regressed at different rates in evolution, and then provide an adaptive explanation for that phenomenon.)

Sometimes even the correct topology of description is unknown. The brain is divided into anatomical divisions corresponding to certain separable nervous functions that can be localized, but memory is not one of those functions. The memory of specific events seems to be stored diffusely over large regions of the cerebrum rather than being localized microscopically. As one moves from anatomy to behavior the problem of a correct description becomes more acute and the opportunities to introduce arbitrary constructs as if they were evolutionary traits multiply. Animal behavior is described in terms of aggression, division of labor, warfare, dominance, slave-making, cooperation—and yet each of these is a category that is taken directly from human social experience and is transferred to animals.

The decision as to which problem is solved by each trait of an organism is equally difficult. Every trait is involved in a variety of functions, and yet one would not want to say that the character is an adaptation for all of them. The green turtle *Chelonia mydas* is a large marine turtle of the tropical Pacific. Once a year the females drag themselves up the beach with their front flippers to the dry sand above the high-water mark. There they spend many hours laboriously digging a deep hole for their eggs, using their hind flippers as trowels. No one who has watched this painful process would describe the turtles' flippers as adaptations for land locomotion and digging; the animals move on land and dig with their flippers because nothing better is available. Conversely, even if a trait seems clearly adaptive, it cannot be assumed that the species would suffer in its absence. The fur of a polar bear is an adaptation for temperature regulation, and a hairless polar bear would certainly freeze to death. The color of a polar bear's fur is another matter. Although it may be an adaptation for camouflage, it is by no means certain that the polar bear would become extinct or even less numerous if it were brown. Adaptations are not necessary conditions of the existence of the species.

For extinct species the problem of judging the adaptive status of a trait is made more difficult because both the trait and its function must be reconstructed. In principle there is no way to be sure whether the dorsal plates of *Stegosaurus* were heat-regulation devices, a defense mechanism, a sexual recognition sign or all these things. Even in living species where experiments can be carried out a doubt remains. Some modern lizards have a brightly colored dewlap under the jaw. The dewlap may be a warning sign, a sexual attractant or a species-recognition signal. Experiments removing or altering the dewlap could decide, in principle, how it functions. That is a different question from its status as an adaptation, however, since the assertion of adaptation implies a historical argument about natural selection as the cause of its establishment. The large dorsal plates of *Stegosaurus* may have evolved because individuals with slightly larger plates were better able to gather food in the heat of the day than other individuals. If, when the plates reached a certain size, they incidentally frightened off predators, they would be a "preadaptation" for defense. The distinction between the primary adaptation for which a trait evolved and incidental functions it may have come to have cannot be made without the reconstruction of the forces of natural selection during the actual evolution of the species.

The current procedure for judging the adaptation of traits is an engineering analysis of the organism and its environment. The biologist is in the position of an archaeologist who uncovers a machine without any written record and attempts to reconstruct not only its operation but also its purpose. The hypothesis that the dorsal plates of *Stegosaurus* were a heat-regulation device is based on the fact that the plates were porous and probably had a large supply of blood vessels, on their alternate placement to the left and right of the midline (suggesting cooling fins), on their large size over the most massive part of the body and on the constriction near their base, where they are closest to the heat source and would be inefficient heat radiators.

Ideally the engineering analysis can be quantitative as well as qualitative and so provide a more rigorous test of the

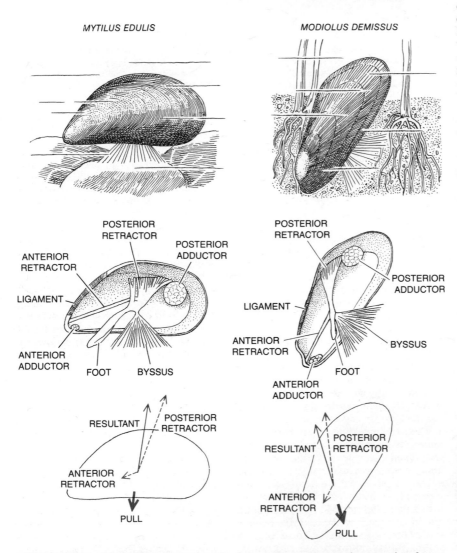

MYTILUS EDULIS

MODIOLUS DEMISSUS

FUNCTIONAL ANALYSIS indicates how the shape and musculature of two species of mussels are adapted to their particular environments. *Mytilus edulis* (*left*) attaches itself to rocks by means of its byssus, a beardlike group of threads (*top*). Its ventral, or lower, edge is flattened; the anterior and posterior retractor muscles are positioned (*middle*) so that their resultant force pulls the bottom of the shell squarely down to the substratum (*bottom*). *Modiolus demissus* (*right*) attaches itself to debris in marshes. Its ventral edge is sharply angled to facilitate penetration of the substratum; its retractor muscles are positioned to pull its anterior end down into the marsh. The analysis was done by Steven M. Stanley of Johns Hopkins University.

adaptive hypothesis. Egbert G. Leigh, Jr., of the Smithsonian Tropical Research Institute posed the question of the ideal shape of a sponge on the assumption that feeding efficiency is the problem to be solved. A sponge's food is suspended in water and the organism feeds by passing water along its cell surfaces. Once water is processed by the sponge it should be ejected as far as possible from the organism so that the new water taken in is rich in food particles. By an application of simple hydrodynamic principles Leigh was able to show that the actual shape of sponges is maximally efficient. Of course, sponges differ from one another in the details of their shape, so that a finer adjustment of the argument would be needed to explain the differences among species. Moreover, one cannot be sure that feeding efficiency is the only problem to be solved by shape. If the optimal shape for feeding had turned out to be one with many finely divided branches and protuberances rather than the compact shape observed, it might have been argued that the shape was a compromise between the optimal adaptation for feeding and the greatest resistance to predation by small browsing fishes.

Just such a compromise has been suggested for understanding the feeding behavior of some birds. Gordon H. Orians of the University of Washington studied the feeding behavior of birds that fly out from a nest, gather food and bring it back to the nest for consumption ("central-place foraging"). If the bird were to take food items indiscriminately as it came on them, the energy cost of the round trip from the nest and back might be greater than the energy gained from the food. On the other hand, if the bird chose only the largest food items, it might have to search so long that again the energy it consumed would be too great. For any actual distribution of food-particle sizes in nature there is some optimal foraging behavior for the bird that will maximize its net energy gain from feeding. Orians found that birds indeed do not take food particles at random but are biased in the direction of an optimal particle size. They do not, however, choose the optimal solution either. Orians' explanation was that the foraging behavior is a compromise between maximum energy efficiency and not staying away from the nest too long, because the young are exposed to predation when they are unattended.

The example of central-place foraging illustrates a basic assumption of all such engineering analyses, that of ceteris paribus, or all other things being equal. In order to make an argument that a trait is an optimal solution to a particular problem, it must be possible to view the trait and the problem in isolation, all other things being equal. If all

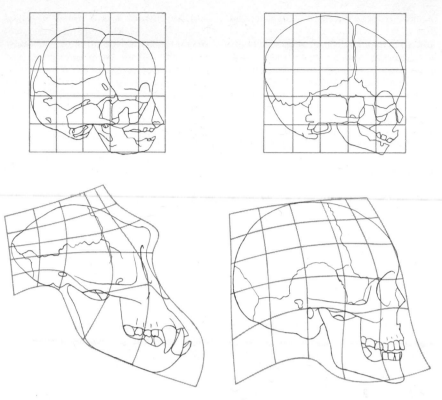

NEOTENY OF HUMAN SKULL is evident when the growth of the chimpanzee skull (*left*) and of the human skull (*right*) is plotted on transformed coordinates, which show the relative displacement of each part. The chimpanzee and the human skulls are much more similar at the fetal stage (*top*) than they are at the adult stage (*bottom*). The adult human skull also departs less from the fetal form than the adult chimpanzee skull departs from its fetal form, except in the case of the chin, which becomes relatively larger in human beings. The chin is a mental construct, however: the result of allometry, or differential growth, of different parts of human jaw.

other things are not equal, if a change in a trait as a solution to one problem changes the organism's relation to other problems of the environment, it becomes impossible to carry out the analysis part by part, and we are left in the hopeless position of seeing the whole organism as being adapted to the whole environment.

The mechanism by which organisms are said to adapt to the environment is that of natural selection. The theory of evolution by natural selection rests on three necessary principles: Different individuals within a species differ from one another in physiology, morphology and behavior (the principle of variation); the variation is in some way heritable, so that on the average offspring resemble their parents more than they resemble other individuals (the principle of heredity); different variants leave different numbers of offspring either immediately or in remote generations (the principle of natural selection).

These three principles are necessary and sufficient to account for evolutionary change by natural selection. There must be variation to select from; that variation must be heritable, or else there will be no progressive change from gen-

eration to generation, since there would be a random distribution of offspring even if some types leave more offspring than others. The three principles say nothing, however, about adaptation. In themselves they simply predict change caused by differential reproductive success without making any prediction about the fit of organisms to an ecological niche or the solution of ecological problems.

Adaptation was introduced by Darwin into evolutionary theory by a fourth principle: Variations that favor an individual's survival in competition with other organisms and in the face of environmental stress tend to increase reproductive success and so tend to be preserved (the principle of the struggle for existence). Darwin made it clear that the struggle for existence, which he derived from Thomas Malthus' *An Essay on the Principle of Population*, included more than the actual competition of two organisms for the same resource in short supply. He wrote: "I should premise that I use the term Struggle for Existence in a large and metaphorical sense.... Two canine animals in a time of dearth, may be truly said to struggle with each other which shall get food and live. But a plant on the edge of the desert is said to struggle for life against the drought."

The diversity that is generated by various mechanisms of reproduction and mutation is in principle random, but the diversity that is observed in the real world is nodal: organisms have a finite number of morphologies, physiologies and behaviors and occupy a finite number of niches. It is natural selection, operating under the pressures of the struggle for existence, that creates the nodes. The nodes are "adaptive peaks," and the species or other form occupying a peak is said to be adapted.

More specifically, the struggle for existence provides a device for predicting which of two organisms will leave more offspring. An engineering analysis can determine which of two forms of zebra can run faster and so can more easily escape predators; that form will leave more offspring. An analysis might predict the eventual evolution of zebra locomotion even in the absence of existing differences among individuals, since a careful engineer might think of small improvements in design that would give a zebra greater speed.

When adaptation is considered to be the result of natural selection under the pressure of the struggle for existence, it is seen to be a relative condition rather than an absolute one. Even though a species may be surviving and numerous, and therefore may be adapted in an absolute sense, a new form may arise that has a greater reproductive rate on the same resources, and it may cause the extinction of the older form. The concept of relative adaptation removes the apparent tautology in the theory of natural selection. Without it the theory of natural selection states that fitter individuals have more offspring and then defines the fitter as being those that leave more offspring; since some individuals will always have more offspring than others by sheer chance, nothing is explained. An analysis in which problems of design are posed and characters are understood as being design solutions breaks through this tautology by predicting in advance which individuals will be fitter.

The relation between adaptation and natural selection does not go both ways. Whereas greater relative adaptation leads to natural selection, natural selection does not necessarily lead to greater adaptation. Let us contrast two evolutionary scenarios. We begin with a resource-limited population of 100 insects of type A requiring one unit of food resource per individual. A mutation to a new type a arises that doubles the fecundity of its bearers but does absolutely nothing to the efficiency of the utilization of resources. We can calculate what happens to the composition, size and growth rate of the population over a period of time [see illustration below]. In a second scenario we again begin with the population of 100 individuals of type A, but now there arises a different mutation a, which does nothing to the fecundity of its bearers but doubles their efficiency of resource utilization. Again we can calculate the population history.

In both cases the new type a replaces the old type A. In the case of the first mutation nothing changes but the fecundity; the adult population size and the growth rate are the same throughout the process and the only effect is that twice as many immature stages are being produced to die before adulthood. In the second case, on the other hand, the population eventually doubles its adult members as well as its immature members, but not its fecundity. In the course of its evolution the second population has a growth rate greater than 1 for a while but eventually attains a constant size and stops growing.

In which of these populations, if in either, would the individuals be better

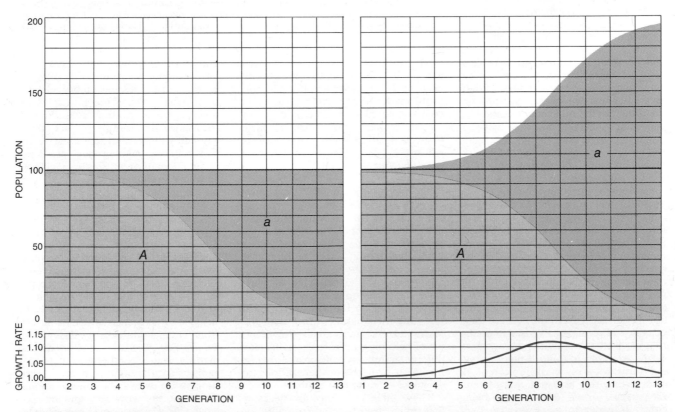

TWO DIFFERENT MUTATIONS have different demographic results for a resource-limited population of 100 insects. In one case (*left*) a mutation arises that doubles the fecundity of its bearers. The new type (*a*) replaces the old type (*A*), but the total population does not increase: the growth rate (*bottom*) remains 1.00. In the other case (*right*) a mutation arises that doubles the carrier's efficiency of resource utilization. Now the new population grows more rapidly, but only for a short time: eventually the growth rate falls back to 1.00 and the total population is stabilized at 200. The question is: Has either mutation given rise to a population that is better adapted?

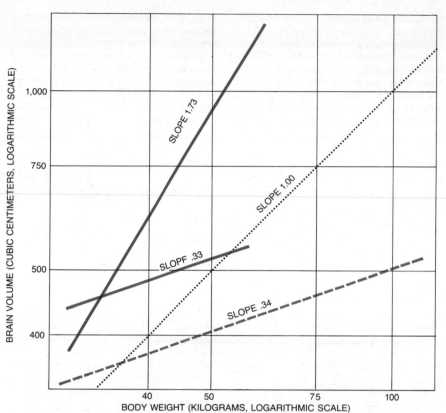

ALLOMETRY, or differential growth rates for different parts, is responsible for many evolutionary changes. Allometry is illustrated by this comparison of the ratio of brain size to body weight in a number of species of the pongids, or great apes (*broken black curve*), of *Australopithecus*, an extinct hominid line (*solid black*), and of hominids leading to modern man (*color*). A slope of less than 1.00 means the brain has grown more slowly than the body. The slope of more than 1.00 for the human lineage indicates a clear change in the evolution of brain size.

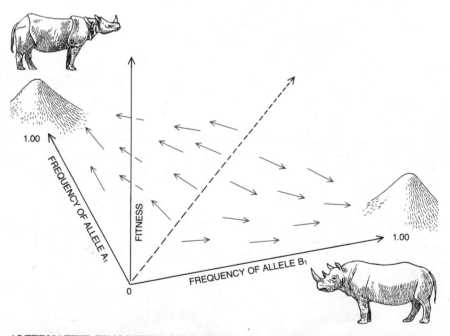

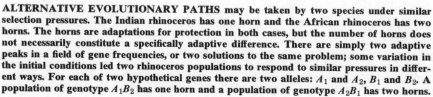

ALTERNATIVE EVOLUTIONARY PATHS may be taken by two species under similar selection pressures. The Indian rhinoceros has one horn and the African rhinoceros has two horns. The horns are adaptations for protection in both cases, but the number of horns does not necessarily constitute a specifically adaptive difference. There are simply two adaptive peaks in a field of gene frequencies, or two solutions to the same problem; some variation in the initial conditions led two rhinoceros populations to respond to similar pressures in different ways. For each of two hypothetical genes there are two alleles: A_1 and A_2, B_1 and B_2. A population of genotype A_1B_2 has one horn and a population of genotype A_2B_1 has two horns.

adapted than those in the old population? Those with higher fecundity would be better buffered against accidents such as sudden changes in temperature since there would be a greater chance that some of their eggs would survive. On the other hand, their offspring would be more susceptible to the epidemic diseases of immature forms and to predators that concentrate on the more numerous immature forms. Individuals in the second population would be better adapted to temporary resource shortages, but also more susceptible to predators or epidemics that attack adults in a density-dependent manner. Hence there is no way we can predict whether a change due to natural selection will increase or decrease the adaptation in general. Nor can we argue that the population as a whole is better off in one case than in another. Neither population continues to grow or is necessarily less subject to extinction, since the larger number of immature or adult stages presents the same risks for the population as a whole as it does for individual families.

Unfortunately the concept of relative adaptation also requires the ceteris paribus assumption, so that in practice it is not easy to predict which of two forms will leave more offspring. A zebra having longer leg bones that enable it to run faster than other zebras will leave more offspring only if escape from predators is really the problem to be solved, if a slightly greater speed will really decrease the chance of being taken and if longer leg bones do not interfere with some other limiting physiological process. Lions may prey chiefly on old or injured zebras likely in any case to die soon, and it is not even clear that it is speed that limits the ability of lions to catch zebras. Greater speed may cost the zebra something in feeding efficiency, and if food rather than predation is limiting, a net selective disadvantage might result from solving the wrong problem. Finally, a longer bone might break more easily, or require greater developmental resources and metabolic energy to produce and maintain, or change the efficiency of the contraction of the attached muscles. In practice relative-adaptation analysis is a tricky game unless a great deal is known about the total life history of an organism.

Not all evolutionary change can be understood in terms of adaptation. First, some changes will occur directly by natural selection that are not adaptive, as for example the changes in fecundity and feeding efficiency in the hypothetical example I cited above.

Second, many changes occur indirectly as the result of allometry, or differential growth. The rates of growth of different parts of an organism are different,

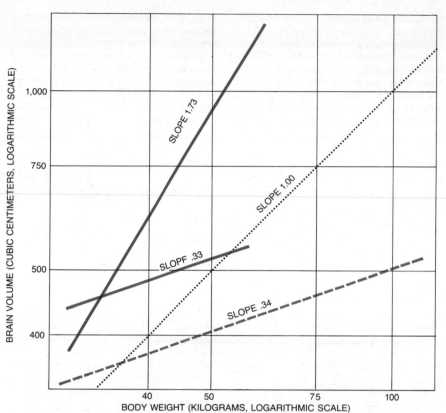

so that large organisms do not have all their parts in the same proportion. This allometry shows up both between individuals of the same species and between species. Among primate species the brain increases in size more slowly than the body; small apes have a proportionately larger brain than large apes. Since the differential growth is constant for all apes, it is useless to seek an adaptive reason for gorillas' having a relatively smaller brain than, say, chimpanzees.

Third, there is the phenomenon of pleiotropy. Changes in a gene have many different effects on the physiology and development of an organism. Natural selection may operate to increase the frequency of the gene because of one of the effects, with pleiotropic, or unrelated, effects being simply carried along. For example, an enzyme that helps to detoxify poisonous substances by converting them into an insoluble pigment will be selected for its detoxification properties. As a result the color of the organism will change, but no adaptive explanation of the color per se is either required or correct.

Fourth, many evolutionary changes may be adaptive and yet the resulting differences among species in the character may not be adaptive; they may simply be alternative solutions to the same problem. The theory of population genetics predicts that if more than one gene influences a character, there may often be several alternative stable equilibriums of genetic composition even when the force of natural selection remains the same. Which of these adaptive peaks in the space of genetic composition is eventually reached by a population depends entirely on chance events at the beginning of the selective process. (An exact analogy is a pinball game. Which hole the ball will fall into under the fixed force of gravitation depends on small variations in the initial conditions as the ball enters the game.) For example, the Indian rhinoceros has one horn and the African rhinoceros has two. Horns are an adaptation for protection against predators, but it is not true that one horn is specifically adaptive under Indian conditions as opposed to two horns on the African plains. Beginning with two somewhat different developmental systems, the two species responded to the same selective forces in slightly different ways.

Finally, many changes in evolution are likely to be purely random. At the present time population geneticists are sharply divided over how much of the evolution of enzymes and other molecules has been in response to natural selection and how much has resulted from the chance accumulation of mutations. It has proved remarkably difficult to get compelling evidence for changes in enzymes brought about by selection, not to speak of evidence for adaptive changes; the weight of evidence at present is that a good deal of amino acid substitution in evolution has been the result of the random fixation of mutations in small populations. Such random fixations may in fact be accelerated by natural selection if the unselected gene is genetically linked with a gene that is undergoing selection. The unselected gene will then be carried to high frequency in the population as a "hitchhiker."

If the adaptationist program is so fraught with difficulties and if there are so many alternative explanations of evolutionary change, why do biologists not abandon the program altogether?

WHALES

SEALS

PENGUINS

FISH

SEA SNAKES

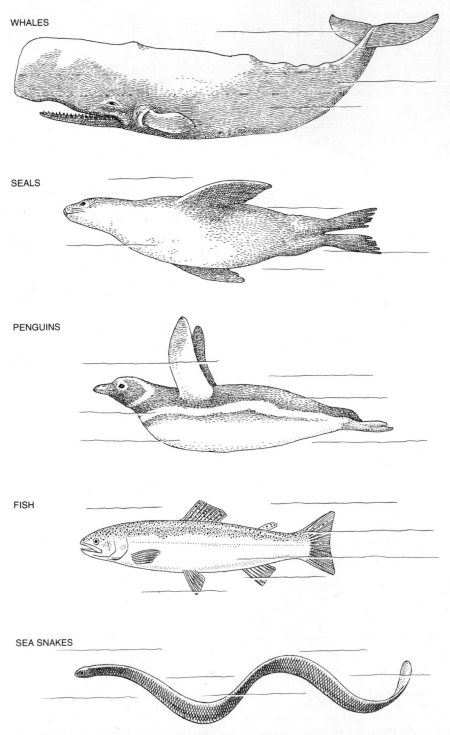

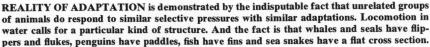

REALITY OF ADAPTATION is demonstrated by the indisputable fact that unrelated groups of animals do respond to similar selective pressures with similar adaptations. Locomotion in water calls for a particular kind of structure. And the fact is that whales and seals have flippers and flukes, penguins have paddles, fish have fins and sea snakes have a flat cross section.

There are two compelling reasons. On the one hand, even if the assertion of universal adaptation is difficult to test because simplifying assumptions and ingenious explanations can almost always result in an ad hoc adaptive explanation, at least in principle some of the assumptions can be tested in some cases. A weaker form of evolutionary explanation that explained some proportion of the cases by adaptation and left the rest to allometry, pleiotropy, random gene fixations, linkage and indirect selection would be utterly impervious to test. It would leave the biologist free to pursue the adaptationist program in the easy cases and leave the difficult ones on the scrap heap of chance. In a sense, then, biologists are forced to the extreme adaptationist program because the alternatives, although they are undoubtedly operative in many cases, are untestable in particular cases.

On the other hand, to abandon the notion of adaptation entirely, to simply observe historical change and describe its mechanisms wholly in terms of the different reproductive success of different types, with no functional explanation, would be to throw out the baby with the bathwater. Adaptation is a real phenomenon. It is no accident that fish have fins, that seals and whales have flippers and flukes, that penguins have paddles and that even sea snakes have become laterally flattened. The problem of locomotion in an aquatic environment is a real problem that has been solved by many totally unrelated evolutionary lines in much the same way. Therefore it must be feasible to make adaptive arguments about swimming appendages. And this in turn means that in nature the ceteris paribus assumption must be workable.

It can only be workable if both the selection between character states and reproductive fitness have two characteristics: continuity and quasi-independence. Continuity means that small changes in a characteristic must result in only small changes in ecological relations; a very slight change in fin shape cannot cause a dramatic change in sexual recognition or make the organism suddenly attractive to new predators. Quasi-independence means that there is a great variety of alternative paths by which a given characteristic may change, so that some of them will allow selection to act on the characteristic without altering other characteristics of the organism in a countervailing fashion; pleiotropic and allometric relations must be changeable. Continuity and quasi-independence are the most fundamental characteristics of the evolutionary process. Without them organisms as we know them could not exist because adaptive evolution would have been impossible.

25

Repeated Segments of DNA

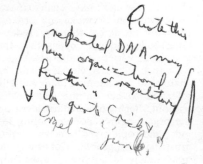

Prote this
repeated DNA may
have organizational or regulatory
Function i
↓ the quote Crick?
Orgel — junk?

by Roy J. Britten and David E. Kohne
April 1970

In cells of higher organisms a significant fraction of the genetic material appears in as many as a million identical or very similar copies. The origin and function of repetitive DNA remain unknown

DNA was conclusively identified as the genetic material in 1944 and its structure was established in 1953. In the next few years it became clear how that structure is critical to DNA's replication, its transcription into RNA and its subsequent translation into particular chains of amino acids to form enzymes and other proteins. Everything that had been learned about DNA by the early 1960's emphasized the significance of the precise and unique sequence of nitrogenous bases that constitutes a gene and suggested that DNA consisted of chains of different genes. It was therefore a major surprise when it was discovered in 1964 that much of the DNA in the cells of the mouse consists of multiple copies of the same or very similar base sequences.

Repeated DNA has now been identified in all higher species that have been examined. It constitutes as little as 20 percent of all the DNA in the cell nuclei of some species and as much as 80 percent in others. The precision of repetition is often imperfect, so that members of a "family" of repeated DNA are usually closely related rather than identical. The number of related base sequences in a family ranges from 50 to two million. The wide distribution of repeated DNA, its persistence through millions of years of evolution and the fact that at least some of it is "expressed," or transcribed into RNA, all suggest that it is important to cell function and the survival of the organism. Yet it is still not known if the repeated segments are genes or parts of genes or if they carry out some role other than the specification of gene products. At this stage it seems likely that the most significant phenomenon may not be at the level of gene repetition; the repeated DNA may have an organizational or regulatory function. More insight into the evolutionary history and the present role of repeated DNA will surely lead to a new understanding of evolution and of the regulation of genetic functions in the living cell.

Before telling the story of the discovery of repeated DNA by our group in the Department of Terrestrial Magnetism of the Carnegie Institution of Washington, it will be well to briefly review the structure and behavior of DNA. Most DNA consists of sequences of only four nitrogenous bases: adenine (*A*), thymine (*T*), guanine (*G*) and cytosine (*C*). Together these bases form the genetic alphabet, and long ordered sequences of them contain in coded form much, if not all, of the information present in the genes. The DNA molecule in its native state (as it is found in the cell) is made up of two strands wound helically around each other. The bases face inward and each is specifically bonded to a complementary base on the other strand; *T* is always linked with *A* and *G* is always linked with *C*. These complementary bases have an affinity for each other such that, when they are paired, they contribute to the stability of the entire double-strand molecule.

In native DNA every base is paired with its complement and the molecule is quite stable. In isolated DNA the bonds between the bases can be broken by heating, and the two strands of the DNA can be completely separated from each other. This dissociation is usually accomplished simply by boiling the DNA solution for a few minutes. Left in solution, the single strands collide with complementary partners and, if the conditions are right, double-strand helixes are formed again. The requirements for such reassociation of DNA are surprisingly simple: the salt concentration must be fairly high, the temperature should be about 25 degrees Celsius below the dissociation temperature and enough time must be allowed for effective collisions to occur between complementary strands.

The extent of reassociation is the critical observation in most experiments with repeated DNA. Reassociation can be measured in a variety of ways, each depending on some easily detected difference between single-strand and double-strand DNA. Some time ago it was discovered that single-strand DNA absorbs more ultraviolet radiation than

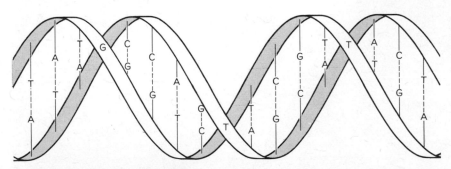

DNA in its native state is a double helix composed of two single strands. Each strand has a backbone of sugar and phosphate groups and a series of nitrogenous bases: adenine (*A*), thymine (*T*), guanine (*G*) and cytosine (*C*). The bases project inward and are linked by hydrogen bonds, which hold the two strands together like the rungs of a ladder; *A* always bonds with *T* and *G* bonds with *C*. The sequence of bases encodes genetic information.

double-strand DNA does; with a spectrophotometer one can measure changes in ultraviolet absorption and thus in the amount of reassociation. In another method single-strand DNA labeled with radioactive atoms is incubated over other strands fixed in agar, and the quantity of radioactive DNA that is reassociated, and thereby bound to the agar, is measured. Our own preferred method for the study of repeated DNA utilizes hydroxyapatite, a crystalline form of calcium phosphate that was originally used for the fractionation of proteins. A number of years ago it was discovered that single-strand and double-strand DNA have different affinities for hydroxyapatite crystals. We have helped to bring this technique to the point where the solution of DNA used for reassociation can simply be passed through the hydroxyapatite. The single-strand DNA passes through the column of crystals; the double-strand DNA is absorbed on the hydroxyapatite and can be removed for assay. The solution and its temperature (about 60 degrees C.) are such that the amount of single-strand DNA that is absorbed is low but the double-strand DNA is efficiently bound. The temperature is only about 25 degrees C. below the dissociation temperature, which prevents the formation of strand pairs held together by very short or very imprecise complementary regions.

When DNA strands derived from different but related sources, such as two related animals, are reassociated, their bases will not be perfectly complementary. The best means now known for recognizing and measuring the extent of differences between similar DNA sequences is to form hybrid-strand pairs between the two DNA's and measure the stability of the resulting helixes. The presence of a few mismatched bases reduces the stability of an otherwise complementary strand pair and leads to its dissociation at a lower temperature. The best current estimate is that for every 1.5 percent of the base pairs that are mismatched the dissociation temperature is reduced one degree. In order to determine the stability of reassociated DNA one need only absorb hybrid-strand pairs on hydroxyapatite and wash the column free of single-strand DNA at 60 degrees. Then the temperature is raised in steps and the DNA that dissociates at each temperature is washed out and assayed.

In 1962 Ellis T. Bolton and Brian J. McCarthy of our biophysics group at the Carnegie Institution had devised the DNA-agar method for measuring the relatedness between different DNA's

REASSOCIATION of strands of DNA is the experimental procedure that gives information about repeated DNA. The schematic drawings show the process of dissociation and reassociation for a hypothetical genome, or gene complement, composed of six copies of a repeated DNA sequence (*color*) and four different (*black*) sequences (*a*). Heating the DNA above the dissociation temperature breaks the hydrogen bonds between complementary base pairs, thus separating the strands (*b*). After the temperature is lowered complementary regions of DNA strands that collide may match up, forming reassociated double strands (*c*). Collision of complementary regions is more likely in the case of repeated DNA than for single-copy DNA. Speed of reassociation is proportional to degree of repetition.

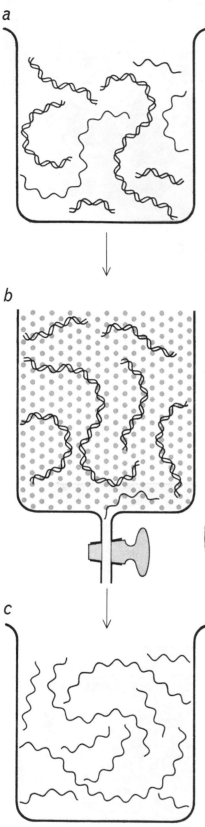

EXTENT OF REASSOCIATION is measured by separating double-strand DNA from single-strand DNA. A solution containing the two (*a*) is passed through a column of hydroxyapatite crystals (*b*). The double strands bind to the hydroxyapatite, whereas the single strands pass through the column (*c*). The double strands may then be collected by raising the column's temperature.

and RNA's and with it had measured the relations among various species of bacteria. With Bill H. Hoyer they undertook to apply the same method to the relations among vertebrate species. The first animal DNA they tried to reassociate was from mouse cells, and they succeeded. They went on to conduct a fascinating series of experiments on the relations between the genes of higher species and the expression of those genes in various tissues.

The trouble was that, according to what was then known about DNA, their experiments should not have worked. The cells of vertebrate animals contain much more DNA than bacteria do. As a result it had been expected that the concentration of each particular DNA sequence from vertebrate cells would be very small, and that after dissociation the complementary strands would rarely collide with each other. Yet the dissociated strands found their partners faster in the vertebrate DNA than they did in the bacterial DNA. Why? The question was set aside for future explanation, but as time passed it increasingly disturbed one of us (Britten). Had the reaction been accelerated by some unknown catalytic action of the agar or by the immobilization of the DNA? An experiment ruled out those possibilities: hybridization proceeded rapidly in free solution. By midsummer of 1964 only one explanation seemed possible and the basic hypothesis was proposed. Certain sequences in the DNA were reiterated again and again. Reiteration, or repetition, would make the concentration of some DNA base sequences much higher than had been expected and would thus account for the rapid reassociation.

On the basis of evidence that these related sequences were not identical with one another, the early hypothesis further suggested that some kind of multiplication process had occurred for selected segments of the DNA, forming large families of related sequences that became integrated into the genetic material and were then passed on to descendants by the usual hereditary mechanisms. It was further supposed that sequence changes, or mutations, altered various members of the families of repeated sequences, so that precise relationship was slowly lost among the member sequences. An essential part of the hypothesis was that new events of multiplication must have occurred at intervals throughout the course of evolution, producing new populations of repeated DNA segments. The evidence obtained since that time has supported this set of hypotheses.

The first direct evidence for repetition came out of an improbable piece of good fortune, again involving the mouse. Michael Waring arrived from the University of Cambridge as a postdoctoral fellow and chose to join in the search for actual pieces of repeated DNA. One of us (Britten) and Waring decided to extract DNA from mouse tissue because that tissue was convenient and available. We were quite unaware that mouse DNA has a singular family of repeated sequences. It is the most obvious one known even today; it makes up a tenth of the total mouse DNA and consists of about a million copies of a short sequence of some 300 base pairs. We had not expected anything so outlandish, but we immediately recognized the highly repetitive component. Because of its special qualities (it is lighter than most mouse DNA and forms a distinct "satellite" band when it is centrifuged in a cesium chloride density gradient), it could be purified in a few weeks and was shown to be an example of repetitive DNA. Soon afterward Ann McClaren and P. M. B. Walker of the University of Edinburgh, who were also working with mouse DNA, observed that 10 percent of it was bound to hydroxyapatite when it was expected to be single-strand and therefore not to bind. They did not imagine that reassociation could be as rapid as their results suggested and they identified their DNA as a stable fraction whose strands either had not separated or had folded back on themselves instantly. Walker's group at Edinburgh continued to investigate rapidly reassociating DNA in rodents. And since 1965 the authors of this article have collaborated on an extensive series of experiments to characterize the repeated sequences in the DNA of many different species.

Let us examine in more detail the paradox of the mouse-DNA reassociation, which led to the discovery of repetitive DNA. As we have noted, the reason animal DNA was expected to reassociate slowly is the large amount of DNA per animal cell [see illustration on opposite page]. If all the base sequences in the DNA of an animal were different from one another, then in a reassociation reaction each sequence would in effect be diluted by all the others. The dilution would lead to a reduced rate of reassociation, and so the rate of reassociation of the DNA from an animal with a large genome size, or DNA content per nucleus, should be less than the rate for DNA from one with a small genome.

This is exactly what we observe if we put aside the effects of the DNA present in multiple copies. In fact, the time required for the reassociation of unrepeated DNA to proceed halfway to completion is proportional to the genome size [see illustration on next page]. The reason is that complementary single strands must collide with each other before they can reassociate. Although most collisions are ineffective, occasionally complementary regions of the two strands are matched up in such a way that a short double-strand region results. If the neighboring regions are also complementary, then the double-strand region will grow into a stable helix; if they are not, the short region is likely to be unstable and therefore to dissociate. Under most practical conditions the rate of collisions controls the rate at which reassociation occurs. Reassociation between DNA strands therefore is normally, in the terminology of chemical kinetics, a second-order reaction. When its time course is plotted on a logarithmic scale, it takes the characteristic form of an S-shaped curve.

Measurement of the rate of reassociation is the best way to identify repetitive DNA. Because the rate of reassociation depends on a number of conditions, the DNA of a standard organism, the bacterium *Escherichia coli*, is taken as a reference. The length of the DNA of the *E. coli* chromosome has been well measured and its genome size is known. Virtually all the nucleotide sequences of its DNA occur only once, that is, it has little repetition. The time course of the reassociation of *E. coli* DNA [see illustration on page 269] reveals the S-shaped curve expected for a single-component reaction, and the precision of agreement with the predicted curve sets a low limit for the presence of repeated DNA. (The agreement also shows that such variables as fragment size have negligible effects and that the hydroxyapatite method adequately separates double-strand DNA from single-strand.)

The time required for half-reassociation is proportional to the DNA content per cell only if each sequence of bases occurs once in the cell's DNA. If a stretch of DNA sequence were to occur twice (if a gene were duplicated, for example), then the concentration of complementary fragments derived from the duplicate region would be twice as great as the concentration of the rest of the DNA. As a result these fragments would take only half as long to reassociate. If there were more copies, the period required for reassociation would be proportionately reduced. This simple rule apparently continues to hold even when as many as a million copies are present,

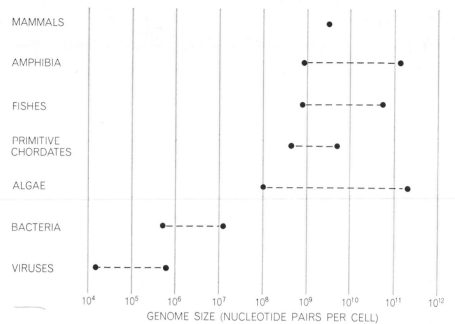

GENOME SIZE, or the amount of DNA per cell, increases with increasing evolutionary complexity. For each group of organisms shown here the minimum and maximum genome size recorded in various species are plotted, expressed as number of nucleotide pairs per haploid cell (a germ cell in higher forms). Mammals all have about the same genome size.

as is the case for the most rapidly associating family in mouse DNA. With this rule we can evaluate the number of copies from measurements of the rate of reassociation.

The DNA that has been most thoroughly studied is the DNA of the calf (because calf thymus glands, which are rich in DNA, are readily available from meat-packers). When its reassociation curve is plotted in the same way as the *E. coli* curve, there are some obvious differences. The reaction starts much sooner and ends much later; it is evidently divided into two quite separate regions. The later region of the curve has been shown to be due to the reassociation of DNA present in single copies. In the earlier part of the curve the DNA on the average reassociates 100,000 times faster, indicating that for each of the segments reassociating in this early region there must be about 100,000 other segments in the genome that are similar enough to form complementary pairs of strands. More detailed measurements show there are two major families of repeated sequences whose reactions overlap. One is present in about a million copies and amounts to only a few percent of the DNA. The other repeated component, accounting for nearly 40 percent of the total DNA of the calf, is present in 66,000 copies, not all of them exactly alike.

The total quantity of DNA in the form of repetitive sequences is quite large:

about 45 percent for calf DNA, according to the hydroxyapatite measurements. Actually this is a somewhat arbitrary figure. It depends on the degree of sequence relationship—the proportion of matching bases in each pair of strands—that is recognized in the particular measurement. We do not yet know how much of the DNA would have to be included if all degrees of sequence relationship greater than chance expectation were taken into account. In several species—Pacific salmon, wheat, onion and the salamander *Amphiuma*—more than 80 percent of the DNA shows repetition that is recognizable under the conditions in which calf shows only 45 percent repetition. Certainly the repeated sequences are not a minor part of the DNA. Why is there so much and what role does it play in the operations of the cell? Before considering the evidence bearing on these unsolved questions we will summarize what is known about the distribution of repeated DNA in various organisms, the frequency of repetition, the apparent evolutionary history of repeated DNA and the distribution of the material in the cell.

In the past few years the DNA of more than 60 plant and animal species has been examined. Significant quantities of repeated sequences have been found in the following groups: primates, other mammals, birds, amphibians, bony and cartilaginous fishes, amphioxus, echino-

mammals particularly
regularity of frequency spectra

derms, brachiopods, insects, other arthropods, mollusks, coelenterates, sponges, protozoans and plants ranging from algae to wheat. Small quantities may be present in the nuclei of fungi, but here a clear separation from DNA outside the nucleus has not yet been made. Bacteria contain a number of copies of the genes that specify ribosomal RNA, but no other nuclear repeated DNA has been observed in the few species tested. (All species appear to have a few copies of certain special genes such as the genes for ribosomal or transfer RNA, and so one can say that repeated DNA probably occurs universally. The large quantities and high frequencies we are describing, however, appear to occur in species higher on the scale of complexity than fungi.)

The number of DNA segments from one cell that have similar base sequences is the repetition frequency for that set. One can regard the set of different repetition frequencies for each species as a spectrum [*see illustration on page 270*]. These spectra are somewhat tentative in detail, since measurements with adequate technique have just been under-

taken; the overall patterns are nonetheless correct, and some of the spectra, such as the spectrum for calf DNA, have been fairly well measured. The repetition spectra show few common features. Calf cells have little or no DNA with frequencies between 10 and 10,000. Toad and snail cells have no families with more than 10,000 members, and their dominant components are in the broad frequency range that is not present in calf DNA. It is too early in the exploration to discern the regularities in frequency spectra that probably exist. For example, it appears that components in the range from 100,000 to a million copies may exist only in the DNA of mammals; each of the mammalian species examined so far has such highly repetitive components, which do not appear even in other vertebrates. Many more measurements are required to establish such generalizations.

It seems likely that repeated DNA originated in past events of DNA replication, each of which produced a family made up of identical copies of some preexisting DNA sequence. Then in the period of time since the original produc-

tion of the copies a number of things must have happened. Segments were translocated and scattered into various parts of the genome; subunits were substituted and deleted. The traces of such events can be found in the present populations, and the result is that in most cases the members of a family of repeated sequences differ somewhat from one another.

Estimates of the degree of sequence difference can be made from the thermal stability of reassociated DNA, since the imperfectly complementary strand pairs have reduced stability compared with perfectly matched strands [*see top illustration, page 271*]. Typically when the entire population of repeated DNA in one organism is examined, a wide range of thermal stability is seen. Some sequences are quite closely related to others, even to the extent of being perfect copies, whereas other sequences differ so much that the strand pairs barely hold together under mild conditions. In the most divergent examples changes appear to have taken place in as many as half of the nucleotides. We believe all degrees of divergence are present, as would be expected if the processes of production and divergence had been active throughout the evolution of higher species.

When did the events that produced these enormous families occur? Some of them happened a very long time ago, perhaps several hundred million years ago. It is possible to date reasonably well the time in evolutionary history when there lived the last common ancestor of two related but now distinct modern species. This time may be considered, if enough fossils have been discovered and examined, to be the time of divergence of the ancestral lines of species that led to the present-day organisms. Any feature, molecular or morphological, that is now exhibited by both species probably originated before the time of divergence. We assume that any family of repetitive DNA held in common by two species probably originated before their ancestral lines diverged. (There are risks in this assumption where gene flow between species is possible, as it is in plants.) The relation between the fraction of the repeated DNA held in common between species and the time since their species lines diverged can be established [*see bottom illustration on page 271*]. Assuming, as seems reasonable, that in each species line there is production of new families of repeated DNA as preexisting families are slowly lost, the curve can be considered a "decay-of-relatedness curve," by analogy

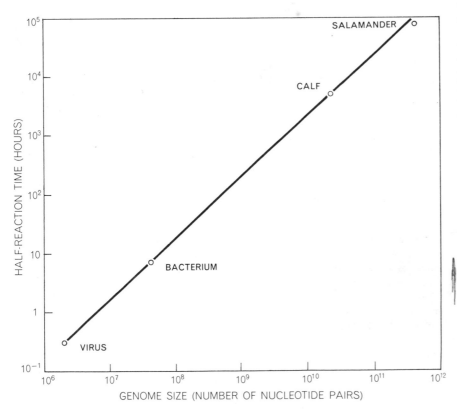

TIME FOR REASSOCIATION of single-copy DNA varies with the size of the genome and therefore increases with the complexity of the organism. Here the time for half-reassociation is plotted against genome size for a virus (T4) and a bacterium (*Escherichia coli*) and for the nonrepetitive DNA of the calf and a salamander, *Amphiuma*. The times shown are for a particular set of conditions; the time can be accelerated by increasing the concentrations of DNA and salt, but for large genomes it is difficult to measure a complete reaction.

with radioactive decay. For the vertebrates shown in the illustration the median age of the repeated families seems to be about 100 million years—the time in which half of the repeated DNA is replaced.

Some of the more abstract but nevertheless fascinating questions about repeated DNA sequences have to do with their organization in the cell. How long are the repeated elements? Do they occur together in the genome? Are they organized in special places in the DNA in keeping with some role in chromosome structure, or are they scattered throughout the DNA?

Several different classes of experiments yield information about the arrangement of the repeated DNA in the genome. One case—perhaps an atypical case—is the mouse satellite DNA, which has recently been studied in much detail by Walker and by William G. Flamm, now at the National Institute of Environmental Health Science, and by others. The sequences of this family are of fairly recent evolutionary origin, having appeared since the ancestral lines leading to the house (or laboratory) mouse and the rat diverged 10 or 20 million years ago. They have been shown to be arrayed in clusters containing as many as 100 copies, and according to recent measurements the clusters appear on many of the chromosomes of the mouse. This family is somewhat exceptional, since it is a very short repeated element present in very many copies and is found in only the one species.

Such detailed evidence is not yet available for any other family of repeated sequences, but there is evidence from a number of species that the repeated sequences are intimately scattered through the DNA. If DNA is prepared in fragments about 20,000 bases long, each fragment is almost certain to include a repeated sequence, and it usually includes more than one. The extent of scattering is known in a little more detail in calf DNA, where three-quarters of the fragments broken down to about 4,000 bases contain segments of repeated DNA.

One approach to the difficult problem of the function of repeated DNA is to assay the extent to which it is expressed. All the genes of an organism's genome are present in all its cells, and yet many of the genes carry out no function in particular tissues or at particular times. Such genes are said to be unexpressed, whereas those that are active are said to be ex-

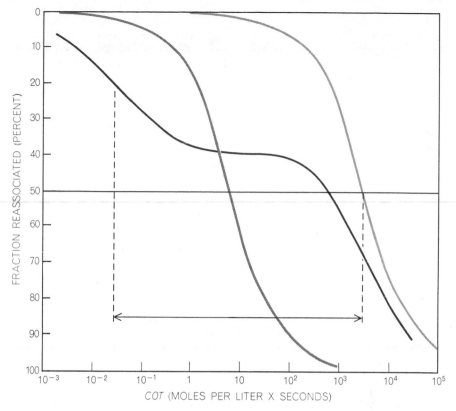

TIME COURSE of reassociation is shown for bacterial and for calf DNA. The amount of reassociation is plotted against the COT, a coined word for the product of DNA concentration and time. The time course for bacterial (E. coli) DNA, almost all of which is single-copy DNA, plots as an S-shaped curve (gray). If calf DNA were also single-copy DNA, it would plot as a similar curve displaced to the right (light color). The actual curve for calf DNA (dark color) has a different shape, however. The shape of the curve indicates that the reaction takes place in two different stages, one early and one late; the midpoints of the two stages (broken vertical lines) are separated by a factor of 100,000. The early stage represents the reassociation of repeated DNA and the later stage that of single-copy DNA.

pressed. The function of many genes is to specify the synthesis of proteins, and the mechanism of these genes' expression is now quite well understood. The first step in the process is the transcription of the base sequence of the DNA into "messenger" RNA by the pairing of complementary RNA bases with the DNA as template. When the messenger RNA reaches the site of protein synthesis, its base sequence supplies information that is translated according to the genetic code to determine the amino acid sequence of the protein. Our interest here is only in the first stage, the synthesis of RNA on the DNA template, as an indication of gene expression. Clearly one can say that a gene is expressed if RNA complementary to one of its DNA strands is synthesized. Therefore a good test for genetic activity is to search for an RNA complementary to the DNA of the gene. In most cases this would be difficult, since preparations of the DNA of particular genes are not ordinarily available. In the case of repeated DNA taken as a whole, however, it is quite easy. The high

frequency of repetition increases the rate of RNA-DNA hybridization just as it increases the rate of DNA reassociation. When proper precautions are taken, RNA that hybridizes rapidly with DNA can be assumed to be RNA that was synthesized on a repeated DNA sequence.

Hybridization experiments have been conducted in a number of laboratories in attempts to learn the mechanism of gene expression and its control. We now know that the limitations of the available techniques restricted the observations to rapidly hybridizing sequences—the sequences involving repeated DNA. The observations therefore made available a large body of measurements of the expression of repeated DNA sequences in a variety of cell types from a number of organisms. These measurements yield two principal results: first, repeated DNA sequences are expressed in every higher organism and cell type that has been examined; second, different families of repeated DNA are expressed in different tissues.

By accident

A particularly significant measurement of differences in patterns of expression can be made with a technique called competition. For this purpose RNA from one tissue is mixed at a high concentration with radioactively labeled RNA from a second tissue in the presence of DNA. To the extent that there are similarities between the RNA populations, the unlabeled RNA "occupies" the DNA and prevents the labeled RNA from hybridizing. In this way RNA populations from different tissues and different stages of embryonic development have been compared. In a number of instances the populations of RNA molecules are entirely dissimilar, indicating that the DNA sequences expressed are also dissimilar, that is, they are members of different families of repeated sequences. Why should a particular family of DNA sequences be expressed in an early embryo and not be expressed at a later stage of development? The answer to this question would probably go a long way toward revealing the role of repeated DNA in the molecular machinery of the cell.

The fact that repetitive DNA is expressed (transcribed into RNA) leaves no doubt in our minds about the significance of some of the repeated sequences. That brings us up against the problem of the hundreds of thousands and millions of copies present in the mammals. What role could such very large numbers of copies have? If these frequencies were observed even throughout the vertebrates, one might consider a structural or organizing role for the highly repetitive DNA, but it is unlikely that 1,000 times more copies would be required for such a purpose in the mouse than in, say, the toad. Mammalian chromosomes do not seem to be that much more complex. They have comparable amounts of DNA per cell; in fact, some amphibian cells contain a great deal more DNA than mammalian cells.

Taking the evidence all in all, the large size of some families of repeated DNA is probably best explained if the production of a family of repeated sequences is taken to be an analogue of a more familiar kind of mutation. Repeated sequences could be produced in large numbers "by accident," that is, by an event not directly related to their ultimate function. In the long run, of course, their fate would presumably be determined by natural selection, but for some time a large number of unexpressed copies might remain in the genome. In other words, we are led to the somewhat unorthodox view that a significant part of the DNA in the cell may not have a current genetic function.

We have called the appearance of a family of repeated sequences "saltatory replication" to imply a sudden event of multiplication. The evidence indicates that on an evolutionary time scale the appearance of a family is sudden, although we cannot tell whether the events occur within the lifetime of an organism or are spread over a few million years. Several steps are obviously required. First a segment of DNA must be multiplied; then the copies must be transmitted by heredity. The set of copies must become intimately associated with the chromosomes and ultimately must be linked directly into the principal DNA strands. If that much is accomplished in a small population within the species, this DNA must be transmitted throughout the species over a large number of succeeding generations, presumably by natural selection. Such a process would be inexplicable in neo-Darwinian terms unless some genetic advantage were associated with the family of repeated DNA sequences. We are unable

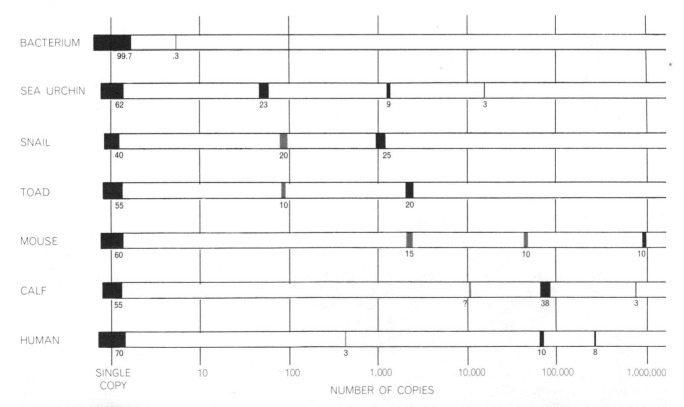

PATTERNS OF REPETITION of DNA sequences are shown in the form of spectra for a number of organisms. Each spectral band represents a repetitive class of DNA that has been identified to date. The width of the band indicates what percent (*given below the band*) of the total DNA is represented by that class. The position of the band indicates the frequency, or number of copies. (*Gray bands are those whose precise frequencies are not known.*) For example, the most repetitive mouse component amounts to 10 percent of its DNA and has about a million copies. Snail and toad data are from Eric H. Davidson of Rockefeller University.

to specify what the advantage might be unless an actual function were already being carried out by the repetitive DNA at the earliest stages of the process.

The time period would be greatly compressed if the saltatory replication were due to a virus infection that swept through a large population. A model for such an event might be the behavior of bacteria that survive a viral infection because they have previously accepted the virus genome into an integrated, or lysogenic, state within their own chromosome and by so doing have achieved immunity to the virus. Cellular resistance to virus infection by this kind of mechanism has not been clearly demonstrated in animals. The transformation of animal cells by viruses, however, is well known, and in many cases it may be due to the integration of the viral genome with the host chromosomes. Even so, there is no reason to expect that an enormous number of copies would be involved. It is therefore merely possible—not likely—that an infection of some kind is responsible for the transmission of a family of repeated sequences to all the surviving members of a species.

Obviously we do not understand how a family of repeated DNA originates, but such families exist, and they must have originated somehow. What can be said of their subsequent history? We know that some of the sequences have become functional because RNA complementary to them is produced. We do not know when in their history the useful attributes were developed, how many of the members of a family of sequences are utilized or what function they carry out. It is clear, however, that the families of repeated sequences evolve and that the sequences change slowly. It is even possible to estimate the rate of change. Estimating the "decay of relatedness" is one approach, and more elegant methods are now available. Fairly good estimates can be made of the actual rate of base substitution in DNA by measuring the stability of strand pairs hybridized between chosen species. Such measurements have only been started and no broad generalizations can yet be drawn, but we should soon know whether or not natural selection operates to conserve the sequences of some of the members of a family while allowing others to change at a more rapid rate. We now know a little about the history of repetitive sequences and we possess techniques to learn a great deal more, but a comprehensive set of measurements is needed that will require much work at many laboratories.

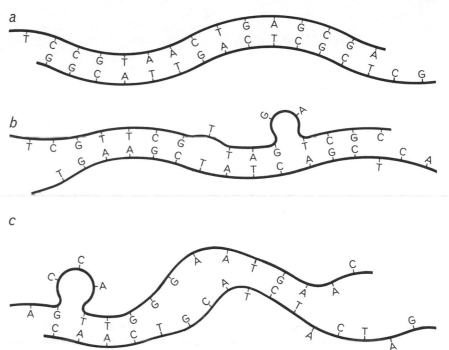

STABILITY of reassociated DNA depends on how well the two strands are hydrogen-bonded, which depends in turn on the extent to which the two strands are composed of complementary base sequences. In native DNA or in two strands from identical copies of a repetitive DNA all the bases match and the strand pair is quite stable (a). If some bases do not match, the stability is reduced (b), and if many are mismatched, the stability is poor (c). The stability of DNA can be determined by measuring its dissociation temperature.

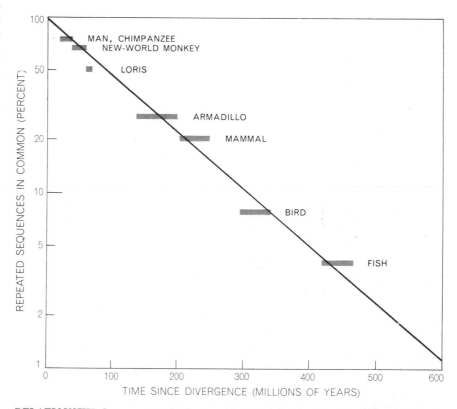

RELATIONSHIP between the similarity of repeated-DNA sequences and evolutionary history is shown, based on measurements made by Bill H. Hoyer, Brian J. McCarthy and Ellis T. Bolton of the Carnegie Institution. Vertical scale shows the fraction of repeated sequences held in common by the DNA of the rhesus monkey and that of other animals. The time of divergence of species lines leading to the rhesus and to other modern species is plotted on the horizontal scale. ("Mammal" refers to nonprimates other than armadillo.)

26

The Neutral Theory
of Molecular Evolution

by Motoo Kimura
November 1979

*It holds that at the molecular level most evolutionary
change and most of the variability within a species are
caused not by selection but by random drift of mutant
genes that are selectively equivalent*

The Darwinian theory of evolution through natural selection is firmly established among biologists. The theory holds that evolution is the result of an interplay between variation and selection. In each generation a vast amount of variation is produced within a species by the mutation of genes and by the random assortment of genes in reproduction. Individuals whose genes give rise to characters that are best adapted to the environment will be the fittest to survive, reproduce and leave survivors that reproduce in their turn. Species evolve by accumulating adaptive mutant genes and the characters to which those genes give rise.

In this view any mutant allele, or mutated form of a gene, is either more adaptive or less adaptive than the allele from which it is derived. It increases in the population only by passing the stringent test of natural selection. For more than a decade now I have championed a different view. I believe most of the mutant genes that are detected only by the chemical techniques of molecular genetics are selectively neutral, that is, they are adaptively neither more nor less advantageous than the genes they replace; at the molecular level most evolutionary changes are caused by the "random drift" of selectively equivalent mutant genes.

The Evolution of Darwinism

The controversy between the neutralist view and the "panselectionist" assumption arises from the way the modern "synthetic" theory of evolution has itself evolved. When Darwin formulated his original theory, the mechanisms of inheritance and the nature of heritable variations were not known. With the rise of Mendelian genetics in this century the way was opened for efforts to supply a genetic base for Darwin's insights. This was achieved largely through the elucidation by H. J. Muller of the fundamental nature of the gene and through

the methods of population genetics developed mainly by R. A. Fisher, J. B. S. Haldane and Sewall Wright. On their foundation subsequent studies of natural populations by Theodosius Dobzhansky, paleontological analyses by George Gaylord Simpson, the "ecological genetics" of E. B. Ford and his school and other investigations built a large and impressive edifice of neo-Darwinian theory.

By the early 1960's there was a general consensus that every biological character could be interpreted in the light of

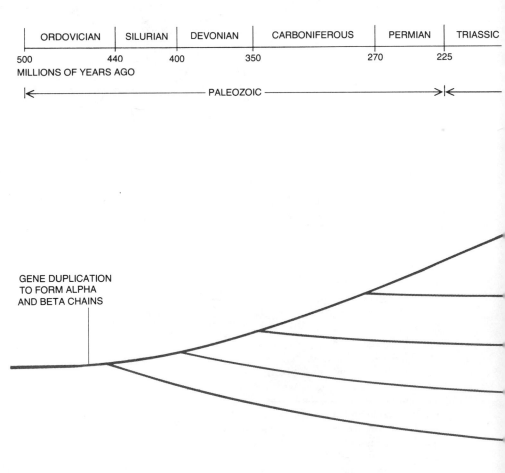

ORDOVICIAN	SILURIAN	DEVONIAN	CARBONIFEROUS	PERMIAN	TRIASSIC

500   440   400   350   270   225

MILLIONS OF YEARS AGO

←—————————————— PALEOZOIC ——————————————→|←————

GENE DUPLICATION
TO FORM ALPHA
AND BETA CHAINS

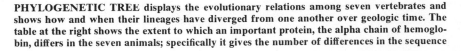

PHYLOGENETIC TREE displays the evolutionary relations among seven vertebrates and shows how and when their lineages have diverged from one another over geologic time. The table at the right shows the extent to which an important protein, the alpha chain of hemoglobin, differs in the seven animals; specifically it gives the number of differences in the sequence

adaptive evolution through natural selection and that almost no mutant genes were selectively neutral. As Ernst Mayr stated the case in 1963, "I consider it... exceedingly unlikely that any gene will remain selectively neutral for any length of time." A great deal was said by many workers about how genes interact, how gene pools of species are organized and how gene frequencies in populations change in the course of evolution. These conclusions, however, were necessarily inferences based on observations at the phenotypic level: the level of the form and function arising from the operation of genes. There was no way of knowing what actually goes on in evolution at the level of the internal structure of the gene.

Meanwhile the mathematical theory of population genetics was becoming quite sophisticated (which is rather unusual in biology). Particularly noteworthy was the theoretical framework provided by the manipulation of partial differential equations called diffusion equations. Diffusion models enable one to describe the behavior of mutant alleles by considering the random changes resulting from random sampling of gametes (germ cells) in reproduction as well as the deterministic changes caused by mutation and selection. Although the diffusion-equation method involves approximation, it yields answers to important but difficult questions that are inaccessible by other methods, such as: What is the probability of fixation for a single mutant appearing in a finite population and having a certain selective advantage, that is, what is the probability that it will eventually spread through the entire population?

The applicability of this method to gene changes in evolution, however, remained rather limited for some time. The reason is that population genetics deals with the concept of gene frequencies (the relative prevalence of various alleles within a population), whereas conventional studies of evolution were conducted at the phenotypic level, and there was no direct way of connecting the two sets of data unambiguously. That obstacle was removed with the advent of molecular genetics. It became possible to compare, in related organisms, individual RNA molecules (the direct products of genes) and proteins (the ultimate products) and so to estimate the rate at which allelic genes are substituted in evolution. It also became possible to study the variability of genes within a species. At last the time was at hand for applying the mathematical theory of population genetics to find out how genes evolve. One might have expected that the principle of Darwinian selection would prove to prevail at that fundamental level. Indeed, many evolutionary biologists found what they expected to find, and they have tended to extend panselectionism to the molecular level.

The Neutral Theory

The picture of evolutionary change that actually emerged from molecular studies seemed to me, however, to be quite incompatible with the expectations of neo-Darwinism. One of my salient findings with regard to evolution was that in a given protein the rate at which amino acids (the subunits of proteins) are substituted for one another is about the same in many diverse line-

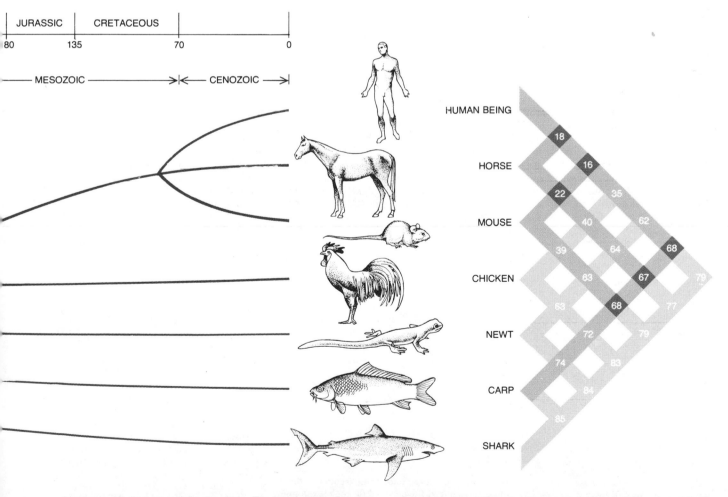

of amino acids that constitutes the chain. The hemoglobin molecule has two alpha chains and two beta chains, which originated through the duplication of a single gene some 450 million years ago. The table reflects the approximate uniformity (predicted by the neutral theory) of the rate of evolution of a given protein in very different organisms. The number of amino acid differences is roughly 20 when any of the three mammals are compared with one another, and it is approximately 70 when the carp is compared with any of the three mammals.

TYPE OF CHANGE	HUMAN ALPHA V. HUMAN BETA	CARP ALPHA V. HUMAN BETA
NO CHANGE	62	61
ONE-NUCLEOTIDE	55	49
TWO-NUCLEOTIDE	21	29
GAP	9	10
TOTAL	147	149

NUMBER OF DIFFERENCES between the amino acid sequences of the alpha chain and the beta chain of human hemoglobin is compared with the number of differences between the sequences of the alpha chain in the carp and the human beta chain. The column at the left categorizes the amino acid sites according to whether there is no change, a change due to a minimum of at least one nucleotide substitution or at least two substitutions in the genetic code at each site, or a "gap": an addition or a deletion of an amino acid. The numbers are similar whether one compares the chains in the same species or in the two species, suggesting that alpha chains in two lineages have accumulated mutations at about same rate for 400 million years.

ages. Another finding was that the substitutions seem to be random rather than having a pattern. A third finding was that the overall rate of change at the level of DNA, the actual genetic material, is very high, amounting to the substitution of at least one nucleotide base (DNA subunit) per genome (total genetic complement) every two years in a mammalian lineage. As for the extent of variability within a species, electrophoretic methods for detecting small differences among proteins suddenly disclosed a wealth of genetic variability; the proteins produced by a large fraction of the genes in diverse organisms were found to be polymorphic, that is, they were present in the species in variant forms. In many cases the protein polymorphisms had no visible phenotypic effects and no obvious correlation with environmental conditions.

In 1967, as I considered these puzzling observations, I decided they suggested two things. One was that a majority of the nucleotide substitutions in the course of evolution must be the result of the random fixation of neutral or nearly neutral mutants rather than the result of positive Darwinian selection. The other was that many protein polymorphisms must be selectively neutral or nearly so and must be maintained in a population by the balance between mutational input and random extinction. I presented these thoughts at a meeting of the Genetics Club in Fukuoka in November, 1967, and in a short paper in *Nature* the following February. In 1969 strong support came from a paper in *Science* by Jack Lester King, now of the University of California at Santa Barbara, and Thomas H. Jukes of the University of California at Berkeley. They had arrived at the same ideas on molecular evolution (although not on protein polymorphisms) independently, and they presented cogent supporting data from molecular biology.

The papers suggesting a neutral theory were severely criticized by evolutionists who believed the new molecular data could be understood in the light of orthodox neo-Darwinian principles, and the neutralist-selectionist controversy continues today. The essential difference between the two schools of thought can be appreciated by comparing their differing explanations of the evolutionary process by which mutant genes come to be substituted in a species. Every substitution involves a sequence of events in which a rare mutant allele appears in a population and eventually spreads through the population to reach fixation, or a frequency of 100 percent. Selectionists maintain that for a mutant allele to spread through a species it must have some selective advantage (although they admit that an allele that is itself neutral may occasionally be carried along by "hitchhiking" on a gene that is selected for and with which it is closely linked, and may thus reach a high frequency).

Neutralists, on the other hand, contend that some mutants can spread through a population on their own without having any selective advantage. If a mutant is selectively equivalent to preexisting alleles, its fate is left to chance. Its frequency fluctuates, increasing or decreasing fortuitously over time, because only a relatively small number of gametes are "sampled," out of the vast number of male and female gametes produced in each generation, and are therefore represented in individuals of the next generation [*see illustration on page 275*].

In the course of this random drift the overwhelming majority of mutant alleles are lost by chance, but a remaining minority of them eventually become fixed in the population. If neutral mutations are common at the molecular level and if the random drift is continuous over a long time (say millions of generations), the genetic composition of the population will change significantly. For any neutral mutant that appears in a population the probability of eventual fixation is equal to its initial frequency. The average length of time until fixation (excluding alleles that are lost) is four times the "effective" population size, or $4N_e$. (The effective size of a population is approximately equal to the number of breeding individuals in one generation, and it is usually much smaller than the total number of individuals in the species.)

The neutral theory, I should make clear, does not assume that neutral genes are functionless but only that various alleles may be equally effective in promoting the survival and reproduction of the individual. If a mutant allele encodes variant amino acids in a protein, the modified protein need function only about as well as the original form; it need not be precisely equivalent. In higher organisms particularly, homeostasis counteracts external environmental changes just as it does internal physiological changes; fluctuations in the environment do not necessarily imply

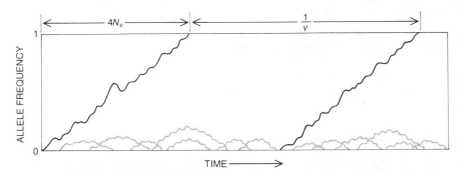

MUTANT ALLELES (variant genes) arise in a population at random. Their frequency fluctuates; in time most of them disappear (*gray lines*), but some of them spread through the population to fixation: a frequency of unity, or 100 percent (*black lines*). Population-genetic studies show that for a neutral allele that is destined for fixation the average number of generations until fixation is four times the effective population size, or $4N_e$. The average number of generations between consecutive fixations is equal to the reciprocal of the mutation rate *v*.

comparable fluctuations in the Darwinian fitness of mutant genes.

Some criticisms of the neutral theory arise from an incorrect definition of "natural selection." The phrase should be applied strictly in the Darwinian sense: natural selection acts through—and must be assessed by—the differential survival and reproduction of the individual. The mere existence of detectable functional differences between two molecular forms is not evidence for the operation of natural selection, which can be assessed only through investigation of survival rates and fecundity. Moreover, a clear distinction should be made between positive (Darwinian)

selection and negative selection. The latter, which Muller showed is the commoner form, eliminates deleterious mutants; it has little to do with the gene substitutions of evolution. A finding of negative selection does not contradict the neutral theory. Finally, the distinction between gene mutation in the individual and gene substitution in the population should be kept in mind; only the latter is directly related to molecular evolution. For advantageous mutants the rate of substitution is greatly influenced by population size and degree of selective advantage (as I shall show below) as well as by the mutation rate.

Two major findings with regard to

molecular evolution demonstrate particularly clearly that its patterns are quite different from those of phenotypic evolution and that the laws governing the two forms of evolution are different. One is the finding, alluded to earlier, that for each protein the rate of evolution in terms of amino acid substitutions per year is approximately constant and about the same in various lineages. The other is that molecules or parts of a molecule subjected to a relatively small degree of functional constraint evolve at a higher rate (in terms of mutant substitutions) than those subjected to stronger constraints do.

Molecular Evolution

The constancy of the evolutionary rate is apparent in the molecule of hemoglobin, which in bony fishes and higher vertebrates is a tetramer (a molecule with four large subunits) consisting of two identical alpha chains and two identical beta chains. In mammals amino acids are substituted in the alpha chain, which has 141 amino acids, at the rate of roughly one substitution in seven million years. This corresponds to about one substitution in a billion years (or 10^{-9} substitution per year) per amino acid site. The rate does not appear to depend on such factors as generation time, living conditions and population size. The approximate constancy of the rate is evident when the number of amino acid differences between the alpha chains of various vertebrates is charted together with the phylogenetic tree showing the relations among the vertebrates and the times when they diverged from one another in evolution [see illustration on pages 272 and 273].

The alpha and beta chains have essentially the same structure, are about the same length and show roughly the same rate of evolutionary amino acid substitution. They arose through gene duplication some 450 million years ago and became differentiated as they accumulated mutations independently. If one compares the divergence between the alpha and the beta chain of man with the divergence between the alpha chain of the carp and the beta chain of man, it is evident that in both cases the alpha and beta chains differ from each other to roughly the same extent. Because the alpha chain of man and that of the carp differ from each other in about half of their amino acid sites this suggests that the alpha chains in two distinct lineages, one leading to the carp and the other to man, have accumulated mutations independently and at practically the same rate over a span of about 400 million years. Moreover, the rate of amino acid substitution observed in these comparisons is very similar to the rates observed in comparisons of the alpha chains in various mammals.

My assertion of constancy of the evo-

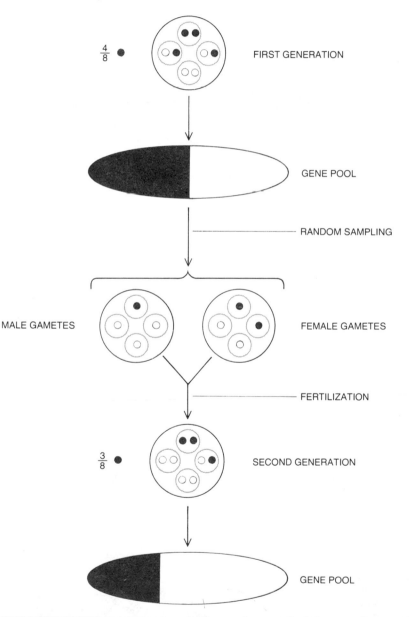

FIRST GENERATION

GENE POOL

RANDOM SAMPLING

MALE GAMETES

FEMALE GAMETES

FERTILIZATION

SECOND GENERATION

GENE POOL

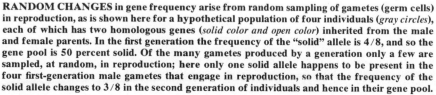

RANDOM CHANGES in gene frequency arise from random sampling of gametes (germ cells) in reproduction, as is shown here for a hypothetical population of four individuals (gray circles), each of which has two homologous genes (solid color and open color) inherited from the male and female parents. In the first generation the frequency of the "solid" allele is 4/8, and so the gene pool is 50 percent solid. Of the many gametes produced by a generation only a few are sampled, at random, in reproduction. Here only one solid allele happens to be present in the four first-generation male gametes that engage in reproduction, so that the frequency of the solid allele changes to 3/8 in the second generation of individuals and hence in their gene pool.

lutionary rate at the molecular level has been criticized by, among others, Richard C. Lewontin of Harvard University, who called the asserted constancy "simply a confusion between an average and a constant" and "nothing but the law of large numbers." The remarks reveal a misunderstanding of the nature of molecular evolution. One is attempting here to compare intrinsic rates of evolution in different lineages. The death rates characteristic of man and of an insect do not become equal by merely being averaged over a long period of time or over a large number of individuals; there is no reason to expect two averages to converge on each other unless the intrinsic factors shaping them are the same. My point is that intrinsic evolutionary rates are essentially determined by the structure and function of molecules and not by environmental conditions.

Evolutionary rates are, to be sure, not precisely constant in the sense that a rate of radioactive decay is constant. My colleague Tomoko Ohta and I showed in 1971 that the variance (the squared standard deviation) of the evolutionary rate observed for hemoglobins and for the protein cytochrome c in different mammalian lines is about 1.5 to 2.5 times larger than the variance to be expected if it were due only to chance. Charles H. Langley and Walter M. Fitch did a more elaborate analysis at the University of Wisconsin School of Medicine, combin-

ing data for the alpha and beta hemoglobin chains, cytochrome c and fibrinopeptide A; they found a variation in rates of mutant substitution about 2.5 times larger than the expected variance due to chance fluctuations, and they took this as evidence against the neutral theory. Yet they also showed that when the estimated number of substitutions between diverging branches of a phylogenetic tree is plotted against the corresponding time of divergence, the points fall on a straight line, which suggests the substantial uniformity of the evolutionary rates. It seems to me to be wrong to emphasize local fluctuations as evidence against the neutral theory while neglecting to inquire into why the rate remains essentially constant.

Rates of Evolution

Turning now to the quantitative relations that determine rates of evolution, consider first the nucleotides constituting a genome: a single (haploid) set of chromosomes. For a human being the number of nucleotides is very large, on the order of 3.5 billion. Because the mutation rate per nucleotide site is low (perhaps 10^{-8} per generation, or one mutation per 100 million generations) one can assume that whenever a mutant appears it is at a new site. This assumption is called the "infinite site" model in population genetics.

Let v represent the mutation rate per

gamete per unit time (generation). Since each individual has two sets of chromosomes, the total number of new mutants introduced into a population of N individuals in each generation is $2Nv$. Now let u be the probability that a single mutant will ultimately reach fixation. Then, in a steady state in which the process of substitution goes on for a very long time, the rate k of mutant substitution per unit time is given by the equation $k = 2Nvu$. That is, $2Nv$ new mutants appear in each generation, of which the fraction u eventually reach fixation, and k represents the rate of evolution in terms of mutant substitutions. The equation can be applied not only to the genome as a whole but also, with good approximation, to a single gene consisting of several hundred nucleotides or to the protein encoded by a gene.

The probability u of ultimate fixation is a well-known quantity in population genetics. If the mutant is selectively neutral, u equals $1/(2N)$. The reason is that any one of the $2N$ genes in the population is as likely as any other to be fixed, and so the probability that the new mutant will be the lucky gene is $1/(2N)$. (This assumes that the process is being viewed over a very long period of time, since the average time for a neutral gene to sweep through the population is $4N_e$.) Substituting $1/(2N)$ for u in the equation for the rate of evolution ($k = 2Nvu$), one gets $k = v$. That is, the rate of evolution in terms of mutant substitutions in the population is simply equal to the rate of mutation per gamete, independent of what the population size may be.

This remarkable relation applies only for neutral alleles. If the mutant has a small selective advantage s, then u equals $2s$ with good approximation, and the equation for the rate of evolution becomes $k = 4Nsv$. That is, the rate of evolution for selectively advantageous genes depends on the size of the population, the selective advantage and the rate at which mutants having a given selective advantage arise in each generation. One should expect, in this case, that the rate of evolution would depend strongly on the environment, being high for a species offered new ecological opportunities but low for those kept in a stable environment. It is highly unlikely, I think, that the product Nsv should be the same in diverse vertebrate lineages, in some of which phenotypic evolution has been very rapid (as in the line leading to man) and in others of which phenotypic evolution has long since practically ceased (as in the line leading to the carp). And yet the observed rates of molecular evolution show remarkable constancy. It seems to me that this constancy is much more compatible with the expectation of the neutral theory, that is, with the equation $k = v$, than it is with the selectionist relation $k = 4Nsv$.

Even more striking than the constan-

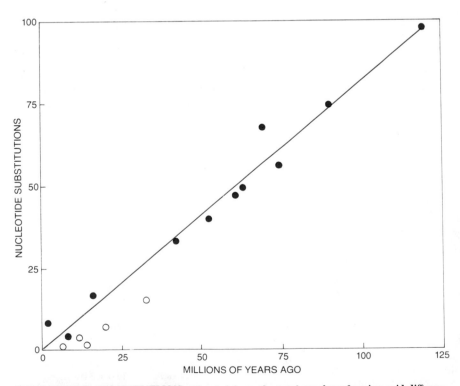

NUCLEOTIDE SUBSTITUTIONS estimated from the total number of amino acid differences observed in seven proteins in 16 pairs of mammals are plotted against the time since members of each pair diverged. Except for lineages involving primates (*open circles*) the points fall close to a straight line, again suggesting approximate uniformity of a protein's molecular evolutionary rate. Data are from Walter M. Fitch of University of Wisconsin School of Medicine.

cy of the rate of evolution is the second major feature of molecular evolution: the weaker the functional constraint on a molecule or a part of a molecule, the higher the evolutionary rate of mutant substitutions. There are regions of DNA between genes, for example, and in the case of higher organisms even within genes, that do not participate in protein formation and must therefore be much less subject to natural selection; some recent research has indicated that nucleotide substitutions are particularly prevalent in such regions of DNA.

Functional Constraint

This relation between a relative lack of selective constraint and a relatively high rate of molecular evolution has been well established for certain proteins. Among the proteins so far investigated the highest evolutionary rate has been found in fibrinopeptides, which appear to have little function, if any, after they become separated from fibrinogen to yield fibrin, a protein that plays a role in blood clotting. The same effect is observed in the case of the C chain of the proinsulin molecule, a precursor of insulin. The C chain, which is cleaved from the precursor to form active insulin, evolves at a rate several times higher than that of the active molecule. The effect of constraint on the evolutionary rate has also been noted for different parts of the hemoglobin molecule. The precise structure of the surface of the molecule is presumably less significant than the structure of the pockets, in the interior of the molecule, that hold the iron-containing heme groups. Ohta and I have estimated that the regions of both the alpha and the beta chain that are at the surface of the protein evolve about 10 times faster than the regions forming the heme pocket.

The genetic code is based on groups of three nucleotides, with each triplet "codon" in a strand of RNA specifying a particular amino acid of the protein chain encoded by the RNA. For example, the codon *GUU* (the letters stand for particular nucleotide bases) specifies the amino acid valine. So does *GUC,* however; the genetic code is "degenerate," with most amino acids being designated by two or more synonyms, which typically differ only in the third position of the triplet. As a result a large fraction (perhaps 70 percent) of all random nucleotide substitutions at the third position are synonymous changes and do not lead to amino acid replacements. There is growing evidence that evolutionary nucleotide substitution goes on at a particularly high rate at the third position. Michael Grunstein of the University of California at Los Angeles and his colleagues compared the RNA sequences encoding the protein histone IV in two species of sea urchin. They found that although the protein has maintained a

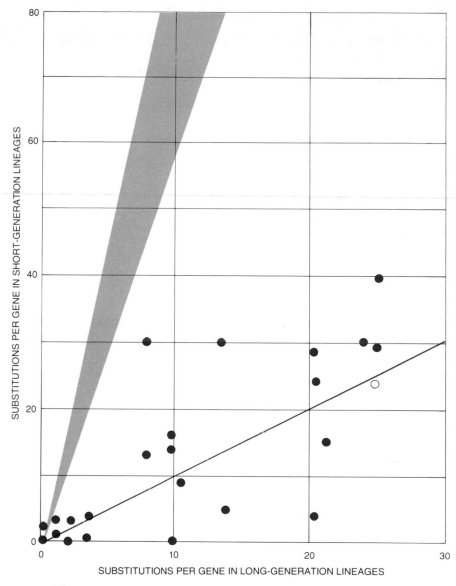

MOLECULAR EVOLUTIONARY RATE in mammals that have a short generation time was compared with the rate in mammals having a long generation time by Allan C. Wilson and his colleagues at the University of California at Berkeley. Each point represents the ratio of the implied nucleotide substitutions in the two animals of a pair since the two diverged from a common ancestor; the open circle, for example, is for the beta chain of hemoglobin in the elephant (*abscissa*) and in the mouse (*ordinate*). **If the rate of change per year were identical in both animals of a pair, the points would fall on a line (*solid color*). Actually the points are close to that absolute-time line and far from sector predicted for generation-time effect (*colored area*). Apparently molecular evolutionary rate is roughly constant per year, not per generation.**

practically unchanged amino acid sequence for about a billion years, a number of synonymous nucleotide differences are found in the RNA sequences of the two species. On the basis of their data and paleontological evidence on the length of time since the species diverged, I have estimated that the rate of nucleotide substitution has been roughly 3.7×10^{-9} per year at the third position, a very high rate. What is remarkable is that there have been so many synonymous mutant substitutions in the histone-IV gene in spite of the very low rate of amino acid changes in the corresponding protein.

These observations can be explained

simply and consistently by the neutral theory. Suppose a certain fraction f_0 of all molecular mutants are selectively neutral and the rest are definitely deleterious. Then the mutation rate v for neutral alleles is equal to the total mutation rate v_T multiplied by f_0, so that the overall rate k of mutant substitution becomes equal to $v_T f_0$. Now assume that the probability that a mutational change is neutral (not harmful) depends strongly on functional constraints. The weaker the constraint is, the larger will be the probability f_0 that a random change is neutral, with the result that the rate of evolution k increases. The maximum evolutionary rate is attained when f_0

PROTEIN	EVOLUTIONARY RATE
FIBRINOPEPTIDES	9.0
PANCREATIC RIBONUCLEASE	3.3
HEMOGLOBIN CHAINS	1.4
MYOGLOBIN	1.3
ANIMAL LYSOZYME	1.0
INSULIN	.4
CYTOCHROME c	.3
HISTONE IV	.006

PROTEINS VARY WIDELY in their rate of evolution. Here the rate is given for several proteins in terms of the number of amino acid substitutions per amino acid site per billion years. The rate is particularly high for fibrinopeptides, which appear to have little function after they are cleaved from a precursor molecule to yield the active blood-clotting protein fibrin. Proteins such as fibrinopeptides are subject to less functional constraint, and evolve faster, than proteins whose precise shape is significant and hence subjects them to stronger constraint.

equals 1, that is, when all the mutations are neutral. In my opinion the high evolutionary rates observed at the third position of the codon are rather near this limit.

The neutral theory, then, predicts that as functional constraint diminishes, the rate of evolution converges to the maximum value set by the total mutation rate. Confirmation of such a convergence, or plateauing, of molecular evolutionary rates by further studies would be strong evidence in support of the neutral theory. This interpretation of the molecular data will not make sense to selectionists. In their view molecules or parts of a molecule that are evolving rapidly in terms of mutant substitutions must have some important but as yet unknown function and must be undergoing rapid adaptive improvement by accumulating beneficial mutants. And they will see no reason to believe the upper limit of the evolutionary rate is related to the total mutation rate.

Polymorphism

Neutralists and selectionists also have diametrically opposed explanations for the mechanisms by which genetic variability is maintained within a species, particularly in the form of protein polymorphism: the coexistence in a species of two or more different forms of a protein. Neutralists maintain that polymorphisms are selectively neutral and are maintained in a population through mutational input and random extinction; in every generation a number of neutral mutants arise and in time either become fixed in the population or are lost, and in the process they contribute to genetic variability in the form of polymorphisms. In the neutral view polymorphism and molecular evolution are not two distinct phenomena; polymorphism is simply one phase of molecular evolution.

Selectionists maintain that polymorphisms are actively maintained by some form of "balancing selection," notable among which are heterotic selection, or "heterozygote advantage," and frequency-dependent selection. At one time the former was enthusiastically proposed as the main agent maintaining polymorphisms. There are instances in which individuals that are heterozygous for a particular gene (that carry a different allele of the gene on each of their two chromosomes) are fitter than individuals that are homozygous for either allele (that carry one or the other allele on both chromosomes). Selection will then tend to preserve both alleles in the population as a balanced polymorphism. In 1973, however, Roger D. Milkman of the University of Iowa found abundant polymorphisms in the bacterium *Escherichia coli*, which is a haploid organism: it has only one set of genes. Heterozygote advantage cannot explain such polymorphisms.

Nowadays many selectionists explain polymorphisms as being the result of frequency-dependent selection, in which the fitnesses of two alleles vary with their relative frequency. This was first proposed by the late Ken-Ichi Kojima of the University of Texas at Austin and his colleagues, who obtained results indicating marked frequency-dependent selection affecting the genes for the enzymes esterase-6 and alcohol dehydrogenase (ADH) in the fruit fly *Drosophila melanogaster*; Bryan Clarke of the University of Nottingham reported that he had confirmed Kojima's results in the case of ADH. On the other hand, experiments done by Tsuneyuki Yamazaki, now of Kyushu University, failed to show any such selection for esterase-5 alleles in *Drosophila pseudoobscura*. A group led by Terumi Mukai of Kyushu carried out extensive studies of selection for several enzymes in *D. melanogaster* and found no evidence for a difference in the fitness of variant forms of the enzymes. And recent careful, large-scale experiments by Mukai and Hiroshi Yoshimaru have failed to find any frequency-dependent selection for ADH in *D. melanogaster*.

If selection is not responsible for maintaining polymorphisms, what neutralist explanation is there for the fact

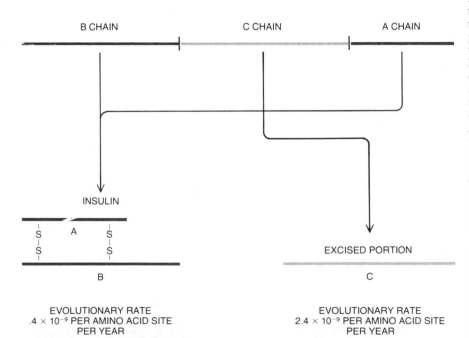

C CHAIN OF PROINSULIN (*light color*) is cleaved from the precursor molecule and discarded, and the A and B chains (*dark color*) are joined by disulfide bridges (*S–S*) to form the active insulin molecule. In keeping with the C chain's relative freedom from functional constraint, it has been found to evolve at a much higher rate than the two chains of the active molecule.

that some proteins are more often polymorphic than others? Recently Richard K. Koehn of the State University of New York at Stony Brook and W. F. Eanes, now of Harvard, and also Masatoshi Nei's group at the University of Texas at Houston, have shown that in various *Drosophila* species there is a significant correlation between the genetic variability (or polymorphism) of proteins and the weight of their molecular subunits. This is easy to explain according to the neutral theory because the larger the size of a subunit is, the higher its mutation rate should be. Harry Harris of the University of Pennsylvania and his colleagues could not find the same correlation when they investigated human polymorphisms, but they did find that single-subunit enzymes are more polymorphic than multiple-subunit ones, something that Eleftherios Zouros of Dalhousie University had earlier reported in *Drosophila*. One finding by Zouros and Harris fits the neutral theory particularly well: multiple-subunit enzymes that form hybrid molecules by combining with enzymes encoded by other genes have a clearly reduced level of polymorphism. The precise interaction among subunits required to form such enzymes would increase the degree of functional constraint and so reduce the possibility that a mutation will be harmless, or neutral.

Neutralists, in other words, consider molecular structure and function to be the major determinants of protein polymorphisms. Selectionists consider environmental conditions to be the major determinants. They have maintained that there should be a correlation between environmental variability and genetic variability. They predicted, for example, that organisms living at the bottom of the deep sea would generally be found to display little genetic variability because their environment is stable and homogeneous, whereas organisms living in the intertidal zone would display a great deal of genetic variability because their environment is a changeable one. The prediction was logical and plausi-

ble, but it failed: genetic variability has been found to be generally extremely high among organisms living at the bottom of the oceans and to be very low among organisms living in the intertidal zone.

Models

In order to carry out quantitative studies based on population genetics one needs mathematical models for the mutational production of new alleles. The first such model was proposed in 1964 by James F. Crow of the University of Wisconsin and me. It is based on the fact that each gene consists of a large number of nucleotides, so that a practically infinite number of alleles can arise; the model therefore assumes that any new mutant arising represents a new allele rather than a preexisting one. The model predicts that variability within a species, in terms of the average heterozygosity H per gene, will be determined essentially by the product of the effective population size N_e and the mutation rate v per gene per generation, rather than by N_e and v separately. Specifically, H equals $4N_ev/(4N_ev + 1)$. For example, if the mutation rate is 10^{-6} and the effective population size is 10^5, the average heterozygosity per gene will be about .286, that is, 28.6 percent of the individuals are heterozygous at each gene locus on the average. The larger either the population size or the mutation rate per gene per generation is, the closer the average heterozygosity will be to unity (or 100 percent).

The model assumes that alleles are identified at the level of the gene in terms of actual nucleotide substitutions. Most observations of variability depend, however, on the electrophoresis of proteins, which has much less resolving power and is far from revealing all nucleotide substitutions (or even all amino acid changes), so that the observed heterozygosity is less than the true amount. Even when the model is modified to take account of this problem, very large populations should, according to the neutral theory, display nearly 100 percent heter-

ozygosity. When observations suggest otherwise, the theory is subject to criticism. For example, Francisco J. Ayala of the University of California at Davis has reported that in the neotropical fruit fly *Drosophila willistoni*, for which he estimates a very large effective population size of 10^9, he has found an observed heterozygosity of roughly 18 percent. He points out that even assuming a very low rate of neutral mutations per generation, 10^{-7}, the predicted heterozygosity is practically 100 percent.

There are at least two ways to deal with this apparent inconsistency. First of all, it is possible that the effective population size of *D. willistoni* has not been as large as 10^9 even if the apparent present size of the population is enormous. One can show mathematically that the genetic variability due to neutral alleles can be greatly reduced by a population "bottleneck" from time to time, after which it takes millions of generations for the variability to build up again to the theoretical level characteristic of a very large population maintained constantly over a long period. In this sense such neotropical species as *D. willistoni* may still show the effects of the bottleneck imposed by the last continental glaciation, between some 30,000 and 10,000 years ago. In addition the local extinction of colonies of a species, which may be fairly frequent, must reduce the effective population size.

A second possibility is that, as Ohta first proposed in 1973, the majority of "neutral" alleles may not actually be strictly neutral but rather may be very slightly deleterious. Adopting Ohta's idea, but retaining room for truly neutral mutations also, I have considered a model in which the selection coefficient s' against the mutant follows a particular distribution (the gamma distribution) [*see illustration on page 280*]. The mutation rate for variants whose negative selection (s') value is smaller than the reciprocal of two times the population size, or $1/(2N)$, can be considered the effectively neutral mutation rate v_e. It can be shown that this effectively neu-

L. PICTUS MESSENGER RNA	GA U	AAC	AUC	CAA	GG A	AU A	AC U	AAA	CCG	GC A	AUC
S. PURPURATUS MESSENGER RNA	GA C	AAC	AUC	CAA	GG U	AU C	AC G	?	?	GC U	AUC
HISTONE IV AMINO ACID SEQUENCE IN BOTH SPECIES	Asp	Asn	Ile	Gln	Gly	Ile	Thr	Lys	Pro	Ala	Ile
	24	25	26	27	28	29	30	31	32	33	34

NUCLEOTIDE SEQUENCES of the messenger RNA encoding the protein histone IV in two sea-urchin species, *Lytechinus pictus* and *Strongylocentrotus purpuratus*, were compared by Michael Grunstein of the University of California at Los Angeles. There are four RNA nucleotides (*A, G, U* and *C*); three-nucleotide codons specify the various amino acids that constitute a protein. Most amino acids are specified by two or more synonymous codons, which usually differ only at their third position. In this short stretch of RNA coding for amino acid sites 24 through 34 of histone IV there are five synonymous differences (*color*) in third-position nucleotides. That is, there has been a high rate of nucleotide substitution at the unconstrained third position, leaving the amino acid sequences unaffected.

tral mutation rate decreases as the population increases; in the case illustrated the rate is proportional to 1 divided by the square root of the population size. In this model the level of heterozygosity increases only slowly as the population increases. Moreover, given a realistic assumption about generation time, the rate of evolution in terms of mutant substitutions would be roughly constant per year for various lineages if the mu-

tation rate per generation is constant. Note that although this explanation invokes natural selection, it is quite different from the selectionist explanation.

A Quantitative Approach

The neutral theory of molecular evolution and polymorphism that I have developed in collaboration with my colleagues Ohta and Takeo Maruyama is

distinguished from most selectionist approaches—in particular from the approach of ecological genetics—in that it aims at a quantitative description of molecular evolution, which we attempt to achieve by manipulating diffusion equations. It is a venture in what might be called molecular population genetics. Nei and his associates at Houston have contributed greatly to this effort, in particular by connecting theoretical predictions with actual observations. They have shown, for example, that the variance of heterozygosity for particular enzymes within a species can be predicted fairly accurately by the neutral theory on the basis of observations of mean heterozygosity.

Because our theory is quantitative it is testable and therefore much more susceptible to refutation when it is wrong than are selectionist theories, which can invoke special kinds of selection to fit special circumstances and which usually fail to make quantitative predictions. To test the neutral theory, however, it is necessary to estimate such quantities as mutation rates, selection coefficients, population sizes and migration rates. Many evolutionary biologists maintain that such population-genetic quantities can never be accurately determined and that consequently any theory dependent on them is a futile exercise. I, on the other hand, believe these quantities must be investigated and measured if the mechanisms of evolution are to be understood. Surely astronomers and cosmologists cannot eschew theories set forth in terms of astronomical quantities simply because such quantities are hard to estimate accurately.

Darwinian selection acts mainly on phenotypes shaped by the activity of many genes. Environmental conditions surely play a decisive role in determining what phenotypes are selected for; Darwinian, or positive, selection cares little how those phenotypes are determined by genotypes. The laws governing molecular evolution are clearly different from those governing phenotypic evolution. Even if Darwin's principle of natural selection prevails in determining evolution at the phenotypic level, down at the level of the internal structure of the genetic material a great deal of evolutionary change is propelled by random drift. Although this random process is slow and insignificant in the time frame of man's ephemeral existence, over geologic time it makes for change on an enormous scale.

People have told me, directly and indirectly, that the neutral theory is not important biologically because neutral genes are not involved in adaptation. My own view is that what is important is to find the truth, and that if the neutral theory is a valid investigative hypothesis, then to establish the theory, test it against the data and defend it is a worthwhile scientific enterprise.

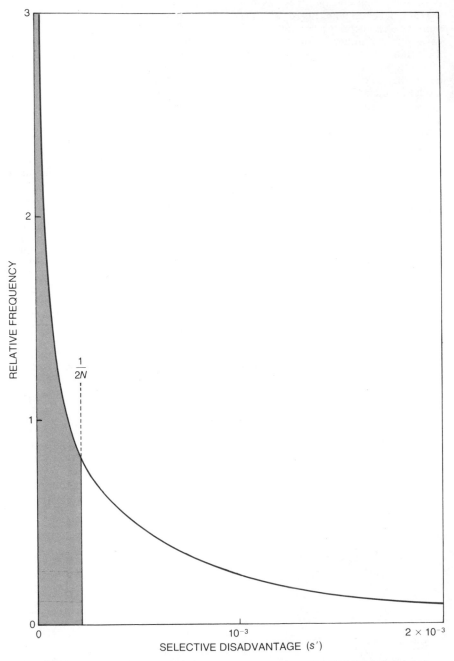

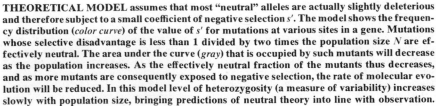

THEORETICAL MODEL assumes that most "neutral" alleles are actually slightly deleterious and therefore subject to a small coefficient of negative selection s'. The model shows the frequency distribution (*color curve*) of the value of s' for mutations at various sites in a gene. Mutations whose selective disadvantage is less than 1 divided by two times the population size N are effectively neutral. The area under the curve (*gray*) that is occupied by such mutants will decrease as the population increases. As the effectively neutral fraction of the mutants thus decreases, and as more mutants are consequently exposed to negative selection, the rate of molecular evolution will be reduced. In this model level of heterozygosity (a measure of variability) increases slowly with population size, bringing predictions of neutral theory into line with observation.

APPLIED GENETICS

APPLIED GENETICS VII

INTRODUCTION

While plant and animal breeding are clearly practical applications of genetic and evolutionary theory, much of the technology developed for these practical pursuits preceded the articulation of the theories that later informed them, and for this reason no articles dealing with these most important and practical fields have been included. I have chosen instead to include three articles about genetic applications in which technology clearly followed fundamental discoveries. One deals with a controversial area where the myth of the Promethean fire is most apt. I refer to Clifford Grobstein's article "The Recombinant-DNA Debate," which, after its original publication in 1977, was used as the pivotal article in a *Scientific American Reader* edited by David Freifelder devoted to the topic of recombinant DNA (W. H. Freeman and Company, 1978).

The second article, "Genetic Amniocentesis" by Fritz Fuchs, is hardly less controversial in its implications, in that a frequent outcome of prenatal diagnosis is the indication for therapeutic abortion. Nevertheless, it will provide an understanding of the rapid progress that is being made in the field of medical genetics.

The third article, "Bacterial Tests for Potential Carcinogens" by Raymond Devoret, describes the advances that have been made since Bruce Ames developed the first inexpensive and quick screening test for chemical carcinogens. The principle of the Ames test is based on the notion that most cancers arise by somatic mutation; thus a chemical that is mutagenic for a tester strain of bacteria should be immediately suspect as a carcinogen in man. The validity of this notion was shown when Ames was able to demonstrate that a number of known carcinogens, which initially did not test out as mutagens for bacteria, did so when they were exposed to tissue extracts prior to testing. Thus it came to be recognized that some form of metabolic derivative of the "carcinogenic" chemical was the actual carcinogen.

However, as Devoret's article indicates, the correlation between mutagenicity for bacteria and carcinogenicity for animals is not perfect. While there may be some trivial reasons for this, Devoret suggests that the criterion of DNA damage may be more critical than mutagenicity. This criterion raises the possibility that carcinogens may bring about other events in addition to mutation that are DNA-damage-related, such as crossing over, and that these events also contribute to carcinogenesis.

The Recombinant-DNA Debate

by Clifford Grobstein
July 1977

*The four-year-old controversy over the potential
biohazards presented by the gene-splicing method and
the effectiveness of plans for their containment is
viewed in a broader context*

The guidelines for research involving recombinant-DNA molecules issued a year ago by the National Institutes of Health were the culmination of an extraordinary effort at self-regulation on the part of the scientific community. Yet the policy debate over recombinant-DNA research was clearly not laid to rest by the appearance of the NIH guidelines. Instead the debate has escalated in recent months both in intensity and in the range of public involvement. A watershed of sorts was reached in March at a public forum held by the National Academy of Sciences in Washington. The forum was in part a repeat performance by scientists arguing fixed positions that were established early in the debate. There were, however, new participants on the scene, and they presented a varied and rapidly shifting agenda. They made it clear that research with recombinant DNA had become a political issue. As one speaker remarked, the Academy forum may have been the last major public discussion of recombinant DNA arranged by the scientists involved in the research. Nonscientists at the forum, by word and deed, reiterated the theme that science has become too consequential either to be left to the self-regulation of scientists or to be allowed to wear a veil of political chastity.

Science of course is crucially consequential to society, precisely because it is an intensifying source of both benefits and risks. Research with recombinant DNA may provide major new social benefits of uncertain magnitude: more effective and cheaper pharmaceutical products; better understanding of the causes of cancer; more abundant food crops; even new approaches to the energy problem. These and other possible outcomes are envisioned in "best-case scenarios" for the future application of recombinant-DNA technology. "Worst-case scenarios" can also be conceived: worldwide epidemics caused by newly created pathogens; the triggering of cat-

astrophic ecological imbalances; new tools for militarists and terrorists; the power to dominate and control the human spirit.

Both the best-case and worst-case scenarios are largely speculative; the gap between them symbolizes the large degree of uncertainty that surrounds this major step forward in molecular genetics. The material basis of biological heredity has been broken into in the past two decades, and it seems as though each of the fragments has acquired a life of its own. In this resulting period of instability fear threatens to override wonder as the implications of the research diffuse more widely. The fear is not so much of any clear and present danger as it is of imagined future hazards. The classic response to such fears is rigid containment: the Great Wall, the Maginot Line, the cold war. All are manifestations of the effort to provide absolute

security against unpredictable risks, and yet each generates its own risk. The escalation of the recombinant-DNA debate has a component of this kind of behavior, but there is a more rational component as well.

The first round of the fateful debate began in 1974, when investigators at the leading edge of work in this field declared a voluntary moratorium on several types of experiment judged to be conceivably risky. A set of techniques had been developed that made it possible to cut the long, threadlike molecules of DNA into pieces with the aid of certain enzymes, to recombine the resulting segments of DNA with the DNA of a suitable vector, or carrier, and to reinsert the recombinant into an appropriate host cell to propagate and possibly to function.

The significance of the new develop-

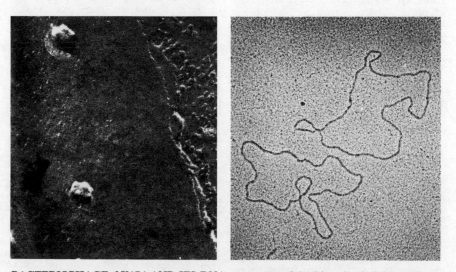

BACTERIOPHAGE ΦX174 AND ITS DNA are portrayed in this pair of electron micrographs. The virus infects the common intestinal bacterium *Escherichia coli*. In the micrograph at left, made by Jack D. Griffith of the Stanford University School of Medicine and Andrew Staehelin of the University of Colorado, two φX174 particles are seen attached to surface of an *E. coli* cell. In micrograph at right, made by Griffith, the DNA molecules of two φX174 viruses are seen in their double-strand form; each molecule is about 18,000 angstroms long.

ments is rooted in the central biological role of DNA as the transmitter of genetic information between generations. The transmission of the encoded genetic message depends on the ability of a cell to generate exact replicas of the parental DNA and to allocate the replicas among the offspring. In addition the success of genetic transmission depends on the ability of the offspring to "express" the encoded information properly by referring to it to control essential life processes. The mechanism of genetic expression in higher organisms is at present only dimly understood, and the discovery of the new recombinant-DNA techniques seemed immediately to open a broad new avenue to increased knowledge in this field.

The detailed mechanisms of genetic replication and expression are enormously complex. The essence of the matter, however, is found in the famous "double helix" structure of DNA. Both of the two long, interwound and complementary strands of the DNA molecule are made up of four kinds of nucleotides, cytosine, guanine, adenine and thymine (abbreviated *C, G, A* and *T*), which are linked end to end like a train of boxcars. The genetic message of each strand is embodied in the particular sequence of nucleotides, any one of which may follow any other. For example, the sequence *CATTACTAG* contains five identifiable English words: *CAT, AT, TACT, ACT* and *TAG*. The genetic message, however, is "written" in triplets: *CAT, TAC* and *TAG*. In general each triplet "codon" determines, through a series of intermediate steps, the position of a specific amino acid in a protein molecule.

Proteins, like nucleic acids, can be visualized as long trains of boxcars coupled end to end; here, however, the subunits are amino acids rather than nucleotides. The sequence of nucleotides in a given DNA molecule determines the sequence of amino acids in a particular protein, with each triple-nucleotide codon placing one of 20 possible amino acids at each successive position in the protein chain. The sequence of amino acids in turn specifically establishes both the structure and the function of the protein. Thus the nucleotide sequence of DNA precisely specifies the protein-building properties of the organism. Moreover, virtually every property of the organism, from enzymatic action to eye color, depends on protein structure in one way or another.

The transmission of the essential genetic information between generations depends on the precise replication of the nucleotide sequences of DNA. The mechanism for replication stems from the complementary relation between the two strands of the DNA molecule. A sequence on one strand (for example *CATTACTAG*) lies immediately opposite a complementary sequence (*GTA-*

ATGATC) on the other strand. The strands are complementary because *C* and *G* are always opposite on the intercoiled strands, as are *A* and *T*. Complementarity depends on the special chemical affinity, or binding, between *C* and *G* on the one hand and *A* and *T* on the other. The sum of these bonds, repeating along the length of the strands, is what holds the strands together in the double helix. Under appropriate conditions affinity is reduced and the two strands can unwind and separate. The single strands can again pair and rewind when conditions for high affinity are restored.

Double-strand DNA replicates by means of an extension of these properties. The unwinding and separation of the strands begins at a localized site along the DNA molecule. In the presence of suitable enzymes and free nucleotides a new chain is formed next to the exposed portion of each unpaired older chain. Each nucleotide lines up next to its opposite number (*C* next to *G*, *A* next to *T*). The complementary sequence thus established is then linked end to end by an enzyme that closes the nucleotide couplings. When the replication process has traveled along the entire length of the original double helix, two new helixes identical with the first one have been formed. The replication of DNA is the most fundamental chemical reaction in the living world. It fully accounts for the classical first principle of heredity: like begets like.

If DNA replication always worked without error, life would be far more homogeneous than it is. Here, however, a second classical principle of heredity intervenes: the principle of mutational variation, or the appearance in the offspring of new hereditary characteristics not present in the progenitors. Mutations arise through error, at least partly in the replication process. For example, the substitution of one nucleotide by another changes the triplet codon and puts a different amino acid in the corre-

sponding position in the resulting protein. Single-nucleotide errors lead to single-amino-acid errors. Thus, a single-nucleotide error is responsible for the human disease sickle-cell anemia. Most mutations are not such simple, single-nucleotide exchanges; nevertheless, they correlate directly with altered, transposed or deleted nucleotide sequences in DNA. When these changes appear in a gene (that is, a segment of DNA that codes the amino acid sequence of a particular protein), a change in the protein and hence in the hereditary properties it controls is the result.

Therein lies the crux of recombinant-DNA technology. It makes possible for the first time the direct manipulation of nucleotide sequences. Changes in nucleotide sequence that are produced by "natural" errors are random, even when their overall frequency is artificially increased. In natural populations Darwinian selection "chooses" among the random errors, increasing the representation in breeding populations of those errors that lead to more offspring in particular environments. Artificial selection, practiced by human beings for millenniums, favors errors that meet human needs (agricultural breeding) or whims (exotic-pet breeding). The success of both natural and artificial selection, however, is dependent on the random occurrence of desirable mutations. There was no way to direct genetic change itself until recombinant-DNA techniques came along. The new techniques enable one to deliberately introduce known and successful nucleotide sequences from one strain or species into another, thereby conferring a desired property.

The recombinant-DNA approach involves experimental ingenuity and detailed knowledge of the DNA molecule. It begins with an attack on DNA by the proteins called restriction enzymes, which are isolated from bacteria. The enzyme attack breaks the double chain

GENETIC CODE of an extremely small bacterial virus, the bacteriophage designated ϕX174, is given by the sequence of letters on the opposite page. The letters stand for the four nucleotides cytosine, guanine, adenine and thymine, which are linked end to end to make up each strand of the normally double-strand DNA molecule. The genetic message embodied in each strand of DNA is represented by the particular sequence of nucleotides, any one of which may follow any other. In the ϕX174 virus the DNA molecule, which has only a single circular strand for part of its life cycle, consists of approximately 5,375 nucleotides; the nucleotides are grouped into nine known genes, which are responsible in turn for coding the amino acid sequences of nine different proteins. For example, the dark-color segment of the molecule, called gene *J*, codes for a small protein that is part of the virus; this segment also happens to be the shortest gene in the ϕX174 genome. The complete nucleotide sequence for the DNA in ϕX174 was worked out recently by Frederick Sanger and his colleagues at the British Medical Research Council Laboratory of Molecular Biology in Cambridge. About 2,000 pages of this type would be required to show the nucleotide sequence for the DNA in the chromosome of a typical single-cell bacterium; roughly a million pages would be needed to similarly display the genetic code embodied in DNA molecules that make up chromosomes of a mammalian cell.

```
GAGTTTTATCGCTTCCATGACGCAGAAGTTAACACTTTCGGATATTTCTGATGAGTCGAAAAATTATCTTGATAAACGAGGAATTACTACTGCTTGTTTA
TCAACTACCGCTTTCCAGCGTTTCATTCTCGAAGAGCTCGACGCGTTCCTATCCAGCTTAAAAGAGTAAAAGGCGGTCGTCAGGTGAAGCTAAATTAAGC
AACGATTCTGTCAAAAACTGACGCGTTGGATGAGGAGAAGTGGCTTAATATGCTTGGCACGTTCGTCAAGGACTGGTTTAGATATGAGTCACATTTTGTT
TGAACTGAGTACTAAAGAATGGATAATCACCAACTTGTCGTAGCCTGAGTCTATCATTAGGTGCGAGAAAATTTTACAGTTGTTCTCTTAGAGATGGTAC
TACTGAACAATCCGTACGTTTCCAGACCGCTTTGGCCTCTATTAAGCTCATTCAGGCTTCTGCCGTTTTGGATTTAACCGAAGATGATTTCGATTTTCTG
TCGCTCCCATAGGATGTTTCAGGTCGCATGGTATTTGCGTTCGGAGTTGCGTCGCTGCTCGTGCTCTCGCCAGTCATCGTTAGGTTTGAAACAATGAGCA
TTCCTGCTCCTGTTGAGTTTATTGCTGCCGTCATTGCTTATTATGTTCATCCCGTCAACATTCAAACGGCCTGTCTCATCATGGAAGGCGCTGAATTTAC
CAGTCATTCTTGCAGTCACAAAGGACGCGCATGTGCGTTCCATTTGCGCTTGTTAAGTCGCCGAAATTGGCCTGCGAGCTGCGGTAATTATTACAAAGG
GCAGAAGAAACGTGCGTCAAAAATTACGTGCGGAAGCAGTGATGTAATGTCTAAAGGTAAAAAACGTTCTGGCGCTCGCCCTGGTCGTCGGCAGCCGTT
TTAAATAGGAGTTCATTCCCCGGCTTCGGGGACGTTAATTTAACAACTGGTGGATGTATGGTTTCTGCTCGCCGGAAATGCGAACGCGAAATCATCGAGCCG
ATGTCTAATATTCAAACTGGCGCCGAGCGTATGCCGCATGACCTTTCCCATCTTGGCTTCCTTGCTGGTCAGATTGGTCGTCTTATTACCATTTCAACTA
TACAGATGTCATCTCAGTTATCGTTCCGGTGCTGCGTTACCTCTTTCTGCCTCTCGCGGTTGCCGCAGGTAGAGCTTCCTCAGCCGGTCGCTATTGGCCTC
TTTTACTTTTTATGTCCCTCATCGTCACGTTTATGGTGAACAGTGGATTAAGTTCATGAAGGATGGTGTTAATGCCACTCCTCTCCCGACTGTTAACCAA
GATATCTATAGTTTATTGGGACTTTGTTTACGAATCCCTAAAATAACCATAGTCCCAATTAGCACGGTTCTTTTCGCCGTACCAGTTATATTGGTCATCA
CGTATTTTAAAGCGCCGTGG--ATGCCTGACCGTACCGAGGCTAACCCTAATGAGCTTAATCAAGATGATGCTCGTTATGGTTTCCGTTGCTGCCATCT
TCGTCGAACGTCTGGGTATTACAGTTATCTACACCATCTTCAGCAGTAAACCGCTCTTTCGAGTCAGAGTCCTCCTTCGCCTCGTCAGGTTTACAAAAAC
TATGCTAATTTGCATACTGACCAAGAACGTGATTACTTCATGCAGCGTTACCATGA-GTTATTTCTTCATTTGGAGGTAAAAACCTCATATGACGCTGACA
ACTTGTGCTGGTCTTTTGACCGGATTGCTGCAAACCAGTCAAGGTAGTTGTAGTATCGGTCTACGGGTCTCTAATCTCGCGTACTGTTCATTTCCTGCCA
ACAGACCTATAAACATTCTGTGCCGCGTTTCTTTGTTCCTGAGCATGGCACTATGTTTACTCTTGCGCTGGTTCGTTTTCCGCCTACTGCGACTAAAGAG
GGAAGTATCTTTAAAGTGCGCCGCCGTTCAACGGTATGTTTTGTCCCAGCGGTCGTTATAGCCATATTCAGTTTCGTGGAAATCGCAATTCCATGACTTA
ATGTTTTCCGTTCTGGTGATTCGTCTAAGAAGTTTAAGATTGCTGAGGGTCAGTGGTATCGTTATGCGCCTTCGTATGTTTCTCCTGCTTATCACCTTCT
TTGACTTGCTGACTTTGTGACCAGTATTAGTACCAACGCTTATTCATGCGCAAGAACGTTTAGTGGTCTTCCGCCAAGGACTTACTTACCCTTCGGAAGT
GTTGCAGTGGATAGTCTTACCTCATGTGACGTTTATCGCAATCTGCCGACCACTCGCGATTCAATCATGACTTCGTGATAAAAGATTGAGTGTGAGGTTA
TTTCAGACTTTGTACTAATTTGAGGATTCGTCTTTTGGATGGCGCGAAGCGAACCAGTTGGGGAGTCGCCGTTTTTAATTTTAAAAATGGCGAAGCCAAT
TATTTCTCGCCACAATTCAAACTTTTTTTTCTGATAAGCTGGTTCTCACTTCTGTTACTCCAGCTTCTTCGGCACCTGTTTTACAGACACCTAAAGCTACA
GGACTAATCGCCGCAACTGTCTACATAGGTAGACTTACGTTACTTCTTTTGGTGGTAATGGTCGTAATTGGCAGTTTGATAGTTTTATATTGCAACTGCT
TTGTTTCAGTTGGTGCTGATATTGCTTTTGATGCCGACCCTAAATTTTTTGCCTGTTTGGTTCGCTTTGAGTCTTCTTCGGTTCCGACTACCCTCCCGAC
TAACGGCCCGCATGCCCCTTCCTGCAGTTATCAGTGTGTCAGGAACTGCCATATTATTGGTGGTAGTACCGCTGGTAGGTTTCCTATTTGTAGTATCCGT
AACGTCTACGTTGGTTTCATGGTTTGGTCTAACTTTACCGCTACTAAATGCCGCGGATTGGTTTCGCTGAATCAGGTTATTAAAGAGATTATTTGTCTCC
AACTGGCGGAGGTTTGTTAAATCTGTACCGCGGTGGTCGTTCTCGTCTTCGTTATGGCGGTCGTTATCGTGGTTTGTATTTAGTGGAGTGAATTCACCGA
AAAGCCGCCTCCGGTGGCATTCAAGGTGATGTGCTTGCTACCGATAACAATACTGTAGGCATGGGTGATGCTGGTATTAAATCTGCCATTCAAGGCTCTA
CTTCACGGTCGGACGTTGCATGGAAGTTCTTCAGGAAATGGTCGAAATCGTTATCGTGTGCTTTGTTTTGATCCCCGCCGGAGTAGTCCCAATCCTTGTA
TGCCGTTTCTGATAAGTTGCTTGATTTGGTTGGACTTGGTGGCAAGTCTGCCGCTGATAAAGGAAAGGATACTCGTGATTATCTTGCTGCTGCATTTCCT
GGTCAACGTAAAATCATTCGAGAAAAACTAAGAGTTTAGGCCGCAGTTGGTATGTGCGTCTCCTTCGTAGTCGTGGTCGTGCGAGGGTTCGTAATTCGAG
ACAATCAGAAAGAGATTGCCGAGATGCAAAATGAGACTCAAAAAGAGATTGCTGGCATTCAGTCGGCGACTTCACGCCAGAATACGAAGACCAGGTATA
CCTTTGGACGACAACGAACCTTTCTAACCACAAAAGGTATTATCTGCGTTGCGTCGTCATCTGAGGAAGACA-CTTATTCGTTCGTAGAGTAAAACACGT
GAGATTATGCGCCAAATGCTTACTCAAGCTCAAACGGCTGGTCAGTATTTTACCAATGACCAAATCAAAGAAATGACTCGCAAGGTTAGTGCTGAGGTTG
GGTCTTCGTCGTAGTCACTGCTGTAATCTTTATAGGAAACGTCATCGCGCGTTATACTCTTCTCGGTATGGCGACTAAGACGCAAACGACTACTTGATTCA
TGTGGTTATATTTTTCATGGTATTGATAAAGCTGTTGCCGATACTTGGAACAATTTCTGGAAAGACGCGGTAAAGCTGATGGTATTGGCTCTAATTTGTCT
TAAGTCTTCCCATTATTCTTGCTTGGTATTTTTTCGGAGGTTCTAAACCTCCGTACTTTTGTATGTTAACCCTCCCACAGTTAGGACTGCCAATAAAGGA
GTCACGCTGATTATTTTGACTTTGAGCGTATCGAGGCTCTTAAACCTGCTATTGAGGCTTGTGGCATTTCTACTCTTTCTCAATCCCCAATGCTTGGCTT
CGGCAGTTGTATGTATAGTGGTAATAGCTTGAGTTGCGGGACGTATGCTTTTCTGTCTTAGAGAAGGTTCTCGAACTACGCCAATAGGTAGACGAATACC
CATAAGGCTGCTTCTGACGTTCGTGATGAGTTTGTATCTGTTACTGAGAAGTTAATGGATGAATTGGCACAATGCTACAATGTGCTCCCCCAACTTGATA
CCCGCAAGTCGTCGGTCGAACGTTTTGACGCATTGGCAGAAGAGCAAGAGATTTTTGGTAAAAAGCAGGGGAAGCCCCGCCACCAGATATCACAATAATT
TCTTAAGGATATTCGCGATGAGTATAATTACCCCAAAAAGAAAGGTATTAAGGATGAGTGTTCAAGATTGCTGGAGGCCTCCACTAAGATATCGCGTAGA
GATTAGCCAGCAGTCGGTTGCACTCTCACAGTTTTTGCTATTTGGTTGGTAGTCGTACTCGGACAGCGTAACGTAAGTAGTTTGCGACTTATCGTTTCGG
AGGCGTTTTATGATAATCCCAATGCTTTGCGTGACTATTTTCGTGATATTGGTCGTATGGTTCTTGCTGCCGAGGGTCGCAAGGCTAATGATTCACACGC
CGATGGACATCCTTCACAGGCGTATTTCACGTGGCGTACCTTTACTTCTGCCGGTAATCGACATGGTATGAGTCCGTGTGTTTTATGACTATCGTCAGC
GTTGACCCTAATTTTGGTCGTCGGGTACGCAATCGCCGCCAGTTAAATAGCTTGCAAATACGTGGCCTTATGGTTACAGTATGCCCATCGCAGTTCGCT
TGTATCTTTGGTTGTCGGTATATTGACCATCGAAATTCGCCGAGTGGAAATCGTAGTTGTCCGGTGTTGGTTGGTCTTGCACTTTTTCGCAGGACGCACA
GGCTAAATACGTTAACAAAAGTCAGATATGGACCTTGCTGCTAAAGGTCTAGGAGCTAAAGAATGGAACAACTCACTAAAAACCAAGCTGTCGCTACTT
CGAACCATTCAACCTAATTCGTGAGGCACCTGTCTAAACAGTAACACTCGTAAAAGTAGGGCTTCAACGCCGAGTAAGACTAAGACTTGTCGAAGAACCC
TGGGTTACGACGCGACGCCGTTCAACCAGATATTGAAGCAGAACGCAAAAAGAGAGGATGAGATTGAGGCTGGGAAAAGTTACTGTAGCCGACGTTTTGGC
ACGTCCAACCTATGCCGGTTAGTAAAAATAGCTTCGCGCGTATTTAAACTCGTCTAAACAGCAGTGTCCAACGCGCG
```

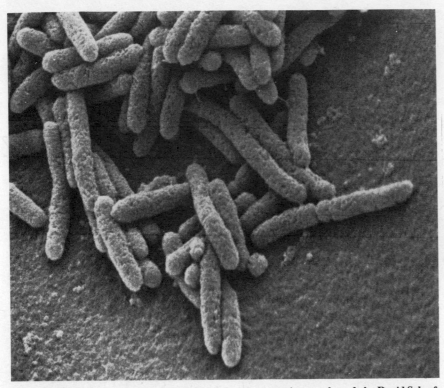

PILE OF E. COLI CELLS appears in this scanning electron micrograph made by David Scharf. Some of the cells have been caught in the act of asexual reproduction (cell division); a few appear to be transferring their DNA by means of the threadlike connection characteristic of the process known as conjugation. *E. coli* bacteria are considered by most investigators to be most suitable host cells for recombinant-DNA experiments. Magnification is 11,000 diameters.

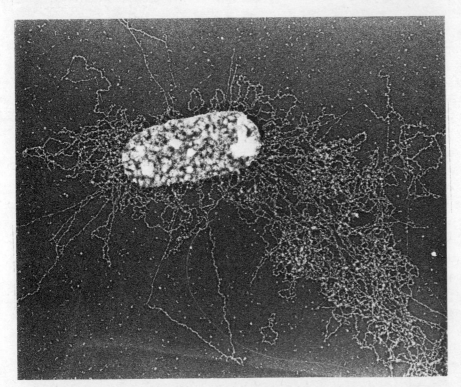

E. COLI SPEWS OUT DNA through its chemically disrupted cell wall in this electron micrograph by Griffith. Most of the DNA is in the form of a single large molecule of double-strand DNA, which constitutes the chromosome of this simple prokaryotic organism. In addition the *E. coli* bacterium may have one or more of the independently replicating loops of DNA known as plasmids; one of these smaller extrachromosomal DNA molecules can be seen near the bottom. Plasmids derived from *E. coli* cells play an important role in recombinant-DNA research, since they form one class of vectors, or carriers, into which segments of "foreign" DNA can be spliced prior to their being reinserted into an appropriate host cell to propagate, thereby duplicating not only their own native nucleotide sequence but also the foreign sequence.

of DNA at particular sequences, say at the sequence *CATTAC,* which is opposite the complementary sequence *GTAATG.* The break does not always occur at the same point on the two strands. It may, for example, be between the two *T*'s in the first strand but just to the right of the *ATG* in the second strand. On separation one piece therefore ends in *TAC,* whereas the other ends in *ATG.* Since the single-strand ends are complementary, they will under suitable conditions stick side by side, and they can then be coupled together end to end. If the same restriction enzyme is used on the DNA from two different sources, both of which have the appropriate target sequence, then sequences with the same "sticky" ends will result. By taking advantage of this stickiness two sequences from any source can be recombined into a single DNA molecule.

The only further step necessary is to put the recombinant DNA into a suitable host organism. The recombinant must have the ability to penetrate the host and become part of its genetic system. An effective way to accomplish this has been developed for the common intestinal bacterium *Escherichia coli.* In addition to its single large circular chromosome the *E. coli* bacterium may have one or more independently replicating, smaller loops of DNA known as plasmids. The plasmids can be isolated from the bacteria, broken open by restriction enzymes and used as one component of a recombinant. After linking up the plasmid DNA with the "foreign" DNA the circular form of the plasmid can be restored and the structure returned to a whole cell. There it can resume replication, duplicating not only its own native sequence but also the foreign one. A strain of bacteria is thus obtained that will yield an indefinite number of copies of the inserted nucleotide sequence from the foreign source.

Standing alone, none of this appears to be particularly momentous or threatening; it is only a new and intriguing kind of chemistry applied to living organisms. Given the complexity of living organisms and the still more complex world of social phenomena, however, this new chemistry quickly builds into varied new potentials, both speculative and real. Suppose, for example, one were to isolate the nucleotide sequence necessary to produce a potent toxin and to transfer it to *E. coli,* usually a harmless inhabitant of every human intestinal tract. Would a dangerous new pathogen be created? Would the transformed *E. coli* release a toxin in the human gut? Might such a new pathogen escape from control and induce epidemics? Questions of this kind have answers, but they take time to find. To gain some time for reflection investigators in 1974 called for a partial and temporary moratorium on those experiments thought to be potentially the riskiest. The separation of the certainly safe experiments from the less certainly safe ones became the chief function of the guidelines released by the NIH in June, 1976. The guidelines, which replaced the temporary moratorium, were derived from worst-case analyses of various kinds of experiments; the object was to evaluate the possible range of hazards and to prescribe appropriate matching safeguards in order to minimize the unknown risks. The guidelines assigned heavy responsibility to individual investigators, and they buttressed this responsibility with special monitoring committees in the sponsoring institutions and in the funding agency.

If such regulations have been adopted, why is debate continuing? Briefly, it is because the matching of estimated risk and prescribed containment adopted by the guidelines is regarded by critics as being inadequate in dealing with potential biohazards and incomplete in failing to address other important issues. The most vocal critics have presented their own worst-case analyses in the scientific and general press. These accounts have led to widespread alarm and to public-policy deliberations at the level of local communities, states and the Federal Government. The expressed concerns of the critics have generated a revised agenda for what is now emerging as a broadened second round of policymaking.

Potential biohazards and estimated degrees of risk continue to dominate the debate. The NIH guidelines balance the estimated risk of a given experiment and recommend specific measures for containing the risks. (Risk, it must be remembered, means possible danger, not demonstrated danger.) Those experiments judged to present an excessive risk are entirely proscribed. At the other end of the spectrum experiments judged to present an insignificant risk require only the safeguards of good laboratory practice. Between these extremes the guidelines establish various levels of estimated risk and prescribe combinations of suitably increasing physical and biological containment. The release into the environment of any recombinant organisms is forbidden.

Unfortunately, given the growing but still limited state of knowledge, wide disagreement is possible, both as to estimated degrees of risk and as to the efficacy of the proposed containment. Some critics project fragmentary information into the inevitable spread of dangerous, newly created organisms, threatening both the public health and the environment. Some defenders project the same fragmentary information to the conclusion that the NIH guidelines are already overly cautious. They believe the actual hazard under existing precautions will turn out to be no greater than that routinely faced in the use

of automobiles, jet aircraft and other accepted technologies. The wide range of estimates is possible because of the multiplicity of conceivable experiments and because experience and critical data are inadequate for certainty on many points. One fact that is certain is that no known untoward event has yet resulted from recombinant-DNA research.

What emerges on the new policy agenda, then, is the need for effective policy-oriented research to reduce the current uncertainty as to the risk of particular kinds of experiments. For example, there is dispute over the use of *E. coli* as a host for recombinant DNA. One side argues that scientists must be mad to pick a normal human inhabitant (and a sometime human pathogen) to serve as a host for recombinant DNA. This view, in extreme form, demands the suspension of all recombinant-DNA research until an organism safer than *E. coli* can be found. The other side argues (1) that the vast amount of information available on *E. coli* makes it invaluable, (2) that the *K*-12 strain of *E. coli* actually used in laboratory research has been so modified genetically in adapting to laboratory conditions that it survives only with difficulty in the human intestine and (3) that new strains of *K*-12 have been developed with additional genetic deficiencies that will make survival outside of laboratory conditions essentially impossible. The use of such genetically deficient strains is what is meant by the term "biological containment." The

concept is supported by proponents of the research as an efficacious new approach to safety and derided by critics as likely to be circumvented by natural recombination.

Such differences of opinion are normally reduced by scientists to experimental questions. For example, the suitability of the *K*-12 strain of *E. coli* as an experimental organism can be judged only from the effect of recombinant genes on the ecological relations of *E. coli* within the human intestine, including the degree of success of recombinant strains in competing with other strains of *E. coli* and with other organisms. Information on these matters is growing. Such questions, however, are not normally subjects of profound scientific interest. They have recently become matters of priority only because they may provide information that would be useful in arriving at a policy decision. Research on policy-oriented questions has never had a very high status among scientists engaged in basic research or even among those engaged in applied research. Therefore policy-oriented research must be encouraged through special funding mechanisms and through suitable new institutional arrangements. A regulatory agency for recombinant-DNA research and other conceivably hazardous kinds of research is urgently needed outside the NIH, and it should include a research component. The Center for Disease Control and its National Institute for Occupational Safety and Health come

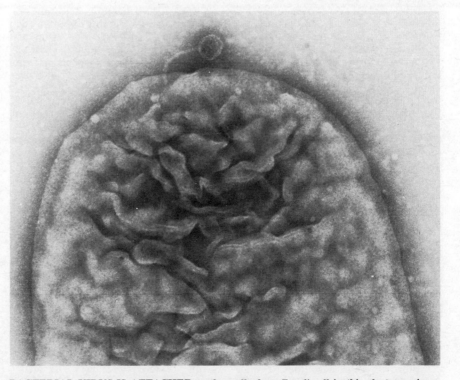

BACTERIAL VIRUS IS ATTACHED to the wall of an *E. coli* cell in this electron micrograph made by Maria Schnoss of the Stanford School of Medicine. This particular virus, named bacteriophage lambda, normally infects the bacterium by injecting its DNA into the host cell through a long taillike appendage. The magnification is approximately 140,000 diameters.

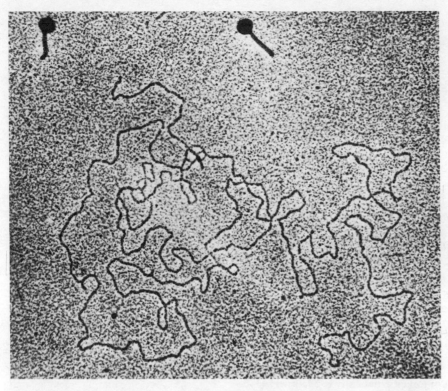

BACTERIOPHAGE LAMBDA AND ITS DNA are both represented in this electron micrograph provided by Griffith. Two complete lambda viruses are at the top; the long doublestrand DNA molecule of a disrupted lambda is below them. DNA from bacteriophage lambda can also serve as a vector for recombinant-DNA experiments involving *E. coli* host cells.

to mind as possible models for such a dual-purpose agency.

Also related to the question of biohazards is a controversy over the desirability of centralizing recombinant-DNA research facilities. Some of those who fear severe dangers from recombinant organisms have urged that the potentially more hazardous research be concentrated in remote places with extremely stringent containment procedures. Those who minimize the hazard are opposed to the concentration concept because it would tend to separate the research from the intellectual mainstream and would be unnecessarily expensive in facilities. The argument has been particularly strenuous with respect to experiments requiring *P*3 facilities, which are defined as those necessary to contain "moderate risk" experiments. *P*4 facilities for "high risk" experimentation are expected to be fewer in number because of their high cost; generally speaking they are likely also to be comparatively isolated. The current NIH guidelines provide little direction in these matters. A decision on a firmer policy belongs on the discussion agenda. Particularly urgent is careful consideration of such intermediate possibilities as the use of centralized, high-risk facilities for making particular recombinations for the first time. These activities, together with preliminary testing of new recombinants for possible hazards, might also be

carried out by the proposed new regulatory agency.

A special case that emphasizes the advantages of initial testing in a central facility is provided by what are called "shotgun" experiments. These experiments, which offer special advantages to the investigator, may also present special hazards. Shotgun experiments involve exposing the total DNA of a given organism to restriction enzymes in order to obtain many DNA fragments. The fragments are then each recombined with DNA from a suitable vector and the recombinants are randomly reinserted into *E. coli* host cells. The next step is to spread the *E. coli* cells on a nutrient substrate so that each recipient cell, containing a particular inserted foreign sequence, grows into a colony. If the experiment is successful, the yield is a "library" of all the nucleotide sequences of a particular organism, each sequence growing in a separate strain and accessible to manipulation and cross-combination at will.

This experimental approach is laborious but far less so than anything else available for the exploration of the complex genetic systems of higher organisms. There is, however, a risk of unknown magnitude that portions of the DNA with unknown or repressed functions might duplicate and create unanticipated hazards. The result might be

particularly unfortunate if the original DNA preparation were to contain genetic material from parasites or from viruses associated with the species under study. Under the NIH guidelines, therefore, shotgun experiments are regarded as being more dangerous than those involving purified and characterized DNA. Experiments in this category are treated as being increasingly more dangerous as the test organism under study is biologically more like the human organism. Thus experiments with primate DNA are considered to be more dangerous than experiments with mouse DNA. This approach appears to represent a reasonable precaution with respect to human health hazards, but it is less reasonable with respect to potential ecological effects. For example, shotgun recombinants involving DNA from plant sources could conceivably lead to ecologically dangerous effects if they were to escape into the environment. Shotgun procedures might therefore be best conducted first in special centralized facilities that could also act as storage and distribution centers for the recombinant products once they had been tested for safety.

These examples suggest several advantages for the creation of a Center for Genetic Resources. The center might not only carry out DNA recombinations suspected to be hazardous but also function to preserve genetic information contained in threatened natural species and in special strains of cells or organisms developed for research and other purposes. Stored genetic information can be expected to be increasingly important in the future. For example, new genetic infusions into domesticated stocks of plants and animals from their wild progenitors have long been used to strengthen the response of the domesticated stocks to changing conditions of husbandry. The sources of wild progenitors are threatened by the reduction of wild habitats all over the world.

The possibility of a biohazard need not arise only as a by-product of basic research. The practical applications of recombinant-DNA techniques, together with the applied research and development leading to them, are at least equally likely sources. For example, recombinant techniques may enormously expand the use of bacteria (and other microorganisms) for the production of certain proteins and other pharmacological products. Microorganisms have long played an essential role in the food, beverage, pharmaceutical and chemical industries, and more precise genetic control of their characteristics has already yielded large benefits. The recombinant-DNA techniques not only offer advances on current practice but also suggest a new realm of "bacterifacture" in which the rapid, controlled growth of microorganisms is coupled to the production of specific products normally made only by higher organisms. Included among the possibilities are the production of insulin, blood-clotting factors and immunological agents. The probability of those possibilities ever being realized is no more easily assessed than the risks, but success in realizing them clearly could provide substantial economic and social benefits. Accordingly entrepreneurial interests have been aroused.

The NIH guidelines are silent on the matter of commercial applications other than stipulating that large-scale experiments (beyond production batches of 10 liters) with recombinants "known to make harmful products" be prohibited unless specially sanctioned. The guidelines also require detailed reporting of proposed recombinant-DNA experiments, a provision that runs counter to the protection of proprietary interest. There have been discussions of these matters between the NIH and representatives of industry. In addition industry spokesmen have testified at Congressional hearings. It is known that some industrial research already is under way and that representatives of industry generally endorse the precautionary approach of the NIH guidelines, but they are resistant to limitations on proprietary rights and on the size of batch production. Moreover, patent policy has come up as an issue and there has been some uncertainty in the Department of Commerce as to how it should be handled. Indeed, the possible commercial applications of recombinant-DNA techniques have yet to be publicly evaluated as a serious policy question, and they must be high on the agenda of the next round of discussions.

The problems of commercial applications lead from immediate issues to broader ones and to a larger time frame. Recombinant-DNA techniques have revived the debate over "genetic engineering" and have once again raised questions about the applications of fundamental biomedical research to technology, to the quality of life and to the future of society. Recombinant DNA has now joined nuclear fission, overpopulation, famine and resource shortages in the

doomsday scenarios of "creative pessimism." These issues are even more difficult to deal with objectively than those related to potential biohazards, but they are plainly apparent in the general public discussion and in the public statements of respected scientists.

For example, Robert L. Sinsheimer of the California Institute of Technology has persistently raised issues that are in part practical and in part philosophical. Along with George Wald of Harvard University and Erwin H. Chargaff of the Columbia University College of Physicians and Surgeons, he suggests that the entire recombinant-DNA approach to gaining an understanding of the complexities of higher genetic systems is misbegotten. The argument is not that the approach may not work but that its alleged huge risks are unnecessary because less risky, although slower, means are available. Sinsheimer emphasizes the fundamental difference between simple prokaryotic organisms such as bacteria and complex eukaryotic organisms, including human beings. Prokaryotes, typically one-cell organisms, have a single, comparatively simple chromosome floating freely within the cell body, whereas eukaryotic cells have a nucleus that is bounded by a membrane and contains a number of far more complex chromosomes. The paleontological record suggests that prokaryotes existed on the earth for a billion or more years before the more complex eukaryotes arrived on the scene. Sinsheimer proposes that throughout the evolution of the eukaryotes there has been a genetic barrier between them and the prokaryotes, behind which eukaryotes have developed their more complex mechanisms of genetic control. To transfer these mechanisms, which are possibly the key to the evolutionary success and enormous diversity of eukaryotes, to prokaryotes may introduce, he says, incalculable evolutionary dangers. The prokaryotes may be made far more effective, both as competitors and as parasites, negating an ancient evolutionary strategy.

Sinsheimer's argument has won only a few vocal adherents among biologists, and he himself concedes that it is speculative. Nevertheless, his argument

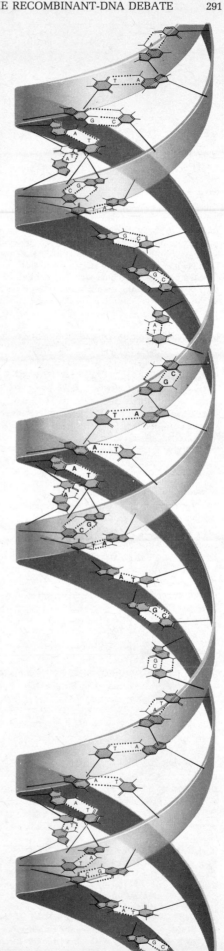

DOUBLE-HELIX STRUCTURE OF DNA is evident in this simplified diagram of a short segment of the deoxyribonucleic acid (DNA) molecule. The sugar and phosphate groups that are linked end to end to form the outer structural "backbones" of the double-strand molecule are represented schematically here by the two helical colored bands. The inner portion of each polynucleotide chain, drawn in somewhat greater detail, consists of a variable sequence of four kinds of bases: two purines (adenine and guanine, or *A* and *G*) and two pyrimidines (thymine and cytosine, or *T* and *C*). The two chains, which run in opposite directions, are held together by hydrogen bonds (*dotted black lines*) between pairs of bases. Adenine is always paired with thymine, and guanine is always paired with cytosine. The planes of the bases are perpendicular to the common axis of the two helixes. The diameter of the double helix is 20 angstroms. Adjacent bases are separated by 3.4 angstroms along the axis and are displaced successively around the axis by an angle of 36 degrees. The structure therefore repeats after 10 bases on each chain (360 degrees), or at intervals of 34 angstroms. The genetic information is stored in the sequence of bases along each chain. In this case the sequence *CATTACTAG* on one strand is identified in boldface type opposite complementary sequence *GTAATGATC* on other strand.

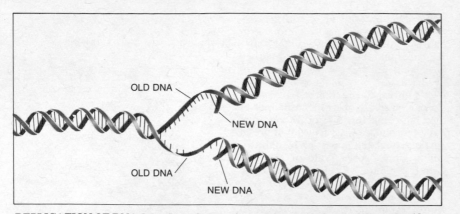

REPLICATION OF DNA depends on the complementary relation between the nucleotide sequences on the two strands of the DNA molecule. Under appropriate chemical conditions the hydrogen bonds between the bases are weakened and the two strands can unwind and separate. In the presence of suitable enzymes and free nucleotides a new chain can be formed next to the exposed portion of each unpaired older chain. The complementary sequence that is formed by each nucleotide lining up next to its opposite is then linked end to end by an enzyme that "zips up" the nucleotide couplings. In this way two new helixes identical with the first can be formed.

has attracted significant public attention, and it is widely cited to support opposition to continued recombinant-DNA research. Bernard D. Davis, a Harvard Medical School microbiologist, has provided a rebuttal, particularly with respect to the concept of a genetic barrier between prokaryotes and eukaryotes. He believes there has been an ample and continuous opportunity for the exchange of DNA between the two groups. He points out that bacteria can take up naked DNA from their immediate environment and that *E. coli* would be exposed to such DNA arising from dead human cells in the human intestine. Microorganisms might similarly take up DNA in the process of decomposing dead animals. Therefore, Davis argues, most recombinants probably have already been tried in the natural evolutionary arena and have been found wanting. Reasoning on analogy with extensive information on pathogenic bacteria, Davis concludes that under the existing NIH guidelines the probability for survival in nature of laboratory-produced prokaryote-eukaryote recombinants is vanishingly small.

This clash of opinion on a major biological issue illustrates the difficulty of assessment of even comparatively value-free questions when critical information is fragmentary. The controversy over the risk-benefit ratio becomes even more intense when issues involve substantial value judgments as well. Here again a concern of skeptics and opponents of recombinant techniques is sharply articulated by Sinsheimer. He asks: "Do we want to assume the basic responsibility for life on this planet? To develop new living forms for our own purposes? Shall we take into our own hands our own future evolution?" Since the questions include such concepts as responsibility, purpose and control of the future, they clearly involve considerations beyond science alone.

The human species has, of course, been altering life on this planet from the beginnings of human culture. When hunting and gathering gave rise to animal husbandry and agriculture, human choice and purpose began to influence the evolution of selected species. Unconscious human selection was replaced by deliberate plant and animal breeding, and the further development of human culture is now clearly altering the entire ecosystem. Moreover, the biocultural progression of the human species, based partly on human purpose, is undoubtedly altering the human gene pool and will slowly modify the species in unpredictable ways. Nevertheless, the advent of recombinant-DNA techniques has obviously enhanced the prospects for genetic engineering and has restressed the need to assess its implications. Can it be assumed that success in introducing recombinant DNA's into *E. coli* means that there will be similar success in introducing them into the human species? If it can, what is the probable time frame for applying the technique to the human species? Is it accurate and responsible to suggest that we have almost in hand control of "our own future evolution"? These certainly are questions for scientific assessment, and they should have a prominent place on the new policy agenda.

Sinsheimer has gone into still another controversial area, not only for the scientific community but also for the entire

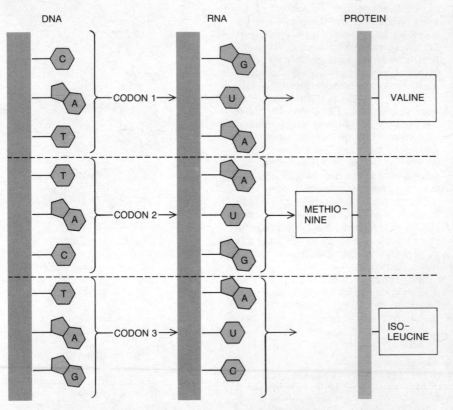

ROLE OF DNA IN PROTEIN SYNTHESIS is suggested by this highly schematic diagram. The genetic message contained in the nucleotide sequence *CATTACTAG*, for example, is "written" in the form of the triplet "codons" *CAT, TAC* and *TAG*. Each codon determines, through a series of intermediate steps involving a molecule of ribonucleic acid (RNA), the position of a specific amino acid in a protein molecule. Thus the sequence of nucleotides in a given DNA molecule specifies the corresponding sequence of amino acids in a particular protein, with each triple-nucleotide codon placing one of 20 possible amino acids at each successive position in the protein chain. Since the sequence of amino acids in turn establishes both the structure and the function of the protein, the nucleotide sequence of DNA determines virtually every property of organism. Letter *U* stands for the pyrimidine uracil, a constituent of RNA.

society. Arguing that time may be needed to "pace" new genetic knowledge to
human capacities for putting nature to
intelligent use, he wonders whether
"there are certain matters best left unknown, at least for a time." This is high
heresy in the scientific community,
whose fundamental premise is that the
growth of knowledge is the driver and
not the captive of other values. The rejection of the concept of "forbidden
knowledge" was part of the heroic period at the beginning of modern science,
when it included willingness to face the
Inquisition and the stake. Having been
seared by the nuclear flame and now
confronting the more subtle implications of the innermost language of life,
20th-century scientists fear not the stake
but the judgment of history. Chargaff, a
pioneer in the investigations that led to
the decipherment of the genetic language, says: "My generation, or perhaps
the one preceding mine, has been the
first to engage, under the leadership of
the exact sciences, in a destructive colo-
nial warfare against nature. The future
will curse us for it."

Sinsheimer and Chargaff, along with a
number of philosophers, historians and
sociologists of science, are clearly suggesting that the possible consequences
of knowing must be consciously included in decisions about the directions of
the search for knowledge itself. No issue
cuts more deeply to the core of modern
science. The self-doubt expressed by
some scientists reflects a general ques-
tioning in the U.S. of the net benefits of
science and technology. Cost-benefit
analysis is a current preoccupation,
and it is being increasingly applied to the
generation of knowledge itself. It is hard
enough to assess what we may gain or
lose from particular new knowledge; it
is even harder to assess the costs of not
having it. This problem is epitomized
by the recombinant-DNA controversy.
The rise of molecular genetics in the
U.S. is the direct product of a series of
decisions made after World War II that
provided funds for biomedical research.
The objective was the conquest of the
"killer" diseases: cancer, heart disease
and stroke. Those diseases are still much
with us, although they are better understood and cared for. Meanwhile, out of
Federally supported research also came
the impetus that led to the discovery of
the double helix, the genetic code, the
structure of proteins and recombinant
DNA. In a classic "double take" the
public is now asking whether it has been
buying health and well-being or chimer-
ic monsters. Is molecular genetics and
all biomedical technology a sorcerer's
apprentice? Are we increasing rather
than lessening our burden of pain and
anxiety?

The last question leads to yet another
issue. Biohazard and ecohazard may
arise inadvertently, but "sociohazard"
may be the product of deliberate malev-

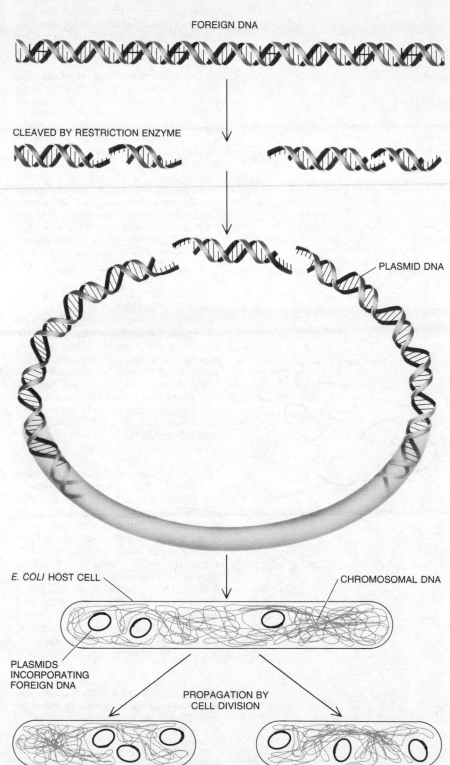

RECOMBINANT-DNA TECHNIQUE makes it possible for the first time to deliberately introduce nucleotide sequences from the DNA of one strain or species of organism into the
DNA of another. The DNA of the "foreign" organism is first treated with restriction enzymes,
which cleave the double-strand molecule at particular nucleotide sequences (typically thousands of base pairs apart) on a random basis. The same enzyme is then used to cleave the DNA
of a suitable vector, in this case a plasmid isolated from *E. coli* bacteria. Since the break caused
by the enzyme does not occur at the same point on both strands, the chemical treatment results
in a mixture of DNA segments that have complementary single-strand ends. Under suitable
conditions the "sticky" ends of two different sequences can be coupled to form a single DNA
molecule. For example, after recombining the foreign DNA with the plasmid DNA the circular form of the plasmid can be restored and the structure can be inserted into a suitable host cell
(in this case *E. coli*), where the plasmid can resume replication, thereby propagating an indefinite number of "cloned" copies of the inserted nucleotide sequence from the foreign source.

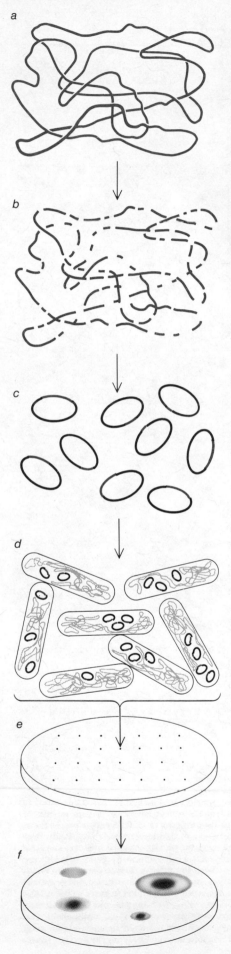

a

b

c

d

e

f

"SHOTGUN" EXPERIMENT is a type of recombinant-DNA experiment in which the total DNA of an organism (a) is exposed to restriction enzymes in order to yield many fragments (b), which are then recombined with the DNA from a suitable vector (c) and randomly reinserted with the vector into the host cells (d). The E. coli hosts are next spread on a nutrient substrate (e) so that each recipient cell, containing a particular inserted foreign nucleotide sequence, can grow into a colony (f). The result, if the experiment is successful, is a "library" of all the nucleotide sequences of the organism. Under the guidelines issued by the National Institutes of Health last year shotgun experiments are regarded as being potentially more hazardous than those involving purified and characterized DNA, since it is not known whether portions of the DNA with unknown or repressed functions might cause unexpected problems.

olence. The U.S. is a signatory to an international legal convention that has renounced biological warfare, including research to produce the necessary agents. Not all countries have taken this step, and public renunciation without adequate inspection cannot ensure that covert activities do not exist. Opponents of recombinant-DNA research see its techniques as being ideally suited to serve malevolent purposes, either as agents of organized warfare or of sabotage and terrorism. The techniques do not require large installations or highly sophisticated instrumentation. Contrary views have not denied this but have noted that recombinant-DNA techniques would not be the first technology to have potential malevolent applications. Explosives have such applications, but society does not completely ban them; it takes prudent precautions against their misuse.

Nevertheless, the issue of the possible misuse of recombinant-DNA technology deserves a place on the policy agenda, because it emphasizes the need for international discussion of the implications and management of recombinant-DNA research and recombinant-DNA applications. It can be argued that the U.S. is not ready for such discussion until its own policies are in better order. It is not too early, however, to begin the internal consideration of how best to approach the international arena.

These are the chief issues that have emerged from the policy debate so far. It is a not inconsiderable list. The debate has not been raging on every street corner, but it became strenuous enough in Cambridge last summer to have repercussions across the continent. For example, an evaluation presented by a panel of nonscientists to the Cambridge City Council was not too different in content from one produced by a task force of the Quality of Life Board of the City of San Diego, where I live. Both groups accepted within their community the continuance of recombi-

nant-DNA research requiring P3 facilities but sought somewhat greater assurances of safety than those provided by the NIH guidelines. Meanwhile the Attorney General of the State of New York held a public hearing, and a joint hearing was conducted by two committees of the Assembly of the State of California Legislature. Legislation regulating DNA research was later introduced in the California Assembly, and it is still under consideration. Congress has also held several hearings and various items of regulatory legislation have been introduced in both the Senate and the House of Representatives. These local, state and Federal initiatives emphasize the necessity to get on with the policy agenda.

The agenda should be viewed in at least two time frames: immediate and longer range. A consensus has been growing that there is an immediate need to give the quasi regulation represented by the NIH guidelines a statutory base. In particular, regulation must be extended to activities not supported by Federal agencies, especially in the industrial sector. However this is to be done, it is important to maintain flexibility, since the problems to be dealt with will change as greater knowledge and experience are acquired.

Moreover, given the complexity of the longer-term issues, immediate legislation probably should be provisional and limited. A mechanism should be included, however, that actively leads toward a more definitive future policy. This requires provision for a new, comprehensive assessment of all the issues raised by recombinant-DNA research, including the probable effectiveness of the regulatory devices put in place under the NIH guidelines.

The need for such a new national assessment is demonstrated by the nature of the critical challenge to the product of the earlier assessment. First, it has been alleged that the 1975 Asilomar conference establishing the pattern for the NIH guidelines was dominated by scientists involved in the research, and therefore it could not yield a broad enough perspective. Second, it is argued that the earlier assessment was devoted primarily to the question of potential biohazards and did not address in any depth other gravely important questions. The passage of time has added several more points: that circumstances already have changed as research has progressed, that experience has grown and that a wider range of opinion has come to bear on the issue. Whatever format is adopted for the reappraisal of the recombinant-DNA issue, the public must be assured that the process is a comprehensive and objective one.

Whoever undertakes this new national review should first carefully examine the current situation, including the actual effectiveness of the regulatory mecha-

		BIOLOGICAL CONTAINMENT (FOR *E. COLI* HOST SYSTEMS ONLY)		
		*EK*1	*EK*2	*EK*3
PHYSICAL CONTAINMENT	P1	DNA from nonpathogenic prokaryotes that naturally exchange genes with *E. coli* Plasmid or bacteriophage DNA from host cells that naturally exchange genes with *E. coli*. (If plasmid or bacteriophage genome contains harmful genes or if DNA segment is less than 99 percent pure and characterized, higher levels of containment are required.)		
	P2	DNA from embryonic or germ-line cells of cold-blooded vertebrates DNA from other cold-blooded animals and lower eukaryotes (except insects maintained in the laboratory for fewer than 10 generations) DNA from plants (except plants containing known pathogens or producing known toxins) DNA from low-risk pathogenic prokaryotes that naturally exchange genes with *E. coli* Organelle DNA from nonprimate eukaryotes. (For organelle DNA that is less than 99 percent pure higher levels of containment are required.)	DNA from nonembryonic cold-blooded vertebrates DNA from moderate-risk pathogenic prokaryotes that naturally exchange genes with *E. coli* DNA from nonpathogenic prokaryotes that do not naturally exchange genes with *E. coli* DNA from plant viruses Organelle DNA from primates. (For organelle DNA that is less than 99 percent pure higher levels of containment are required.) Plasmid or bacteriophage DNA from host cells that do not naturally exchange genes with *E. coli*. (If there is a risk that recombinant will increase pathogenicity or ecological potential of host, higher levels of containment are required.)	
	P3	DNA from nonpathogenic prokaryotes that do not naturally exchange genes with *E. coli* DNA from plant viruses Plasmid or bacteriophage DNA from host cells that do not naturally exchange genes with *E. coli*. (If there is a risk that recombinant will increase pathogenicity or ecological potential of host, higher levels of containment are required.)	DNA from embryonic primate-tissue or germ-line cells DNA from other mammalian cells DNA from birds DNA from embryonic, nonembryonic or germ-line vertebrate cells (if vertebrate produces a toxin) DNA from moderate-risk pathogenic prokaryotes that do not naturally exchange genes with *E. coli* DNA from animal viruses (if cloned DNA does not contain harmful genes)	DNA from nonembryonic primate tissue DNA from animal viruses (if cloned DNA contains harmful genes)
	P4		DNA from nonembryonic primate tissue DNA from animal viruses (if cloned DNA contains harmful genes)	

"SHOTGUN" EXPERIMENTS USING *E. COLI K*-12 OR ITS DERIV-
ATIVES AS THE HOST CELL AND PLASMIDS, BACTERIOPHAGES
OR OTHER VIRUSES AS THE CLONING VECTORS

EXPERIMENTS IN WHICH PURE, CHARACTERIZED "FOREIGN"
GENES CARRIED BY PLASMIDS, BACTERIOPHAGES OR OTHER
VIRUSES ARE CLONED IN *E. COLI K*-12 OR ITS DERIVATIVES

SOME EXAMPLES of the physical and biological containment requirements set forth in the NIH guidelines for research involving recombinant-DNA molecules, issued in June, 1976, are given in this table. The guidelines, which replaced the partial moratorium that limited such research for the preceding two years, are based on "worst case" estimates of the potential risks associated with various classes of recombinant-DNA experiments. Certain experiments are banned, such as those involving DNA from known high-risk pathogens; other experiments, such as those involving DNA from organisms that are known to exchange genes with *E. coli* in nature, require only the safeguards of good laboratory practice (physical-containment level *P*1) and the use of the standard *K*-12 laboratory strain of *E. coli* (biological-containment level *EK*1). Between these extremes the NIH guidelines prescribe appropriate combinations of increasing physical and biological containment for increasing levels of estimated risk. (In this table containment increases from upper left to lower right.)

Thus physical-containment levels *P*2, *P*3 and *P*4 correspond respectively to minimum isolation, moderate isolation and maximum isolation. Biological-containment level *EK*2 refers to the use of new "crippled" strains of *K*-12 incorporating various genetic defects designed to make the cells' survival outside of laboratory conditions essentially impossible. Level *EK*3 is reserved for an *EK*2-level host-vector system that has successfully passed additional field-testing. Because of the very limited availability of *P*4 facilities and because no bacterial host-vector system has yet been certified by the NIH as satisfying the *EK*3 criteria, the recombinant-DNA experiments now in progress in the U.S. with *E. coli* host systems are with a few exceptions limited to those in the unshaded boxes. Experiments with animal-virus host systems (currently only the polyoma and SV40 viruses) require either the *P*3 or the *P*4 level of physical containment. Experiments with plant-virus host systems have special physical-containment requirements that are analogous to the *P*1-to-*P*4 system.

nisms provided by the NIH guidelines. Particular attention needs to be paid to the local institutional biohazards committees mandated by the NIH guidelines. Beyond the responsibility assigned to the principal investigator these committees are the only source of local surveillance and standard-setting. Their composition and charge are unique, yet their authority and procedures are stipulated only generally in the NIH guidelines. They may well need the stimulus and support of external interests to carry out their important task. Moreover, no provision has been made for budgeting what may turn out to be their considerable cost for technical surveillance, personnel training and medical monitoring. Like all insurance, security against biohazard must be bought. The cost should be borne as an additional expense of the research, not as a competitor for existing funds.

Similarly, the actual performance of the NIH study sections, which are mandated by the guidelines to be independent evaluators of biohazards and containment, needs to be examined. Study sections are already heavily overloaded with the job of evaluating scientific quality. Yet these part-time peer groups are asked to assume another difficult function. If the responsibility is to be taken seriously, it too will entail additional costs.

Of special importance for early attention is an effective monitoring system for following the actual directions of recombinant-DNA research. The techniques involved are so rich in possibilities, whether for fundamental research or applications, for benefit or risk, that "early warning" is essential. Systematic following of the directions of investigators' interests, from applications for support through informal communication to formal publication, is essential to the early detection and assessment of either risks or opportunities. Needless to say, monitoring is particularly difficult in industrial research. It might therefore be desirable to limit or postpone certain development efforts pending closer study and greater knowledge of the underlying problems.

Equally urgent is a determined effort toward a more effective assessment of risks and their limitation. The specific assignment of responsibility for this kind of policy-oriented research should be an early recommendation of the body undertaking the reassessment. Given the differing perspectives required by regulation and the NIH mission to promote health-related research, the regulatory function probably belongs elsewhere in the long run. On the other hand, given the need for careful study of the implications of relying on existing agencies or of establishing a new one, the temporary continued assignment of this responsibility to the NIH may be desirable. This interim solution, if it is adopted, must be accompanied by additional funding to carry it out effectively.

Considerations of biohazards and physical and biological containment have necessarily had a high priority in this early phase of recombinant-DNA research. Many informed observers believe, however, that these concerns will decline in importance as research continues and experience grows. Therefore although the current furor makes a rational approach to the biohazards question an essential part of any successful recombinant-DNA policy, this approach does not exhaust the longer-term requirements and may even distort them. More crucial in the long run may be several other issues that have been raised directly or indirectly.

For example, in investing in fundamental genetic research that can profit from recombinant-DNA techniques, what relative priorities should be assigned to potential applications? In the past the national strategy in biomedical research has been to invest directly in basic research, without declared objectives, while also investing in specific objectives, allowing some of the latter support to "trickle down" to basic research. Thus an investigator of the interaction of viruses and cells, say, might be alternatively or simultaneously supported by funds for fundamental investigation and by funds intended for promoting the development of an effective therapy for cancer. What should be the priorities among possible practical applications of molecular genetics? Competing lines of inquiry include the microbiological synthesis of drugs, specific human gene therapies, the improved efficiency of photosynthesis, nitrogen fixation by food crops, enhanced agricultural production and so on. There are quite different potential risks and benefits in each of these directions, and all are unlikely to be maximally supported at once. In the new areas that are opening up is a new research strategy called for? If it is, by what procedures should it be formulated and how should it be implemented?

It is widely recognized that there is a logical continuum running from basic research through applied research and development to technological application. It is also recognized that movement along this continuum is neither smooth nor fully predictable and that varying motivations and institutional arrangements operate along its length. Recombinant-DNA techniques are the product of fundamental investigation, supported almost entirely by the partnership of the Federal Government and the universities. For the moment, at least, the techniques are likely to remain useful primarily in that area. The techniques may also be useful for various industrial purposes, however. Given the nature of the original investment as well as the complex issues raised, should technological uses, at least for a time, be kept under Federal control? Should some of the return from successful applications be employed to recycle the original investment of Federal resources? Should this promising new technology be a prototype for establishing a revolving capital fund to support a more stably financed basic-research effort?

The possibilities of genetic engineering and evolutionary control illustrate the fundamental dilemmas raised by the new capabilities conferred by scientific knowledge. Society has entered an age of intervention, in which the automatic operation of natural processes is increasingly, through informed intervention, brought consciously into the orbit of human purpose. Many events that humanity formerly could regard only as a boon or a scourge—an act of God or of nature—are now the partial product of human decision and intervention. If human beings do not have the capability today to invent new organisms or to initiate life itself, they may soon have that capability. If they cannot today consciously and fully control the behavior of large ecosystems, that power is not far beyond what has already been achieved. The humility of individuals understandably shrinks from awesome powers that were earlier assigned to divine will. It was not, however, the humility of individuals that conferred these emerging capabilities or is called on to control them today. It was the social interaction of individuals, operating through social institutions, that brought us to the present fateful decision making. Imperfect though they are, our social institutions built the platform for the age of intervention.

The policy challenge we face, refracted in the exquisite structure and potential of the double helix, is whether we can create institutions able to transform the fruits of an age of reason into the achievements of an age of intervention. There are voices today urging us not only to eschew conscious intervention but also to distrust and limit the uses and consequences of reason itself. Perhaps it needs to be restated that it was, after all, natural selection that evoked the double helix and all it conveys. Included among the products are human knowledge and judgment, to which has now passed the duty of designing social processes and structures that can cope with the manipulability of the double helix itself.

The concept and control of the double helix signal a new frontier of biocultural progression. A stereoscopic vision that includes both "creative pessimism" and "creative optimism" is now required. Neither alone can do justice to the profound revelations human beings have recently experienced. A single eye is particularly limited in yielding depth and perspective. For the age of intervention at least two are needed.

Genetic Amniocentesis

28

June 1980

The fluid that surrounds a fetus in the uterus can now be examined for the prenatal detection of genetic disorders, yet the procedure is being performed for only a tenth of the parents most in need of it

Amniocentesis, from the Greek *amnion,* the fetal membrane, and *kentesis,* a pricking or puncture, signifies the insertion of a needle into the membranous sac surrounding the fetus in the womb of a pregnant woman and the withdrawal of some of the fluid in which the fetus is suspended. The fluid is sometimes withdrawn late in a pregnancy to remove the excess that accumulates in the pathological condition called polyhydramnios. Currently it is sampled far more often with an entirely different intent.

In 1955 Povl Riis and I, working at the University of Copenhagen, demonstrated that the sex of a fetus can be determined by staining the nuclei of cells found floating in the amniotic fluid, because the cells are from the fetus and in cells from the tissues of a female the chromatin (the material in the nucleus that accepts the stain) includes a condensation that can be seen as a dark mass: the Barr body. We also were able to determine the *ABO* blood group of the fetus by adding cells from the amniotic fluid to solutions containing red blood cells whose blood group is known and then adding to the solution the antibody to cells of such a group. In the solution in which the two populations of cells are of the same group (both types of cells have *ABO* marker molecules arrayed on their surface) the antibody causes a clumping of cells in which both populations participate.

Such findings led us to predict as early as 1956 that many other features of the genetic makeup of the fetus could be identified by examining the amniotic fluid and the cells floating in it, and that it might therefore be possible to diagnose a number of hereditary disorders well before the birth of an afflicted infant. What we surely did not predict was the rapidity with which the prenatal diagnosis of genetic disorders has progressed since those early years.

The advantage of prenatal diagnosis is clear. Among the numerous fetal abnormalities that can now be detected many lead to debilitating disease and many cause death at an early age. Others cause mental retardation so severe that it precludes a normal life. For most of the abnormalities no treatment now exists, but if they can be detected early in gestation, the pregnancy can be ended. Even those prospective parents who do not consider the termination of a pregnancy acceptable may think it important to know the diagnosis so that they can prepare for the birth of an afflicted infant. Moreover, a large proportion of parents who know themselves to be at risk of bearing a child with a genetic disease no doubt would never undertake a pregnancy if prenatal diagnosis were not available, or would choose abortion automatically if a pregnancy occurred.

In the U.S. and a number of other countries amniocentesis and the related laboratory procedures are now available in many medical centers. Obstetricians are aware of its benefits and its risks, and public discussion has brought amniocentesis to the attention of at least part of the population. At present, however, only an estimated 10 percent of the women in this country who are substantially at risk of having children with genetic disorders that can be detected by amniocentesis actually have the procedure performed.

The amniotic fluid usually is sampled 16 to 17 weeks after the pregnant woman's last menstrual period. This corresponds to the 14th or 15th week of the pregnancy, a time that is found to be optimal for the procedure in part because the amniotic fluid has then attained a volume of 175 to 225 milliliters (six to eight ounces), from which as much as 25 milliliters can be removed without substantially collapsing the fluid space. The fluid has several sources. Some of it is secreted from the upper respiratory tract of the fetus. Some of it diffuses through the fetal skin. Some of it diffuses through the membranes of the amniotic sac. The most constant source of the fluid is the urine excreted by the fetus.

Suspended in the fluid are live and dead cells. Some slough off from the skin of the fetus. Some are in the urine and in the respiratory-tract secretions. Others slough off from the umbilical cord, from the inner surface of the amniotic sac and from the placenta, to which the umbilical cord is attached. The placenta is an organ interposed between the maternal and the fetal bloodstream. In genetic content the cells from all these sources in essence are fetal cells, because the cord, the sac and the placenta, in addition to the fetus itself, derive from the fertilized ovum.

The fluid is best removed immediately after examination of the image of the fetus in a pulse-echo ultrasonogram. The essence of this technique is that when vibrations with frequencies of from 20,000 to several million cycles per second meet the interface between tissues of different densities, they are reflected. The reflections yield an image on which can be measured the size of various structures. A notably useful measurement is that of the size of the fetus's head, which is well correlated with its age.

Sonography makes it possible to diagnose major structural abnormalities. Oscillations in the sonogram reveal the fetal heartbeat several weeks before it can be detected with an ordinary stethoscope. The technique therefore makes

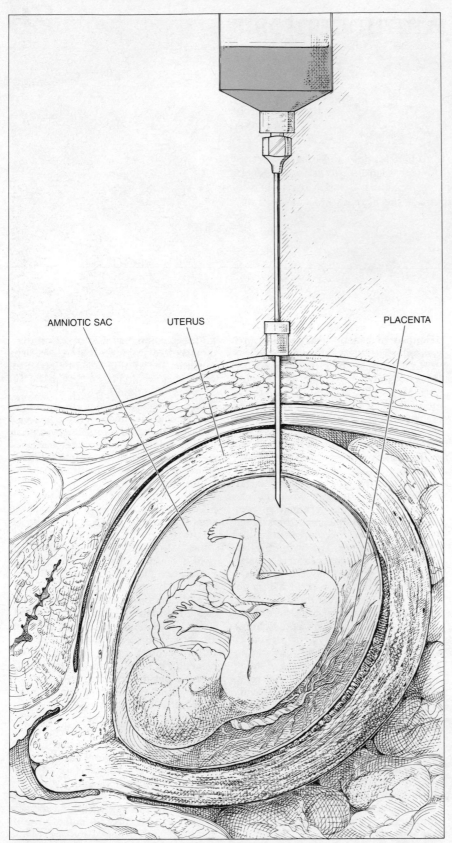

AMNIOTIC SAC   UTERUS   PLACENTA

NEEDLE IS INSERTED through a pregnant woman's abdominal wall to remove fluid from the sac in which the fetus is suspended. The procedure, called amniocentesis, provides samples of the fluid (and of the cells floating in it) for the prenatal diagnosis of genetic disease. The illustration, which is actual size, shows the fetus in the 15th week of pregnancy. It is the time at which the amniocentesis is best performed, in part because the fluid has sufficient volume so that the sample can be taken and in part because the completion of the tests of the fluid and its cells, two to four weeks after the fluid is sampled, leaves open to the prospective parents the choice of having the pregnancy ended if a severe disease is diagnosed. Many of the fetal abnormalities that can be detected by amniocentesis lead to crippling, life-shortening disorders.

possible the early diagnosis of a fetal death. With the aid of sonography a site for the puncture of amniocentesis is selected so that the needle used for the procedure avoids the head of the fetus, and if possible the placenta, and enters an accessible pool of fluid. A problem for which sonography is particularly valuable is the detection of twins and other multiple gestations. With rare exceptions each fetus in a multiple gestation has its own amniotic sac, and so it is necessary to sample the fluid in each sac.

In any case the fluid always is removed by means of a long, thin needle that is inserted through the mother's abdominal wall. Small quantities of the fluid are used for biochemical assays The remaining fluid is placed in tissue-culture flasks and incubated so that the living cells in the fluid multiply. This requires two to four weeks. If the examination of the fluid and its cells then reveals a fetal abnormality, the pregnancy can still be terminated well before the fetus is viable.

One reason for the culturing is that a large number of cells can then be caught as they divide and their genetic material is condensed into chromosomes. At other times the genetic material is a disorganized tangle that cannot be analyzed. Typically the chromosomes are photographed under a high-resolution microscope and identified according to their size. In a newer technique the chromosomes are stained with dyes that bind selectively to certain parts of individual chromosomes to produce characteristic patterns of banding. The normal human complement of chromosomes is 46. All but two of them are found in pairs. The remaining two are sex chromosomes: they bear the genes that make the child male or female. In the cells of a male one of the chromosomes (designated X) is large. The other one (Y) is small. In the cells of a female both of them are X. It turns out that one of the X chromosomes in each cell of a female condenses to form the Barr body.

The group of hereditary diseases that were the first to be diagnosed by amniocentesis consists of the sex-linked disorders, in which the disease is caused by a defective gene that is part of an X chromosome. In males the presence of such a defect results in manifestation of the disease because the defective version of the gene is the only copy in each cell. In females, however, a defective X chromosome is masked by a normal X. The disease can affect a female only if both of her X chromosomes are defective, and that is quite unlikely; it requires that the genetic inheritance from her mother include the defective gene and that her father have the disease. It is all the more unlikely in view of the severity of the diseases, most of which are crippling and greatly shorten life. Two examples are hemophilia, in which the blood lacks certain substances that make it coagulate and thereby arrest bleeding, and Duchenne's muscular dystrophy, a progressive muscle weakness that by early adolescence confines its victim to a wheelchair.

Hence a woman with a defective *X* chromosome is a carrier who will pass the sex-linked defect on to half of her children. On the average half of her male children will suffer from the disease and half of her female children will be carriers. A woman at risk of bearing a child with a sex-linked disease usually knows it, because for many generations some of her male ancestors have been afflicted by the disease. The examination of the chromosomes in cells from the amniotic fluid then reveals the sex of the fetus. The cells can be stained to demonstrate the presence or absence of the Barr body, or the cells can be cultured so that the sex chromosomes can be identified. Since the disease has a 50 percent incidence in the sex at risk (the male), the indication for amniocentesis is certainly quite strong. (It should be added that recent efforts to find a correlation between the sex-linked disorders and some feature of the fetus other than the sex may ultimately make possible the prenatal diagnosis of the sex-linked disorders themselves.)

A second group of disorders that can be detected by amniocentesis consists of disorders caused by gross chromosomal abnormalities, usually an abnormal number of chromosomes but sometimes a major structural defect in a single chromosome. The commonest of these disorders is Down's syndrome, in which the cells of the fetus each have three rather than the usual two of the chromosome designated No. 21. The chromosomal abnormality is called trisomy 21. The consequences are mental retardation and internal malformations that often include congenital heart disease and lead to early death. Typically the infant's head is small and flattened and has mongoloid features such as slanted eyes. Certain other chromosomal abnormalities have clinical features that are equally severe. Still others are compatible with a normal life except that they cause sterility.

Chromosomal abnormalities that are disastrous to a fetus can be the result of an innocuous change in the chromosomes in either parent. That change is a translocation: the displacement of a chromosome, or part of a chromosome, onto a second chromosome. Such a defect may exist in all the parent's cells; if it does, the defect existed in the fertilized ovum from which all the cells arose. Alternatively, the defect may exist in only some of the parent's cells. The cells that have the abnormality are those that arose from a cell in the embryo in which the translocation first occurred. In either case the translocation causes neither loss

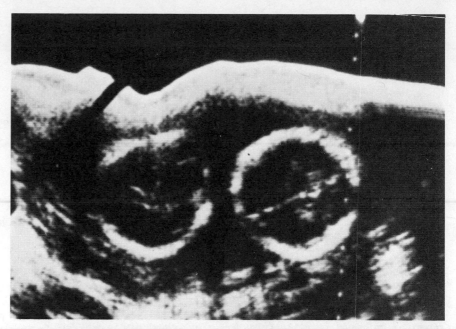

IMAGE OF A PREGNANCY is made by ultrasonography and is viewed before amniocentesis is performed so that the needle reaches an accessible pool of fluid in the fetal sac. The technique constructs the image from high-frequency vibrations, which are reflected at the interface between tissues of different densities. The sonogram shown here is of a pregnancy at 21 weeks. It reveals the gestation of twins. The white ovals defining their heads are caused by reflections of ultrasound at the skull. The white line at the midplane of the head at the right is caused by reflections of ultrasound at the falx cerebri, a thick sheet of tissue between the left and right hemispheres of the brain. A dotted line coursing diagonally downward at the right is placed on the image by an electronic measuring device and is used to measure the size of the fetal head, from which the age of the fetus is judged. The dots are at one-centimeter intervals. The sonogram was made by the Division of Ultrasound at the New York Hospital–Cornell Medical Center.

CHROMOSOMES from a cell in the amniotic fluid are identical with those in the cells of the fetus. The set shown here is from the progeny of a cell in the fluid. The set includes an extra chromosome No. 13; hence the cells of the fetus also have that defect, which is called trisomy 13 and almost always causes spontaneous miscarriage. The presence of two *X* chromosomes (*bottom right*) shows that the fetus is a female. If the child is born, however, her head and brain will be severely deformed and it is likely that she will die before she is six months old. The chromosomes in the illustration were prepared by the Cytogenetics Laboratory of the New York Hospital–Cornell Medical Center. The cell from which they were taken was about to divide, and so each chromosome is present in two copies that join at the structure called the centromere.

nor gain of chromosomal material in any one cell of the body. Hence the parent does not manifest any disease. The parent's germ cells form, however, by meiosis, a process in which cells divide but their chromosomes do not replicate. Instead they are simply allocated among daughter cells, so that if a chromosome has translocated, one germ cell may get too much chromosomal material and another may get too little. Both conditions are grave. Some fetuses with cells that include 45 or 47 chromosomes are miscarried spontaneously. Others are carried to term and then are born with severe defects.

Trisomies can also occur by mutation. The risk of its happening increases with advancing maternal age, perhaps because all germ cells in the ovaries were formed before the mother herself was born. Germ cells released from the ovary during later years of the mother's fertile period have therefore been exposed to mutagens (mutation-causing agents) longer than those that were ovulated early.

A third group of disorders detectable by amniocentesis consists of diseases caused by the biochemical defects known as inborn errors of metabolism. The inborn error is usually the deficiency of a certain enzyme. It is caused by a mistake in the gene that specifies the structure of the enzyme. Although the incorrect gene itself cannot yet be detected (in most cases the site of the gene, normal or abnormal, in the chromosomal material is not yet known), the defect is revealed by assays of the amniotic fluid or of its cells that show abnormal rates of enzyme activity, abnormal metabolites or normal metabolites in abnormal concentrations. In a few instances the knowledge that a certain disease is caused by a particular inborn error has led to the development of a therapy, but most of the inborn errors entail severe, debilitating diseases that currently are incurable and have symptoms that cannot be alleviated. About 75 inborn errors of metabolism, each of them quite rare but collectively the source of much human suffering, can now be diagnosed by amniocentesis.

An example is Tay-Sachs disease, in which the fatty substance called ganglioside accumulates abnormally and progressively in cells of the nervous system. The consequences include blindness, dementia and paralysis. Usually the child dies before attaining the age of five. The inborn error that causes Tay-Sachs disease is a deficiency of the enzyme hexosaminidase A. When heat is applied to the enzyme, it loses its efficacy; with the aid of an assay that is based on this property an examination of skin cells, blood plasma or white blood cells taken from prospective parents reveals carriers of the defective gene. As it happens, carriers are quite rare in the general population, but the incidence is considerable among Jews of eastern European origin. There the incidence may be as great as one person in 30.

Although the precise location of the defective gene is not yet known, it does not lie on a sex chromosome. Moreover, the defective gene is recessive, so that in the chromosomes of the fetus both versions of the gene must be defective to bring out Tay-Sachs disease. To put it another way, if both prospective parents are carriers, the chances are one in four that their child will manifest the disease. Thus amniocentesis is indicated if both of the partners are carriers. An assay of cells from the amniotic flu-

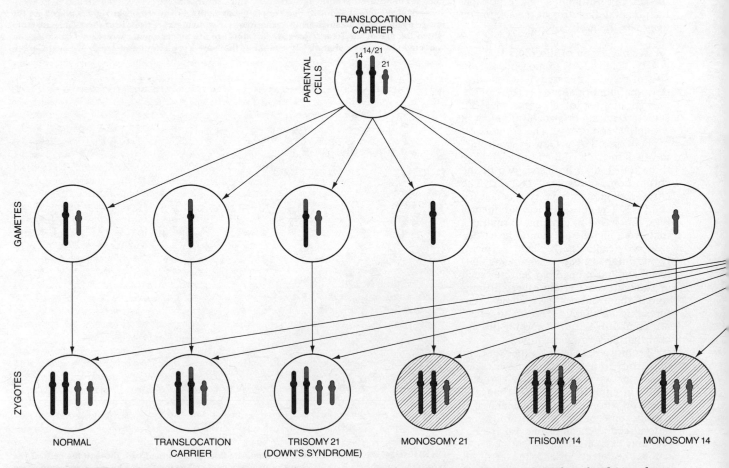

CHROMOSOMAL ABNORMALITY (such as the one shown in the bottom illustration on the preceding page) sometimes is inherited in the manner diagrammed here. The parental cell at the right includes a normal complement (two versions) of chromosome No. 14 (*black*) and chromosome No. 21 (*color*). The parental cell at the left is abnormal. One version of chromosome No. 14 and one of chromosome No. 21 have broken near the centromere. The short arms of the chromosomes have disappeared and the long arms have fused to produce a hybrid 14/21. The defect is called a translocation; it may produce in the parent no sign of disease. When the cell divides to produce germ cells, however, the chromosomes are allocated unevenly. The six possible outcomes for fertilized ova appear at the bottom of the illustration. The three at the right are not found in live births; it appears that the abnormality always induces miscarriage. Of the three at the left

id will then determine whether the fetus has inherited the disease.

The discovery by David J. H. Brock and Roger Sutcliffe of Western General Hospital in Edinburgh that the incidence of certain fetal malformations is correlated with elevated levels of the protein called alpha-fetoprotein (AFP) both in the amniotic fluid and in the mother's blood opened the way for the prenatal diagnosis of a fourth group of disorders. Most of them are defects of the neural tube, the structure in the embryo that becomes the central nervous system. The defects include anencephaly, in which the brain of the infant is partially or completely absent, along with the upper part of the skull, and spina bifida, a defect that leaves the spinal cord covered only by membranes or only by skin and membranes and not by the bones of the vertebral column. Certain other malformations involving the kidneys or the gastrointestinal tract and the abdominal wall have also been associated with abnormally high AFP levels in the amniotic fluid.

AFP is manufactured in the liver of the fetus, from which it is secreted into the fetal circulation in amounts that in-

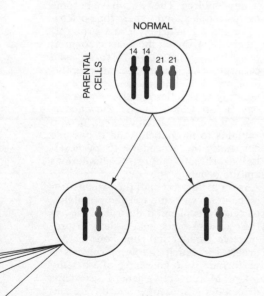

one outcome is normal and one has the translocation. In the third the presence of an extra chromosome No. 21 (a condition called trisomy 21) causes Down's syndrome, characterized by mental retardation and various malformations. Two-thirds of all fetuses whose cells have trisomy 21 miscarry spontaneously.

crease until the 20th week of the pregnancy. From there it passes into the amniotic fluid; it is probably excreted in the urine of the fetus, although it also may diffuse through the skin. Small amounts then diffuse through the placenta and the amniotic sac and enter the maternal circulation.

For anencephaly it is hypothesized that a further path exists: the AFP diffuses through the thin walls of developing blood vessels in the brain of the fetus, which is exposed to the amniotic fluid by the congenital malformation. AFP does not itself cause the neural-tube defects; it is produced in normal amounts in the liver of a malformed fetus. What makes AFP important is that it is the only protein known to exist in the fetus that is not produced in adults. Hence its presence in the mother at levels that seem to be abnormal can be a diagnostic sign.

In pregnant women who have previously delivered an infant with neural-tube defects or who have family members with such defects the level of AFP in the amniotic fluid should be measured. Such women are known to be at greater risk than the general population of bearing an afflicted child. In women not known to be at risk the level of AFP in the maternal blood can be measured in early pregnancy, and amniocentesis can then be performed if the level in the maternal blood is elevated in repeated samples.

There is, however, no sharp demarcation between normal and abnormal maternal levels. Moreover, there are several possible reasons for misinterpreting the measurement, such as the undetected death of the fetus or the miscalculation of its age. The latter is a source of error because the level of AFP considered normal changes with the age of the fetus. In cases of maternal levels deemed to be borderline it therefore becomes important to apply sonography to determine the fetal age and make certain that the fetus is alive before amniocentesis is done.

In Britain and in the Scandinavian countries large-scale AFP screening programs for pregnant women have already been instituted. In the U.S. the Food and Drug Administration and the National Institutes of Health recommend further studies before large-scale screening programs are undertaken. Our own experience at the New York Hospital–Cornell Medical Center suggests that screening programs have realized their promise. In our opinion the principal reason for not yet instituting large-scale programs is a lack of widespread expert knowledge and facilities.

It was recognized from the beginning that the insertion of a needle into the amniotic sac and the withdrawal of a sample of amniotic fluid can be associated with certain hazards, particularly to

the fetus. It is therefore important to weigh the risks of a genetic disorder against the hazards of the procedure in each individual case.

Two major hazards are known. They are the increased risk of spontaneous abortion or rupture of the amniotic sac, which can occur for several weeks following amniocentesis, and the risk of causing in the mother an immune reaction that is directed at the fetus. The reaction is one in which an *Rh*-negative mother (a woman whose red blood cells lack the cell-surface molecules called rhesus factor) develops antibodies that attack the red blood cells in an *Rh*-positive fetus. The possibility arises because the puncture of the placenta by the needle used for amniocentesis is sometimes unavoidable and may enable fetal red blood cells to enter the maternal circulation. The hazard can be prevented, however, by injecting anti-*Rh* globulin into the maternal circulation of an *Rh*-negative mother. This substance is an antibody that destroys the *Rh*-positive blood cells that have found their way into the maternal circulation before they can trigger the immune response. If both the mother and the father are *Rh*-negative, the treatment need not be performed, because in that case the fetus cannot be *Rh*-positive, and so the immune reaction cannot occur.

The risk of amniocentesis, then, is essentially an increased risk of miscarriage. It augments a natural risk of miscarriage, which is estimated to be 1 to 2 percent at the stage of the pregnancy when the amniocentesis is done. In a study published by the National Institutes of Health the total fetal loss (including stillbirths) for 1,040 subjects of amniocentesis and for 992 controls was 3.5 and 3.2 percent respectively. In 1,440 instances of amniocentesis performed at the New York Hospital–Cornell Medical Center from 1972 to 1978 the total fetal loss was only 1.5 percent, and undoubtedly there would have been some losses even if amniocentesis had not been performed. Temporary symptoms that did not entail the death of the fetus occurred in 16 percent of the patients. Such symptoms included lower abdominal pain, leakage of amniotic fluid and vaginal bleeding.

How do these figures compare with the risk of giving birth to a severely diseased child? When amniocentesis is undertaken because of the previous birth of a diseased child, because the mother or the father is a translocation carrier of a chromosomal defect or because the level of AFP in the maternal blood is abnormally high, the chances of detecting a serious abnormality are substantially greater than the increased risk of fetal loss or of serious but nonfatal complications. When it is undertaken because of the mother's age, only if she is 40 or older does the chance of finding a defect definitely outweigh the increased

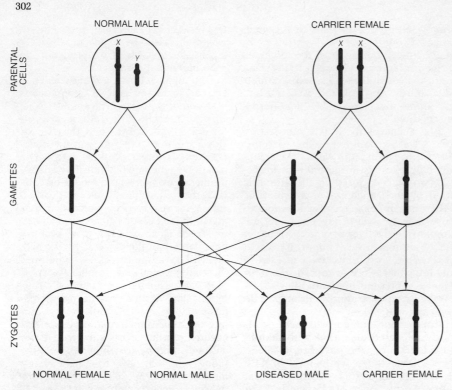

PARENTAL CELLS

NORMAL MALE   CARRIER FEMALE

GAMETES

ZYGOTES

NORMAL FEMALE   NORMAL MALE   DISEASED MALE   CARRIER FEMALE

SEX-LINKED DISEASE such as hemophilia is caused by a defective gene (*color*) on an *X* chromosome. In the cells of a female the presence of two *X* chromosomes ensures that a normal version of the gene is present. The female is thus a carrier; she manifests no disease. In cells of a male the sex chromosomes are an *X* and a *Y*. Hence if a defective gene is on an *X* chromosome, no normal version is present. Women who are carriers usually know that some of their male ancestors had the disease, and inspection of the chromosomes in cells from the amniotic fluid reveals fetal sex. The chance that a male child born to a carrier will have the disease is one in two.

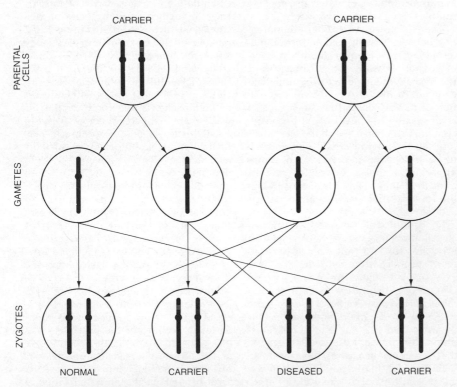

PARENTAL CELLS

CARRIER   CARRIER

GAMETES

ZYGOTES

NORMAL   CARRIER   DISEASED   CARRIER

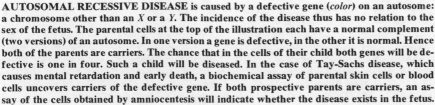

AUTOSOMAL RECESSIVE DISEASE is caused by a defective gene (*color*) on an autosome: a chromosome other than an *X* or a *Y*. The incidence of the disease thus has no relation to the sex of the fetus. The parental cells at the top of the illustration each have a normal complement (two versions) of an autosome. In one version a gene is defective, in the other it is normal. Hence both of the parents are carriers. The chance that in the cells of their child both genes will be defective is one in four. Such a child will be diseased. In the case of Tay-Sachs disease, which causes mental retardation and early death, a biochemical assay of parental skin cells or blood cells uncovers carriers of the defective gene. If both prospective parents are carriers, an assay of the cells obtained by amniocentesis will indicate whether the disease exists in the fetus.

risk to the fetus. If she is between 35 and 40, the risks probably balance.

Most women in that age group decide in favor of amniocentesis, and if they are properly counseled and the local facilities allow, it is certainly reasonable to offer them the opportunity of having it done. For public health, however, amniocentesis raises a difficult question. It can be estimated that not performing it in women whose only apparent risk factor is that they are between 35 and 40 would eliminate two-thirds of the total number of amniocenteses that are currently performed. This would enable the present U.S. health-care facilities to provide amniocentesis to three times as many women who are more substantially at risk. Such a policy, however, simply cannot be instituted within the current system of private medical care.

Some changes are occurring. In the past, for example, it was often pediatricians who had to explain genetic mechanisms and risk factors to parents after the birth of a child with a congenital disease. Now it is obstetricians who have to provide information, because many parents want to know the risk of genetic disorders before they give birth to any children. The issues have become more complex, and to help the obstetricians there has arisen a new group of medical personnel: the genetic counselors. Although their training varies, such counselors have taken special programs designed to give them some knowledge not only of genetics but also of psychology. Before amniocentesis is now undertaken it is often a genetic counselor who explains to the patient (and if possible her mate) the procedure, its purposes, its hazards, its failures, its limitations and its benefits.

In 95 to 98 percent of the patients amniocentesis will rule out the possibility of certain congenital disorders. Nevertheless, it must be made clear to prospective parents that ruling out the disorders for which the amniocentesis is performed is no guarantee of a healthy infant. Moreover, if a genetic disorder is detected, the subsequent need for counseling and support is even greater than before. By the time the results of amniocentesis are available the age of the fetus is 17 to 21 weeks. It is not surprising, therefore, that the psychological effects of an abortion performed for a genetic indication can be severe, particularly in a first pregnancy. In one study done at the University of California School of Medicine in San Francisco depression developed in 92 percent of the women and 82 percent of the men in couples who elected to interrupt a pregnancy, and in four out of 13 families the husband and wife separated. Yet most of the families studied would repeat their course of action and elect abortion over the birth of a defective child.

Failure to tell patients about the pos-

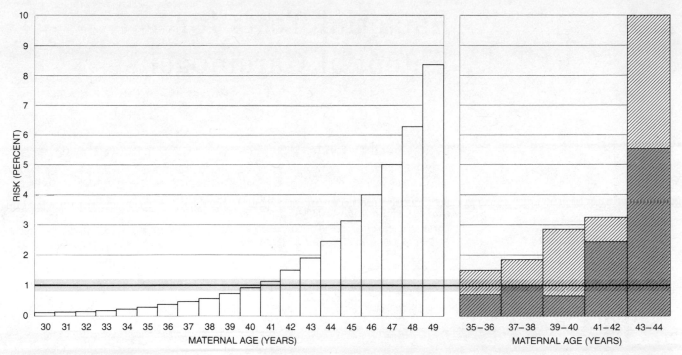

RISK OF CHROMOSOMAL ABNORMALITY in a fetus is known to increase with the age of the mother. The chart at the left shows the empirical risk that a woman will give birth to an infant whose cells have the commonest abnormality: an extra chromosome No. 21. Such infants are born with Down's syndrome. In deciding to do amniocentesis the risk of such a birth must be weighed against the risk of miscarriage, estimated at 1 percent, that the procedure itself entails. The two risks are in balance if the mother is in her late thirties. If she is 40 or older, the risk of miscarriage becomes the lesser of the two. The data are from a study by Ernest B. Hook and Agneta Lindsjö. Colored bars in the chart at the right show empirical probabilities, again by the age of the mother, that the presence of an extra chromosome No. 21 will be discovered by amniocentesis. Many such fetuses are miscarried spontaneously. Hence the bars are higher than those at the left. The bars with hatching at the right show empirical probabilities that any excess of chromosomes (or any deficit) will be discovered by amniocentesis. Those bars are higher still. The data in the chart at the right are from a study by Mitchell S. Golbus and his colleagues.

sibility of genetic diagnosis by amniocentesis has brought legal action for malpractice against obstetricians in recent years. This is not the place to discuss the legal aspects of amniocentesis in detail. Still, it is ironic to contemplate that a small number of plaintiffs in the U.S. have been awarded considerable sums in settlement of some of these cases. Meanwhile approximately 90 percent of the mothers substantially at risk of bearing children with disorders that can be detected by amniocentesis did not receive the benefits of the procedure.

A second legal issue involves the borderline between medical research and clinical practice. If the state of New York, for example, decides to allow and perhaps even encourage the screening of fetuses for neural-tube defects by means of AFP assays and the state of New Jersey does not, will obstetricians in New Jersey leave themselves vulnerable to lawsuit if they fail to tell their patients that the test can be done on the other side of the Hudson River in New York?

The most important aspect of the future of amniocentesis (and prenatal diagnosis in general) is to make the present techniques available to every mother at risk of bearing an abnormal child. Nevertheless, recent discoveries show the ways in which prenatal diagnosis can be expected to develop. For one thing, continuing work will surely increase the number of inborn errors of metabolism that can be detected by amniocentesis. At the same time the list of inborn errors known to exist can be expected to grow as the biochemical disorders underlying other genetic diseases are clarified.

Inborn errors, however, are rare. A more significant prospect is the prenatal diagnosis of diseases of higher incidence. Some examples are thalassemia and sickle-cell anemia, both of which are due to abnormalities in hemoglobin. One approach to the prenatal diagnosis of pathologies of hemoglobin requires the detection of the abnormal hemoglobin itself in samples of fetal blood. Such samples are obtained by inserting into the amniotic sac a needle thicker than the one used for amniocentesis. Through the needle is inserted a slender fiber-optic device that includes in turn a long, thin needle. Blood can then be taken from a fetal artery that the clinician actually can see on the surface of the placenta. The technique, which is called fetoscopy, requires great skill and experience. Even so, it carries a much greater risk to the fetus than amniocentesis.

This approach may, however, become obsolete. In 1978 Yuet Wai Kan and Andrée M. Dozy reported that sickle-cell anemia can be diagnosed prenatally by means of amniocentesis. Their method is based on the use of enzymes that cleave DNA strands at the sites of specific sequences of the nucleotides composing them. The strands are taken from the nuclei of cells in the amniotic fluid. Some of the strands incorporate the gene that encodes the structure of hemoglobin. It turns out that the cleavage of those strands by a particular enzyme produces fragments of one size if the gene is normal and fragments of another size if the gene is defective. The difference reveals the defect. Similar techniques should eventually make possible the prenatal diagnosis of many genetic diseases for which there is currently no method.

To be sure, genetic disorders at present cannot be cured. At the time the fetus first becomes visible by sonography, and hence might conceivably be accessible for genetic engineering, it consists of millions of cells, and copies of the genes, and of any defects in them, are present in every one. Nevertheless, certain inborn errors of metabolism have symptoms that can be alleviated after birth by dietary measures or by the administration of drugs. It is imaginable that such treatment might be applicable to the fetus. Moreover, it is a truism in medicine that the first step in rational treatment is an exact diagnosis, and this is what amniocentesis provides in many cases. The successes of amniocentesis therefore suggest the hope that at some time in the future measures to alter the expression of genes and thereby cure disease might be administered in utero.

29

Bacterial Tests for Potential Carcinogens

by Raymond Devoret
August 1979

*New short-term tests can identify environmental agents
that cause damage to DNA, the primary event in
chemical carcinogenesis. The tests are also valuable for
clarifying the mechanism of DNA damage*

The fact that physical and chemical agents in the environment enhance the incidence of cancer has become of great concern. It is clear that an adequate limitation of human exposure to carcinogens would save lives. To identify potential carcinogens in the environment is therefore an urgent task. In the case of chemical carcinogens that is no easy matter. It is estimated that more than 50,000 different man-made chemicals are currently in commercial and industrial use; between 500 and 1,000 new chemicals are put on the market every year. The standard animal tests for potential carcinogenicity take a long time and cost a great deal of money.

Fortunately there is an alternative to the classical animal tests. One can take advantage of the profound unity of living matter and resort to bacteria as the test organisms. A bacterial assay for carcinogenicity takes a few hours or days rather than the two or three years required for an animal test, and it costs far less. An effective bacterial test has been developed by Bruce N. Ames of the University of California at Berkeley, based on the ability of a chemical to cause mutations in bacteria. More recently my colleagues and I at the Centre National de la Recherche Scientifique in Gif-sur-Yvette have devised a group of tests based on a chemical's ability to induce the development of a dormant virus in bacteria. The Ames test and our tests not only provide means of identifying dangerous chemicals but also are powerful new tools for learning to understand the primary events of the carcinogenic process initiated by chemicals.

Cancer is a disease of highly evolved multicellular organisms such as human beings, whereas bacteria are minute single cells at the opposite end of the evolutionary scale. It may therefore seem paradoxical that bacteria can serve to identify substances that cause cancer. Actually, however, there is no paradox.

One tends to think of a cancer as a tumor that can spread through the body to form multiple tumor colonies (metastases). That is a clinical, macroscopic view of cancer at a multicellular stage, since just one gram of malignant tumor already contains a million cancer cells. Cancer begins at the level of the single cell. A cell in an adult tissue evolves in such a way that it departs from conformity with the strict physiological rules governing the set of identical cells that constitute a tissue; it becomes a unique and distinguishable defect in an otherwise monotonous structure. The cell begins to divide and a tumor grows. Some of the daughter cells may break the tissue barrier, invading adjacent tissues and usually metastasizing to distant sites. Cancer cells have a great selective advantage, since they escape the programmed fate of most normal cells: to age and die. For cancer cells the entire body is a culture medium in which they thrive, ultimately to die with the body they kill.

Physical and chemical agents in the environment cause cancer by damaging DNA, the cell's hereditary material. DNA damage initiates a complex cellular process that in mammalian cells can eventually lead to transformation into a cancerous state. Agents that damage DNA are therefore potential carcinogens. DNA is the hereditary material of all living cells, and both DNA lesions and the cellular processes that repair them are remarkably similar in bacteria and in human cells; what is detrimental to bacterial DNA is likely to harm human DNA. That is the theoretical justification for substituting bacteria for mammalian cells in tests to detect damage to DNA.

The theoretical justification is supported by experimental and practical results: bacterial tests distinguish with more than 90 percent reliability between known carcinogens and known noncarcinogens, and they have identified as potential carcinogens new chemicals that have subsequently been shown to be carcinogenic in animal tests. Of course, the manifestations of DNA damage are very different in bacteria from the transformation to the cancerous state that is observed in mammalian cells. In compensation the bacterial tests are so much faster and less expensive as to finally make comprehensive screening for potential carcinogens a feasible objective.

The carcinogenic potential of physical agents, and notably of ionizing radiations, is much better understood than that of chemical agents. Broad popular awareness that radiations can cause cancer has come only in the era of nuclear weapons and reactors, but the danger actually became apparent soon after the discovery of X rays by Wilhelm Konrad Röntgen in 1895. Only four years later it was reported that a technician who checked newly manufactured X-ray tubes by fluoroscoping his own hand was afflicted with a skin cancer; he eventually died of it. The warning was ignored, and most of the first generation of radiation therapists died of cancer. The carcinogenic effect of the ultraviolet radiation in sunlight has also been known for some time. As long ago as 1905 a French physician named Dubreuilh observed that skin cancers of the back of the neck were particularly prevalent among workers who were exposed to the sun as they tended the vineyards and harvested grapes in the Bordeaux region.

Although X rays, gamma rays and other radiations hit all the components of cells in a random manner, it was recognized quite early that the genetic material must be the most radiation-sensitive cellular target. DNA constitutes only a minor fraction of the chemical components of a cell, but direct or indirect damage to it has great impact on the

cell's future, whereas damage to proteins and other cellular components has much less effect. That is because DNA is the memory of the cell. DNA replicates to beget DNA, and so errors in DNA are transmitted from cell generation to cell generation.

The double helix of DNA consists of two chains of sugar (deoxyribose) and

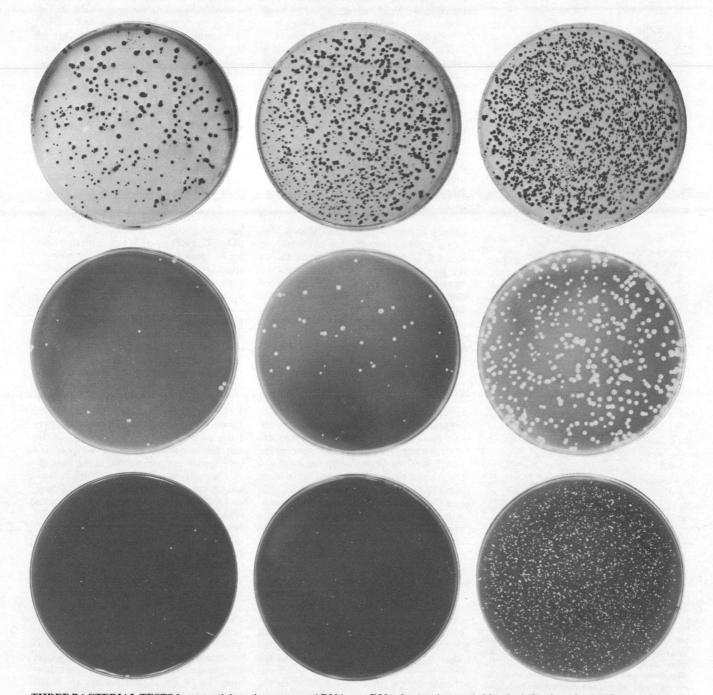

THREE BACTERIAL TESTS for potential carcinogens reveal DNA damage: the Ames test (*top*), the inductest (*middle*) and the lambda mutatest (*bottom*). In each case the culture plate at the left is an untreated control; tester bacteria on the center plate were treated with a moderate dose and those at the right with a higher dose. The Ames test shows the extent of reverse mutations in histidine-deficient *Salmonella typhimurium* that enable the revertant bacteria to proliferate. A background of spontaneous mutant colonies (*red stain*) is seen at the left. Many more colonies grow when tester bacteria are exposed to 250 nanograms (*center*) and 750 nanograms (*right*) of the potent mutagen (and carcinogen) nitrosoguanidine. In the inductest (*middle*)

DNA damage is revealed by the induction of a prophage, a dormant bacterial virus integrated in the DNA of "lysogenic" *Escherichia coli;* mature phage particles burst out of the tester bacteria and create plaques on a lawn of indicator bacteria of strain *A* (*red stain*). Here the DNA damage was caused by the antitumor drug mitomycin C, 10 nanograms of it on the center plate, 200 nanograms at the right. In the mutatest (*bottom*) a modified form of the prophage makes plaques on a lawn of *E. coli* of strain *B* (on which nonmutated phage cannot form plaques) when it undergoes mutation in its "operator" regions and can no longer remain dormant. Again the treatment was 10 nanograms of mitomycin C (*center*) and 200 nanograms (*right*).

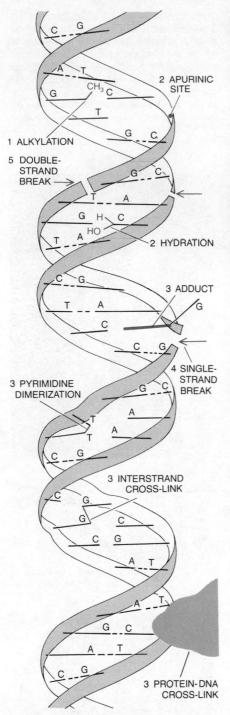

1 ALKYLATION

5 DOUBLE-STRAND BREAK

2 APURINIC SITE

2 HYDRATION

3 ADDUCT

3 PYRIMIDINE DIMERIZATION

4 SINGLE-STRAND BREAK

3 INTERSTRAND CROSS-LINK

3 PROTEIN-DNA CROSS-LINK

STRUCTURAL ALTERATIONS are imposed by radiations or by chemical agents on the double helix of DNA, two chains of sugar and phosphate groups (helical bands) linked by paired bases: either adenine (A) and thymine (T), or guanine (G) and cytosine (C). The alterations can be classified in five categories, examples of which are illustrated: negligible helix distortions (1), as by alkylation of one of the bases; minor distortions (2) caused by hydration or the absence of a base; major distortions (3) caused by insertion of an "adduct," linking of two bases to form a dimer, or cross-linking between the two strands or between a strand and a protein; breaks in a single strand (4) or in both strands (5). Any structural alteration affects DNA's function as a template for replication, but some negligible alterations are not sensed by a cell as damage to DNA.

phosphate groups linked, as by the rungs of a twisted ladder, by paired nitrogenous bases: two purines (adenine and guanine) and two pyrimidines (thymine and cytosine). Adenine always pairs with thymine and guanine with cytosine, but the sequence of the bases along a strand of DNA is variable and carries a particular coded message. Any alteration, even a slight one, in the structure of the double helix affects the functions of the DNA, one of which is to serve as a template for its own replication.

Not every DNA alteration is sensed in the cell as DNA damage. Defined precisely, DNA damage is an alteration that constitutes a stumbling block for the replication machinery and hence hampers the replication of DNA, endangering the survival of the cell. Once incurred, DNA damage calls for repair, which is accomplished by the interplay of at least a score of enzymes whose action is governed by as many genes. Repair is never totally efficient, and so many cells die. Some cells may survive, however, even though the lesions are not totally removed from their DNA, because a repair process has bypassed the lesions. Replication then reconstitutes an undamaged double helix, but one bearing a coded message different from the original one. The scars left on the DNA by such a process are mutations.

The correlation between radiation-induced cancer and radiation-induced DNA damage has long been apparent to radiation biologists, but it is only recently that molecular evidence has been found that DNA damage is a direct cause of cancer. The evidence comes from patients suffering from xeroderma pigmentosum, who are extremely sensitive to sunlight and, while they are still very young, develop skin cancers of which they may eventually die. James E. Cleaver of the University of California School of Medicine in San Francisco and Dirk Bootsma of Erasmus University in Rotterdam have demonstrated that xeroderma patients suffer from a well-characterized genetic defect: their cells cannot carry out a particular DNA-repairing process.

At noontime on a sunny day the flow of ultraviolet radiation that reaches the earth is strong enough to generate pyrimidine dimers in the DNA of exposed cells by linking two laterally adjacent thymines or a thymine and a cytosine. Most such DNA lesions in skin cells are repaired in normal people by an excision process strikingly similar to the "cut and patch" repair process in bacteria exposed to the same radiation. In xeroderma patients, however, the lesions go unrepaired, and their accumulation appears to bring on cell transformation: DNA damage breeds cancer.

Appreciation of the carcinogenic role of chemicals in the environment has come slowly, even though instances of cancer caused by occupational exposure

have long been observed. As early as 1775 the British physician Sir Percival Pott correlated the incidence of cancer of the scrotum in men who had once been chimney sweeps with the accumulation of soot in their groin area many years before. Experimental findings on chemical carcinogens date back at least to 1918, when two Japanese investigators, K. Yamagiwa and K. Ichikawa, showed that skin cancers could be induced by repeated applications of coal tar to the ear skin of rabbits. One of the chemicals responsible for causing such cancers is benzo[a]pyrene, which is also present in soot, cigarette smoke and charred meat.

Progress toward understanding chemical carcinogenesis has been slow for at least three reasons. First of all, most chemical carcinogens are not biologically active in their original form, so that testing them in that form does not reveal their carcinogenic nature. Only about a decade ago did it become clear, as a result in particular of the investigations of James A. Miller and Elizabeth C. Miller of the University of Wisconsin, that normal metabolic processes, which convert food into substances the body can absorb and eliminate and convert harmful compounds into harmless ones, transform environmental chemicals into metabolites capable of inducing cancer. Those metabolites had to be characterized in order to show just how their parent substances are threats.

A second delaying factor was that the actively carcinogenic metabolites react with various cell components, including RNA and proteins as well as DNA. Since there are a lot of proteins, performing important functions, it was often assumed that damage to proteins might play the major role in the cancerous transformation initiated by chemical carcinogens. The key role of damage to DNA in carcinogenesis by chemicals was recognized only recently, and the results of the bacterial tests have provided strong (although indirect) evidence for such a mechanism.

Finally, the understanding of chemical carcinogenesis has been obscured by the fact that DNA damage, although it is the essential event in the initiation of carcinogenesis, does not usually in itself lead to cancerous transformation; additional factors are apparently required to promote the complex chain of cellular events culminating in the transformation. DNA-damaging agents in themselves are therefore only potential carcinogens.

As I pointed out at the beginning of this article, only a small fraction of the flood of chemicals reaching the market every year can be tested accurately by means of the standard animal assays. In order to obtain results with statistical significance a great many animals must be (or at least should be) exposed to each tested chemical; for that reason alone a comprehensive screening of new chemicals would be impractical (other than the few, such as food additives, drugs and cosmetics, for which testing is mandated). Even when animal tests are feasible, manufacturers need low-cost, fast tests if they are to identify DNA-damaging substances while new products are still under development and alternative ones can still be sought.

Studies of the epidemiology of human cancers have provided much information on environmental carcinogens, but such studies are of little immediate value in detecting new potential carcinogens because most human cancers appear only some 20 or 30 years after the exposure that gives rise to them. Epidemiological analysis has most frequently been successful in detecting a chemical to which an identifiable subpopulation, such as workers in a particular industry, are exposed and because of which they show a high incidence of a particular type of cancer. For some widely distributed carcinogens there are no clearly identifiable subpopulations for statistical analysis.

There is no doubt, then, that the present situation calls for simple, fast, inexpensive methods of detecting potential carcinogens, which is to say DNA-damaging agents. One possibility is to measure DNA damage directly in mammalian cells by determining biochemically the incidence of particular forms of DNA damage in cells exposed to a chemical. Such studies are being done by molecular biologists, and their results provide valuable standards of reference. For screening purposes, however, it is cheaper and faster to detect and measure the extent of DNA damage by scoring its manifestations in bacteria.

One of the great advantages of assays done with bacteria is the enormous biological amplification implicit in bacterial manipulations. It is easy to grow as many as a billion (10^9) bacteria per milliliter of culture medium. A mutational event such as a change in a single base pair in the bacterial DNA, which is impossible to detect by standard biochemical methods, will be revealed as a mutant bacterium. That single bacterium can be selected from among 10^9 cells because its daughter cells, and only they, will proliferate and form a colony visible to the unaided eye on an agar nutrient plate. Since a colony consists of about a million (10^6) bacteria, a rare single mutational event with a probability of, say, one in 100 million (a probability of 10^{-8}) would thus be amplified by a factor of 100 trillion (10^{14}).

The series of cellular events that leads to mutation, and to several other manifestations of DNA damage in bacteria, has recently been somewhat clarified. The bacterium *Escherichia coli*,

which inhabits the colon of a number of mammals including human beings, is genetically programmed to divide and (under optimal conditions for growth) form two daughter bacteria in 30 minutes. When the DNA of a bacterium is damaged, the bacterium may need more than two hours to resume its cycle of division (if it is not killed by the damage). During that time there is a sequence of cellular events.

Immediately after the primary DNA lesions are incurred a first round of repair is effected: an excision repair, in which the damaged segment is cut out of the DNA and replaced by an undamaged DNA sequence. Some residual lesions will remain, however. If they are bulky or clustered, the residual lesions hamper the DNA-replication machinery, and replication stops abruptly. Unless the arrest of replication is transitory

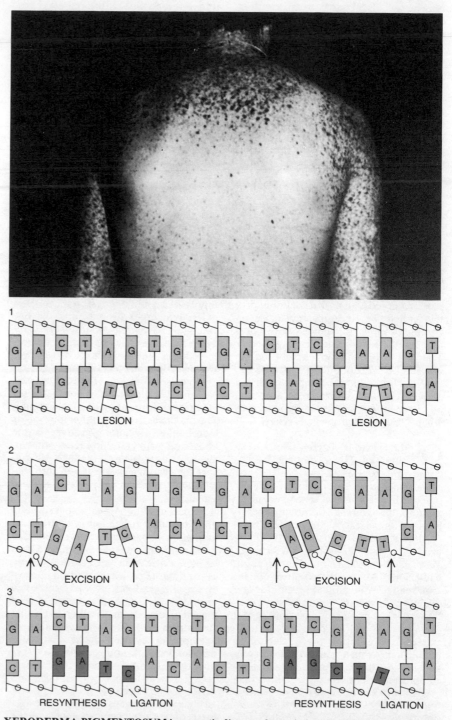

XERODERMA PIGMENTOSUM is a genetic disease whose victims develop skin cancer as a result of normal exposure to sunlight, typically on uncovered parts of the body, as is shown in the photograph at the top. These patients lack a DNA-repair process that in normal individuals removes thymine dimers, the lesions created by the ultraviolet radiation present in sunlight (1). In the normal repair process the bases in the lesion areas are first excised (2). Then the excised stretches of DNA are resynthesized and the new segments are ligated to undamaged stretches (3).

the cell's survival is at risk. As Miroslav Radman of the University of Brussels and Evelyn M. Witkin of Rutgers University first suggested, the following cellular adaptive mechanisms come into play to cope with the emergency:

1. A second round of repair is induced. This inducible and error-prone repair (nicknamed "SOS repair") tends to restore the structure of the DNA even though there are errors in the coded message; indeed, this adaptive response may be successful partly because it does not "bother" to follow the base-pairing rules of normal DNA replication. At any rate, the price paid for cell survival appears often to be mutagenesis.

2. Cell partition ceases. The elongation of the cell that ordinarily precedes division is protracted, and the cells may form filaments. The elongation may be adaptive in that it facilitates recombination between the two sets of damaged chromosomes in the cell, the intact segments of each combining to yield an intact chromosome.

3. If a prophage, or dormant bacterial virus, is present in the cell, it is induced to develop into a large number of mature progeny phage that burst out of the cell. This adaptive response evolved by the prophage ensures its survival when a host cell appears to be doomed to die: the rats leave the sinking ship.

Each of these adaptive responses is in part promoted by a multifunctional protein called the RecA protein (for "recombination," since defects in the protein impair recombination in general). The RecA protein appears to be required in E. coli not only for recombination but also for the responses listed above. The RecA protein is activated and its synthesis dramatically increased when DNA replication is blocked.

Of the various bacterial manifestations of DNA damage, Ames chose mutagenesis as the basis of his pioneering work to develop a test for potential carcinogens. His Salmonella–mammalian-liver assay, known generally as the Ames test, is currently the standard test and by far the most widely used. The tester organism is a strain of Salmonella typhimurium, another colon bacterium, bearing a mutation (his−) that renders it unable to manufacture one of the enzymes required for the synthesis of the amino acid histidine, a necessary component of proteins. As a result of the mutation the bacterium is unable to grow in a mineral nutrient medium unless the medium is supplemented with an external supply of histidine.

On very rare occasions a his− mutation undergoes reversion: a back mutation restores the DNA's normal coding sequence for the needed enzyme and thereby restores the internal supply of histidine. The reversion can be scored because only the revertant bacteria form colonies on a medium that lacks histidine. Obviously the spontaneous rate of reversion, which is ordinarily very low, will be considerably enhanced if the his− bacteria are exposed to a chemical that induces mutations. This is the theoretical basis of the Ames test.

Actually Ames and his colleagues had to introduce three important modifications into the original his− strain to make it a sensitive and versatile tester bacterium. Bacteria such as E. coli and S. typhimurium have a rather impermeable envelope that reduces or even prevents the penetration of many chemicals into the cell. (This bacterial armor has evolved because the bacteria must usually survive in a hostile environment such as an intestine or a sewer.) Ames and his colleagues overcame the envelope barrier by introducing a mutation that gives rise to defects in the envelope. They went on to make the strain more sensitive to DNA-damaging agents by eliminating its capacity for excision repair, so that most of the primary lesions remain unhealed. And they introduced into the bacterium a plasmid, a foreign genetic element that makes DNA replication more error-prone. By means of these three modifications a strain was constructed in which just a few molecules of a carcinogen are able to create DNA lesions, each of which is likely to engender a mutation; of those mutations, some will be such that the internal supply of histidine is restored.

The real breakthrough, and the one that made the Salmonella test truly effective, was Ames's idea of mixing the tester bacteria with an extract of rat liver and thereby subjecting the tested chemical to mammalian metabolic processes. As I pointed out above, it is usually not the original form of a chemical carcinogen that is active but rather one of its metabolites. Because the liver is the body's major metabolic factory the enzymes of a rat-liver extract should convert the chemical being tested into metabolites that react with DNA—if there are any.

In practice the Ames test is usually done by adding the chemical to be examined to his− tester bacteria immersed in a rat-liver extract and plating the mixture on a solid nutrient medium devoid of histidine. (For demonstration purposes the chemical can instead be spotted on a disk of filter paper, which is then placed on a medium on which the bacteria, mixed with the liver preparation, have previously been plated.) After two days of incubation any cells that have undergone the reversion mutation will give rise to revertant colonies. The number of such colonies per mole of the tested chemical provides a quantitative estimate of the mutagenic potency of the chemical.

The simplicity, sensitivity and accuracy of the Salmonella test for screening large numbers of environmental sources of potential carcinogens has resulted in its current application in more than 2,000 governmental, industrial and academic laboratories throughout the world; it is estimated that 2,600 chemicals have been subjected to the test. Ames and his collaborator Joyce McCann have themselves validated the procedure by testing more than 300 chemicals that were previously reported, on the basis of animal experiments, to be either carcinogens or noncarcinogens. About 90 percent of the reputed carcinogens turned out to be mutagenic and about the same proportion of the noncarcinogens were negative for mutagenicity. Other mutagenicity tests have since been devised with E. coli as the tester bacteria, and their efficiency is about as high as that of the original Salmonella test.

Two impressive accomplishments of mutagenicity tests can be mentioned to give an idea of their value. In Japan the chemical furyl furamide, known as AF-2, was added to a broad range of common food products for some years as an antibacterial agent. It had not shown any carcinogenic activity in standard tests on rats in 1962 or on mice in 1971. Then in 1973 T. Sugimura and his colleagues at the National Cancer Centre in Tokyo found that AF-2 was highly mutagenic in bacteria; they could easily demonstrate the mutagenic activity of the additive contained in just one slice of fish sausage! The discovery prompted a new round of more thorough animal tests for carcinogenicity, which showed that AF-2 was indeed a carcinogen. It was withdrawn from the market. If it were not for the bacterial test, AF-2—which had passed two approved animal tests and been declared negative for carcinogenicity—would presumably still be a component of fish sausage and other Japanese food products.

In 1975 Ames and his colleagues reported that 89 percent of the major class of hair dyes sold on the U.S. market contained mutagenic compounds. Since then the cosmetic industry has modified the composition of most hair dyes. It is estimated that several tens of millions of people dye their hair in the U.S. alone, which suggests that the discontinued components had presented a considerable risk.

In spite of all their advantages the mutagenicity tests have some technical and even theoretical limitations. Since mutations are revealed in the Ames assay by a restoration of enzyme activity, any mutation that does not happen to reconstruct the precise DNA sequence that codes for the histidine-making enzyme is not observed. To take one example, the antitumor drug bleomycin, which does its therapeutic work by damaging the DNA of tumor cells (as do about half of all antitumor agents), fails to induce the mutation that is scored in the Salmonella test. A false-negative re-

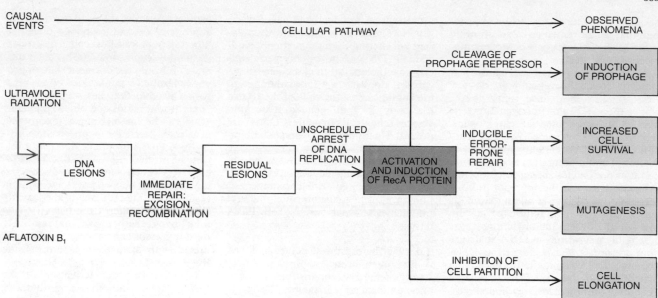

CAUSAL EVENTS — CELLULAR PATHWAY — OBSERVED PHENOMENA

SEQUENCE OF CELLULAR EVENTS follows DNA damage imposed on *E. coli* by ultraviolet radiation or by the carcinogen aflatoxin B_1. Immediate repair processes mend most of the lesions but leave residual damage, causing an unscheduled arrest of DNA replication. This threat to cell survival activates the RecA protein to become a protein-cleaving enzyme and also induces the synthesis of more RecA. Various forms of the protein are involved in three processes that give rise to the four observed phenomena shown at right.

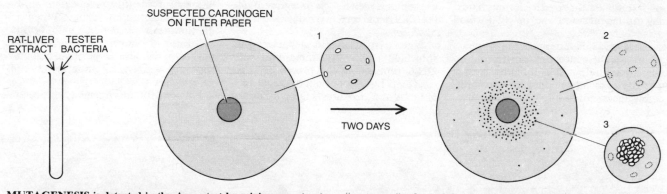

MUTAGENESIS is detected in the Ames test by mixing an extract of rat liver (which supplies mammalian metabolic functions) with tester bacteria (which cannot grow because a mutation makes them unable to manufacture histidine, a necessary nutrient) and plating the mixture on an agar medium so that a thin layer of bacteria covers the medium evenly, as is shown on a microscopic scale (*1*). In this "spot assay" a dose of the chemical to be tested is placed on a disk of filter paper on the tester bacteria. After two days most of the *his⁻* bacteria have died for lack of histidine (*2*), but DNA damage caused by the chemical diffusing out from the disk has given rise to mutations, some of which result in reversion of the *his⁻* mutation. The histidine-making revertant bacteria proliferate, forming visible colonies (*3*).

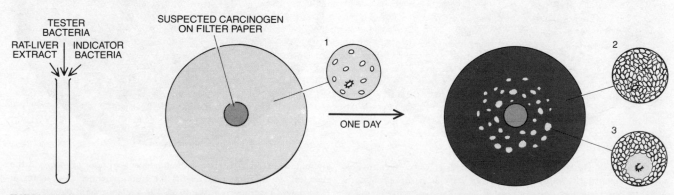

INDUCTION of a dormant bacterial virus, prophage lambda, is detected in the inductest. Lysogenic tester bacteria are mixed with rat-liver extract and then with indicator bacteria. The mixture is plated; the medium is covered with a thin layer of indicator bacteria interspersed with a few lysogenic bacteria (*1*). After a day most of the plate is covered by a thick lawn of indicator bacteria (*2*). Where the chemical that is being tested has diffused from the filter-paper disk the DNA damage it causes leads to the induction of mature lambda phage. The phage particles burst out of the lysogenic cells and kill indicator bacteria in the vicinity, making visible plaques on the lawn (*3*).

sponse of this kind is a technical problem that can be remedied by substituting other tester bacteria or a complementary short-term test.

False-positive responses are more significant from a theoretical point of view because they may raise some doubts about the validity of mutagenicity tests for identifying potential carcinogens. The point is that some chemical reactions with DNA are highly mutagenic in bacterial and mammalian cells without, as far as one can tell, being carcinogenic. Among these are the incorporation of an analogue of one of the nitrogenous bases and the methylation of certain sites on the bases. Such reactions cause negligible alterations in DNA structure that are not sensed in the cell as DNA damage; DNA replication proceeds on schedule and the new DNA carries a readable—although wrong—coded message. This process is termed direct mutagenesis, and in scoring for mutations it should be clearly distinguished from the more frequent indirect mutagenesis. In the latter process, described above, DNA damage brings about a transient arrest in replication; replication is resumed with the help of the RecA protein on a template that carries lesions, causing mutagenesis of the newly formed strand.

Since it is DNA damage that appears to initiate the cancerous transformation of a mammalian cell, a chemical is a potential carcinogen if it is a DNA-damaging agent, not simply because it causes

mutations. A mutagen can have strong genetic effects on a biological population by altering the information encoded in the DNA of individuals' germ cells but nonetheless fail to cause cancer because it does no real DNA damage to the somatic cells: the cells of the rest of the body. Not all genetic toxic substances (chemicals that affect DNA) are potential carcinogens, whereas chemical carcinogens—because they damage DNA—are all indirect mutagens. In other words, mutagenic activity in bacteria is correlated with carcinogenesis in mammals primarily through indirect mutagenesis, which results from DNA damage.

In 1953, before the structure of DNA had even been established, André Lwoff of the Pasteur Institute in Paris anticipated that "inducible lysogenic bacteria might become a good test for carcinogenic and perhaps anticarcinogenic activity." A bacterium is said to be lysogenic when it carries, in a dormant state, the DNA of a "temperate" bacterial virus, which in this dormant state is called a prophage. One such temperate virus is phage lambda, which becomes a prophage when it is integrated into the DNA of certain lysogenic strains of E. coli.

When lysogenic E. coli bacteria are subjected to any treatment that halts DNA replication, the prophage is induced: its DNA loops out of the bacterial DNA and also directs the synthesis of

proteins forming the virus particle, and a progeny of mature phage develops, bursting out of the host cell. The process is called lysogenic induction. Under normal conditions the dormant state of prophage lambda is maintained by a repressor, a protein that lies on the DNA's "operator" regions and blocks the operation of the lambda genes (except the gene that directs the synthesis of repressor to keep the prophage dormant). Jeffrey W. Roberts and Christine W. Roberts of Cornell University discovered that the induction of prophage lambda results from the cleavage of the lambda repressor; together with Nancy Craig they have recently shown that relatively pure repressor can be cleaved when it is mixed with an activated form of the RecA protein.

By triggering the activation of the RecA protein into the form that can cleave a viral repressor, DNA damage results in the induction of a dormant virus. The induction of prophage lambda can therefore serve as a test for DNA damage. Even before the mechanism of prophage development was understood at a molecular level, lysogenic induction was applied in the pharmaceutical industry to identify prospective antibiotics and antitumor drugs. When induction tests were done with ordinary strains of lysogenic E. coli, however, they did not give a positive response for such known carcinogens as benzo[a]pyrene.

Patrice Moreau, Adriana Bailone and

SPOT ASSAYS visualize the efficacy of bacterial tests qualitatively, although quantitative assays are preferable (*see illustration on page 305*). In the Ames test (*left*) the tested chemical was ethyl methane sulfonate, an alkylating agent and potent mutagen. A dense halo of revertant *S. typhimurium* colonies is seen around the disk from which the mutagen diffused. (Close to the disk there is a zone in which a toxic concentration of the chemical killed all bacteria.) The larger colonies are mutant bacteria that arose spontaneously without exposure to the chemical. In the inductest (*right*) the tested chemical was aflatoxin B_1, a potent carcinogen of the liver. The aflatoxin doses placed on the four test disks were (*clockwise from top right*) 0, 20, 200 and 2,000 nanograms. In the background a few of the lysogenic *E. coli* tester bacteria have spontaneously produced mature phage that have killed the nearby indicator bacteria, producing plaques at random locations.

I therefore set out five years ago to renovate the lambda-induction test to make it capable of identifying all kinds of DNA-damaging agents. We reasoned that because the molecular mechanism of lysogenic induction was understood better than that of mutagenesis, a prophage-induction test should provide more insights than mutagenicity tests could into the precise effect of chemical carcinogens on DNA and should provide a supplementary tool for screening as well.

Following Ames's lead, we constructed new tester bacteria that are permeable to chemicals and deficient in excision-repair enzymes, and we assayed potential carcinogens in the presence of a rat-liver metabolizing mixture. The prophage-induction test, or inductest, has turned out to discriminate effectively between carcinogens and noncarcinogens in our laboratory and in several others.

The inductest has certain advantages over mutagenicity tests. Mutations can be scored only in bacteria that survive exposure to a chemical treatment that is often toxic as well as mutagenic, and no more than one bacterium out of 1,000 is likely to be detected as a histidine revertant. In contrast, lysogenic induction can be observed, when it takes place, in the bulk of a cell population. Prophage induction is a mass effect, largely independent of whether or not cells would survive the chemical treatment; a cell that is induced to produce a progeny of phage would die in any case. The inductest can therefore continue to give a positive response for a highly toxic potential carcinogen at dose levels that would kill the tester bacteria in a mutagenicity test.

The fact that in the inductest most cells undergo a dramatic change led us to design a biochemical assay of lysogenic induction. Sankar Adhya, Maxwell E. Gottesman and Asis Das of the National Cancer Institute in the U.S. constructed a bacterial strain in which the gene for the enzyme galactokinase is hooked up to the lambda DNA in such a way that the lambda repressor blocks the synthesis of the enzyme. Alain Levine, Moreau, Steven Sedgwick and I demonstrated that when a DNA-damaging chemical activates the RecA protein in this strain, the repressor is cleaved and galactokinase is synthesized, and the amount of enzyme activity that can be detected reveals the extent of DNA damage. This biochemical assay may be the shortest of the short-term tests; it takes only half a day to test for a potential carcinogen.

Moreau and I have also constructed strains, containing a new form of prophage lambda, in which one can identify chemicals that induce mutagenesis without damaging DNA. This lambda mu-

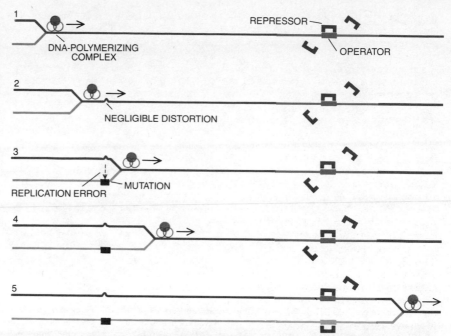

DIRECT MUTATION is a result of DNA replication on a minor distortion of a DNA strand. This diagram and the one below show the replication of only one strand of bacterial DNA (*blue*) carrying a piece of viral DNA, prophage lambda (*red*), which is kept in a dormant state by a repressor protein that blocks an operator region. The DNA strand is replicated by a polymerizing complex (*1*). A negligible distortion of the bacterial DNA (*2*) does not obstruct replication, but id does give rise to an error in replication, so that the newly formed DNA strand (*light blue*) carries a mutation (*3*). The negligible distortion is not sensed as DNA damage, RecA protein is not induced, repressor stays in place and prophage is replicated in dormant state (*4, 5*).

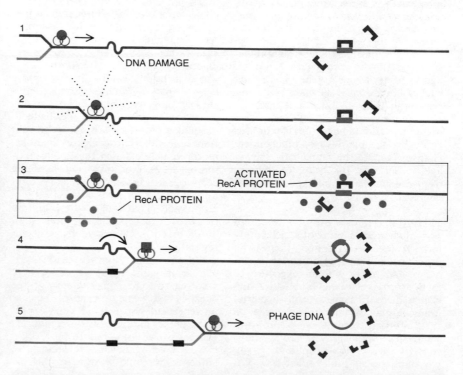

INDIRECT MUTATION results from the events that follow DNA damage (*1*), which blocks DNA replication and produces an "SOS" signal (*2*). A large amount of RecA protein (*orange*) is synthesized (*3*). The RecA somehow facilitates DNA replication on a damaged template, apparently by temporarily modifying the polymerizing complex (*4*). This replication is highly mutagenic not only at the lesion site but also elsewhere (*5*). Activated RecA cleaves the repressor (*4, right*), freeing the operator region of the prophage DNA. The prophage is thereby induced to form phage DNA (*5*), which in turn forms virus particles that burst out of bacterium.

312

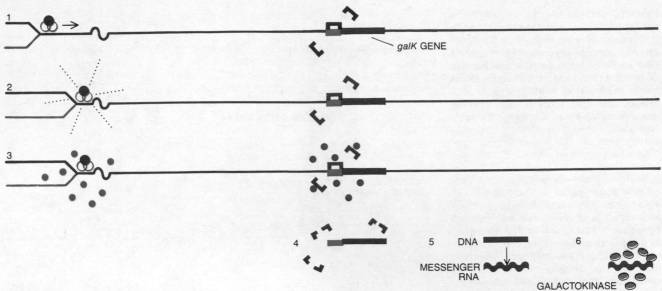

ENZYME-INDUCTION TEST is a biochemical counterpart of the inductest. The gene, *galK*, for synthesis of galactokinase is hooked up to prophage-lambda DNA in such a way that enzyme synthesis is blocked by lambda repressor. When DNA is damaged, activated RecA protein cleaves repressor. Liberated *galK* DNA (4) is transcribed into messenger RNA (5), which is translated into galactokinase (6).

tatest, as we call it, reveals whether a chemical gives rise to mutations directly (that is, without causing DNA damage and therefore without inducing prophage) or indirectly (by damaging the DNA and producing active virus).

Although neither the inductest nor the lambda mutatest has yet been validated as extensively as the Ames mutagenicity test, the two appear to be valuable complementary assays, in particular for determining the nature of DNA damage and for establishing a correlation between potency in producing DNA damage in bacteria and potency in causing cancer in mammals. Perhaps more important in the long run, a biological system comprising both a cell and a dormant virus should be a valuable tool for analyzing the pathway of biochemical changes, genetic or nongenetic (cleavage of repressor, for example), that are triggered when cells are exposed to carcinogens.

Given that man is exposed to a flood of chemicals, some of which surely increase the risk of cancer, there is an urgent need to identify potential carcinogens and lower human exposure to them to a level of negligible risk and keep it there. In such an effort bacterial tests can play a major role. In order to ensure broad acceptance of uniform standards, some investigators have strongly favored the adoption of a single, standardized mutagenicity test. Often, however, it is advisable to adapt a screening effort to the particular physical and chemical properties of the substance that is under study. It would therefore seem to be sensible to subject each chemical to a battery of tests. Any test has its foibles, which give rise to false-negative or false-positive respon-

ses. Several independent and complementary determinations of a chemical's DNA-damaging capacity can provide a balanced and hence more compelling assessment of carcinogenic risk.

The pronouncement that a particular chemical is a potential carcinogen has obvious social, economic and even political implications. The chemical may present a risk to a very large population, as was the case with the food additive AF-2 in Japan. That situation was dealt with rather easily: AF-2 could be replaced by an innocuous substitute. It was dispensable, and so it was simply banned from the market. Some chemical carcinogens cannot easily be dispensed with and cannot be banned; particular problems are raised by substances as different as cigarette smoke, motor-vehicle exhaust fumes and antitumor drugs. For most indispensable carcinogens it is clearly necessary to reduce human exposure to the safest levels possible. That calls for the adoption of a broadly accepted set of safety standards to be implemented by stringent regulations.

We stand today, with regard to chemical carcinogens, about where we were with regard to ionizing radiations three decades ago. The advent of nuclear power reactors made it necessary to protect workers in the industry from direct sources of radiation and to protect everyone from exposure to radioactive effluents and solid wastes. An international agency sponsored by the United Nations worked out a set of standards and regulations designed to minimize human exposure to various sources of radiation. The regulations were broadly accepted and implemented, and they have stood the test of time. They should provide a valuable reference for evolv-

ing standards for chemical carcinogens.

The experts who set the standards for radiation safety have proceeded from two basic facts. One is that mankind is exposed to a natural background of radiations whose biological effect is not perceptible (even though in principle there is no threshold for the effects of ionizing radiations). The other is that biological damage is proportional to the energy released within the living system being considered, and that this linear relation is more or less independent of the particular source of radiation. The experts assumed that for the known sources an absorbed dose only a few times higher than the dose delivered by background radiation, and less than the level that would cause a doubling of the mutation rate in a mammalian population, would constitute a negligible risk; in general such doses were adopted as being permissible for human exposure. In the case of newly devised or newly discovered radioactive materials for which there were no accumulated experimental data, permissible values were calculated on the basis of the linear relation between biological damage and the energy released within the organism.

Since chemical pollution is worldwide, an international committee of experts should propose protective regulations for adoption by individual governments. Some principles should gain broad acceptance rather easily: for example, all unnecessary potential carcinogens should be banned. A list of carcinogens that cannot be dispensed with but to which exposure should be strictly limited will also have to be drawn up. It may be difficult, however, to reach a consensus on such a list and on the permissible environmental concentrations of each chemical. One would like to begin by defining permissible concentrations of indispensable carcinogens in relation to a "natural" background. The trouble is that cultural differences, social habits and other factors have considerably distorted notions of what is a natural background level of negligible risk. (For geographical, economic and religious reasons the carcinogenic background is surely lower in Salt Lake City than it is in New York.)

If a negligible-risk level can be agreed on, the next step would be to estimate, for each indispensable carcinogen, the concentrations that should be allowed for workers in the industry and (at a much lower level) for the general population. Can bacterial tests help in setting such levels? Yes, if there is proportionality between potency in causing DNA damage in bacteria and potency in causing cancer in mammals. Matthew S. Meselson and Kenneth Russell of Harvard University have attempted to demonstrate a direct relation between mutagenic potency in the *Salmonella* test and carcinogenic potency in laboratory animals. For 10 of the 14 known chemical carcinogens they considered there was a correlation.

The many exceptions to the rule of correlation were to be expected because in a mammal the extent of DNA damage in a target cell depends on factors (which may be different in bacteria) such as the specific metabolism of carcinogens in the organ involved, the cells' permeability to particular metabolites and tendency to accumulate them and the extent of residual DNA damage after the cellular repair mechanisms have had their effect. Each of these variables must be determined for each individual carcinogen if one is to set valid levels for human exposure. Each chemical—in contrast to individual sources of radiation—requires a thorough specific study to determine its carcinogenic potency. Bacterial tests have good indicative value.

Even though establishing regulations will be much more difficult for chemical carcinogens than it was for radiations, the regulations should be established soon. It is wiser and safer to have safety regulations than not to have them. A large part of the human population contracts cancer; one person in five dies of it. It is important to determine the precise mechanism by which each DNA-damaging agent acts on DNA, not only to prevent cancer but also to develop better ways of curing it. About half of all the drugs administered in an effort to arrest cancer are DNA-damaging agents, and so a significant number of people are being regularly exposed, for therapeutic purposes, to such agents. By establishing, through studies in bacteria, the mechanism whereby antitumor agents affect DNA, investigators may be able to help design more drugs that are effective in curing cancer without also causing cancer.

Bacterial tests for potential carcinogens should help to answer two questions now being asked by industry, government and people at large. What chemicals are DNA-damaging agents? How much of each—in our air, our water and our food, in the products we manufacture and consume—constitutes a biological risk? These two questions will be answered satisfactorily, however, only if a third question is asked: What is the mechanism whereby each chemical affects DNA? Only basic knowledge can breed safety.

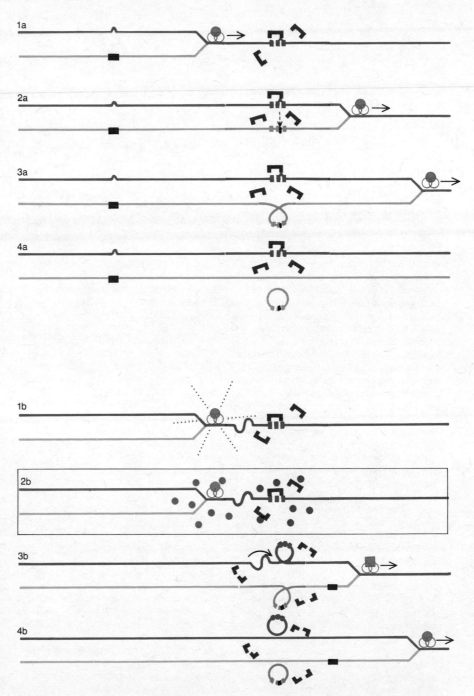

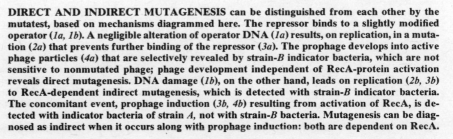

DIRECT AND INDIRECT MUTAGENESIS can be distinguished from each other by the mutatest, based on mechanisms diagrammed here. The repressor binds to a slightly modified operator (*1a, 1b*). A negligible alteration of operator DNA (*1a*) results, on replication, in a mutation (*2a*) that prevents further binding of the repressor (*3a*). The prophage develops into active phage particles (*4a*) that are selectively revealed by strain-*B* indicator bacteria, which are not sensitive to nonmutated phage; phage development independent of RecA-protein activation reveals direct mutagenesis. DNA damage (*1b*), on the other hand, leads on replication (*2b, 3b*) to RecA-dependent indirect mutagenesis, which is detected with strain-*B* indicator bacteria. The concomitant event, prophage induction (*3b, 4b*) resulting from activation of RecA, is detected with indicator bacteria of strain *A*, not with strain-*B* bacteria. Mutagenesis can be diagnosed as indirect when it occurs along with prophage induction: both are dependent on RecA.

BIBLIOGRAPHIES

I PRINCIPLES OF HEREDITY

1. Experiments in Plant Hybridization

Reprint of "Versuche über Pflanzen-Hybriden" by Gregor Mendel (1865). In *Classic Papers in Genetics*. Edited by James A. Peters. Prentice-Hall, Inc., 1959.

2. Genes That Violate Mendel's Rules

MECHANISMS OF MEIOTIC DRIVE. S. Zimmering. L. Sandler and B. Nicoletti in *Annual Review of Genetics*, Vol. 4, pages 409–436; 1970.

SEGREGATION DISTORTION AFTER FIFTEEN YEARS. D. L. Hartl and Y. Hiraizumi in *The Genetics of Drosophila melanogaster*, edited by E. Novitski and M. Ashburner. Academic Press, 1976.

DYNAMICS OF SPERMIOGENESIS IN DROSOPHILA MELANOGASTER, VII: EFFECTS OF SEGREGATION DISTORTER (*SD*) CHROMOSOME. K. T. Tokuyasu, W. J. Peacock, and R. W. Hardy in *Journal of Ultrastructure Research*, Vol. 68, No. 1, pages 96–107; January, 1977.

EXPERIMENTAL POPULATION GENETICS OF MEIOTIC DRIVE SYSTEMS, I: PSEUDO-*Y* CHROMOSOMAL DRIVE AS A MEANS OF ELIMINATING CAGE POPULATIONS OF *Drosophila Melanogaster*. Terrence W. Lyttle in *Genetics*, Vol. 86, pages 413–445; June, 1977.

ON THE COMPONENTS OF SEGREGATION DISTORTION IN *Drosophila Melanogaster*. Barry Ganetzky in *Genetics*, Vol. 86, pages 321–355; June, 1977.

3. The Genes of Men and Molds

GENES AND THE CHEMISTRY OF ORGANISM. G. W. Beadle In *American Scientist*, Vol. 34, pages 31–53; 1946.

THE PRINCIPLES OF HEREDITY, Third Edition. Laurence H. Snyder. D. C. Heath, 1946.

THE PHYSIOLOGY OF THE GENE. S. Wright in *Physiological Reviews*, Vol. 21, pages 487–527; 1941.

II THE CHEMICAL BASIS OF HEREDITY

4. Transformed Bacteria

THE GENETIC CHEMISTRY OF THE PNEUMOCOCCAL TRANSFORMATIONS. Rollin D. Hotchkiss in *The Harvey Lectures (1953–54)*, Series 49, pages 124–144. Academic Press Inc., 1955.

HELICAL STRUCTURE OF DEOXYPENTOSE NUCLEIC ACID. M. H. F. Wilkins and others in *Nature*, Vol 172, No. 4382, pages 759–762; October 24, 1953.

SYMPOSIUM PAPERS ON THE NUCLEIC ACIDS. Proceedings of the National Academy of Sciences *(in press)*.

5. The Structure of the Hereditary Material

THE BIOCHEMISTRY OF THE NUCLEIC ACIDS. J. N. Davidson. Methuen & Co., Ltd., 1954.

6. The Duplication of Chromosomes

THE BIOCHEMISTRY OF THE NUCLEIC ACIDS. J. N. Davidson. John Wiley & Sons, Inc., 1953.

THE NUCLEIC ACIDS. Edited by Erwin Chargaff and J. N. Davidson. Academic Press, Inc., 1955.

NUCLEIC ACIDS. F. H. C. Crick in *Scientific American,* Vol. 197, No. 3, pages 188–200; September, 1957. (Offprint 84.)

THE ORGANIZATION AND DUPLICATION OF CHROMOSOMES AS REVEALED BY AUTORADIOGRAPHIC STUDIES USING TRITIUM-LABELED THYMIDINE. J. Herbert Taylor, Philip S. Woods and Walter L. Hughes in *Proceedings of the National Academy of Sciences,* Vol 43, No. 1, pages 122–128; January, 1957.

PHYSICAL TECHNIQUES IN BIOCHEMICAL RESEARCH. VOL. III: CELLS AND TISSUES. Edited by Gerald Oster and Arthur W. Pollister. Academic Press, Inc., 1956.

A SYMPOSIUM ON THE CHEMICAL BASIS OF HEREDITY. William D. McElroy and Bentley Glass. Johns Hopkins Press, 1957.

7. The Bacterial Chromosome

THE BACTERIAL CHROMOSOME AND ITS MANNER OF REPLICATION AS SEEN BY AUTORADIOGRAPHY. John Cairns, in *Journal of Molecular Biology,* Vol. 6, No. 3, pages 208–213; March, 1963.

COLD SPRING HARBOR SYMPOSIA ON QUANTITATIVE BIOLOGY, VOLUME XXVIII: SYNTHESIS AND STRUCTURE OF MACROMOLECULES. Cold Spring Harbor Laboratory of Quantitative Biology, 1963.

MOLECULAR BIOLOGY OF THE GENE. James D. Watson. W. A. Benjamin, Inc., 1965. See pages 255–296.

III   GENETIC ANALYSIS

8. "Transduction" in Bacteria

BACTERIAL TRANSDUCTION. Norton D. Zinder in *Journal of Cellular and Comparative Physiology.* Vol. 45, Supplement 2, pages 23–49; May, 1955.

GENETIC EXCHANGE IN SALMONELLA. Norton D. Zinder and Joshua Lederberg in *Journal of Bacteriology,* Vol. 64, No. 5, pages 679–699; November, 1952.

GENETIC STUDIES WITH BACTERIA. M. Demerec, Zlata Hartman, Philip E. Hartman, Takashi Yura, Joseph S. Gots, Haruo Ozeki and S. W. Glover. Publication 612, Carnegie Institution of Washington, 1956.

"TRANSDUCTION IN *Escherichia coli* k-12." M. L. Morse, Esther M. Lederberg and Joshua Lederberg in *Genetics,* Vol. 41, No. 1, pages 142–156; January, 1956.

"TRANSDUCTION OF FLAGELLAR CHARACTERS IN *Salmonella.*" B. A. D. Stocker, N. D. Zinder and J. Lederberg in *The Journal of General Microbiology,* Vol. 9, No. 3, pages 410–433; December, 1953.

"TRANSDUCTION OF LINKED GENETIC CHARACTERS OF THE HOST BY BACTERIOPHAGE P1." E. S. Lennox in *Virology,* Vol. 1, Nos. 1–5, pages 190–206; 1955.

9. Viruses and Genes

THE CONCEPT OF VIRUS. A. Lwoff in *The Journal of General Microbiology,* Vol. 17, No. 2, pages 239–253; October, 1957.

GENETIC CONTROL OF VIRAL FUNCTIONS. François Jacob in *The Harvey Lectures,* Series LIV, 1958–1959, pages 1–39. Academic Press Inc., 1960.

MICROBIAL GENETICS. Tenth Symposium of the Society for General Microbiology. Cambridge University Press, 1960.

PHYSIOLOGICAL ASPECTS OF BACTERIOPHAGE GENETICS. S. Brenner in *Advances in Virus Research,* Vol. 6, pages 137–158; 1959.

A SYMPOSIUM ON THE CHEMICAL BASIS OF HEREDITY. Edited by William D. McElroy. Johns Hopkins Press, 1957.

VIRUSES AS INFECTIVE GENETIC MATERIAL. S. E. Luria in *Immunity and Virus Infection,* edited by Victor A. Najjar, pages 188–195. John Wiley and Sons, Inc., 1959.

THE VIRUSES: BIOCHEMICAL, BIOLOGICAL, AND BIOPHYSICAL PROPERTIES. Edited by F. M. Burnet and W. M. Stanley. Academic Press Inc., 1956.

10. The Genetics of a Bacterial Virus

BACTERIOPHAGE REPRODUCTION. Sewell P. Champe in *Annual Review of Microbiology, Vol. 17, 1963.*

MOLECULAR BIOLOGY OF BACTERIAL VIRUSES. Gunther S. Stent. W. H. Freeman and Company, 1963.

"PHYSIOLOGICAL STUDIES OF CONDITIONAL LETHAL MUTANTS OF BACTERIOPHAGE T4D." R. H. Epstein, A. Bolle, C. M. Steinberg, E. Kellenberger, E. Boy de la Tour, R. Chevalley, R. S. Edgar, M. Susman, G. H. Denhardt and A. Lielausis in *Cold Spring Harbor Symposia on Quantitative Biology, Vol. XXVIII.* 1963.

11. The Fine Structure of the Gene

THE ELEMENTARY UNITS OF HEREDITY. Seymour Benzer in *The Chemical Basis of Heredity,* edited by William D. McElroy and Bentley Glass, pages 70–93. The Johns Hopkins Press, 1957.

GENETIC RECOMBINATION BETWEEN HOST-RANGE AND PLAQUE-TYPE MUTANTS OF BACTERIOPHAGE IN SINGLE BACTERIAL CELLS. A. D. Hershey and Raquel Rotman in *Genetics*, Vol. 34, No. 1, pages 44–71; January, 1949.

INDUCTION OF SPECIFIC MUTATIONS WITH 5-BROMOURACIL. Seymour Benzer and Ernst Freese in *Proceedings of the National Academy of Sciences*, Vol. 44, No. 2, pages 112–119; February, 1958.

THE STRUCTURE OF THE HEREDITARY MATERIAL. F. H. C. Crick in *Scientific American*, Vol. 191, No. 4, pages 54–61; October, 1954.

ON THE TOPOGRAPHY OF THE GENETIC FINE STRUCTURE. Seymour Benzer in *Proceedings of the National Academy of Sciences*, Vol. 47, No. 3, pages 403–415; March, 1961.

12. Hybrid Cells and Human Genes

HYBRIDIZATION OF SOMATIC CELLS. B. Ephrussi. Princeton University Press, 1972.

LINKAGE ANALYSIS IN MAN BY SOMATIC CELL GENETICS. Frank H. Ruddle in *Nature*, Vol. 242, No. 5394, pages 165–167; March 16, 1973.

ASSIGNMENT OF THREE GENE LOCI (PGK, HGPRT, G6PD) TO THE LONG ARM OF THE HUMAN X CHROMOSOME BY SOMATIC CELL GENETICS. Florence C. Ricciuti and Frank H. Ruddle in *Genetics*, Vol. 74, No. 4, pages 661–678; August, 1973.

LOCALIZATION AND INDUCTION OF THE HUMAN THYMIDINE KINASE GENE BY ADENOVIRUS 12. Raju Kucherlapati, James K. McDougall and Frank H. Ruddle in *Nature New Biology*, Vol. 245, No. 145, pages 170–173; October 10, 1973.

IV GENE EXPRESSION AND REGULATION

13. How Do Genes Act?

THE ABNORMAL HUMAN HEMOGLOBINS. Harvey A. Itano, William R. Bergren and Phillip Sturgeon in *Medicine*, Vol. 35, No. 2, pages 121–159; May, 1956.

THE BIOCHEMISTRY OF GENETICS. J. B. S. Haldane. George Allen & Unwin Ltd., 1954.

A SYMPOSIUM ON THE CHEMICAL BASIS OF HEREDITY. William D. McElroy and Bentley Glass. Johns Hopkins Press, 1957.

14. The Genetic Code

THE FINE STRUCTURE OF THE GENE. Seymour Benzer in *Scientific American*, Vol. 206, No. 1, pages 70–84; January, 1962. (Offprint 120.)

GENERAL NATURE OF THE GENETIC CODE FOR PROTEINS. F. H. C. Crick, Leslie Barnett, S. Brenner and R.J. Watts-Tobin in *Nature*, Vol. 192, No. 4809, pages 1227–1232; December 30, 1961.

MESSENGER RNA. Jerard Hurwitz and J. J. Furth in *Scientific American*, Vol. 206, No. 2, pages 41–49; February, 1962. (Offprint 119.)

THE NUCLEIC ACIDS: VOL. III. Edited by Erwin Chargaff and J. N. Davidson, Academic Press Inc., 1960.

15. The Genetic Code: III

THE GENETIC CODE, Vol. XXXI: 1966 COLD SPRING HARBOR SYMPOSIA ON QUANTITATIVE BIOLOGY. Cold Spring Harbor Laboratory of Quantitative Biology, in press.

MOLECULAR BIOLOGY OF THE GENE. James D. Watson W. A. Benjamin, Inc., 1965.

RNA CODEWORDS AND PROTEIN SYNTHESIS, VII: ON THE GENERAL NATURE OF THE RNA CODE. M. Nirenberg, P. Leder, M. Bernfield, R. Brimacombe, J. Trupin, F. Rottman and C. O'Neal in *Proceedings of the National Academy of Sciences*, Vol. 53, No. 5, pages 1161–1168; May, 1965.

STUDIES ON POLYNUCLEOTIDES, LVI: FURTHER SYNTHESIS, IN VITRO, OF COPOLYPEPTIDES CONTAINING TWO AMINO ACIDS IN ALTERNATING SEQUENCE DEPENDENT UPON DNA-LIKE POLYMERS CONTAINING TWO NUCLEOTIDES IN ALTERNATING SEQUENCE. D. S. Jones, S. Nishimura, and H. G. Khorana in *Journal of Molecular Biology*, Vol. 16, No. 2, pages 454–472; April, 1966.

16. Gene Structure and Protein Structure

CO-LINEARITY OF β-GALACTOSIDASE WITH ITS GENE BY IMMUNOLOGICAL DETECTION OF INCOMPLETE POLYPEPTIDE CHAINS. Audree V. Fowler and Irving Zabin in *Science*, Vol. 154, No. 3752, pages 1027–1029; November 25, 1966.

CO-LINEARITY OF THE GENE WITH THE POLYPEPTIDE CHAIN. A. S. Sarabhai, A. O. W. Stretton, S. Brenner and A. Bolle in *Nature*, Vol. 201, No. 4914, pages 13–17; January 4, 1964.

THE COMPLETE AMINO ACID SEQUENCE OF THE TRYPTOPHAN SYNTHETASE A PROTEIN (α SUBUNIT) AND ITS COLINEAR RELATIONSHIP WITH THE GENETIC MAP OF THE A GENE. Charles Yanofsky, Gabriel R. Drapeau, John R. Guest and Bruce C. Carlton in

Proceedings of the National Academy of Sciences, Vol. 57, No. 2, pages 296–298; February, 1967.

MUTATIONALLY INDUCED AMINO ACID SUBSTITUTIONS IN A TRYPTIC PEPTIDE OF THE TRYPTOPHAN SYNTHETASE A PROTEIN. John R. Guest and Charles Yanofsky in *Journal of Biological Chemistry,* Vol. 240, No. 2, pages 679–689; February, 1965.

ON THE COLINEARITY OF GENE STRUCTURE AND PROTEIN STRUCTURE. C. Yanofsky, B. C. Carlton, J. R. Guest, D. R. Helinski and U. Henning in *Proceedings of the National Academy of Sciences,* Vol. 51, No. 2, pages 266–272; February, 1964.

17. The Nucleotide Sequence of a Viral DNA

A RAPID METHOD FOR DETERMINING SEQUENCES IN DNA BY PRIMED SYNTHESIS WITH DNA POLYMERASE. Frederick Sanger and Alan R. Coulson in *Journal of Molecular Biology,* Vol. 94, No. 3, pages 441–448; May 25, 1975.

OVERLAPPING GENES IN BACTERIOPHAGE φX174, B. G. Barrell, G. M. Air and C. A. Hutchison III in *Nature,* Vol. 264, No. 5581, pages 34–41; November 4, 1976.

NUCLEOTIDE SEQUENCE OF BACTERIOPHAGE φX174 DNA. Frederick Sanger, G. M. Air, B. G. Barrell, N. L. Brown, Alan R. Coulson, John C. Fiddes, C. A. Hutchison III, P. M. Slocombe and M. Smith in *Nature,* Vol. 265, No. 5595, pages 687–695; February 24, 1977.

18. A DNA Operator-Repressor System

STRUCTURE OF THE λ OPERATORS. Tom Maniatis and Mark Ptashne in *Nature,* Vol. 246, No. 5429, pages 133–136; November 16, 1973.

THE NUCLEOTIDE SEQUENCE OF THE LAC OPERATOR. Walter Gilbert and Allan Maxam in *Proceedings of the National Academy of Sciences,* Vol. 70, No. 12, pages 3581–3584; December, 1973.

CONTROL ELEMENTS IN THE DNA OF BACTERIOPHAGE λ. Tom Maniatis, Mark Ptashne and Russell Maurer in *Cold Spring Harbor Symposia on Quantitative Biology,* Vol. XXXVIII, pages 857–869; 1974.

REPRESSOR, OPERATORS, AND PROMOTERS IN BACTERIOPHAGE LAMBDA. Mark Ptashne in *The Harvey Lectures: Series 69.* Academic Press, 1975.

V GENETIC TRANSACTIONS

19. The Molecule of Infectious Drug Resistance

EXTRACHROMOSOMAL INHERITANCE IN BACTERIA. Richard P. Novick in *Bacteriological Reviews,* Vol. 33, No. 2, pages 210–235; June, 1969.

THE PROBLEMS OF DRUG-RESISTANT PATHOGENIC BACTERIA. Conference of the New York Academy of Sciences, edited by Eugene L. Dulaney and Allen I. Laskin in *Annals of the New York Academy of Sciences,* Vol. 182, June 11, 1971.

MOLECULAR STRUCTURE OF BACTERIAL PLASMIDS. Royston C. Clowes in *Bacteriological Reviews,* Vol. 36, No. 3, pages 361–405; September, 1972.

DRUG-RESISTANCE FACTORS AND OTHER BACTERIAL PLASMIDS. G. G. Meynell. The MIT Press, 1973.

20. How Viruses Insert their DNA into the DNA of the Host Cell

THE LINEAR INSERTION OF A PROPHAGE INTO THE CHROMOSOME OF *E. COLI* SHOWN BY DELETION MAPPING. Naomi C. Franklin, William F. Dove and Charles Yanofsky in *Biochemical and Biophysical Research Communications,* Vol. 18, No. 5–6, pages 910–923; March 12, 1965.

EPISOMES. Allan M. Campbell. Harper & Row, Publishers, 1969.

THE BACTERIOPHAGE LAMBDA. Edited by A. D. Hershey. Cold Spring Harbor Laboratory, 1971.

NOTE ON THE STRUCTURE OF PROPHAGE λ. Phillip A. Sharp, Ming-Ta Hsu and Norman Davidson in *Journal of Molecular Biology,* Vol. 71, No. 2. pages 499–501; November 14, 1972.

INTEGRATIVE RECOMBINATION OF BACTERIOPHAGE LAMBDA DNA *IN VITRO.* Howard A. Nash in *Proceedings of the National Academy of Sciences of the United States of America,* Vol. 72, No. 3, pages 1072–1076; March, 1975.

DNA INSERTION ELEMENTS, PLASMIDS AND EPISOMES. Edited by A. I. Bukhari, J. Shapiro and S. Adhya. Cold Spring Harbor Laboratory, 1977.

21. RNA-Directed DNA Synthesis

ONCOGENIC RIBOVIRUSES. Frank Fenner in *The Biology of Animal Viruses: Vol. II.* Academic Press, 1968.

CENTRAL DOGMA OF MOLECULAR BIOLOGY. Francis Crick in *Nature,* Vol. 227, No. 5258, pages 561–563; August 8, 1970.

MECHANISM OF CELL TRANSFORMATION BY RNA TUMOR VIRUSES. Howard M. Temin in *Annual Review of Microbiology: Vol. XXV*. Annual Reviews, Inc. 1971.

THE PROTOVIRUS HYPOTHESIS: SPECULATIONS ON THE SIGNIFICANCE OF RNA-DIRECTED DNA SYNTHESIS FOR NORMAL DEVELOPMENT AND FOR CARCINOGENESIS. Howard M. Temin in *Journal of the National Cancer Institue*, Vol 46, No. 2, pages III–VII; February, 1971.

22. Transposable Genetic Elements

THE CONTROL OF GENE ACTION IN MAIZE. Barbara McClintock in *Brookhaven Symposia in Biology*. No. 18, pages 162–184; 1965.

IS-ELEMENTS IN MICROORGANISMS. Peter Starlinger and Heinz Saedler in *Current Topics in Microbiology and Immunology*, Vol 75, pages 111–152; 1976.

TRANSPOSABLE GENETIC ELEMENTS AND PLASMID EVOLUTION. Stanley N. Cohen in *Nature*, Vol. 263, No. 5580, pages 731–738; October 28, 1976.

DNA INSERTION ELEMENTS, PLASMIDS AND EPISOMES. Edited by A. I. Bukhari, J. A. Shapiro and S. L. Adhya. Cold Spring Harbor Laboratory, 1977.

MOLECULAR MODEL FOR THE TRANSPOSITION AND REPLICATION OF BACTERIOPHAGE MU AND OTHER TRANSPOSABLE ELEMENTS. James A Shapiro in *Proceedings of the National Academy of Sciences of the United States of America*, Vol. 76, No. 4, pages 1933–1937; April 1979.

VI EVOLUTION

23. Evolution

ON THE ORIGIN OF SPECIES. Charles Darwin. Facsimile Edition. Harvard University Press, 1964.

THE NATURE OF THE DARWINIAN REVOLUTION. Ernst Mayr in *Science*, Vol. 176, No. 4038, pages 981–989; June 2, 1972.

DARWIN AND NATURAL SELECTION. Ernst Mayr in *American Scientist*, Vol. 65, No. 3, pages 321–327; May–June, 1977.

24. Adaptation

ADAPTATION AND NATURAL SELECTION: A CRITIQUE OF SOME CURRENT EVOLUTIONARY THOUGHT. George C. Williams. Princeton University Press. 1966.

EVOLUTION IN CHANGING ENVIRONMENTS: SOME THEORETICAL EXPLORATIONS. Richard Levins. Princeton University Press, 1968.

ADAPTATION AND DIVERSITY: NATURAL HISTORY AND THE MATHEMATICS OF EVOLUTION. E. G. Leigh, Jr. Freeman, Cooper and Company, 1971.

25. Repeated Segments of DNA

A MOLECULAR APPROACH IN THE SYSTEMATICS OF HIGHER ORGANISMS. B. H. Hoyer, B. J. McCarthy and E. T. Bolton in *Science*, Vol. 144, No. 3621, pages 959–967; May 22, 1964.

NUCLEOTIDE SEQUENCE REPETITION: A RAPIDLY REASSOCIATING FRACTION OF MOUSE DNA. Michael Waring and Roy J. Britten in *Science*, Vol. 154, No. 3750, pages 791–794; November 11, 1966.

REPEATED SEQUENCES IN DNA. R. J. Britten and D. E. Kohne in *Science*, Vol. 161, No. 3841, pages 529–540; August 9, 1968.

HIGHLY REPETITIVE DNA IN RODENTS. P. M. B. Walker, W. G. Flamm and A. McClaren in *Handbook of Molecular Cytology*, edited by A. Lima-De-Faria. North-Holland Publishing, 1969.

26. The Neutral Theory of Molecular Evolution

MOLECULAR POPULATION GENETICS AND EVOLUTION. Masatoshi Nei. North-Holland Publishing Co., 1975.

STATISTICAL STUDIES ON PROTEIN POLYMORPHISM IN NATURAL POPULATIONS, I: DISTRIBUTION OF SINGLE LOCUS HETEROZYGOSITY. Paul A Fuerst, Ranajit Chakraborty and Masatoshi Nei in *Genetics*. Vol. 86, No. 2, Part 1, pages 455–483; June 1977.

LACK OF EXPERIMENTAL EVIDENCE FOR FREQUENCY-DEPENDENT SELECTION AT THE ALCOHOL DEHYDROGENASE LOCUS IN DROSOPHILA MELANOGASTER. Hiroshi Yoshimaru and Terumi Mukai in *The Proceedings of the National Academy of Sciences of the United States of America*, Vol. 76, No. 2, pages 876–878; February, 1979.

MODEL OF EFFECTIVELY NEUTRAL MUTATIONS IN WHICH SELECTIVE CONSTRAINT IS INCORPORATED. Motoo Kimura in *The Proceedings of the National Academy of Sciences of the United States of America*, Vol. 76, No. 7, pages 3440–3444; July, 1979.

THEORETICAL ASPECTS OF POPULATION GENETICS. Motoo Kimura and Tomoka Ohta. Princeton University Press, 1971.

VII APPLIED GENETICS

27. The Recombinant-DNA Debate

THE MANIPULATION OF GENES. Stanley N. Cohen in *Scientific American*, Vol. 233, No. 1, pages 24–33; July, 1975. (Offprint 1324.)

MOLECULAR BIOLOGY OF THE GENE. J. D. Watson. Addison-Wesley Publishing Co., Inc. 1977.

NATIONAL INSTITUTES OF HEALTH GUIDELINES FOR RESEARCH INVOLVING RECOMBINANT DNA MOLECULES. Public Health Service, U.S. Department of Health, Education, and Welfare, June 23, 1976.

RECOMBINANT DNA RESEARCH: BEYOND THE NIH GUIDELINES. Clifford Grobstein in *Science*, Vol. 194, No. 4270, pages 1133–1135; December 10, 1976.

NOTES OF A BIOLOGY-WATCHER: THE HAZARDS OF SCIENCE. Lewis Thomas in *The New England Journal of Medicine*, Vol. 296, No. 6, pages 324–328; February 10, 1977.

RECOMBINANT DNA: FACT AND FICTION. Stanley N. Cohen in *Science*, Vol. 195, No. 4279, pages 654–657; February 18, 1977.

28. Genetic Amniocentesis

EARLY DIAGNOSIS OF HUMAN GENETIC DEFECTS: SCIENTIFIC AND ETHICAL CONSIDERATIONS. Edited by Maureen Harris. Government Printing Office, 1971.

ALPHA-FETOPROTEIN IN THE ANTENATAL DIAGNOSIS OF ANENCEPHALY AND SPINA BIFIDA. D. J. H. Brock and R. G. Sutcliffe in *The Lancet*, Vol. 2 for 1972, No. 7770, pages 197–199; July 29, 1972.

MIDTRIMESTER AMNIOCENTESIS FOR PRENATAL DIAGNOSIS: SAFETY AND ACCURACY. The NICHD National Registry for Amniocentesis Study Group in *The Journal of the American Medical Association*. Vol. 236, No. 13, pages 1471–1476; September 27, 1976.

ANTENATAL DIAGNOSIS OF SICKLE-CELL ANAEMIA BY DNA ANALYSIS OF AMNIOTIC-FLUID CELLS. Yuet Wai Kan and Andrée M. Dozy in *The Lancet*, Vol. 2 for 1978, No. 8096, pages 910–911; October 28, 1978.

29. Bacterial Tests for Potential Carcinogens

METHODS FOR DETECTING CARCINOGENS AND MUTAGENS WITH THE *SALMONELLA*/MAMMALIAN-MICROSOME MUTAGENICITY TEST. Bruce N. Ames, Joyce McCann, and Edith Yamasaki in *Mutation Research*, Vol. 31, No. 6, pages 347–364; December, 1975.

PROPHAGE λ INDUCTION IN *ESCHERICHIA COLI* K12 ENVA UVRB: A HIGHLY SENSITIVE TEST FOR POTENTIAL CARCINOGENS. Patrice Moreau, Adriana Bailone and Raymond Devoret in *Proceedings of the National Academy of Sciences of the United States of America*, Vol. 73, No. 10, pages 3700–3704; October, 1976.

ULTRAVIOLET MUTAGENESIS AND INDUCTIBLE DNA REPAIR IN *ESCHERICHIA COLI*. Evelyn M. Witkin in *Bacteriological Reviews*, Vol. 40, No. 4, pages 869–907; December, 1976.

Scientific American Articles on Genetics
(to July, 1980)

Allison, Anthony C., "Sickle Cells and Evolution." Aug. 1956 (Offprint 1065).

Ayala, Francisco J., "The Mechanisms of Evolution." Sept. 1978 (Offprint 1407).

Bauer, W. R., F. H. C. Crick, and James H. White, "Supercoiled DNA." July 1980 (Offprint 1474).

Beadle, George W., "Ionizing Radiation and the Citizen." Sept. 1959 (page 219).

Beadle, George W., "Genes of Men and Moulds." Sept. 1948 (Offprint 1).

Bearn, A. G., and James L. German III, "Chromosomes and Disease." Nov. 1961 (Offprint 150).

Bearn, A. G., "The Chemistry of Hereditary Disease." Dec. 1956 (page 126).

Beerman, Wolfgang, and Ulrich Clever, "Chromsome Puffs." April 1964 (Offprint 180).

Benzer, Seymour, "The Fine Structure of the Gene." Jan. 1962 (Offprint 120).

Benzer, Seymour, "The Genetic Dissection of Behavior." Dec. 1973 (Offprint 1285).

Bishop, J. A., and Laurence M. Cook, "Moths, Melanism and Clean Air." Jan. 1975 (Offprint 1314).

Bodmer, Walter F., and Luigi Luca Cavalli-Sforza, "Intelligence and Race." Oct. 1970 (Offprint 1199).

Brady, Roscoe, "Hereditary Fat-metabolism Diseases." Aug. 1973 (page 83).

Britten, Roy J., and David E. Kohne, "Repeated Segments of DNA." April 1970 (Offprint 1173).

Brown, Donald D., "The Isolation of Genes." Aug. 1973 (Offprint 1278).

Buzzati-Traverso, A., "The State of Genetics." Oct. 1951 (page 22).

Cairns, John, "The Bacterial Chromosome." Jan. 1966 (Offprint 1030).

Cairns, John, "The Cancer Problem." Nov. 1975 (Offprint 1330).

Campbell, Allan M., "How Viruses Insert Their DNA into the DNA of the Host Cell." Dec. 1976 (Offprint 1347).

Cavalli-Sforza, Luigi, " 'Genetic Drift' in an Italian Population." Aug. 1969 (page 30).

Cavalli-Sforza, L. L., "The Genetics of Human Populations." Sept. 1974 (page 80).

Clarke, Bryan, "The Causes of Biological Diversity." Aug. 1975 (Offprint 1326).

Clowes, Royston C., "The Molecule of Infectious Drug Resistance." July 1975 (Offprint 1269).

Cohen, Stanley N., "The Manipulation of Genes." July 1975 (Offprint 1324).

Cohen, S. N., and J. A. Shapiro, "Transposable Genetic Elements." Feb. 1980 (Offprint 1460).

Crick, F. H. C., "Nucleic Acids." Sept. 1957 (Offprint 54).

Crick, F. H. C., "The Structure of the Hereditary Material." Oct. 1954 (page 54).

Crick, F. H. C., "The Genetic Code." Oct. 1962 (Offprint 123).

Crick, F. H. C., "The Genetic Code: III." Oct. 1966 (Offprint 1052).

Croce, Carlo M., and Hilary Koprowski, "The Genetics of Human Cancer." Feb. 1978 (Offprint 1381).

Crow, James, "Ionizing Radiation and Evolution." Sept. 1959 (Offprint 55).

Crow, James, "Genes That Violate Mendel's Rules." Feb. 1979 (Offprint 1418).

Curtis, Byron C., and David R. Johnston, "Hybrid Wheat." May 1969 (page 21).

Darlington, C. D., "The Origin of Darwinism." May 1959 (page 60).

Davidson, Eric H., "Hormones and Genes." June 1965 (Offprint 1013).

Dayhoff, Margaret Oakley, "Computer Analysis of Protein Evolution." July 1969 (page 86).

Delbruck, Max, and Mary Bruce, "Bacterial Viruses and Sex." Nov. 1948 (page 46).

Devoret, Raymond, "Bacterial Tests for Potential Carcinogens." Aug. 1979 (Offprint 1433).

Dickerson, Richard E., "The Structure and History of an Ancient Protein." April 1972 (Offprint 1245).

Dickerson, Richard E., "Chemical Evolution and the Origin of Life." Sept. 1978 (Offprint 1401).

Dobzhansky, Theodosius, "Genetics." Sept. 1950 (page 55).

Dobzhansky, Theodosius, "The Genetic Basis of Evolution." Nov. 1950 (Offprint 6).

Dobzhansky, Theodosius, "The Present Evolution of Man." Sept. 1960 (page 206).

Dulbecco, R., "The Induction of Cancer by Viruses." April 1967 (Offprint 1069).

Dunn, L. C., "Genetic Monsters." June 1950 (page 16).

Eckhardt, Robert B., "Population Genetics and Human Origins." Jan. 1972 (Offprint 676).

Edgar, R. S., and R. H. Epstein, "The Genetics of a Bacterial Virus." Feb. 1965 (Offprint 1004).

Eiseley, Loren C., "Alfred Russel Wallace." Feb. 1959 (page 70).

Eiseley, Loren C., "Charles Darwin." Feb. 1956 (Offprint 108).

Ephrussi, Boris, and Mary C. Weiss, "Hybrid Somatic Cells." April 1969 (Offprint 1137).

Fiddes, John C., "The Nucleotide Sequence of a Viral DNA." Dec. 1977 (Offprint 1374).

Frankel-Conrat, Heinz, "The Genetic Code of a Virus." Oct. 1964 (page 46).

Friedmann, Theodore, "Prenatal Diagnosis of Disease." Nov. 1971 (Offprint 1234).

Fuchs, Fritz, "Genetic Amniocentesis." June 1980 (page 47).

Gamow, George, "Information Transfer in the Living Cell." Oct. 1955 (page 70).

Glass, H. Bentley, "The Genetics of the Dunkers." Aug. 1953 (Offprint 1062).

Glass, H. Bentley, "Maupertius, A Forgotten Genius." Oct. 1955 (page 100).

Goldschmidt, Richard B., "Phenocopies." Oct. 1949 (page 46).

Goodenough, Ursula W., and R. P. Levine, "The Genetic Activity of Mitochondria and Chloroplasts." Nov. 1970 (Offprint 1203).

Gorini, Luigi, "Antibiotics and the Genetic Code." April 1966 (page 102).

Gurdon, John B., "Transplanted Nuclei and Cell Differention." Dec. 1968 (Offprint 1128).

Grobstein, Clifford, "The Recombinant DNA Debate." July 1977 (Offprint 1362).

Hadorn, Ernst, "Fractionating the Fruit Fly." April 1962 (Offprint 1166).

Hanawalt, Philip C., and Robert H. Haynes, "The Repair of DNA." Feb. 1967 (page 63).

Hannah-Alava, Aloha, "Genetic Mosaics." May 1960 (page 118).

Hollaender, Alexander, and George E. Stapleton, "Ionizing Radiation and the Living Cell." Sept. 1959 (page 94).

Holley, Robert W., "The Nucleotide Sequence of a Nucleic Acid." Feb. 1966 (Offprint 1033).

Horowitz, Norman H., "The Gene." Oct 1956 (Offprint 17).

Hotchkiss, R. D., and E. Weiss, "Transformed Bacteria." Nov. 1956 (Offprint 18).

Hurwitz, Jerard, and J. J. Furth, "Messenger RNA." Feb. 1962 (Offprint 119).

Ingram, Vernon M., "How Do Genes Act?" Jan. 1958 (Offprint 104).

Jacob, François, and Elie L. Wollman, "Sexuality in Bacteria." July 1956 (page 109).

Jacob, Francois, and Elie L. Wollman, "Viruses and Genes." June 1961 (Offprint 89).

Joravsky, David, "The Lysenko Affair." Nov. 1962 (page 41).

Kalmus, Hans, "Inherited Sense Defects." May 1952 (Offprint 406).

Kimura, M., "The Neutral Theory of Molecular Evolution." Nov. 1979 (Offprint 1451).

Kellenberger, Edouard, "The Genetic Control of the Shape of a Virus." Dec. 1966 (Offprint 1058).

Kettlewell, H. B. D., "Darwin's Missing Evidence." March 1959 (Offprint 842).

Knight, C. A., and Dean Fraser, "The Mutation of Viruses." July 1955 (Offprint 59).

Kornberg, Arthur, "The Synthesis of DNA." Oct. 1968 (Offprint 1124).

Lack, David, "Darwin's Finches." April 1953 (Offprint 22).

Ledley, Robert S., and Frank H. Ruddle, "Chromosome Analysis by Computer." April 1966 (Offprint 1040).

Lewontin, Richard C., "Adaptation." Sept. 1978 (Offprint 1408).

Luria, Salvador E., "The Recognition of DNA in Bacteria." Jan. 1970 (page 88).

Lwoff, Andre, "The Life Cycle of a Virus." March 1954 (page 34).

MacAlpine, Ida, and Richard Hunter, "Porphyria and King George III." July 1969 (Offprint 1149).

McKusick, Victor A., "The Royal Hemophilia." Aug. 1965 (page 88).

McKusick, Victor A., "The Mapping of Human Chromosomes." April 1971 (Offprint 1220).

Manglesdorf, Paul C., "Hybrid Corn." Aug. 1951 (Offprint 1150).

Manglesdorf, Paul C., "The Mystery of Corn." July 1950 (Offprint 26).

Manglesdorf, Paul C., "Wheat." July 1953 (page 50).

Maniatis, Tom, and Mark Ptashne, "A DNA Operator-Repressor System." Jan. 1976 (Offprint 1333).

Margulis, Lynn, "Symbiosis and Evolution." Aug. 1971 (Offprint 1230).

May, Robert M., "The Evolution of Ecological Systems." Sept. 1978 (Offprint 1404).

Mayr, Ernst, "Evolution." Sept. 1978 (Offprint 1400).

Melnick, Joseph L., and Fred Rapp, "The Footprints of Tumor Viruses." March 1966 (page 34).

Miller, O. L., Jr. "The Visualization of Genes in Action." March 1973 (page 34).

Mirsky, Alfred E., "The Chemistry of Heredity." Feb. 1953 (Offprint 28).

Mirsky, Alfred E., "The Discovery of DNA." June 1968 (Offprint 1109).

Mittwoch, Ursula, "Sex Differences in Cells." July 1963 (page 54).

Muller H. J., "Radiation and Human Mutation." Nov. 1955 (Offprint 29).

Nirenberg, Marshall W., "The Genetic Code, II." March 1963 (Offprint 153).

Ptashne, Mark, and Walter Gilbert, "Genetic Repressors." June 1970 (Offprint 1179).

Puck, Theodore T., "Radiation and the Human Cell." April 1960 (page 142).

Ruddle, Frank H., and Raja S. Kucherlapati, "Hybrid Cells and Human Genes." July 1974 (Offprint 1300).

Ryan, Francis J., "Evolution Observed," Oct. 1953 (page 78).

Sager, Ruth, "Genes Outside the Chromosome." Jan. 1965 (Offprint 1002).

Schopf, J. William, "The Evolution of the Earliest Cells." Sept. 1978 (page 110).

Sigurbjornsson, Bjorn, "Induced Mutations in Plants." Jan. 1971 (Offprint 1210).

Sinsheimer, Robert L., "Single-stranded DNA." July 1962 (Offprint 128).

Smith, John Maynard, "The Evolution of Behavior." Sept. 1978 (page 176).

Sonneborn, T. M., "Partner of the Genes." Nov. 1950 (Offprint 39).

Southern, H. N., "A Study in the Evolution of Birds." May 1957 (page 124).

Spector, Deborah H., and David Baltimore, "The Molecular Biology of Polio Virus." May 1975 (page 24).

Spiegelman, S., "Hybrid Nucleic Acids." May 1964 (page 48).

Stebbins, Ledyard G., Jr., "Cataclysmic Evolution." April 1951 (page 54).

Stein, Gary S., and Janet Swinehart Stein, "Chromosomal Proteins and Gene Regulation." Feb. 1975 (Offprint 1315).

Stern, Curt, "Man's Genetic Future." Feb. 1952 (page 68).

Strong, Leonell C., "Genetics and Cancer." July 1950 (page 44).

Taylor, Herbert J., "The Duplication of Chromosomes." June 1958 (page 36).

Temin, Howard, "RNA-Directed DNA Synthesis." Jan. 1972 (Offprint 1239).

Tomasz, Alexander, "Cellular Factors in Genetic Transformation." Jan. 1969 (page 38).

Valentine, James W., "The Evolution of Multicellular Plants and Animals." Sept. 1978 (page 140).

Washburn, Sherwood L., "The Evolution of Man." Sept. 1978 (page 144).

Watanabe, Tsutomu, "Infectious Drug Resistance." Dec. 1967 (page 19).

Wills, Christopher, "Genetic Load." March 1970 (Offprint 1172).

Wollman, E. L., and Francois Jacob, "Sexualtiy in Bacteria." July 1956 (Offprint 50).

Wood, William B., and Robert S. Edgar, "Building a Bacterial Virus." July 1967 (Offprint 1079).

Yanofsky, Charles, "Gene Structure and Protein Structure." May 1967 (Offprint 1074).

Zahl, Paul A., "The Evolution of Sex." April 1949 (page 52).

Zinder, Norton D., " 'Transduction' in Bacteria." Nov. 1958 (Offprint 106).

Zobel, Bruce, J., "Genetic Improvement of Southern Pines." Nov. 1971 (page 94).

Zuckerkandl, Emile, "The Evolution of Hemoglobin." May 1965 (Offprint 1012).

NAME INDEX

SUBJECT INDEX